Advanced Retrieval-Augmented Generation

Advanced Retrieval-Augmented Generation

Bridging Large Language Models and Knowledge Graphs

Wendy Ran Wei

Huijun Wu

IEEE Press
WILEY

Published by John Wiley & Sons, Inc., Hoboken, New Jersey.

Library of Congress Cataloging-in-Publication Data:

Names: Wei, Wendy Ran author | Wu, Huijun (Writer on cloud computing) author
Title: Advanced retrieval-augmented generation : bridging large language models and knowledge graphs / Wendy Ran Wei, Huijun Wu.
Description: Hoboken, New Jersey : Wiley, [2026] | Includes bibliographical references and index.
Identifiers: LCCN 2026003339 | ISBN 9781394374687 hardback | ISBN 9781394374700 adobe pdf | ISBN 9781394374694 epub
Subjects: LCSH: Semantic networks (Information theory) | Natural language processing (Computer science) | Generative artificial intelligence | Information retrieval—Automation
Classification: LCC Q387.5 .W45 2026
LC record available at https://lccn.loc.gov/2026003339

Cover Design: Wiley
Cover Image: © Wendy Ran Wei

To my beloved cats—Coco, Milk, Pudding, and Cream—for the purrs and company through every page. And to my family, for igniting my dreams.

—Wendy

For wanderers who found their compass, may you discover light in exploration.

—Huijun

Contents

Foreword

This book is a rare achievement in technical literature: it teaches not by presenting isolated concepts, but by guiding readers through a carefully orchestrated intellectual journey. From the opening pages, one senses it was written with exceptional discipline, foresight, and genuine care for how deep understanding develops. Each chapter serves dual purposes—functioning both as a rigorous, comprehensive survey of its domain and as a foundation stone supporting the next stage of learning. Reading it feels like ascending a series of carefully designed terraces, where each level broadens your view, deepens your perspective, and prepares you for the challenges ahead.

The book's architecture reflects this pedagogical intent. Part I establishes our foundations in the long arc of information retrieval, tracing its evolution from keyword search through neural methods, then positioning large language models and their inherent limitations before introducing retrieval-augmented generation (RAG). Part II pivots to graphs and knowledge graphs, revealing how structured representations provide the rigor, grounding, and explanatory power that pure language models lack. Part III synthesizes these approaches, demonstrating how RAG can be seamlessly integrated with graph-based knowledge into coherent, production-ready systems. Finally, Part IV expands the horizon—surveying advanced frameworks including RAPTOR, G-Retriever, Microsoft GraphRAG, and HippoRAG, while thoughtfully examining the ethical considerations and research frontiers that will shape our field's future.

What strikes me most profoundly is the balance the authors achieve throughout. Each chapter delivers not merely exposition but also rich bibliographies and practical, working code examples. This approach transforms abstract concepts into tangible implementations, ensuring that theoretical insights are always grounded in executable reality. This pattern of theory-to-practice integration continues seamlessly through the final chapters, with supplemental materials and appendices providing readers with the hands-on foundation needed to fully internalize the concepts they've absorbed.

For those of us advancing distributed and agentic systems, this book transcends the role of a reference manual—it provides the core constructs of task orientation and knowledge integration that will underpin the next generation of intelligent systems. Its methodology—progressively detailed, successively deeper, yet always maintaining coherence—mirrors the very architecture of the sophisticated systems it describes.

I want to acknowledge the exceptional authors behind this work. Wendy Ran Wei recently joined my Relevance and Personalization team at Airbnb, bringing the same intellectual depth and clarity that illuminate every page of this book. I'm excited to see her apply this expertise to advance our work in relevance, personalization, agentic systems, and matching as we push Airbnb toward new frontiers. Her co-author, Huijun Wu, contributes complementary insights and technical depth that clearly strengthen the entire work. Together, they have created a resource that will serve as both a foundation and a catalyst for our rapidly evolving field.

Sanjeev Katariya
Head of Relevance and Personalization, Airbnb

About the Authors

Wendy Ran Wei

Wendy Ran Wei is a Machine Learning Engineer at Airbnb, where she works on large-scale search, recommendation systems, and LLM-driven relevance and personalization. She specializes in building production-grade AI systems that bridge information retrieval, knowledge graphs, and retrieval-augmented generation.

Wendy holds a Ph.D. in Statistics from The Ohio State University, with a minor in Computer Science, and a Bachelor's degree in Mathematics and Statistics from Zhejiang University. Her academic training provides a strong foundation in statistical modeling, machine learning, and data-driven systems. She has extensive industry experience across leading technology companies, including Twitter, Pinterest, Meta, and Coupang, developing large-scale machine learning systems for social networks and e-commerce platforms. Her work spans recommendation systems, ads relevance, search retrieval, and user modeling, with a focus on delivering high-impact, real-world applications.

Wendy is also a co-author of the book *Data Science: Methods and Practice* (China Machine Press, 2025). Her recent work focuses on integrating large language models with traditional retrieval systems, advancing techniques in RAG, knowledge-enhanced AI, and generative applications.

Huijun Wu

Huijun Wu is an Engineer at Samsung Electronics America (SEA), where he builds large-scale data platforms, advanced AI model infrastructure, and LLM-driven solutions for Samsung Ads. He holds a PhD in Computer Science from Arizona State University (ASU) and graduated from Huazhong University of Science and Technology (HUST).

Huijun's career began in 2007 with the development of CacheDB, an in-memory real-time database kernel adopted by China Southern Power Grid, earning him two patents. In 2011, he pursued his PhD at ASU, focusing on mobile cloud computing; during this time, he co-authored the book *Mobile Cloud Computing* and developed POEM, a framework that won the Best Student Paper Award at IEEE MS 2015.

After graduating in 2016, Huijun joined Twitter to work on Apache Heron, an open-source distributed stream processing engine built for petabyte-scale real-time analytics, and co-authored the book *Heron Streaming*. His collaborative research on forecasting SQL query costs at Twitter also won the Best Industry Track Paper Award at IEEE IC2E 2021. He then moved to Samsung Research America (SRA) to design large-scale AI infrastructure and LLM-powered privacy guardrails. An experienced technical author, his expertise spans distributed systems, AI model infrastructure, enterprise LLM deployment, and generative AI applications.

Preface

Over the past few years, the world has witnessed a profound shift in how machines understand, process, and generate language. With the rise of large language models (LLMs), natural language processing (NLP) has evolved from task-specific pipelines into general-purpose engines capable of reasoning, dialog, and creative generation. Yet, even as these models impress with their linguistic fluency, they continue to struggle with grounding—relying solely on pretrained knowledge, which leads to hallucinations, stale information, and a lack of verifiability. Retrieval-augmented generation (RAG) emerged as a critical solution to this problem, allowing models to dynamically retrieve external data to augment their responses. When combined with structured knowledge representations like graphs and knowledge graphs, RAG systems become even more powerful, capable of contextualized reasoning, accurate referencing, and rich semantic understanding. This fusion, which we call Graph-RAG, forms the central focus of this book.

As artificial intelligence (AI) practitioners and researchers, we have witnessed firsthand the limitations of traditional information retrieval systems and the challenges of maintaining factual accuracy in AI-generated content. Through our combined experience at leading technology companies, we recognized the urgent need for a comprehensive resource that addresses these challenges. This book is the result of years of practical experience, research, and collaboration with the AI community. It represents not just a technical manual, but a roadmap for the future of intelligent information systems. We wrote it with the conviction that the integration of knowledge graphs with RAG will fundamentally transform how AI systems understand and interact with human knowledge.

Why This Book Exists

While the fusion of RAG with knowledge graphs represents the new frontier of intelligent information systems, there is a lack of resources that detail the practical implementation of such systems. Most existing literature focuses on either RAG or knowledge graphs in isolation, leaving practitioners and researchers without a clear roadmap for their integration. This book fills that gap, offering both theoretical insights and actionable techniques for creating systems that are not only smart but also explainable and robust.

This book serves as a comprehensive guide for readers interested in building intelligent systems that merge retrieval, generation, and knowledge representation. By the end of this book, readers will:

- Understand the core principles of RAG and knowledge graphs.
- Learn practical techniques to integrate these technologies for search and recommendation tasks.
- Explore real-world use cases where this integration can transform application performance and user experience.
- Gain hands-on experience through step-by-step examples and case studies, enabling them to implement these concepts in their own domains.

Who This Book Is For

We designed this book to serve a diverse audience, including developers, researchers, data scientists, and AI enthusiasts who are interested in building advanced search and recommendation systems.

- **For Beginners**: If you have a general understanding of data science and software engineering but are new to domain-specific concepts (such as NLP, LLM, and knowledge graphs), Parts I and II provide a gentle introduction. These sections are rich with definitions, examples, and illustrations that build your understanding from the ground up. You'll find these chapters suitable as university course materials for introductory classes.
- **For AI and Machine Learning Practitioners**: If you're already familiar with information retrieval and LLMs but new to RAG, you can start directly with Chapter 3. This approach allows you to leverage your existing knowledge while quickly getting up to speed with RAG concepts.
- **For Advanced Readers**: Those who have already mastered NLP, LLM, RAG, and Graph concepts can dive straight into Part III, where we explore the integration of RAG with graph technologies. This intermediate-level content bridges the gap between traditional approaches and cutting-edge graph-based solutions.
- **For Researchers and Graduate Students**: Part IV is designed for those at the frontier of AI research. These chapters read like research seminar discussions, exploring state-of-the-art frameworks and future directions. This content is ideal for graduate-level courses and those looking to contribute to the field's advancement.

While the concepts and techniques are explained in a practical and accessible manner, a basic understanding of computer science principles, such as algorithms and data structures, will be helpful. Readers should also have a working knowledge of Python, as the book includes hands-on examples and implementations using Python-based libraries and frameworks. Familiarity with foundational AI and machine learning concepts is a plus but not mandatory, as relevant ideas are introduced and explained as needed throughout the book.

How to Use This Book

This book is designed to be accessible while covering advanced topics. To help you gauge your readiness and identify areas for review:

- **Parts I–II (Foundations)**: Require only basic programming experience in Python and familiarity with fundamental computer science concepts (algorithms, data structures).
- **Part III (RAG and Graph Integration)**: Assumes comfort with the material in Parts I–II, or equivalent background in basic machine learning concepts, neural network fundamentals, and transformer architectures.
- **Part IV (Advanced Topics)**: Designed for readers with graduate-level understanding of deep learning, graph neural networks, and current research literature in NLP and information retrieval.

Throughout the book, we provide brief background recaps before introducing advanced concepts. The book's modular structure allows you to chart your own learning path based on your background and objectives. The chapter dependency and reading order are shown in Figure 1.

- **For Self-study**: Each part builds upon the previous one. Beginners can follow Parts I through IV sequentially. Advanced readers may explore targeted chapters for implementation recipes or research discussions.

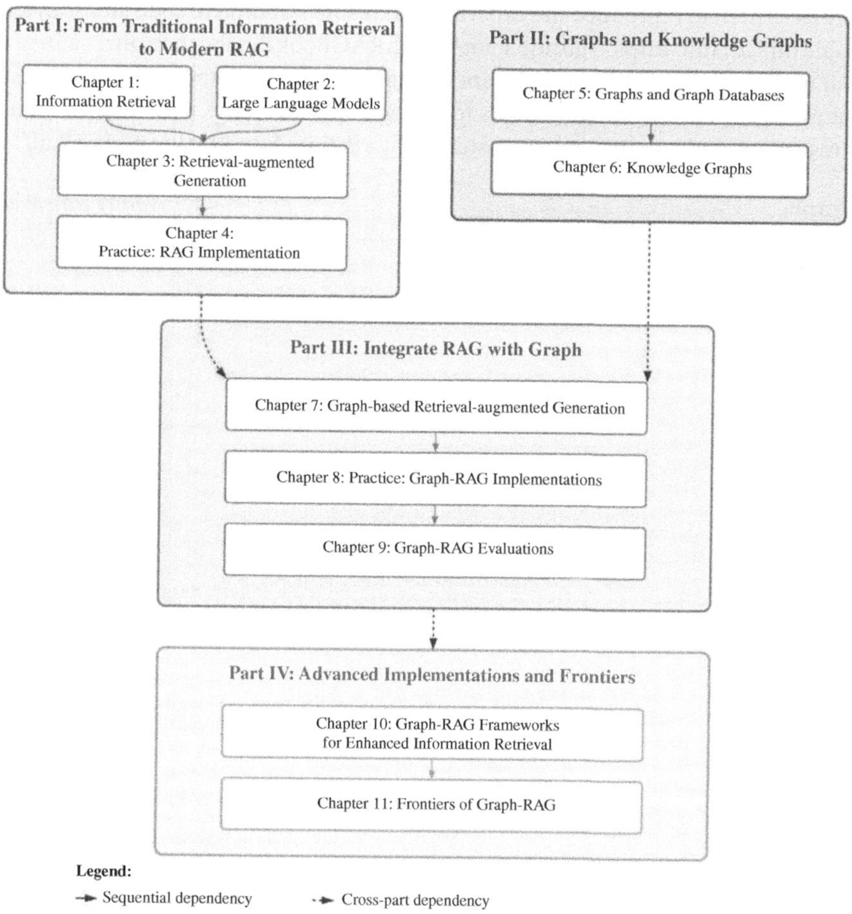

Figure 1 Reading order and chapter dependency.

- **As a Course Textbook**: Parts I and II are suitable for undergraduate or early graduate-level courses in information retrieval, NLP, or knowledge graphs. Part III is appropriate for more advanced courses on applied RAG or semantic systems. Part IV may serve as reading material in graduate seminars or research reading groups.
- **As a Practical Guide**: Throughout the book, we include hands-on tutorials using frameworks like LlamaIndex, LangChain, Neo4j, and real-world case studies in e-commerce, QA, and recommendation. These resources make the book ideal for industry practitioners.

We encourage you to experiment with the provided code repositories, engage with the exercises, and apply these concepts to your own projects. The field of Graph-RAG is rapidly evolving, and this book aims to provide you with both the foundational knowledge and the innovative thinking needed to contribute to its advancement.

Resources and Code Examples

All sections with titles starting with "Practice:" or "Example:" are designed for readers to follow the provided code exactly as written to reproduce the experimental results. Complete code and Jupyter notebooks are available on GitHub: https://github.com/GraphRAGBook/book_code.git.

We will post regular updates on the book's website, including new examples, framework updates, and links to relevant research. We encourage readers to periodically check the website for errata, updates, additional resources, and supplementary materials by visiting: http://www.ragbook.ai/.

San Francisco, California
September, 2025

Wendy Ran Wei

Acknowledgments

The authors would like to express their deepest gratitude to everyone who contributed to the creation of this book. We are especially thankful to our colleagues, advisors, and the broader artificial intelligence (AI) research community for their invaluable insights and feedback throughout the process. We also extend our thanks to the publishing team for their guidance and support, ensuring this book reaches readers in the best possible form.

The first author, Wendy, would like to thank her family and friends for their unwavering encouragement and understanding during countless hours of research and writing. Special thanks go to her mentors and peers—Huiji Gao, Xiaowei Liu, Weiwei Guo, Qing Zhang, Dillon Davis, Chun How Tan, Mihajlo Grbovic, Kedar Bellare, Liwei He, Malay Haldar, Stephanie Moyerman, and Sanjeev Katariya—whose guidance, insights, and thoughtful suggestions greatly enriched this book and strengthened its vigor and vision.

The second author, Huijun, sincerely thanks his family for their unwavering support throughout the writing process, especially his beloved wife for her patience and encouragement. He also extends his gratitude to his peers and collaborators for inspiring the practical applications explored in this work. In particular, he is deeply appreciative of his teammates—Keewook Na, San Tan, Austin Jeong, Zheng Li, Maosong Fu, Mansi Kapoor, Sungwook Ethan Byun, Prashanth Rajendran, Dayoon Woo, Joonwoo (Junu) Jeon, Donggeon Kim, and Jihyeon Ha—for igniting his passion for integrating knowledge graphs with AI systems.

Wendy and Huijun

Introduction

In an era where artificial intelligence (AI) increasingly shapes how we access and process information, a critical challenge has emerged: how can we ensure that AI systems provide accurate, contextual, and verifiable information while maintaining the fluency and accessibility that users expect? This book presents a transformative solution to this challenge through the integration of large language models (LLMs) with knowledge graphs, culminating in what we call graph retrieval-augmented generation (Graph-RAG)—a paradigm that represents the next evolution in intelligent information systems.

In traditional search and recommendation systems, there is often a trade-off between relevance and explainability. While AI-driven methods excel at generating contextually rich outputs, they struggle to ground their results in reliable, interpretable data. On the other hand, knowledge graphs provide structured and semantically rich information but lack the flexibility to adapt to rapidly changing user queries and contexts. The disconnection between these paradigms leads to critical gaps in applications ranging from e-commerce and content discovery to healthcare and legal research. Users are left with either incomplete results or opaque answers that lack accountability. This book aims to address these challenges by demonstrating how to leverage knowledge graphs within RAG frameworks to create systems that are both accurate and transparent.

The Convergence of Two Worlds LLMs have demonstrated remarkable capabilities in understanding and generating human-like text. From generative pretrained transformer (GPT) to Claude, these models have revolutionized natural language processing, enabling applications that seemed impossible just a few years ago. However, they come with inherent limitations: they can hallucinate facts, their knowledge becomes outdated after training, and they struggle with providing structured, verifiable information.

On the other hand, Knowledge Graphs have long served as the backbone of structured information representation. From Google's Knowledge Graph to enterprise knowledge management systems, these structures excel at capturing relationships, maintaining factual accuracy, and enabling complex reasoning over interconnected data. Yet, they often lack the natural language understanding and generation capabilities that make LLMs so accessible.

This book explores how RAG serves as the bridge between these two worlds, and how its evolution into Graph-RAG creates systems that are greater than the sum of their parts.

Why Graph-RAG Matters Now The timing of this book is not coincidental. Several converging trends make Graph-RAG not just relevant but essential:

1. **The Hallucination Crisis:** As LLMs become more prevalent in critical applications—from healthcare to finance—the cost of inaccurate information has never been higher. Organizations need AI systems they can trust.
2. **The Knowledge Update Challenge:** In our rapidly changing world, static knowledge embedded in model parameters quickly becomes obsolete. Dynamic knowledge integration is no longer optional—it's imperative.
3. **The Context Complexity:** Modern applications require understanding not just facts but the intricate relationships between them. Simple retrieval is insufficient; we need systems that can navigate complex knowledge networks.
4. **The Scalability Imperative:** As data volumes explode, traditional retrieval methods struggle. Graph-based approaches offer more efficient ways to organize and access vast knowledge repositories.

What You'll Learn This book takes you on a comprehensive journey through the landscape of advanced retrieval-augmented generation:

Part I: From Traditional Information Retrieval to Modern RAG establishes the foundational concepts. You'll understand how information retrieval evolved from simple keyword matching to sophisticated neural approaches, explore the architecture and capabilities of LLMs, and discover how RAG emerged as a solution to their limitations. Through practical implementations, you'll build your first RAG systems.

Part II: Graphs and Knowledge Graphs introduces the structured world of graph-based knowledge representation. You'll learn about graph databases, algorithms, and the construction of knowledge graphs. This part demystifies how structured knowledge can be captured, stored, and utilized effectively.

Part III: Integrate RAG with Graph is where the magic happens. You'll discover how to combine the retrieval capabilities of RAG with the structured knowledge of graphs, implement Graph-RAG systems using popular frameworks, and evaluate their performance using sophisticated metrics.

Part IV: Advanced Implementations and Frontiers looks toward the future. You'll explore state-of-the-art frameworks like RAPTOR, G-Retriever, Microsoft GraphRAG, and HippoRAG, understanding their unique approaches and applications. The final chapter examines emerging trends, ethical considerations, and research directions that will shape the field's future.

A Practical Approach While this book provides comprehensive theoretical foundations, its heart lies in practical application. Each concept is illustrated with real-world examples, from e-commerce search systems to biomedical knowledge extraction. You'll find:

- **Hands-on Code Examples:** Every major concept is accompanied by implementation examples using popular frameworks like LlamaIndex and LangChain.
- **Real Dataset Applications:** We use datasets like WANDS (e-commerce) throughout the book, showing how concepts apply to actual business problems.
- **Evaluation Frameworks:** Learn not just how to build Graph-RAG systems, but how to measure and improve their performance.
- **Industry Case Studies:** Discover how leading organizations are already leveraging these technologies to transform their operations.

The Road Ahead As you embark on this journey, remember that Graph-RAG represents not just a technical advancement but a fundamental shift in how we think about AI systems. It's about creating AI that is not only intelligent but also accountable, not only fluent but also factual, not only powerful but also transparent.

Whether you're building the next generation of search engines, developing AI assistants for specialized domains, or researching the frontiers of artificial intelligence, the concepts and techniques in this book will equip you with the tools and understanding needed to push the boundaries of what's possible.

Welcome to the future of intelligent information systems. Welcome to the world of Graph-RAG.

About the Companion Website

This book is accompanied by a companion website.

www.wiley.com/go/advretrievalaugmentedgeneration1e

To access the website materials, click on the link under Related Resources on the book's page on wiley.com. You can also find this by searching www.wiley.com for the ISBN 9781394374687.

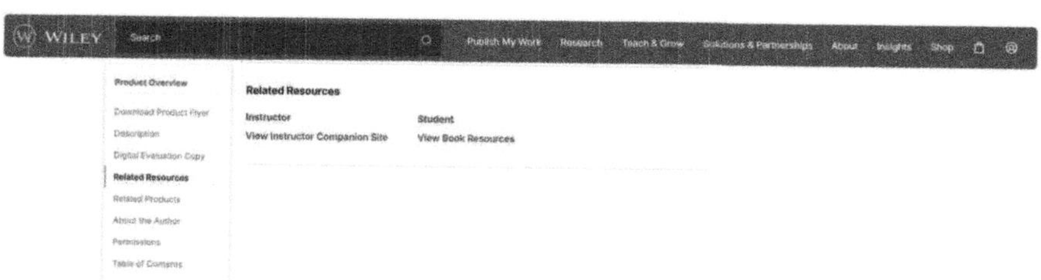

This website includes:

- Source code and implementation examples from the book.
- Jupyter notebooks demonstrating Retrieval-Augmented Generation (RAG), Graph-RAG, and Knowledge Graph integration techniques.
- Sample datasets and benchmark examples used throughout the book.
- Companion materials and supplementary resources for selected chapters.

Part I

From Traditional Information Retrieval to Modern RAG

Information retrieval (IR) is the scientific discipline concerned with organizing, searching, and retrieving relevant information from vast collections of data. At its core, IR addresses the fundamental challenge of connecting users with the information they need, precisely when they need it. While encompassing diverse applications, search systems and recommendation systems stand as the two most prominent and impactful manifestations of IR technology in our digital world.

Search systems empower users to actively seek information by formulating queries, returning results explicitly matching or relevant to their request. Recommendation systems, conversely, operate proactively, anticipating user needs and preferences based on implicit signals and historical interactions to suggest potentially relevant items before an explicit request is made. Both paradigms rely on sophisticated underlying components: ingesting and indexing data (pre-retrieval), employing models like keyword matching (sparse retrieval) or semantic understanding (dense retrieval) to find candidates (retrieval), and then refining results through ranking and evaluation (post-retrieval).

The advent of **large language models (LLMs)** has brought transformative change to the field of IR. Built on the groundbreaking Transformer architecture, which processes sequential data via self-attention mechanisms, models like generative pretrained transformer (GPT), bidirectional encoder representations from transformers (BERT), and T5 have revolutionized how machines process and understand language. Leveraging massive corpora of pretraining data, LLMs can not only understand language contextually but also incorporate a vast amount of general knowledge, making them capable of providing informed and contextually relevant responses. This combination of deep language understanding and extensive knowledge has drastically improved the accuracy, relevance, and user experience of modern IR applications, making interactions more natural and contextually aware.

Despite their advancements, LLMs face challenges such as hallucinations (generating inaccurate or false information) and knowledge staleness (relying on outdated data). To address these limitations, a powerful hybrid approach known as **retrieval-augmented generation (RAG)** has emerged. RAG combines the capabilities of LLMs with external knowledge sources, enabling systems to provide more accurate, context-rich, and up-to-date responses. By leveraging both retrieval and generation, RAG bridges the gap between traditional IR methods and the innovative potential of LLMs.

As illustrated in Figure PI.1, the ever-growing volume of information—whether in the form of raw text, documents, or structured data—there are multiple strategies to extract useful knowledge. Traditionally, IR techniques have been used to identify and return relevant documents. More recently, LLMs have emerged as powerful tools that can be trained on these large corpora to

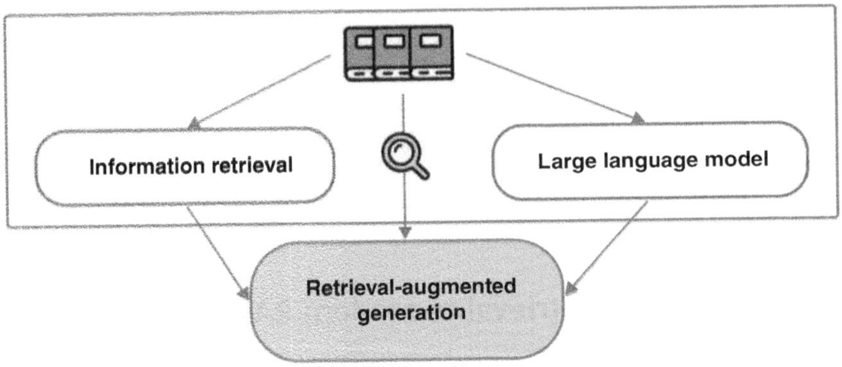

Figure PI.1 Knowledge corpus and knowledge management systems.

perform reasoning, generation, and understanding tasks. Bridging these two approaches is RAG, a hybrid method that combines the precision of IR with the generative capabilities of LLMs.

This first part of the book explores the evolution of IR and its transformation through the integration of LLMs and RAG techniques. The structure of Part I is as follows:

- Chapter 1: Introduces the core principles and evolution of IR, with a focus on key techniques and applications in search engines and recommendation systems.
- Chapter 2: Explores the foundations of LLMs, including their architecture, training paradigms, and how they have redefined the landscape of information access and understanding.
- Chapter 3: Presents an in-depth discussion of RAG, examining its underlying architecture, components, and real-world use cases.
- Chapter 4: Provides practical implementation frameworks for building RAG systems.

Together, these chapters equip readers with the essential knowledge to understand and build modern, LLM-enhanced IR applications. By the end of this section, readers will gain a comprehensive understanding of how IR has progressed—from classical approaches to modern RAG-based systems—empowered by the integration of LLMs and external knowledge retrieval mechanisms.

1

Information Retrieval

In the first chapter of this book, we are going to learn the fundamentals of information retrieval (IR) systems. We are going to dive into details of the definition of IR, its core components, and major applications in **search** and **recommendation** systems, which establish the foundation for subsequent chapters by focusing exclusively on these two application domains.

1.1 Definition and Historical Evolution

IR broadly refers to the process of obtaining relevant information to meet a user's needs. While its essence remains constant—connecting people with the information they seek—its scope and meaning have expanded significantly over time, reflecting the evolution of technology and the growing complexity of information systems.

In this section, we will explore its foundational definition and trace its evolution to build a comprehensive understanding of the IR landscape.

1.1.1 Definition

IR is the science and practice of obtaining relevant information resources that satisfy a user's information needs from a large collection of data. As illustrated in Figure 1.1, IR systems serve as the bridge between users and the vast pool of information, facilitating the retrieval of relevant responses based on user inputs or inferred preferences.

The diagram shows the basic workflow of an IR system, where a user's request (implicit or explicit) is processed to retrieve relevant documents, and these documents are then transformed into meaningful responses. This simplified illustration highlights the core components:

- **User Request**: Represents the user's input or inferred need. This could be an explicit query, such as a search phrase or natural language question, or an implicit signal, such as browsing behavior or preferences.
- **Documents**: The source of information, which could include structured data (e.g., product catalogs, relational databases), semi-structured data (e.g., XML, JSON), or unstructured data (e.g., articles, reviews, multimedia). The IR system organizes and stores this data in formats that allow for efficient search and retrieval.
- **Retrieval Engine**: At the heart of the IR process is the retrieval engine, which matches the user request against a vast collection of data to find relevant information. This step relies on

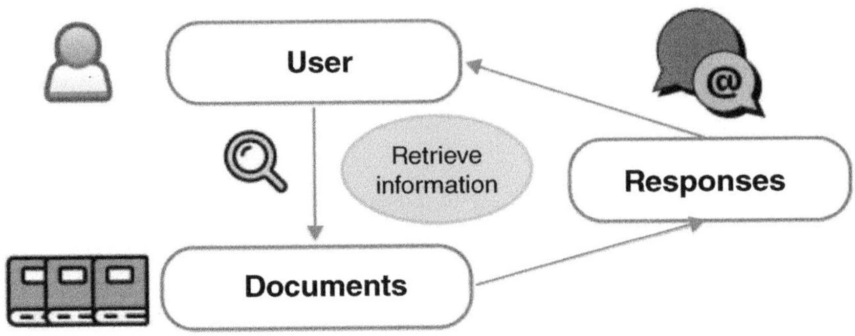

Figure 1.1 Information retrieval systems overview.

a combination of indexing, matching algorithms, and ranking mechanisms to identify and prioritize results.

- **Responses**: The system delivers the retrieved information in a form that meets the user's needs. Responses can take various formats, such as ranked lists of documents, summaries, or curated suggestions, depending on the application.

This concept of IR has evolved significantly over time, from traditional library systems to recent search engines and personalized recommendation systems. The primary purpose of IR is to deliver the right information at the right time, catering to user queries or inferred needs.

1.1.2 Historical Evolution

IR began as a manual process in libraries and archives, transitioning through several key phases:

- Manual Retrieval (pre-1950s): Catalog systems and index cards in libraries.
- Early Days and Text-based Era (1950s–1980s): Introduction of early retrieval systems, including Boolean search systems, vector space models, and probabilistic approaches.
- Web-era Information Retrieval (1990s–2000s): Emergence of link analysis and large-scale distributed systems.
- Neural IR (2010s–present): Integration of deep learning and contextual understanding.

As shown in Table 1.1, each era brought significant improvements in search effectiveness.

1.1.2.1 Manual Retrieval (Pre-1950s)

Libraries and archives relied on catalog systems and index cards to help users locate resources. **Cataloging**, which involves creating records and descriptions of the items in a collection to identify and access them, laid the foundation for today's IR systems by giving structure and findability to data.

Librarians played a central role, manually matching queries with relevant materials. In ancient Greek times, the Library of Alexandria used an early form of card catalogs that allowed scholars to search the scrolls and find important works of the day in this prominent center of learning and knowledge. Figure 1.2 illustrates a card catalog in a library where users can search for books by flipping through drawers of cards categorized by title, author, and subject.

Table 1.1 Evolution of search effectiveness.

Era	Search time	Accuracy (%)	Scale	Understanding level
Manual retrieval (pre-1950s)	Hours to days	10–15	Hundreds	Human-based classification
Early computerization (1950s–1960s)	Minutes to hours	20–30	Thousands	Boolean logic only
Text-based era (1970s–1980s)	Seconds to minutes	40–50	Millions	Term frequency based
Web era (1990s–2000s)	Sub-second	60–70	Billions	Link-based relevance
Neural era (2010s–present)	Milliseconds	80–90	Trillions	Semantic and contextual

Figure 1.2 Woman at main reading room card catalog, [1] / Library of Congress / Public Domain.

1.1.2.2 Early Computerization (1950s–1980s)

The foundation of IR as a field was laid in the 1950s when computer scientist Calvin Mooers first coined the term "information retrieval" in his paper published in the *Proceedings of the International Congress of Mathematicians* [2]. This period marked the transition from manual indexing and searching methods to the earliest forms of automated retrieval systems.

1.1.2.2.1 The Birth of Boolean Search

One of the earliest and simplest approaches to data modeling in IR is based on set theory, where both documents and queries are treated as collections (or "buckets") of words. For example, Figure 1.3 illustrates the intersection of two sets A and B.

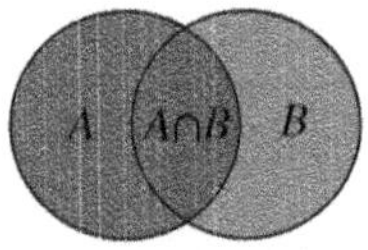

Figure 1.3 A set theory Boolean AND query [3] / Cepheus / Wikimedia Commons / Public Domain.

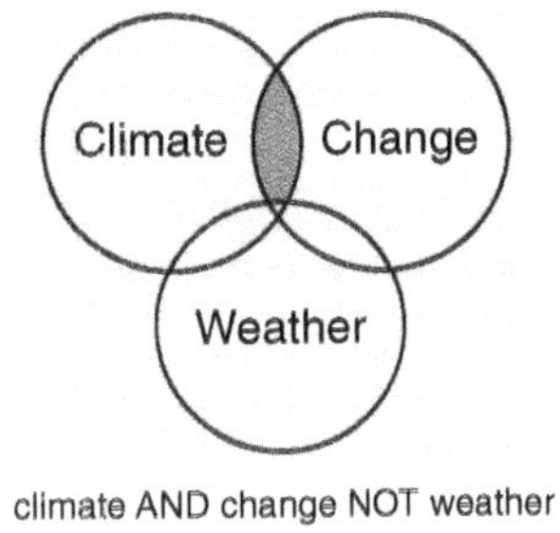

Figure 1.4 Boolean search on "climate AND change NOT weather."

This foundational approach gives rise to the use of **Boolean search systems**. These systems introduced three fundamental logical operators:

- AND requiring all specified terms to be present.
- OR requiring at least one of the specified terms.
- NOT excluding documents with certain terms.

These systems allowed users to construct queries with precision, targeting documents that matched specific criteria. For example, as illustrated in Figure 1.4, a query such as "climate AND change NOT weather" would return documents containing both "climate" and "change" while excluding those containing "weather."

1.1.2.2.2 Fuzzy Set Theory for Relevance Ranking

While traditional Boolean models operate on a strict match/no-match basis, **fuzzy set theory** refines this approach to allow for degrees of membership. In the context of IR, fuzzy logic introduces the concept of weighted relevance, enabling systems to rank documents based on how closely they match a query rather than just whether they match.

Key features of fuzzy set theory in IR include the following:

- Partial Matches: Documents are assigned a relevance score based on their degree of match with the query, rather than being strictly included or excluded.
- Relevance Ranking: This scoring allows documents to be sorted by relevance, improving the user experience by presenting the most relevant results first.

1.1.2.2.3 Vector Space Models

In the 1960s, Gerard Salton at Cornell University developed the "SMART system" [4], which was one of the earliest experimental systems to introduce advanced techniques that later became the cornerstone of modern IR. One pioneering concept proposed in the SMART system is the "vector space model," introducing a mathematical framework for representing documents and queries as vectors in a multidimensional space. This model was a major step forward in IR, moving beyond simple keyword matching to capture the relative importance of terms, significantly improving retrieval precision and ranking.

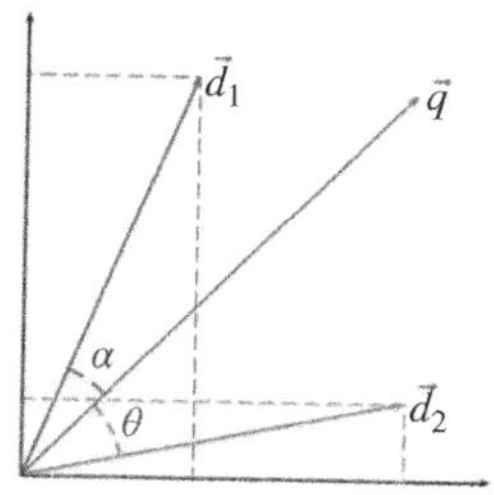

Figure 1.5 Cosine similarity between query and document vectors.

Each document is represented as a vector, where dimensions correspond to unique terms, and values represent their weights. Similarly, queries are also represented as vectors in the same space. The relevance of a document to a query is determined by calculating the cosine similarity between their respective vectors. A higher cosine similarity score indicates greater alignment and relevance. For example, a query such as "machine learning algorithms" would rank documents by how closely their vectors align with the query vector.

Figure 1.5 demonstrates this concept with a concrete example in a two-dimensional (2D) vector space. Consider a query q about "AI algorithms" and two documents: d_1 containing "machine learning algorithms" and d_2 containing "graph algorithms." The vectors represent their positions in the term space, where the angle between vectors determines their similarity. Document d_1 has a smaller angle with the query vector, indicating higher relevance and better alignment with the user's information need, while d_2 shows a larger angle, suggesting lower relevance.

1.1.2.2.4 Term Weights and the TF-IDF Examples

In vector space models, each term in a document is assigned a weight, which quantifies its importance within the document and across the entire collection. These weights form the components of the vector representation of a document. Term weights play a crucial role in determining the similarity between a query and a document, directly influencing the ranking of results.

Term frequencies and document frequencies are commonly used terms for the term weighting. Each term in a document is assigned a weight based on both:

- Term Frequency (TF)—how often the term appears in a document.
- Inverse Document Frequency (IDF)—how rare the term is across the corpus.

The **term frequency-inverse document frequency (TF-IDF)** score for a term t in document d is computed as:

$$\text{TF-IDF}(t, d) = \text{TF}(t, d) \times \text{IDF}(t)$$

To illustrate this in action, suppose we have a query "machine learning algorithms" and three candidate documents. For each document, we compute TF-IDF for the terms, construct vectors, and then rank the documents by their cosine similarity to the query vector. This ranks documents by relevance, which is the fundamental goal of IR: to order results based on how well they satisfy the user's intent. While here we briefly introduce the definition of TF-IDF, we are going to dive into more details of TF-IDF in Section 1.2.3 of the book.

Term weights are the cornerstone of the vector space model, providing the quantitative foundation for measuring the relevance of documents to user queries. These weighting schemes have evolved significantly, leading to more accurate and intuitive IR systems.

1.1.2.2.5 Probabilistic Models and BM25

The probabilistic approach to IR represents a sophisticated evolution beyond vector space models, focusing on estimating the probability that a document is relevant to a given query. Unlike deterministic approaches such as TF-IDF, probabilistic models explicitly handle uncertainty in the retrieval process through statistical inference.

At its core, the probabilistic model attempts to answer a fundamental question: given a user query Q, what is the probability $P(R|D)$ that a document D is relevant (R)? This framework is based on the probability ranking principle (PRP), which states that documents should be ranked according to their probability of relevance to the query.

The **binary independence model (BIM)** serves as a foundational example. It makes several key assumptions:

- Term Independence: The presence/absence of one term doesn't affect another.
- Binary Weighting: Terms are either present or absent.
- Document relevance is independent of other documents.

One of the most successful instantiations of probabilistic ranking is **Best Match 25 (BM25)**, which improves upon TF-IDF by introducing term saturation and length normalization.

BM25 refines relevance estimation by considering how often terms appear (TF), how informative they are (IDF), and how long the document is. As a probabilistic model, it remains one of the most widely used scoring functions in modern retrieval engines. While we briefly introduce the definition of BM25 here, we will discuss it in more details of BM25 in Section 1.2.3 of the book.

A more sophisticated approach is the **Bayesian network model**, which represents documents and queries as nodes in a directed acyclic graph (DAG). As illustrated in Figure 1.6, the network captures: (1) document nodes representing individual documents, (2) term nodes representing index terms, (3) query nodes representing user information needs, and (4) edges representing probabilistic dependencies between nodes.

In Bayesian network models, the relevance of a document is inferred through probabilistic dependencies between query terms, documents, and latent relevance variables. Given a query Q and document D, the system estimates:

$$P(R = 1 \mid Q, D)$$

by propagating beliefs through a graphical model encoding term–document relationships.

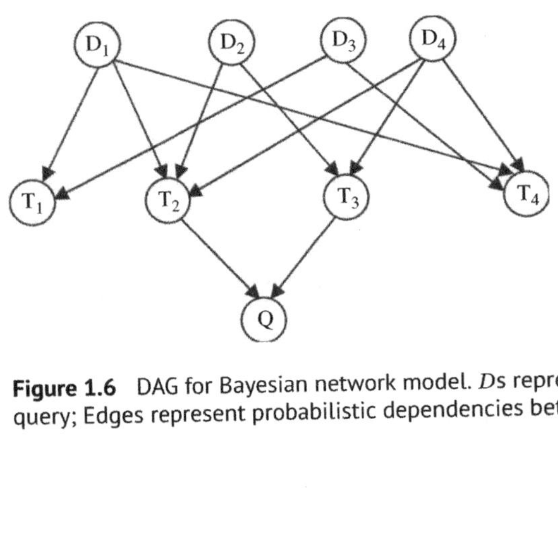

Figure 1.6 DAG for Bayesian network model. Ds represent documents; Ts represent terms; Q represents query; Edges represent probabilistic dependencies between nodes.

As machine learning (ML) continues to advance, probabilistic approaches remain fundamental to modern IR systems, often working in conjunction with neural networks and other AI techniques.

1.1.2.3 Web-era Information Retrieval (1990s–2000s)

The 1990s marked a transformative era for IR with the advent of the World Wide Web (WWW) and the emergence of early search engines. The web revolutionized how individuals accessed and shared information, enabling global-scale connectivity among users, organizations, and machines. This explosive growth in content volume and diversity led to a pressing need for scalable, efficient retrieval systems capable of delivering relevant information from billions of web documents—far beyond the scale of traditional IR systems.

1.1.2.3.1 Web Search as a Retrieval Paradigm

Web search evolved as a distinct and large-scale retrieval method, building upon core IR principles while introducing new architectural and algorithmic innovations. Compared to traditional IR, web search differs in several ways:

- Scale: Web collections include billions of heterogeneous documents spanning text, images, and structured data.
- Noise and Adversarial Content: Web content includes spam, duplicated pages, and low-quality or misleading documents.
- User Interaction: Queries tend to be short (2–3 keywords), with users expecting instant and high-quality results.
- Architecture: Large-scale distributed infrastructure is essential to support indexing, crawling, ranking, and query response.

While web search inherits foundational concepts from vector space models, probabilistic retrieval, and Boolean logic, it diverges significantly in implementation due to its scale and complexity. As illustrated in Figure 1.7, it introduced new components such as web crawling, link analysis, distributed indexing, and real-time ranking.

1.1.2.3.2 Early Search Engines and System Architecture

In the early days of the Web, search engines like Yahoo! initially relied on human-curated directories to organize content (Figure 1.8). These hierarchical listings worked well for a small and structured web but quickly became inadequate as the number of websites exploded.

To address this, search engines began building automated indexing systems powered by web crawlers and distributed processing pipelines. Pioneers like AltaVista were among the first to offer full-text search over large-scale web documents, leveraging parallel architectures for crawling, storage, and retrieval. These systems laid the groundwork for modern search infrastructure, which partitions the web across data centers, supports real-time updates, and handles millions of concurrent queries.

1.1.2.3.3 PageRank and Link-based Retrieval

A major advancement in web IR came in the late 1990s when Larry Page and Sergey Brin at Stanford introduced PageRank, a link-based ranking algorithm that measured the relative importance of web pages based on their hyperlink structure.

Unlike traditional keyword-based scoring methods, PageRank treated hyperlinks as votes of confidence. The core idea was that a page is important if it is linked to by other important pages. This approach was inspired by academic citation analysis. Figure 1.9 illustrates this with two example

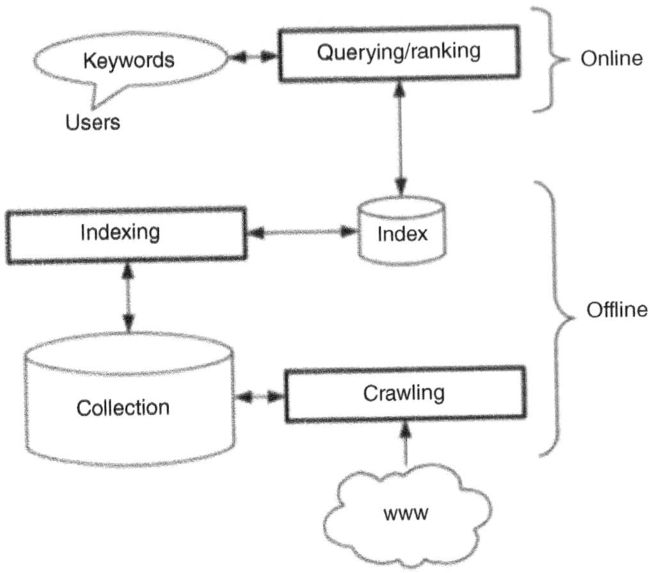

Figure 1.7 Phases of web search [5] / Taylor & Francis Group.

Figure 1.8 Yahoo search page in 2005.

nodes: Node B (a highly cited paper) receives many incoming links and is considered authoritative; Node F (a rarely cited paper) has only a few links and is considered less important.

Thus, PageRank captures the notion of authority propagation across the web graph. The algorithm computes a score for each page using the structure of the graph, assigning higher scores to pages with more influential incoming links.

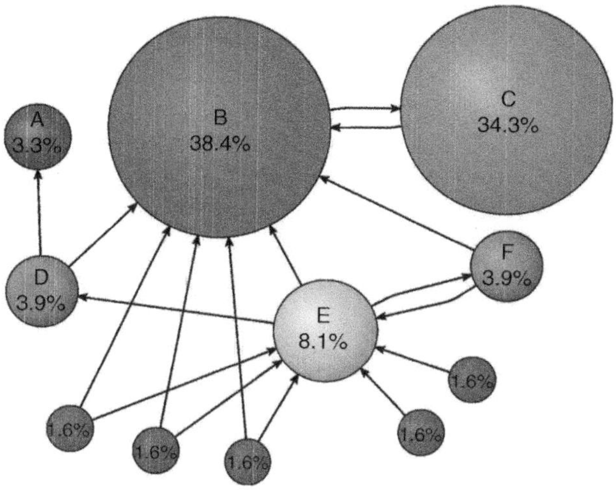

Figure 1.9 Simplified illustration of PageRank with citation-based analogy. Percentages represent perceived importance; arrows represent hyperlinks [6] / 345Kai / Wikimedia Commons / Publuc Domain.

While PageRank represented an important innovation, it was only one part of Google's success story on search engines (Figure 1.10). Other crucial factors included:

- Comprehensive Indexing: Google's aggressive web crawling and indexing strategy captured a larger portion of the web than competitors, providing more comprehensive search results.
- System Performance: Superior infrastructure and algorithms enabled faster query processing and more responsive user experiences.
- Algorithm Integration: PageRank was combined with traditional IR techniques and other ranking factors to create a more robust relevance assessment.
- Continuous Innovation: Google consistently refined its algorithms, incorporating user behavior signals, content quality metrics, and contextual factors.

These complementary components together contributed to Google's ability to consistently deliver relevant, high-quality results, helping it surpass competitors like Yahoo and AltaVista.

1.1.2.3.4 Bridging Web Search and Traditional IR

Despite their architectural differences, web search systems are deeply rooted in traditional IR frameworks. For example, vector space models are still used for scoring document-query relevance; probabilistic models, like BM25, remain foundational in ranking functions; Boolean filters are applied to support structured constraints (e.g., site: filters, date ranges); link analysis (e.g., PageRank) added a graph-based dimension to IR, enriching the representation and evaluation of document authority; user interaction data, such as click-through rates (CTRs) and dwell time, further extend the retrieval model by integrating behavioral signals.

Thus, web search represents an evolutionary integration of existing IR methods with new techniques designed for web-scale data, heterogeneous content, and real-time responsiveness.

Figure 1.10 Google search page in 2001.

1.1.2.4 Big Data and ML (2000s–present)

The early twenty-first century ushered in the big data era, characterized by a massive influx of structured and unstructured data from sources such as social media, mobile devices, e-commerce platforms, Internet of Things (IoT) sensors, and cloud services. This explosion of data volume, velocity, and variety exposed the limitations of traditional keyword-based retrieval approaches, which were often brittle, syntactic, and context-insensitive.

To meet these new demands, IR systems evolved to become artificial intelligence (AI)-powered, leveraging ML, deep learning (DL), and natural language processing (NLP) to understand user intent, rank results by relevance, and deliver personalized responses. These technologies are key components within the broader field of AI, as illustrated in Figure 1.11, which shows the conceptual relationship between these fields:

- AI: The broadest category, encompassing all techniques that enable machines to mimic human intelligence, including reasoning, perception, and decision-making.
- ML: A subset of AI focused on algorithms that learn patterns from data and improve over time without being explicitly programmed.
- DL: A specialized area of ML that uses artificial neural networks with multiple layers to model complex patterns in high-dimensional data.
- NLP: An application domain of AI and ML focused on understanding, generating, and interacting with human language.

1.1.2.4.1 Machine Learning

ML enables systems to learn from user interactions, past behavior, and content characteristics to improve search relevance, recommendation quality, and query understanding. Traditional IR systems used manually engineered features and fixed rules. In contrast, ML-based systems automatically discover useful features, adapt to new data, and optimize ranking functions through training.

One commonly used family of algorithms in IR is learning-to-rank, which uses labeled relevance data to train models that rank documents given a query. These models outperform traditional TF-IDF and BM25 methods in many real-world settings.

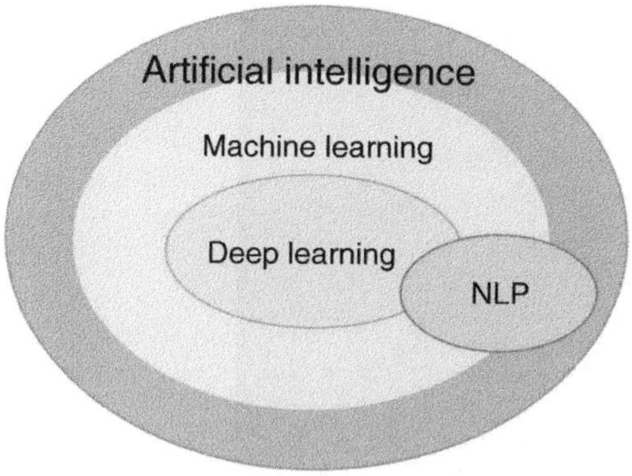

Figure 1.11 Relationships among AI, ML, DL, and NLP (Note: NLP intersects with AI, ML, and DL—it is an application domain that uses techniques from all three, not a subset of any single field).

1.1.2.4.2 Deep Learning

DL has significantly advanced IR by allowing systems to learn hierarchical representations from raw inputs such as text, audio, and images. Unlike shallow models that rely on manual feature extraction, DL models learn distributed, abstract representations that capture semantic meaning.

Two foundational architectures in DL are:

- Convolutional Neural Networks (CNNs): Originally developed for image recognition, CNNs have also been applied to text classification and matching tasks in IR. They work by learning local patterns via convolutional filters and aggregating them for global understanding.
- Recurrent Neural Networks (RNNs): Designed for sequential data, RNNs are used in tasks such as query suggestion, click prediction, and document summarization. They maintain a memory of past inputs, making them suitable for modeling word sequences and capturing context over time. Variants such as long short-term memory (LSTM) and gated recurrent units (GRU) address the limitations of standard RNNs in long-sequence learning.

These models enabled neural IR systems that go beyond bag-of-words and TF, allowing systems to rank based on semantic similarity, not just lexical overlap.

1.1.2.4.3 Natural Language Processing

NLP applies ML and DL to understand and generate human language. Historically, NLP originated from rule-based symbolic AI and only later incorporated statistical and DL methods. For example, early systems such as ELIZA [7], a pattern-matching chatbot that simulated a psychotherapist, and SHRDLU [8], which understood natural language commands within a blocks-world environment, relied entirely on hand-crafted rules and symbolic reasoning rather than statistical learning. Thus, NLP is best viewed as an overlapping domain, not a subset of ML.

In IR, some examples of how NLP techniques can help systems: (1) parse user queries to identify intent and disambiguate meaning; (2) extract entities and relations; (3) match queries to semantically related documents; (4) generate readable answers or summaries. For example, a traditional search for "date" might return ambiguous results related to fruit, calendar dates, or social interactions. NLP-powered systems analyze surrounding context (e.g., in "best date ideas"), detect that the user likely means romantic outings, and retrieve accordingly. We are going to introduce more details of the NLP tasks in Section 2.1.1.2.

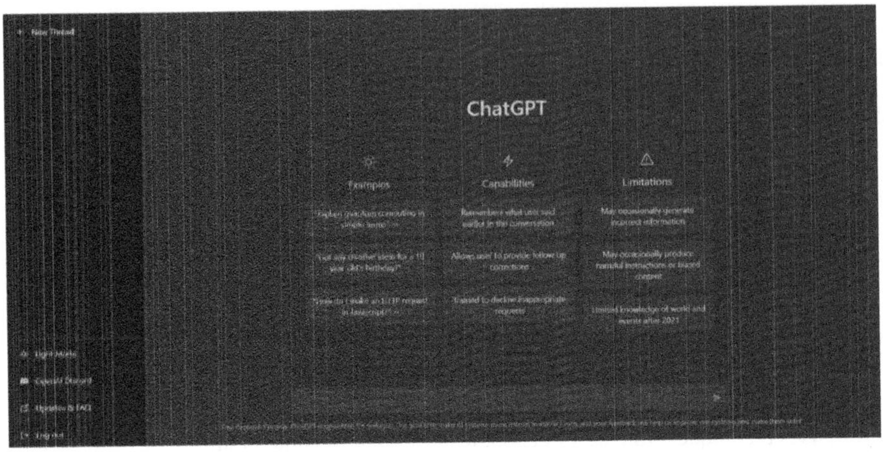

Figure 1.12 OpenAI's ChatGPT UI in 2023. Generated with AI using ChatGPT-OpenAI

1.1.2.4.4 Toward Conversational and Personalized IR

Modern AI-powered IR systems consider user preferences, behavioral signals, and contextual information to tailor results for each individual. E-commerce platforms use these techniques for recommendation systems, suggesting relevant products based on previous behavior, clicks, and natural language queries.

In parallel, conversational AI has emerged as a natural evolution of retrieval technology. Systems now support interactive dialog, allowing users to refine queries, ask follow-up questions, and receive dynamic, multi-turn responses.

One prominent example is OpenAI's GPT (generative pre-trained transformer) family, including ChatGPT (Figure 1.12), which can interpret and respond to user input in a human-like, context-aware manner. These large language models (LLMs) bring together DL, massive data, and transformer architectures to push the boundaries of intelligent search and retrieval. From personalizing e-commerce recommendations to answering complex questions, these innovations redefine how we access and interact with information in a data-rich world.

1.2 Information Retrieval Components

1.2.1 Overview

Modern IR systems have evolved beyond the basic elements shown in Figure 1.1 (users, documents, and responses) into a sophisticated, systematic framework. There are essential components that form the backbone of contemporary IR systems: document indexing, user input processing, retrieval, ranking, and evaluation. As shown in Figure 1.13, these components can be categorized into three different stages: "pre-retrieval," "retrieval," and "post-retrieval."

1.2.1.1 User Input Handling

Modern IR systems account for diverse user inputs, including explicit queries and implicit signals such as browsing history or contextual interactions. This component involves two key tasks:

- Query Understanding: Interpreting and refining explicit user queries (e.g., keywords) for better alignment with document representations.

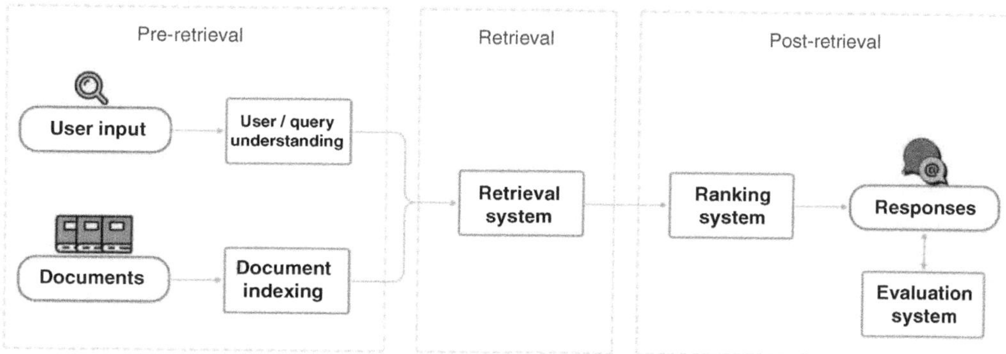

Figure 1.13 Pre-retrieval, retrieval and post-retrieval.

- User Understanding: Incorporating user behavior (e.g., browsing history), preferences, and context to personalize results, particularly for recommendations.

1.2.1.2 Document Handling

This is the process of transforming raw documents into a structured format that facilitates efficient retrieval. Indexing organizes content in a way that enables quick and accurate matching with both explicit queries and implicit signals. This foundational step creates the data structures necessary for efficient retrieval operations.

1.2.1.3 Retrieval

Based on the processed user inputs—whether explicit queries or implicit signals—the system retrieves relevant documents from the indexed repository. A dual retrieval architecture is used to ensure comprehensive coverage of relevant documents:

- Sparse Retrieval: Efficiently handles traditional keyword-based matching.
- Dense Retrieval: Enables semantic understanding and matching.

1.2.1.4 Ranking

Once candidate documents are retrieved, they are ranked based on their relevance to the user input. The ranking system includes:

- Ranker: Initial ranking based on relevance signals such as content similarity and query-document matches.
- Re-ranker: A secondary, more sophisticated ranking layer that refines the order using advanced techniques such as ML models. The initial ranker performs coarse screening based on quickly calculated signals (such as keyword matching), while the re-ranker uses more computationally expensive models (such as neural networks) for fine sorting.

1.2.1.5 Evaluation

The effectiveness of the IR system is assessed through metrics such as precision, recall, and user satisfaction. For implicit inputs, additional metrics, such as CTR or engagement duration, are used to measure system performance and refine its recommendations. For example, if a recommended product has a lower CTR than average, the system will lower its priority in similar queries.

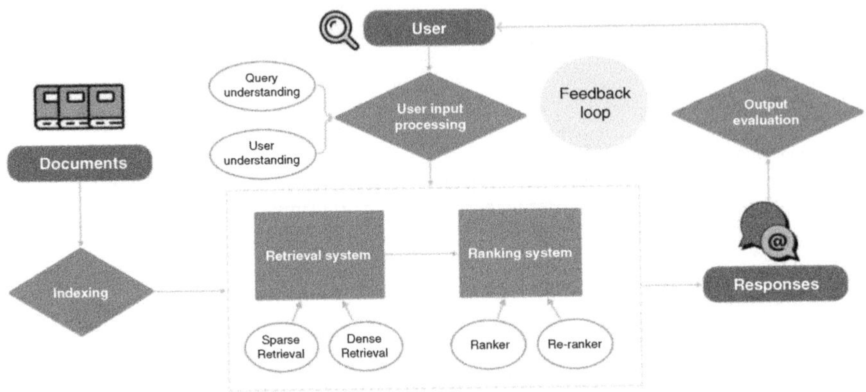

Figure 1.14 Key components of modern IR systems.

The diagram in Figure 1.14 shows how these components form an integrated pipeline, with the feedback loop creating a dynamic system that learns and improves from user interactions. Documents flow through indexing to the retrieval-ranking subsystem (highlighted in the dotted box), while user input undergoes sophisticated processing before entering this core subsystem. The final responses are evaluated, completing the feedback cycle.

This systematic approach enables modern IR systems to cater to diverse use cases, from search engines to personalized recommendation platforms, by seamlessly integrating user signals, retrieval mechanisms, and adaptive evaluation techniques. In the next sections, we are going to dive deeply into each component.

1.2.2 Pre-retrieval

The pre-retrieval phase in modern IR systems consists of two critical components: user input handling and document handling. These two components operate independently: document processing prepares the corpus **offline**, while user input processing runs **online** to adapt to real-time queries. These components lay the groundwork for effective IR by preparing both the query and document corpus for optimal matching.

1.2.2.1 User Input Handling

User input handling transforms raw user queries or implicit signals into a format that can be effectively matched against indexed documents. This process involves several sophisticated steps:

1.2.2.1.1 Query Processing

At the core of user input handling is query processing, which aims to enhance the query's structure, clarity, and expressiveness. The goal is to bridge the gap between what the user types and what they actually mean, increasing the likelihood of retrieving relevant documents. This includes:

- Intent Classification: Determining the type of query (informational, navigational, or transactional) to better align retrieval strategies.
- Query Parsing: Breaking down the input query into meaningful components, identifying key terms, phrases, and potential entities.
- Query Expansion: Enriching the original query with synonyms, related terms, or contextual information to improve recall. Imagine searching for "iPhone charger": query expansion

might add "apple lightning cable" to broaden results, while contextual processing could prioritize sellers near your current location.

- Query Normalization: Normalizing and reformulating problematic queries to improve their likelihood of matching relevant documents. For example, correcting misspellings ("reciepe" → "recipe") or resolving ambiguous terms ("apple" → "Apple Inc.") to improve relevance.

1.2.2.1.2 Contextual Processing

In addition to query refinement, contextual processing seeks to personalize and ground the query using surrounding signals. This makes the system more adaptive to individual users and their situations. It includes:

- Session Context: Incorporating information from the user's current search session to better understand their immediate needs.
- Historical Context: Using the user's past interactions and preferences to personalize query interpretation.
- Environmental Context: Considering factors like location, time, and device type to provide more relevant results.

We are going to cover more details of query processing and contextual processing in Section 1.3 for the search and recommendations system applications.

1.2.2.2 Document Handling

While user input processing adapts queries in real time, document handling forms the offline backbone of IR systems. This component prepares the corpus for fast and accurate retrieval by converting raw documents into structured representations. The two primary steps involved are document preprocessing and index construction. Together, they transform unstructured content into efficient data structures that support millisecond-level query response times in production search and recommendation systems.

1.2.2.2.1 Document Preprocessing

Before any index can be built, documents must first be normalized and cleaned. Document preprocessing consists of a series of text normalization techniques that reduce noise and standardize content, thereby improving both retrieval speed and accuracy. These operations ensure that semantically relevant terms are retained while redundant or irrelevant information is filtered out.

- **Stopword Removal** eliminates functional words that carry little semantic meaning, such as articles ("the," "a"), prepositions ("in," "on"), and common verbs ("is," "are"). This process reduces index size by 20–40% while often improving retrieval precision by focusing on content-bearing terms.
- **Stemming** extracts morphological roots from words, allowing retrieval systems to match related word forms. A search for "housing" can retrieve documents containing "house," "houses," or "housing," significantly improving recall without manual query expansion.
- **Tokenization** separates continuous text into discrete word units. While straightforward for space-separated Western languages, this process becomes significantly more complex for languages like Chinese, Japanese, or Arabic, where word boundaries are not explicitly marked or where morphological complexity requires sophisticated analysis. We are going to discuss more details of tokenization in a language model context in Section 2.1.2.

1.2.2.2.2 Index Construction

Once documents are preprocessed, the next step is to build indexes—the core data structures that make fast and scalable retrieval possible. Without indexing, searching through massive corpora would be prohibitively slow. Index construction converts the preprocessed documents into optimized structures that support efficient lookups and sophisticated ranking algorithms.

The indexing method consists of several key operations:

- Inverted Index Creation: Maps terms to their document locations, forming the backbone of fast term-based search.
- Forward Index Creation: Maps documents to term features, useful for ranking, summarization, or filtering.
- Positional Information: Stores word positions within documents to support phrase and proximity searches.
- Metadata Integration: Incorporates additional document attributes (e.g., author, date, type) into the index to enable faceted search or filters.

An **inverted index** is the dominant data structure in traditional search engines. It allows constant-time access to all documents containing a given term. An inverted index consists of two primary components:

- Vocabulary (Dictionary): A sorted list of all unique terms appearing in the document collection, serving as the primary access point for retrieval operations.
- Posting Lists: For each vocabulary term, a list containing references to all documents where that term appears. Posting lists may also store:
  - TF in each document
  - Term position for phrase matching
  - Document frequency (DF) for ranking algorithms like TF-IDF

There are advanced indexing techniques, including:

- Positional Information: For phrase queries or proximity searches, indexes must store the exact positions where terms appear within documents, significantly increasing storage requirements but enabling sophisticated query types.
- Distributed Indexing: Large-scale systems achieve high performance through parallelization, requiring index partitioning strategies. Two primary approaches exist: document partitioning (each machine handles a subset of documents) and term partitioning (each machine handles a subset of vocabulary terms).
- Compression Techniques: Modern indexes employ various compression methods to reduce storage requirements while maintaining query performance, including delta compression for posting lists and front-coding for vocabulary storage.

These indexing innovations are designed to meet a series of critical system-level requirements. Scalability ensures that the indexing infrastructure can accommodate growing datasets without compromising performance. Latency must be minimized to support real-time applications, such as autocomplete or web search, where users expect sub-second responses. The system must also be flexible, able to handle various document formats and query types—from short keywords to long natural language questions. Just as important is maintainability: indexes must support updates, deletions, and rebuilds with minimal disruption. Lastly, resource efficiency is paramount. Index construction and retrieval should make optimal use of CPU, memory, and storage to reduce operational costs and energy consumption.

1.2.2.2.3 *Example*

Here we use an example to demonstrate the document processing and indexing workflow. Let's say we have a list of AI research papers consisting of six documents:

```
Document 1: "Machine learning algorithms improve natural language processing tasks"
Document 2: "Deep learning revolutionizes computer vision and image processing"
Document 3: "Natural language models enhance machine translation accuracy"
Document 4: "Computer vision algorithms detect objects in real-time processing"
Document 5: "Machine learning applications span computer science and data processing"
Document 6: "Deep neural networks advance artificial intelligence and learning
            systems"
```

The process of constructing the index is as follows: Extract Terms → Build Vocabulary → Create Posting Lists → Sort & Index

Step 1: Document Preprocessing: Before building an index, documents are tokenized and converted into a "bag of words" (BoW) representation: an unordered collection of terms, often paired with frequency counts. Each document is converted into a dictionary of term frequencies. Figure 1.15 shows the BoW results of each document, where each document is represented as an unordered collection of words, with frequency counts but no positional information.

Step 2: Document Indexing: The resulting inverted index is listed in Table 1.2.

Step 3: Boolean Query Using Constructed Index: Inverted indexes make query evaluation efficient via set operations. For example, for the Boolean query: "learning AND computer," here are the steps to use inverted index to retrieve documents:

1. Term Lookup:
   - Retrieve posting list for "learning": {1, 2, 5, 6}
   - Retrieve posting list for "computer": {2, 4, 5}
2. Set Intersection
   - Compute intersection: $\{1, 2, 5, 6\} \cap \{2, 4, 5\} = \{2, 5\}$
3. Result Interpretation
   - Document 2: "Deep **learning** revolutionizes **computer** vision and image processing."
   - Document 5: "Machine **learning** applications span **computer** science and data processing."

This example demonstrates how Boolean queries are efficiently resolved through mathematical set operations, enabling rapid retrieval even from large document collections.

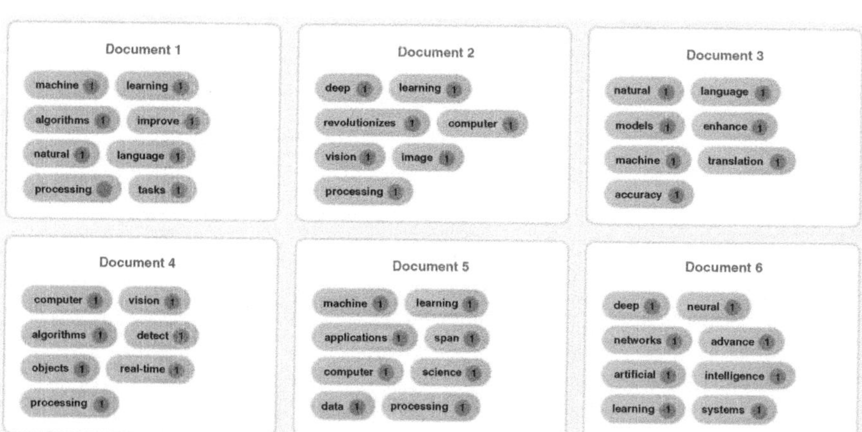

Figure 1.15 Bag of words representation.

Table 1.2 Inverted index table.

Term (vocabulary)	Posting list (Document IDs)	Document frequency
Accuracy	[3]	1
Advance	[9]	1
Algorithms	[1, 4]	2
Applications	[5]	1
Artificial	[6]	1
Computer	[2, 4, 5]	3
Data	[5]	1
Deep	[2, 6]	2
Detect	[4]	1
Enhance	[3]	1
Image	[2]	1
Improve	[1]	1
Intelligence	[6]	1
Language	[1, 3]	2
Learning	[1, 2, 5, 6]	4
Machine	[1, 3, 5]	3
Models	[3]	1
Natural	[1, 3]	2
Networks	[6]	1
Neural	[6]	1
Objects	[4]	1
Processing	[1, 2, 4, 5]	4
Real time	[4]	1
Revolutionizes	[2]	1
Science	[5]	1
Span	[5]	1
Systems	[6]	1
Tasks	[1]	1
Translation	[3]	1
Vision	[2, 4]	2

The pre-retrieval phase is crucial for system performance as it directly impacts both retrieval accuracy and efficiency. Well-processed user inputs and optimized document indexes create the foundation for sophisticated matching and ranking in subsequent phases.

1.2.3 Retrieval Models

IR systems employ a dual retrieval architecture that combines the efficiency of sparse retrieval with the semantic understanding capabilities of dense retrieval. This section explores both approaches and their integration in contemporary IR systems.

Table 1.3 lists the comparisons of sparse retrieval and dense retrieval. The combination of sparse and dense retrieval techniques in modern search systems allows for a comprehensive approach to document retrieval. Sparse methods excel at precise keyword matching and handling straightforward queries, while dense methods capture semantic nuances and handle more complex, context-dependent searches. This dual approach ensures that search systems can effectively bridge the gap between user intent and relevant content across a wide range of query types and complexities.

1.2.3.1 Sparse Retrieval

Sparse retrieval models form the backbone of classical IR systems. They represent documents and queries as high-dimensional sparse vectors, where most dimensions are zero, and focus on exact token matching rather than semantics. The term "sparse" refers to the characteristic that most positions in these high-dimensional vectors contain zero values, corresponding to terms that do not appear in a given document or query. This sparsity enables both computational efficiency and storage optimization, making sparse retrieval particularly suitable for large-scale systems where resource efficiency is paramount.

We introduced basic concepts of keyword search (Boolean retrieval), TF-IDF, and BM25 in Section 1.1.2.2. This section provides comprehensive details of these three foundational sparse retrieval methods, demonstrating their evolution from simple exact matching to sophisticated statistical ranking algorithms. To illustrate the progression and differences between sparse retrieval methods, we'll use a consistent example throughout the section.

The setup of the query and document collections for the retrieval problem setting is as follows: The query is: "machine learning optimization" and the total number of documents is: 10,000 computer science research papers.

Table 1.3 Comparison of sparse retrieval vs. dense retrieval.

Feature	Sparse retrieval	Dense retrieval
Representation	Sparse (term frequency)	Dense (neural embeddings)
Matching method	Lexical (token overlap)	Semantic (vector similarity)
Index type	Inverted Index	Vector index (e.g., FAISS)
Vocabulary mismatch	Problematic	Robust to synonyms and paraphrases
Scalability	Highly efficient	Requires vector indexing and ANN
Deep context handling	None	Enabled via transformers
Query–doc interaction	None (inverted lists)	Implicit (two-tower) or explicit (cross-attention)

The document collection includes papers on various Computer Science topics, with a sample of documents:

```
Document 1: "Machine Learning Optimization in Deep Networks" (200 words)
Document 2: "Advanced Machine Learning Techniques" (120 words)
Document 3: "Database Systems and Query Optimization" (180 words)
Document 4: "Computer Vision Applications" (150 words)
Document 5: "Optimization Algorithms for Machine Learning" (220 words)
```

1.2.3.1.1 Basic Keyword Search

Keyword search represents the simplest form of IR, operating on the principle of lexical matching. Early search systems counted term occurrences, ranking documents by the number of query words they contained.

The naive form of keyword-based search is Boolean retrieval, which evaluates queries using logical operators such as AND, OR, and NOT. It strictly matches query terms against documents without considering TF or semantic meaning.

For our query "ML optimization," we can construct different Boolean expressions as listed in Table 1.4.

Let's examine which documents match each Boolean query variation in Table 1.5:

Table 1.4 Query variations for "ML optimization."

Query variation	Boolean expression	Requirements	Approach
Variation 1	`"machine" AND "learning" AND "optimization"`	Requires ALL three terms to be present	Most restrictive approach
Variation 2	`("machine" AND "learning") OR "optimization"`	Documents must contain either both "machine" and "learning" OR contain "optimization"	More flexible approach
Variation 3	`"machine" OR "learning" OR "optimization"`	Documents need only contain ANY of the three terms	Most permissive approach

Table 1.5 Document query results for search terms.

Document	Contains "machine"	Contains "learning"	Contains "optimization"	Query_1 (AND)	Query_2 (OR)	Query_3 (OR)
Doc 1	✓	✓	✓	✓	✓	✓
Doc 2	✓	✓	×	×	✓	✓
Doc 3	×	×	✓	×	✓	✓
Doc 4	×	×	×	×	×	×
Doc 5	✓	✓	✓	✓	✓	✓

So the Boolean retrieval results are:

- Query 1 (AND): Documents 1, 5 (most restrictive)
- Query 2 (Mixed): Documents 1, 2, 3, 5 (balanced)
- Query 3 (OR): Documents 1, 2, 3, 5 (most permissive)

1.2.3.1.2 TF-IDF

To improve upon binary logic, as introduced in Section 1.1.2.2, TF-IDF provides a numerical weight to each term, balancing how important a word is to a specific document versus the entire corpus. This approach addresses the fundamental question: how important is a term for characterizing a document within a collection?

The definition of TF-IDF is as follows:

- TF measures how often a term appears in a document. It gives more weight to terms that are frequent in the document, assuming they are more important.

$$TF(t, d) = \frac{\text{Number of occurrences of the term “t” in the document “d”}}{\text{Total number of terms in the document “d”}}$$

- IDF measures the rarity of a term across the entire document collection. It reduces the weight of terms that occur in many documents, highlighting terms that are more unique to specific documents.

$$IDF(t, D) = \frac{\text{log (Total number of documents in the collection “D”}}{\text{Number of documents containing the term “t”)}}$$

- TF-IDF combines TF and IDF to provide a balanced measure of term importance within a document and across the collection. Terms with high TF-IDF values are both frequent in a specific document and rare in the overall collection, making them highly relevant for distinguishing that document in the context of a query.

$$\mathrm{TF-IDF(t, d, D) = TF(t, d) \times IDF(t, D)}$$

For the collection of documents listed in the example, here are the statistics for the query terms:

- “machine”: appears in 3,000 of 10,000 documents
- “learning”: appears in 800 of 10,000 documents
- “optimization”: appears in 150 of 10,000 documents

So thc IDF values for these three terms are computed as follows listed in Table 1.6:

Table 1.6 IDF computation for query terms.

Term	Docs containing term	IDF (log(10,000/df)) computation	IDF score
“machine”	3,000	log(10,000/3,000)	0.52
“learning”	800	log(10,000/800)	1.10
“optimization”	150	log(10,000/150)	1.82

Table 1.7 TF-IDF computation for three documents.

Term	IDF	Document 1 (200 words)	Document 2 (120 words)	Document 5 (220 words)
		Count/TF/TF-IDF	Count/TF/TF-IDF	Count/TF/TF-IDF
"machine"	0.52	8/0.040/0.021	4/0.033/0.017	6/0.027/0.014
"learning"	1.1	12/0.060/0.066	8/0.067/0.074	10/0.045/0.050
"optimization"	1.82	15/0.075/0.137	0/0.000/0.000	18/0.082/0.149
Total TF-IDF Score		0.224	0.091	0.213

The TF-IDF retrieval results are listed in Table 1.7. Here are the ranked results by the total TF-IDF score:

```
Document 1: 0.224 - "Machine Learning Optimization in Deep Networks"
Document 5: 0.213 - "Optimization Algorithms for Machine Learning"
Document 2: 0.091 - "Advanced Machine Learning Techniques"
Other documents: < 0.05 (not containing all query terms)
```

This example demonstrates how TF-IDF naturally identifies terms that are both locally important and globally distinctive, making them valuable for retrieval and ranking.

1.2.3.1.3 BM25

Based on TF-IDF, BM25 introduces tunable parameters that prevent TF from growing unboundedly and account for document length normalization. This prevents artificially long documents from dominating search results simply due to term repetition.

In the mathematical formulation of BM25, parameters k_1 and b control TF saturation and document length normalization, respectively:

$$\text{BM25}(q, d) = \sum \text{IDF}(q_i) \times \frac{f(q_i, d) \times (k_1 + 1)}{f(q_i, d) + k_1 \times \left(1 - b + b \times \frac{|d|}{\text{avgdl}}\right)}$$

where:

- $f(q_i, d)$ is the frequency of query term q_i in document d
- $|d|$ is the length of document d
- avgdl is the average document length in collection
- k_1 is the term frequency saturation parameter (typically 1.2–2.0)
- b is the document length normalization parameter (typically 0.75)

In the example listed, we use the below parameters setting:

```
k_1 = 1.5 (term frequency saturation)
b = 0.75 (document length normalization)
avgdl = 180 (average document length in collection)
```

Table 1.8 lists the BM25 computation results for three documents.

Table 1.8 BM25 computation for three documents.

	Document 1 (200 words)	Document 2 (120 words)	Document 5 (220 words)
Length normalization	1.11	0.75	1.17
Term	$f(t,d)$/TF Comp/BM25	$f(t,d)$/TF Comp/BM25	$f(t,d)$/TF Comp/BM25
"machine"	8/1.89/0.98	4/1.54/0.80	6/1.73/0.90
"learning"	12/2.14/2.35	8/2.11/2.32	10/2.12/2.33
"optimization"	15/2.25/4.10	0/0.00/0.00	18/2.29/4.17
Total BM25 score	7.43	3.12	7.4

Table 1.9 Comparison of Boolean retrieval, TF-IDF and BM25.

Method	Boolean	TF-IDF	BM25
Matching type	Exact	Statistical	Probabilistic
Ranking	No	Yes	Yes
Length norm	No	No	Yes
TF saturation	No	Linear	Nonlinear
Best for	Precise filtering	General search	Web-scale search

The retrieved results ranked by the BM25 scores are:

```
Document 1: 7.43 - "Machine Learning Optimization in Deep Networks"
Document 5: 7.40 - "Optimization Algorithms for Machine Learning"
Document 2: 3.12 - "Advanced Machine Learning Techniques"
Other documents: < 2.0 (lower relevance scores)
```

For the above three sparse methods, Table 1.9 lists their comparisons. While the sparse retrieval approach works well for straightforward factual queries, it struggles with semantic nuances and synonymous expressions. The limitations become apparent when users search for "car maintenance," but relevant documents discuss "automobile servicing" or "vehicle upkeep." Basic keyword search fails to recognize these semantic relationships, often missing highly relevant content due to vocabulary mismatch.

1.2.3.2 Dense Retrieval

Dense retrieval represents a fundamental shift from traditional lexical matching to semantic understanding through continuous vector representations. Unlike sparse methods that rely on exact-term matching, dense retrieval models encode both queries and documents into fixed-dimensional dense vectors, which capture semantic relationships through neural network architectures.

1.2.3.2.1 Embeddings and Granularities

As one of the major methods for dense vector representations, **embeddings** are continuous, fixed-size vector representations of linguistic units (words, phrases, sentences, or documents). They map semantically similar items to nearby points in a high-dimensional space, enabling models to reason about meaning numerically.

Word embeddings are vector representations of words that capture syntactic and semantic information based on their context in large corpora. Each word is mapped to a continuous vector in a high-dimensional space (e.g., more than 100 dimensions), where semantically similar words lie close together.

For example, we can obtain the word embeddings for words "king," "queen," "man," and "woman." As shown in the heatmap of Figure 1.16, each word can be represented by a 10-dimensional numerical vector. Using embeddings allows the model to understand similarities and relationships between different words based on their proximity in this vector space. Embeddings support vector arithmetic for analogies. For instance, as shown in Figure 1.17, in a well-trained embedding space, "king" minus "man" plus "woman" might result in a vector close to "queen," with a cosine similarity value of the first five dimensions being equal to 0.8561.

As listed in Table 1.10, embeddings are of multiple levels of granularity, capturing semantic information at varying scopes:

- Word Embeddings: Encode individual words, reflecting their semantic meanings and relationships. These embeddings capture the basic semantic units that can be combined to understand larger contexts.
- Sentence Embeddings: Capture contextual nuances of whole sentences. They aggregate word-level information to represent the meaning of sentences, considering word order, and syntactic structure.
- Document Embeddings: Represent entire documents, summarizing overall themes, and information. Document embeddings provide a holistic view of the content, making them useful

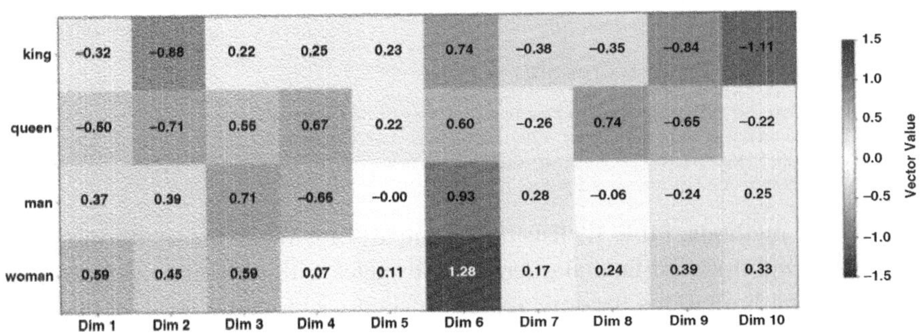

Figure 1.16 Word embeddings (first 10 dimensions) for four words: king, queen, man, and woman.

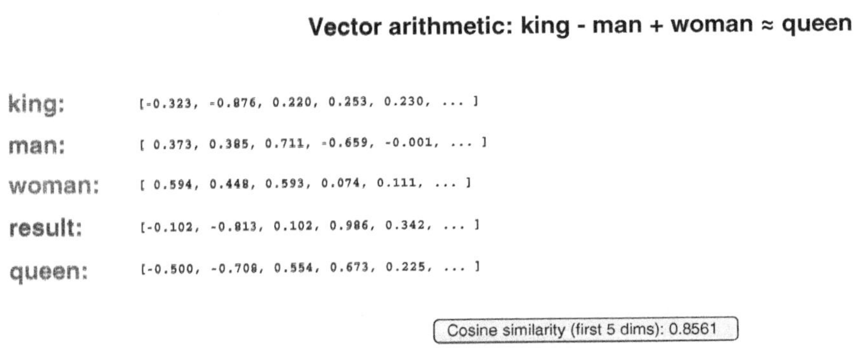

Figure 1.17 Vector arithmetic and similarity metrics.

Table 1.10 Different levels of embedding granularities.

Level	Description
Word embeddings	Represent individual words (e.g., "king," "queen"). Captures basic semantics
Sentence embeddings	Encodes full sentences. Useful for tasks like paraphrase detection or retrieval
Document embeddings	Aggregates multiple sentences. Suitable for long-form retrieval tasks

for retrieving documents that match the overall query intent. Sentence embeddings combine word-level semantics into a unified representation, while document embeddings aggregate sentence-level embeddings to capture global meaning.

Some popular pretrained embeddings include the following:

- OpenAI's Text-embedding-3-small/large: General-purpose, high-accuracy, LLM-aligned vectors.
- BAAI's BGE Models: Multilingual, sentence-level embeddings that support retrieval and reranking across diverse languages and topics.

We are going to discuss more details of how to obtain embeddings in Section 2.1.2.2 of this book.

1.2.3.2.2 Embedding-based Retrieval

Embedding-based retrieval (EBR) is a retrieval paradigm where documents and queries are represented as dense vectors (embeddings) in a latent space. Retrieval is then performed by comparing these vector representations using a similarity function, such as cosine similarity or inner product. This approach allows the system to capture semantic similarity between queries and documents beyond exact keyword matching, enabling more flexible and meaningful retrieval. Modern EBR systems rely on neural networks to generate the embeddings and often use specialized vector indexes to efficiently find nearest-neighbor vectors corresponding to relevant documents.

By encoding queries and documents into dense vectors, EBR facilitates relevance ranking based on semantic similarity rather than exact term matches. This section explores the key architectures, methodologies, and trade-offs in EBR.

Basic Vector Index Retrieval is the foundational approach in EBR, where documents and queries are encoded into dense embeddings using neural networks, and retrieval is performed by searching for the nearest document embeddings to a given query embedding in the vector space. This process typically involves the following steps:

1. Embedding Generation:
   - Query Embedding: Let $q \in \mathbb{R}^d$ be the embedding of the query, computed as $q = f_q(\text{query})$, where f_q is a neural network (e.g., BERT).
   - Document Embedding: Let $d_i \in \mathbb{R}^d$ be the embedding of document D_i, computed as $d_i = f_d(D_i)$, where f_d is a similar or identical neural network.
2. Similarity Function: Cosine similarity between query and document embeddings:

$$\text{sim}(q, d_i) = \frac{q \cdot d_i}{\| q \| \cdot \| d_i \|}$$

3. Retrieval: Return the top-k documents with the highest similarity scores. We can use an index (e.g., FAISS) to approximate the nearest neighbors (ANN) efficiently.

$$\text{Top-}k = \arg \max_{i=1,\ldots,n} \text{sim}(q, d_i)$$

For efficiency considerations, a straightforward dense vector search can be accelerated by vector indexes like inverted file indexes (IVF) or product quantization [9], which cluster or quantize embeddings to limit comparisons. Another approach is using graph-based structures. For example, hierarchical navigable small world (HNSW) [10] has emerged as the de facto standard for high-dimensional approximate nearest neighbor search, combining navigable small world graphs with hierarchical structures inspired by skip lists. These methods dramatically reduce search latency compared to exhaustive comparison, making real-time retrieval feasible even for millions of documents. However, basic vector indexes may still struggle as data scales, which motivates more structured or hierarchical retrieval strategies.

1.2.3.2.3 Two-tower Architecture

The two-tower model is the most widely adopted neural architecture for EBR due to its balance of efficiency, scalability, and performance. It consists of two independent neural networks, or "towers," that encode queries and documents separately into a shared embedding space. The architecture is designed to optimize both real-time query processing and offline document indexing, making it ideal for large-scale retrieval systems.

As illustrated in Figure 1.18, the Two-tower Architecture comprises:

1. Query Tower:
   - Purpose: Encodes user queries into dense embeddings that capture user intent and context.
   - Design Considerations: (1) query understanding: incorporates techniques like query expansion (e.g., synonym augmentation) and contextual analysis to enhance semantic understanding. (2) Low latency: optimized for real-time processing to ensure fast response times in interactive applications. (3) Input features: typically processes raw query text, metadata (e.g., user context), or reformulated queries.
2. Document Tower:
   - Purpose: Encodes documents into embeddings that represent their content and semantics comprehensively.
   - Design Considerations: (1) content understanding: Captures the full context of the document, including text, structure, and metadata (e.g., titles, tags). (2) Pre-computation: designed for offline processing, allowing embeddings to be computed and indexed in advance. (3) Scalability: handles large document collections by leveraging batch processing and distributed computing.

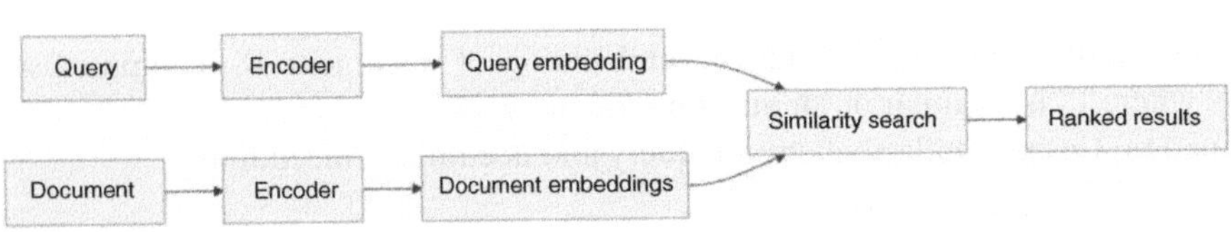

Figure 1.18 Two-tower architecture.

The two-tower model is typically trained using contrastive learning, where the objective is to maximize the similarity between embeddings of relevant query-document pairs while minimizing the similarity for irrelevant pairs. Common loss functions include:

- Contrastive Loss: Encourages positive pairs (relevant query-document) to be closer in the embedding space than negative pairs (irrelevant query-document).
- Triplet Loss: Uses triplets of (query, relevant document, irrelevant document) to enforce ranking.
- In-batch Negatives: Leverages other documents in the same batch as negative examples to improve training efficiency.

1.2.3.2.4 Cross-attention Models

While two-tower models prioritize efficiency, cross-attention models focus on maximizing relevance by enabling deep interactions between query and document representations. These models process query-document pairs jointly, allowing the model to capture fine-grained relationships at the token level.

As shown in Figure 1.19, the cross-attention architecture follows this structure: Query + Document → Joint Encoder → Relevance Score

- Joint Encoder: A single neural network (typically a transformer) processes the concatenated query and document text, using cross-attention mechanisms to model interactions between query tokens and document tokens.
- Cross-attention Mechanism: Each token in the query attends to all tokens in the document (and vice versa), allowing the model to capture context-specific relationships.
- Output: A relevance score (e.g., a scalar value) indicating the likelihood that the document is relevant to the query.

The choice and implementation of retrieval models significantly impact system performance, both in terms of relevance and efficiency. IR systems often employ multiple retrieval approaches, carefully balanced to achieve optimal results across different query types and use cases.

1.2.3.2.5 Hybrid Methods

The choice and implementation of retrieval models have a profound impact on system performance, influencing both relevance and efficiency. Table 1.11 lists the comparisons of the two-tower architecture vs. the cross-attention architecture.

IR systems therefore often employ multiple retrieval approaches in a coordinated pipeline to achieve optimal results across different scenarios. A common pattern is a two-stage (or multistage) retrieval process:

- Stage 1—High-recall Retrieval: Use a fast but less expensive method (like a two-tower EBR model with an ANN index, or even a traditional keyword search) to retrieve a broad set of candidates. This stage prioritizes recall—getting in as many potentially relevant items as possible with minimal latency. The two-tower model is ideal here because it can handle

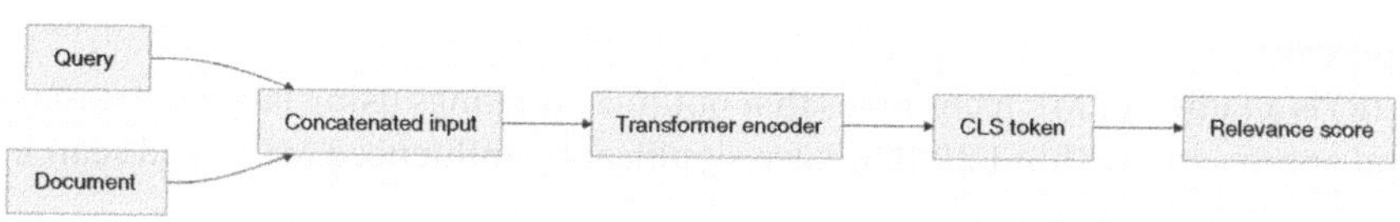

Figure 1.19 Cross-attention architecture.

Table 1.11 Trade-off of two-tower vs. cross-attention models.

Factor	Two-tower	Cross-attention
Latency	1–10 ms	100–1000 ms
Accuracy	Good	State of the art
Scalability	Billions of docs	Thousands of docs
Freshness	Instant	Retraining needed

massive candidate sets quickly, outputting embeddings that are matched via ANN search in milliseconds.

- Stage 2—Precise Reranking: Take the candidate set from Stage 1 and apply a more powerful but slower model to refine the ordering. This could be a cross-attention neural reranker or another model that considers richer features and interactions. This stage focuses on precision—ensuring the very top results are truly the best answers for the query. Because this stage deals with far fewer items (hundreds or less instead of millions), it can afford the computational cost of deep models.

This hybrid approach balances efficiency and accuracy, making it suitable for diverse query types and large-scale deployments.

1.2.4 Post-retrieval

While retrieval models efficiently identify potentially relevant documents, the post-retrieval phase ensures that results are not only technically relevant but also contextually meaningful and personalized to user intent. The post-retrieval stage transforms these raw candidates into a refined, ranked, and user-aligned output. This phase comprises three critical components:

- **Ranking**, which decides the order of the results based on predicted relevance.
- **Result representation**, which determines how results are communicated to users.
- **Evaluation**, which ensures the system's performance meets expectations via metrics and feedback.

Through iterative refinement driven by user feedback, post-retrieval processes enable IR systems to adapt dynamically, ensuring precision and relevance in diverse contexts.

1.2.4.1 Ranking

While retrieval models identify potentially relevant documents, ranking systems determine the order in which they are presented to the user. This process is crucial for ensuring that the most relevant documents appear at the top of the search results. The goal is to produce a preliminary ranked list that captures broadly relevant documents while maintaining scalability.

1.2.4.1.1 Learning-to-rank

Learning-to-rank (L2R) is a family of ML techniques that optimize rankings using labeled datasets [11]. There are several approaches within L2R that have significantly influenced ML-based search methods:

- Pointwise Approaches: These convert ranking problems into regression or classification tasks on individual documents. By treating each document independently, pointwise approaches aim to assign a direct relevance score based on the document's content and the query. For example, a linear regression model predicts a relevance score with this loss function:

$$\mathcal{L}_{\text{pointwise}} = \sum_{i} (s_i - y_i)^2$$

where s_i is the predicted score, and y_i is the true relevance.

- Pairwise Approaches: These treat ranking as a classification problem between pairs of documents. This method evaluates which of two documents is more relevant to a specific query, refining the accuracy of rankings by learning from these comparisons. For example, RankSVM [12] minimizes a hinge loss:

$$\mathcal{L}_{\text{pairwise}} = \sum_{(i,j)\in\text{pairs}} \max(0, 1 - (s_i - s_j))$$

RankNet [13] minimizes a cross-entropy loss:

$$\mathcal{L}_{\text{pairwise}} = -\sum_{(i,j)\in\text{pairs}} \log \sigma(s_i - s_j)$$

Other notable pairwise ranking methods include LambdaRank [14].

- Listwise Approaches: These consider the entire list of documents directly. By modeling the ranking process in a holistic manner, listwise approaches aim to overcome the limitations of pointwise and pairwise methods, offering a more nuanced and effective way to manage rankings in search results. For instance, ListNet [15] minimizes a loss based on the permutation probability of the ranked list:

$$\mathcal{L}_{\text{listwise}} = -\sum_{\pi} P(\pi|y) \log P(\pi|s)$$

where π is a permutation, $P(\pi|y)$ is the ground-truth permutation probability, and $P(\pi|s)$ is the predicted probability. Other notable listwise ranking methods include ListMLE [16] and LambdaMART [17].

Overall, the Pointwise method is efficient but ignores the relative relationship between documents; Pairwise improves local accuracy by comparing document pairs, while Listwise optimizes globally but has a higher computational cost.

Modern IR systems commonly employ a two-stage ranking architecture, consisting of an initial ranking phase followed by a reranking phase. Each stage plays a distinct and complementary role in balancing efficiency and accuracy in large-scale systems.

1.2.4.1.2 Initial Ranking: Fast, Scalable Candidate Selection

The initial ranking stage is responsible for rapidly generating a shortlist of candidate documents or items from a potentially massive index—often millions or billions of entries. The focus here is on speed and recall, ensuring that the set of retrieved documents contains as many relevant candidates as possible.

While fast and scalable, initial ranking models typically rely on limited feature sets, avoid complex computations, and may not fully model fine-grained user intent or context. This can lead to relevant results being improperly ranked lower or less relevant results appearing near the top.

1.2.4.1.3 Re-ranking: Precision-oriented Refinement

The reranking stage refines the initial ranked list by reassessing documents using additional signals and more computationally intensive models. Reranking is essential because it addresses the limitations of initial ranking by incorporating contextual, user-specific, and temporal factors that enhance precision and relevance. By focusing on a smaller subset of top-ranked documents (e.g., top 100), reranking balances computational cost with improved ranking quality, making it a critical step in modern IR systems.

Multifactor reranking combines relevance scores with additional signals through learned weighting functions. A typical implementation might use:

$$\begin{aligned}\text{Final\_Score} &= \alpha \times \text{Relevance\_Score} + \beta \times \text{Freshness\_Score} \\ &\quad + \gamma \times \text{Authority\_Score} + \delta \times \text{Personalization\_Score}\end{aligned}$$

where the weights $\alpha, \beta, \gamma, \delta$ are learned from user interaction data (e.g., clicks, dwell time, conversions) and system performance metrics (e.g., NDCG, CTR). These weights allow the ranking function to dynamically prioritize different aspects of document quality, such as content relevance, recency, trustworthiness, and user-specific preferences.

By combining initial ranking and reranking, systems ensure that retrieved documents are not only relevant but also optimally ordered to meet user expectations. However, the effectiveness of ranking depends on how results are presented to users, which we explore in Section 1.2.4.2 on result representation.

1.2.4.2 Result Representation

Once documents are ranked, the result representation stage determines how they are presented to the user, significantly influencing the overall experience. Effective result representation goes beyond listing documents; it involves structuring and visualizing results to enhance usability, interpretability, and engagement. This stage is critical for minimizing the user's cognitive load and helping them quickly assess relevance without clicking on every link.

Key components of result representation include:

1.2.4.2.1 Snippets and Highlighting

Instead of just showing a title, the system generates a short snippet—a few lines of text from the document that contains the user's query terms. To draw the eye, these query terms are often bolded, providing an immediate visual cue of the document's relevance.

1.2.4.2.2 Rich Results and Knowledge Panels

For certain types of queries, the system can display structured data in a more visually appealing and informative format.

1.2.4.2.3 Personalized Formatting

Recommendation systems tailor result layouts based on user preferences. For example, a streaming service might prioritize movie posters for visual learners or detailed synopses for text-focused users.

1.2.4.2.4 Contextual Highlighting

Highlighting query terms or related concepts in results enhances relevance perception. For example, a news aggregator might bold keywords like "climate change" in article summaries.

For example: In a question-answering system, a query like "What is transformer architecture?" might present results as:

- A concise answer extracted from the top-ranked document: "The transformer is a neural network architecture using self-attention..."

- A list of related articles with snippets and links.
- A visual diagram (e.g., a transformer model schematic) to aid understanding.

In summary, effective representation turns a list of documents into a rich, interactive, and efficient user experience, bridging the final gap between the system's output and the user's understanding.

1.2.4.3 Evaluation

After the ranking and presentation phases, the system produces its final output—whether a ranked list of documents, recommended items, or responses in an interactive system. To ensure high-quality results and continuous improvement, a robust evaluation framework is essential. Evaluation serves as the critical feedback mechanism that measures how well the system meets user needs and achieves its intended goals.

The evaluation process is not a one-time activity but an ongoing cycle that feeds back into the system. By analyzing evaluation results, the system can identify areas for improvement and adapt its retrieval and ranking mechanisms accordingly. This feedback loop is a key feature of modern IR systems, enabling them to learn from user interactions and evolve over time.

1.2.4.3.1 Evaluation Metrics

Evaluation metrics fall into two major categories: offline metrics and online metrics, each tailored to specific use cases. Offline metrics are computed using pre-labeled datasets; online metrics capture real-time user feedback via A/B testing or behavioral signals.

1. **Offline Evaluation**: Offline evaluation measures how well a system performs on known test data with labeled relevance. Common metrics include:
   - **Precision**: The proportion of retrieved documents that are relevant:

     $$\text{Precision} = \frac{|\text{Relevant Documents} \cap \text{Retrieved Documents}|}{|\text{Retrieved Documents}|}$$

     Example: For a query retrieving 10 documents, 7 of which are relevant, precision is $\frac{7}{10} = 0.7$.
   - **Recall**: The proportion of relevant documents retrieved:

     $$\text{Recall} = \frac{|\text{Relevant Documents} \cap \text{Retrieved Documents}|}{|\text{Relevant Documents}|}$$

     Example: If there are 12 relevant documents in the collection and 7 are retrieved, recall is $\frac{7}{12} \approx 0.583$.
   - **Normalized Discounted Cumulative Gain (NDCG)**: Evaluates ranking quality by weighting relevance by position:

     $$\text{DCG}_k = \sum_{i=1}^{k} \frac{\text{rel}_i}{\log_2(i+1)}, \quad \text{NDCG}_k = \frac{\text{DCG}_k}{\text{IDCG}_k}$$

     where rel_i is the relevance score of the document at position i, and IDCG_k is the ideal DCG for the top-k documents. Example: For a ranked list with relevance scores [5, 13, 18], DCG is computed and normalized against the ideal ranking [5, 13, 18].

- **Mean Average Precision (MAP)**: Averages precision across all relevant documents for multiple queries:

$$\text{AP} = \frac{\sum_{k=1}^{n}(\text{Precision@}k \cdot \text{rel}_k)}{|\text{Relevant Documents}|}, \quad \text{MAP} = \frac{\sum_{q \in Q} \text{AP}_q}{|Q|}$$

  Example: For a query with three relevant documents retrieved at ranks 1, 3, and 5, AP considers precision at each relevant position.
- **Reciprocal Rank (RR)**: The inverse of the rank of the first relevant document:

$$\text{RR} = \frac{1}{\text{Rank of First Relevant Document}}$$

  Example: If the first relevant document is at rank 2, RR is $\frac{1}{2} = 0.5$.

2. **Online Evaluation:** Live systems require user-centered metrics that capture behavior and engagement.
   - **Click-through Rate (CTR)**: The proportion of users who click on a result:

$$\text{CTR} = \frac{\text{Number of Clicks}}{\text{Number of Impressions}}$$

     Example: If 100 users see a recommended movie and 20 click, CTR is $\frac{20}{100} = 0.2$.
   - **Dwell Time**: The average time users spend engaging with a result. For example: In a news aggregator, users spending 30 seconds on an article indicate higher engagement than those spending five seconds.
   - **Session Duration**: Measures the total time spent interacting with the system.
   - **Conversion Rate**: The proportion of users who take a desired action (e.g., purchasing a product):

$$\text{Conversion Rate} = \frac{\text{Number of Conversions}}{\text{Number of Users}}$$

     Example: In an e-commerce platform, if 50 out of 1000 users buy recommended headphones, the conversion rate is $\frac{50}{1000} = 0.05$.

Modern evaluation approaches also incorporate A/B testing to compare system variants with real users, and online evaluation methods that capture real-time user feedback and behavioral signals.

User feedback collected during evaluation dynamically adjusts ranking models, ensuring continuous improvement in alignment with user intent. The evaluation framework described here forms the basis for the advanced Graph-RAG system evaluations detailed in Chapter 9.

1.3 Information Retrieval Applications

IR powers a diverse ecosystem of modern applications, from everyday web searches to personalized content recommendations. While these applications share fundamental technical components such as document indexing, candidate retrieval, and result ranking (as discussed in Section 1.2.1), they differ dramatically in their approach to understanding and satisfying user needs.

The most prominent IR applications fall into two primary categories: **search systems** and **recommendation systems**. These represent fundamentally different paradigms for connecting users with relevant information. Search systems respond to explicit user queries where individuals clearly articulate their information needs through typed or spoken input. In contrast, recommendation systems operate without explicit queries, instead inferring user preferences from behavioral patterns, contextual signals, and historical interactions to proactively suggest relevant content.

Table 1.12 Core differences of search and recommendation systems.

Aspect	Search systems	Recommendation systems
User input	Explicit queries (e.g., "best hiking boots")	Implicit signals (past behavior, context)
Primary goal	Match documents to query intent	Anticipate user preferences
Key metrics	Precision, NDCG	Engagement, retention
User intent	Immediate information need	Discovery of new interests
Example	Google search	Netflix "Top Picks for You"

This fundamental distinction in user input processing cascades through every aspect of system design, from the algorithms used for understanding user intent to the metrics employed for evaluating success. Understanding these differences is crucial for designing effective IR systems and selecting appropriate techniques for specific application domains. As shown in Table 1.12, each paradigm serves distinct user needs and employs different approaches to IR.

Beyond these primary applications, IR techniques extend to specialized domains, including targeted advertising systems, AI-powered assistants (such as Siri and Alexa), and domain-specific platforms serving healthcare, legal research, and e-commerce contexts. Each of these applications adapts core IR principles to meet specific user needs and operational constraints.

This section explores search and recommendation systems in detail, examining their distinct approaches to user input processing, core architectural components, and unique technical challenges. Through practical examples and evaluation strategies, we demonstrate how these systems translate theoretical IR concepts into real-world solutions that serve billions of users daily.

1.3.1 Search Systems

Search systems represent the most direct application of IR principles, designed to retrieve and rank documents in response to explicit user queries. The availability of clear textual input makes query understanding and relevance matching central to their operation, creating a pipeline that must effectively transform user intentions into relevant document selections.

The explicit nature of search queries provides both opportunities and challenges. While users clearly express their information needs, the system must accurately interpret potentially ambiguous language, handle variations in query formulation, and understand implicit context that users may not explicitly state. This requirement drives sophisticated query processing techniques that form the foundation of modern search systems.

1.3.1.1 Search System Architecture

The search pipeline consists of four interconnected stages, each specifically tailored to handle explicit user input effectively and ensure that results align closely with the query's intent:

1. **Query Processing**: Transforms raw user input into structured representations suitable for retrieval.
2. **Retrieval**: Identifies candidate documents from large collections using both sparse and dense methods.
3. **Ranking**: Scores and orders candidates based on relevance and quality signals.
4. **Result Presentation**: Formats and displays results to optimize user comprehension and task completion.

This architecture reflects the linear nature of search interactions, where each user query triggers a complete processing cycle designed to satisfy an immediate information need.

1.3.1.2 Query Processing: The Foundation of Search Understanding

Query processing serves as the cornerstone of search systems, transforming raw user input into structured forms that enable effective retrieval and ranking. This complex process addresses the inherent challenges of natural language input while preserving user intent throughout multiple transformation stages.

Modern query processing encompasses three interconnected areas that work together to bridge the gap between human expression and machine understanding: intent recognition, query rewriting, and routing with decomposition. Each area addresses specific aspects of the interpretation challenge while contributing to overall system effectiveness.

1.3.1.2.1 Intent Recognition: Understanding User Goals

The first critical step in query processing involves understanding the user's underlying goal, which drives all subsequent processing decisions. Search intent typically falls into three well-established categories, each requiring different retrieval strategies and result presentation formats [18]:

Informational queries seek knowledge or factual information about specific topics. When users ask, "What is the capital of France?" they expect direct, authoritative answers presented clearly and concisely. These queries benefit from integration with knowledge bases, featured snippets, and structured data that can provide immediate answers.

Navigational queries aim to reach specific websites or digital destinations. A search for "Wikipedia" indicates the user wants to access that particular platform rather than learn about wikis in general. These queries require understanding brand names, website entities, and user navigation patterns to surface the intended destination prominently.

Transactional queries express intent to complete specific actions such as purchasing products or downloading files. "Buy plane tickets to New York" signals commercial intent that benefits from integration with e-commerce platforms, pricing information, and actionable interfaces.

Modern search systems employ ML classifiers trained on large query datasets to automatically categorize intent. For example, when processing "capital of France," the system uses NLP to classify the query as informational and prioritizes factual responses from authoritative knowledge sources like Wikidata or educational institutions.

1.3.1.2.2 Query Rewriting: Enhancing Input Quality

Once intent is established, query rewriting techniques refine the user's input to improve retrieval accuracy and address common issues in natural language expression. This process involves several complementary approaches that standardize, expand, and enrich the original query.

Query normalization addresses basic input quality issues through spelling correction and term standardization. Advanced systems use techniques like Levenshtein distance calculations combined with domain-specific dictionaries to correct errors. For instance, "Azuer Devpos tutorials" becomes "Azure DevOps tutorials" through character-level corrections informed by technology terminology patterns [19].

Query compression removes redundant or irrelevant terms, particularly valuable in conversational search contexts where users provide extensive context. A verbose input like "I'm looking for something affordable. What are the best hiking trails near me?" can be compressed to "best hiking trails near me," focusing processing power on the core information need while preserving essential context.

Query expansion addresses vocabulary mismatch by adding synonyms and related terms that capture broader intent. Statistical approaches analyze query logs to identify terms that frequently co-occur in successful searches, while semantic approaches use word embeddings to find conceptually related terms. For example, "best hiking trails" might expand to include "top-rated hiking trails," "scenic walking paths," or "outdoor recreation areas" based on successful historical searches and semantic relationships [20].

Context enrichment enhances queries with additional information derived from user history, session behavior, and external knowledge sources. This process considers previous queries and clicks within the current session, long-term user preferences when available, and relevant structured data from ontologies or knowledge graphs. For instance, the original query "best laptops 2024" might be enriched to "top-reviewed laptops for students with long battery life in 2024" when the system detects educational context from the user's search history.

1.3.1.2.3 Routing and Decomposition: Managing Complex Information Needs

The final aspect of query processing involves directing queries to appropriate resources and breaking down complex information needs into manageable components.

Query routing ensures that search requests reach the most suitable data sources and processing systems based on query characteristics and system architecture. In enterprise environments, queries about customer transactions might route to secure financial databases with appropriate access controls, while general knowledge queries access public web indexes optimized for broad coverage. Effective routing considers data sensitivity, computational requirements, geographic relevance, and regulatory compliance to optimize both performance and security.

Query decomposition addresses complex queries that contain multiple information needs by breaking them into simpler sub-questions that can be processed independently. For example, "Who won the championship recently, Red Sox or Patriots?" can be decomposed into "When did the Red Sox last win a championship?" and "When did the Patriots last win a championship?" This decomposition enables parallel processing of sub-queries and more precise retrieval for each component, with answers later synthesized into a coherent response [21].

These traditional query processing techniques rely on combinations of predefined rules, statistical models, and heuristic-based enrichment methods to infer user intent and prepare queries for effective retrieval. The sophistication of this processing directly impacts the quality of subsequent retrieval and ranking stages.

1.3.1.3 Search System Challenges

Despite decades of advancement, search systems continue to face significant challenges that drive ongoing research and development efforts.

Intent ambiguity remains a persistent challenge when queries like "apple" could refer to fruit, technology companies, or record labels. Robust intent recognition requires considering user context, search history, and real-time signals to disambiguate effectively.

Scalability demands processing millions of documents and queries in real time while maintaining sub-second response times. This requirement drives innovations in efficient retrieval techniques (such as HNSW indexes discussed in Section 1.2.3.2) and distributed ranking systems.

Evaluation complexity involves measuring not just relevance accuracy through metrics like NDCG and MAP (Section 1.2.4.3), but also user satisfaction, task completion rates, and long-term engagement. Capturing nuanced intent and user context in evaluation frameworks requires ongoing refinement and adaptation to changing user behaviors.

1.3.2 Recommendation Systems

Recommendation systems represent a fundamentally different approach to IR, operating without explicit user queries and instead interpreting implicit behavioral signals to suggest relevant content proactively. This paradigm shift from reactive search to proactive discovery creates unique opportunities for content exploration while introducing distinct technical challenges.

Unlike search systems that respond to immediate, articulated information needs, recommendation systems must continuously analyze user behavior patterns, contextual signals, and content characteristics to predict what users might find valuable or engaging. This prediction challenge requires sophisticated user modeling techniques that capture both momentary interests and evolving long-term preferences.

The absence of explicit queries fundamentally changes system design priorities. While search systems optimize for query-document relevance, recommendation systems focus on user engagement, content discovery, and long-term platform retention. These different objectives drive distinct approaches to candidate generation, ranking strategies, and evaluation methodologies.

1.3.2.1 Recommendation System Architecture

The recommendation pipeline mirrors search systems in its use of retrieval and ranking components but operates from a foundation of implicit user input analysis:

1. **User Interest Modeling**: Infers preferences from behavioral signals, demographic information, and contextual factors.
2. **Candidate Retrieval**: Generates potential recommendations using collaborative filtering, content analysis, and hybrid approaches.
3. **Ranking**: Scores candidates based on predicted user preference, engagement likelihood, and strategic objectives.
4. **Result Presentation**: Displays recommendations in formats optimized for discovery and engagement.

This architecture reflects the proactive nature of recommendation systems, where user models continuously evolve and drive content suggestions without explicit user requests.

For example, TikTok's recommendation system analyzes a user's viewing history (such as frequent engagement with fitness videos) to build interest profiles that inform candidate retrieval using embedding-based techniques. The ranking stage prioritizes content with high engagement potential, while reranking ensures diversity by including content from different genres and creators.

1.3.2.2 User Interest Modeling

User interest modeling forms the foundation of effective recommendation systems, replacing the explicit query understanding required in search systems. This modeling process must infer complex, multifaceted user preferences from indirect behavioral signals and contextual information.

Temporal interest dynamics recognize that user preferences operate at multiple time scales. Short-term modeling captures immediate intent from recent interactions, such as recommending running shoes after a user views fitness gear or suggesting related cooking videos during a baking session. Long-term modeling analyzes historical data spanning weeks or months to identify stable preferences, like a consistent interest in science fiction movies or a preference for organic food products.

Multi-time scale models combine both perspectives to balance immediate relevance with personalized recommendations. Spotify, for example, might recommend a new workout playlist based on recent gym check-ins (short-term signal) while prioritizing rock music based on months of listening history (long-term preference).

Multi-interest recognition acknowledges that users typically maintain diverse, sometimes conflicting interests that cannot be captured by single preference vectors. A user might simultaneously enjoy horror movies, cooking shows, and financial news, representing distinct interest clusters that require separate modeling approaches.

Advanced systems employ techniques such as multi-head attention mechanisms and mixture-of-experts models to identify and separately model these diverse interest dimensions. This approach enables more nuanced recommendations that can appropriately switch between different user personas based on current context and interaction patterns.

1.3.2.3 Traditional Recommendation Techniques

Collaborative filtering leverages user behavior patterns to suggest items based on similarities between users or items. User-based collaborative filtering identifies users with similar interaction patterns and recommends items enjoyed by those similar users. For example, if User A and User B both rated science fiction movies highly, the system might recommend to User A other movies that User B enjoyed but User A hasn't yet seen.

Item-based collaborative filtering focuses on relationships between items rather than users, suggesting items similar to those the user has previously engaged with. This approach might recommend science fiction books based on past purchases of similar titles, using co-interaction patterns to identify related content.

While collaborative filtering excels at capturing complex preference patterns without requiring content understanding, it struggles with the cold-start problem, where new users or items lack sufficient interaction data for accurate recommendations [22]. Solutions include leveraging content-based features for new items or social signals for new users.

Content-based filtering recommends items with attributes similar to those previously engaged with by the user. This approach analyzes item features such as genre, actors, and directors for movies, or topics and authors for articles, to create detailed content profiles. User profiles are built by aggregating features from items with which they've interacted, enabling recommendations through profile matching.

While effective for users with consistent, well-defined preferences, content-based systems risk creating filter bubbles by repeatedly recommending similar content, potentially limiting exposure to diverse experiences and new interests [23].

Hybrid methods combine collaborative and content-based approaches to leverage the strengths of different methodologies while mitigating individual weaknesses. Ensemble methods blend predictions from multiple algorithms, such as averaging scores from matrix factorization and neural models. Feature augmentation uses outputs from one recommendation technique as inputs for another, such as incorporating collaborative filtering user embeddings into content-based models.

Switching strategies adapt recommendation approaches based on contextual factors, deploying popular items for new users while providing personalized suggestions for engaged users. Most production systems, including Netflix and Amazon, employ hybrid approaches to achieve robust performance across diverse user segments and content types [24].

Sequential modeling captures meaningful patterns in user behavior over time, recognizing that interactions often follow logical progressions both within individual sessions and across longer periods. Session-based recommendations model the evolution of user interests during single interaction periods, such as suggesting the next logical video during a YouTube watching session or the next product during an e-commerce browsing session.

Sequence-aware recommenders track how preferences evolve across multiple sessions and extended time periods, such as gradually shifting music tastes from pop to indie rock or evolving reading preferences from fiction to nonfiction as life circumstances change [25].

1.3.2.4 Recommendation System Challenges

Recommendation systems face unique challenges that differ significantly from those encountered in search systems.

The cold-start problem affects both new users who lack sufficient interaction history for accurate preference modeling and new items that haven't received enough user feedback for confident recommendations. This challenge requires creative solutions such as onboarding questionnaires, social signal integration, or content-based fallback strategies.

Filter bubbles and echo chambers can result from overreliance on past behavior patterns, potentially limiting user exposure to diverse content and new experiences. Balancing personalization with diversity requires sophisticated techniques that encourage exploration while maintaining engagement.

Evaluation complexity involves measuring success through online metrics such as CTRs, dwell time, and long-term user retention (Section 1.2.4.3). However, optimizing for immediate engagement may conflict with longer-term goals of user satisfaction, content discovery, and platform diversity. This tension requires careful balancing of multiple, sometimes competing objectives.

1.3.3 Example: WANDS Dataset

To illustrate the practical application of IR techniques throughout the remainder of this book, we will utilize the WANDS (wayfair product search relevance dataset) as a concrete example that demonstrates both search and recommendation system principles in an e-commerce context.

The WANDS dataset provides an open-source benchmark specifically designed to evaluate the relevance of search results in real-world e-commerce scenarios. This dataset encompasses the complexities of product search, including diverse query types, varied product attributes, and nuanced relevance judgments that reflect actual user behavior patterns.

By grounding our theoretical discussions in this practical dataset, we can demonstrate how the concepts explored in this chapter—from query processing techniques to user interest modeling—translate into measurable improvements in system performance. The dataset's availability at https://github.com/wayfair/WANDS enables readers to explore these techniques hands-on, bridging the gap between academic understanding and practical implementation.

This practical foundation will prove particularly valuable as we explore advanced topics such as neural retrieval architectures, ranking optimization, and evaluation methodologies in subsequent chapters, providing concrete examples that illustrate both the potential and the challenges of modern IR systems.

1.3.3.1 Dataset Overview and Structure

The WANDS dataset represents a comprehensive e-commerce search benchmark that captures the real-world complexity of product discovery. Understanding its structure provides the foundation for implementing and evaluating IR techniques in commercial contexts.

1.3.3.1.1 Core Dataset Components

The WANDS dataset is organized into three primary files that together form a complete search evaluation framework:

Product catalog (`product.csv`) contains detailed information about 42,994 products, including:

Product ID: Unique identifier for each product
Title: Customer–facing product name
Class: Product category classification
Category Hierarchy: Nested taxonomy path
Description: Detailed product description with specifications
Features: Product attributes and specifications as structured key–value pairs
User Ratings: Average customer rating on a 1–5 scale
Reviews: Collection of user–generated reviews with sentiment indicators

Query collection(`query.csv`) encompasses 480 diverse search queries representing authentic user search behaviors:

Query ID: Unique identifier for each search query
Raw Query Text: Original user input without preprocessing
Associated Class: Primary product category the query targets

Relevance judgments (`label.csv`) provide 233,448 human-annotated relevance assessments using a three-level scale:

Exact: Product perfectly matches query intent and user need
Partial: Product partially satisfies query intent with some relevance
Irrelevant: Product does not match query intent or user need

1.3.3.1.2 Dataset Statistics and Characteristics

Table 1.13 lists the key statistics and characteristics of the datasets.

And here is an example product entry:

```
# Example Query Entry
"query.csv": [
    {
      "query_id": "0",
      "query": "salon chair",
      "query_class": "Massage Chairs"
    },
    {
      "query_id": "1",
      "query": "smart coffee table",
      "query_class": "Coffee & Cocktail Tables"
    }
  ],

# Example Product Entry
"product.csv": [
```

Table 1.13 WANDS dataset file structure and contents.

File	Contents	Records	Key Fields
`product.csv`	Product details	42,994	ID, title, category, description, features
`query.csv`	Customer searches	480	Query ID, text, class
`label.csv`	Relevance labels	233,448	Query ID, Product ID, relevance score

```
    {
      "product_id": 1,
      "product_name": "all-clad 7 qt. slow cooker",
      "product_class": "Slow Cookers",
      "category_hierarchy": "Kitchen & Tabletop / Small Kitchen Appliances / Pressure & Slow Cookers / Slow Cookers / Slow Slow Cookers",
      "product_description": "create delicious slow-cooked meals, from tender meat to flavorful veggies , with this easy-to-use slow cooker . the unit features a nonstick cast-aluminum insert that moves seamlessly from the oven or stovetop to the electric base to the table . you can use the insert alone or with the slow cooker to make a variety of one-pot dishes from soup to desserts , and much more . you can even prepare your ingredients in the morning , place everything in the slow cooker , and walk away to come home to the aroma of a hot , healthy dinner at the end of a busy day . with its sleek stainless-steel finish , the slow cooker not only presents beautifully , but it' s also the perfect size to accommodate the whole family or a large group when entertaining .",
      "product_features": {
        "capacity_quarts": "7",
        "product_type": "slow cooker",
        "programmable_cooking_settings": ["slow cook", "keep warm"],
        "overall_height_top_to_bottom": "14.2",
        "overall_depth_front_to_back": "11.3",
        "programmable_food_settings": ["soup", "stew"],
        "overall_width_side_to_side": "19.9",
        "top_programmable_settings": ["soup & stew"],
        "dishwasher_safe_parts": ["cooking pot/insert", "lid"],
        "features": ["non-stick", "automatic shutoff", "adjustable temperature settings"],
        "housing_heating_base_material": "stainless steel",
        "cooking_pot_insert_material": "aluminum",
        "operation_type": "programmable",
        "finish": "white",
        "overall_product_weight": "13",
        "uniform_packaging_and_labeling_regulations_uplr_compliant": "no",
        "commercial_warranty": "no",
        "indicator_light": "yes",
        "removable_cooking_pot_insert": "yes",
        "programmable": "yes",
        "country_of_origin": "china",
        "supplier_intended_and_approved_use": "residential use"
      },
      "rating_count": 100,
      "average_rating": 2.0,
      "review_count": 98
    }
]

# Example Label Entry
"label.csv": [
    {
      "id": 3001,
      "query_id": 27,
      "product_id": 24531,
      "label": "Exact"
    },
    {
      "id": 3002,
      "query_id": 27,
      "product_id": 40850,
      "label": "Partial"
    },
    {
      "id": 5322,
      "query_id": 42,
      "product_id": 583,
      "label": "Irrelevant"
    }
  ]
```

1.3.3.1.3 Query Type Analysis

The WANDS dataset includes diverse query patterns that reflect real user search behaviors in e-commerce contexts. To better understand the complexity and variety of user information needs, we categorize the queries into five distinct types based on their characteristics and user intent. Table 1.14 summarizes these query categories with representative examples and their implications for retrieval system design.

This query diversity ensures that retrieval systems must handle various levels of specificity, ambiguity, and user intent, making WANDS a comprehensive evaluation benchmark for e-commerce search applications.

1.3.3.2 Dataset Setup and Preprocessing

Effective utilization of the WANDS dataset requires systematic setup and preprocessing to prepare data for retrieval system implementation and evaluation.

1.3.3.2.1 Initial Data Loading

The dataset can be accessed and loaded using standard data processing libraries. Download github repo and read CSV files into pandas dataframes:

```
# Clone the WANDS repository
git clone https://github.com/wayfair/WANDS.git

# Load data into pandas dataframes
def get_wands_dfs(wands_folder="../WANDS/dataset/"):
    df_product = get_product_df(wands_folder + "product.csv")
    df_query = get_query_df(wands_folder + "query.csv")
    df_label = get_label_df(wands_folder + "label.csv")

    mask_query = df_label["query_id"].isin(df_query["query_id"])
    mask_product = df_label["product_id"].isin(df_product["product_id"])
    df_label = df_label[mask_query & mask_product]

    return df_product, df_query, df_label
```

1.3.3.2.2 Train-test Split Strategy

To ensure reliable evaluation, we implement a stratified split that preserves query and product distribution characteristics:

```
from sklearn.model_selection import train_test_split

# Split queries
df_query_train, df_query_test = train_test_split(
```

Table 1.14 WANDS dataset query type categories and examples.

Query type	Example queries	Characteristics
Categorical queries	"living room furniture," "outdoor dining sets"	Broad category exploration, combining location and product type
Product-specific queries	"l shaped sectional sofa," "round glass coffee table"	Detailed specifications, precise material and shape requirements
Feature-oriented queries	"waterproof outdoor rug," "adjustable height desk"	Functional requirements, mechanism-focused attributes
Intent-based queries	"small space dining solutions," "pet-friendly furniture"	Problem-solving orientation, lifestyle-specific needs
Ambiguous queries	"modern lights," "storage solutions"	Style preferences without specificity, broad functional needs

```
    df_query, test_size=0.25, random_state=42
)

# Create corresponding label splits
df_label_train = df_label[df_label["query_id"].isin(df_query_train["query_id"])]
df_label_test = df_label[df_label["query_id"].isin(df_query_test["query_id"])]
```

This split provides:

- **360 queries (75%)** for training and development of retrieval models.
- **120 queries (25%)** for final evaluation and system comparison.

1.3.3.3 Problem Formulation for Search and Recommendation

The WANDS dataset enables investigation of both search and recommendation scenarios that reflect real-world e-commerce challenges.

1.3.3.3.1 Search Problem Definition

Objective: Given a user search query, retrieve and rank the most relevant products from the catalog to maximize user satisfaction and task completion.

Success Criteria: An effective search system should: (1) prioritize products with "Exact" relevance (score 2) at top positions; (2) appropriately rank "Partial" matches (score 1) in middle positions; (3) minimize or eliminate "Irrelevant" results (score 0) from top rankings; (4) maintain computational efficiency for real-time query processing.

1.3.3.3.2 Recommendation Problem Formulation

Objective: Predict user preferences and suggest relevant products based on implicit signals, behavioral patterns, and contextual information without explicit queries.

Success Criteria: An effective recommendation system should: (1) identify products that align with user interests and needs; (2) balance personalization with diversity to encourage exploration; (3) adapt recommendations based on user context and session behavior; and (4) provide explanations that build user trust and understanding.

Synthetic User Generation: Since WANDS lacks user behavioral data, we create synthetic user profiles. The methods we use include: (1) generate virtual users with diverse preference patterns; (2) assign purchase histories based on product categories and features; (3) create realistic browsing and interaction sequences; and (4) simulate cold-start scenarios for new user recommendations.

Refer to Appendix B for details of the synthetic data generation. This synthetic approach enables comprehensive recommendation system testing while respecting privacy constraints and data availability limitations.

By working through these progressive implementations, readers will gain comprehensive experience with modern IR techniques while developing practical skills directly applicable to real-world e-commerce and recommendation system challenges. The WANDS dataset serves dual purposes as both a hands-on learning tool and a rigorous evaluation benchmark, ensuring that theoretical understanding translates effectively into measurable system improvements.

Throughout the remainder of this book, we will consistently employ the WANDS dataset using the experimental setup and preprocessing procedures established in this section.

1.4 Challenges with IR Systems

While both traditional search and recommendation systems have been successful in various domains, they also face several common challenges:

1.4.1 Lack of Context Awareness

Traditional systems focus heavily on exact term matching or predefined user preferences, often missing the broader context or intent behind a user query. For example, they may struggle to connect a user's interest in "apple fruit" to articles on fruit preservation techniques or health benefits.

1.4.2 Cold-start Problem

In recommendation systems, the cold-start problem arises when there is insufficient user data to make accurate recommendations, particularly for new users or items. This limitation makes it difficult for collaborative filtering models to recommend items to users with limited interaction history.

1.4.3 Scalability

As the volume of data grows, traditional indexing and matching techniques can struggle to scale efficiently, particularly when they rely on exact matching or predefined item attributes (category tags or metadata fields).

1.4.4 Difficulty in Handling Complex Queries

Traditional IR systems are limited in their ability to handle complex or multifaceted queries, such as those requiring a deep understanding of a specific domain or context. For example, a query like "What are the environmental benefits of electric cars in cold climates?" would be difficult for a traditional search engine to resolve with a single result, as it would require an understanding of both environmental science and automotive engineering.

With the advent of LLMs, IR systems have evolved to address many of the shortcomings of traditional approaches. LLMs bring a deeper understanding of language, context, and user intent, allowing for more accurate retrieval of relevant information. By leveraging LLMs, IR systems can now understand not just the words used in a query but also the intent behind them, providing results that are more relevant and personalized.

In the next sections, we will explore how LLMs enable semantic search and enhance recommendation systems, transforming IR from a basic keyword or feature-matching process into a sophisticated, context-aware retrieval mechanism. Some examples include: enhanced query understanding through contextual intent detection; improved document relevance ranking via semantic similarity scoring; generating concise summaries and answers using summarization and paraphrasing capabilities.

1.5 Summary

This chapter provided a comprehensive foundation in IR, tracing its remarkable evolution from manual card catalogs to sophisticated AI-powered systems. We began by exploring the historical progression through four distinct eras: manual retrieval methods, early computerization featuring Boolean search and vector space models, the web era that introduced PageRank and large-scale indexing, and the current big data era characterized by ML and DL approaches.

We examined the core components of modern IR systems through a three-stage pipeline: pre-retrieval (including user input processing and document indexing), retrieval models (from sparse approaches like TF-IDF and BM25 to dense retrieval using embeddings and neural architectures), and post-retrieval processes (ranking, reranking, and evaluation). This systematic breakdown revealed how traditional keyword-based methods have evolved to incorporate semantic understanding through EBR and hybrid architectures.

The practical applications of IR were illustrated through search and recommendation systems, with hands-on examples using the WANDS e-commerce dataset. These examples demonstrated real-world implementations, including query expansion, hybrid ranking systems, and various recommendation scenarios from next-item prediction to cold-start recommendations.

Despite these advances, we identified persistent challenges in IR systems, including handling ambiguous queries, scaling to massive datasets, and balancing relevance with diversity. These limitations motivate the integration of generative AI capabilities explored in subsequent chapters.

References

1 Jack Delano. Woman at main reading room card catalog in the library of congress, 1930. [Photograph] Between 1930 and 1950. https://www.loc.gov/item/90712184/.

2 Calvin Mooers. Information retrieval viewed as temporal signaling. In *Proceedings of the International Congress of Mathematicians*, volume 1, pages 572–573, 1950.

3 Wikipedia Contributors. Set theory, 2024. https://en.wikipedia.org/wiki/Set_theory. Accessed: 2025-05-25.

4 Gerard Salton and Michael E Lesk. The smart automatic document retrieval system—an illustration. *Communications of the ACM*, 8(6):391–398, 1965.

5 Marcia Bates. *Understanding Information Retrieval Systems*. Auerbach Publications, 2011.

6 Wikipedia Contributors. PageRank — Wikipedia, the free encyclopedia, 2024. https://en.wikipedia.org/wiki/PageRank. Accessed: 2025-5-25.

7 Joseph Weizenbaum. Eliza—a computer program for the study of natural language communication between man and machine. *Communications of the ACM*, 9(1):36–45, 1966.

8 Terry Winograd. Understanding natural language. *Cognitive Psychology*, 3(1):1–191, 1972.

9 Herve Jegou, Matthijs Douze, and Cordelia Schmid. Product quantization for nearest neighbor search. *IEEE Transactions on Pattern Analysis and Machine Intelligence*, 33(1):117–128, 2010.

10 Yu A Malkov and Dmitry A Yashunin. Efficient and robust approximate nearest neighbor search using hierarchical navigable small world graphs. *IEEE Transactions on Pattern Analysis and Machine Intelligence*, 42(4):824–836, 2018.

11 Tie-Yan Liu. Learning to rank for information retrieval. *Foundations and Trends® in Information Retrieval*, 3(3):225–331, 2009.

12 Thorsten Joachims. Optimizing search engines using clickthrough data. In *Proceedings of the 8th ACM SIGKDD International Conference on Knowledge Discovery and Data Mining*, pages 133–142, 2002.

13 Chris Burges, Tal Shaked, Erin Renshaw, Ari Lazier, Matt Deeds, Nicole Hamilton, and Greg Hullender. Learning to rank using gradient descent. In *Proceedings of the 22nd International Conference on Machine Learning*, pages 89–96, 2005.

14 Christopher Burges, Robert Ragno, and Quoc Le. Learning to rank with nonsmooth cost functions. *Advances in Neural Information Processing Systems*, 19:193–200, 2006.

15 Zhe Cao, Tao Qin, Tie-Yan Liu, Ming-Feng Tsai, and Hang Li. Learning to rank: From pairwise approach to listwise approach. In *Proceedings of the 24th International Conference on Machine Learning*, pages 129–136, 2007.

16 Fen Xia, Tie-Yan Liu, Jue Wang, Wensheng Zhang, and Hang Li. Listwise approach to learning to rank: Theory and algorithm. In *Proceedings of the 25th International Conference on Machine Learning*, pages 1192–1199, 2008.

17 Christopher JC Burges. From RankNet to LambdaRank to LambdaMART: An overview. *Learning*, 11(23–581):81, 2010.

18 Andrei Broder. A taxonomy of web search. In *ACM SIGIR Forum*, volume 36, pages 3–10. New York, NY, USA: ACM, 2002.

19 Karen Kukich. Techniques for automatically correcting words in text. *ACM Computing Surveys (CSUR)*, 24(4):377–439, 1992.

20 Hang Cui, Ji-Rong Wen, Jian-Yun Nie, and Wei-Ying Ma. Probabilistic query expansion using query logs. In *Proceedings of the 11th International Conference on World Wide Web*, pages 325–332, 2002.

21 Uma Sawant and Soumen Chakrabarti. Learning joint query interpretation and response ranking. In *Proceedings of the 22nd International Conference on World Wide Web*, pages 1099–1110, 2013.

22 Andrew I Schein, Alexandrin Popescul, Lyle H Ungar, and David M Pennock. Methods and metrics for cold-start recommendations. In *Proceedings of the 25th Annual International ACM SIGIR Conference on Research and Development in Information Retrieval*, pages 253–260, 2002.

23 Eli Pariser. *The Filter Bubble: What the Internet Is Hiding from You*. UK: Penguin, 2011.

24 Carlos A Gomez-Uribe and Neil Hunt. The netflix recommender system: Algorithms, business value, and innovation. *ACM Transactions on Management Information Systems (TMIS)*, 6(4):1–19, 2015.

25 Balázs Hidasi and Alexandros Karatzoglou. Recurrent neural networks with top-k gains for session-based recommendations. In *Proceedings of the 27th ACM International Conference on Information and Knowledge Management*, pages 843–852, 2018.

2

Large Language Models

In Chapter 1, we explored the foundational principles and applications of information retrieval (IR) systems, emphasizing the differences and commonalities between search and recommendation engines. As these systems grow increasingly complex and capable, a key enabler of their evolution has been the rise of large language models (LLMs). These models, built on the Transformer architecture, have revolutionized the way machines understand and generate human language. In this chapter, we provide a comprehensive overview of LLMs—tracing their evolution, exploring their inner workings, and highlighting how they power modern information systems. We begin with the historical development of machine learning (ML) paradigms and move through essential LLM building blocks such as tokenization, embeddings, and attention mechanisms. We then examine how LLMs are trained, categorized, and adapted for real-world use, followed by a detailed discussion of their deployment in search and recommendation systems. Finally, we conclude with the limitations of LLMs in isolation and introduce retrieval-augmented generation (RAG) as a solution for grounding LLM outputs with external knowledge. This sets the stage for deeper architectural and implementation discussions in the chapters ahead.

2.1 LLMs Overview

This section explores the foundational principles behind LLMs and the technological breakthroughs that have enabled their emergence. We also trace the evolution of ML paradigms that paved the way for LLMs, each marked by a shift in methodology, scale, and intelligence.

2.1.1 Evolution of ML Paradigms

As discussed in Section 1.1.2.4, the explosion of big data and rapid advancements in ML have reshaped the landscape of IR. Traditional keyword-based systems struggled to scale with the volume, variety, and ambiguity of unstructured data. The need for systems capable of understanding context, semantics, and user intent led to a new era in AI—one that would be increasingly defined by LLMs. To understand the significance of LLMs, it's essential to trace the historical evolution of ML while recognizing the modern paradigms that define current LLM development.

2.1.1.1 Four Phases of Machine Learning Evolution

The development of LLMs is part of a broader evolution in ML, marked by four distinct phases. Each phase represents not just technological advancement, but a fundamental paradigm shift, reflecting the growing complexity and sophistication of AI systems.

2.1.1.1.1 *Phase 1. Traditional Machine Learning: Manual Feature Engineering*

In the foundational era of ML, algorithms such as support vector machines (SVMs) [1], decision trees [2], and Naive Bayes [3] dominated the field. The hallmark of this phase was manual feature engineering—a labor-intensive process where domain experts manually crafted and extracted relevant features from raw data.

The success of these models was intrinsically tied to human expertise and intuition. For instance, as illustrated in Figure 2.1, building an email spam classifier required engineers to manually extract features such as keyword frequency, sender reputation scores, email length, and specific linguistic patterns. This approach demanded substantial domain knowledge and engineering effort, as the quality of manually crafted features directly determined model performance.

The primary limitation of this phase was clear: models could only be as sophisticated as the features humans could conceive and implement, creating a performance ceiling that was difficult to transcend without significant expertise and time investment.

2.1.1.1.2 *Phase 2. Neural Networks: Architecture Engineering*

As computational power increased exponentially and datasets grew larger, the field experienced a fundamental shift. Neural networks began to supplant traditional models, ushering in the second phase of ML evolution. This transition marked a move from manual feature extraction to architecture engineering—designing sophisticated network structures capable of automatically discovering and learning complex patterns.

Convolutional neural networks (CNNs) [4] revolutionized computer vision by automatically learning hierarchical image representations, eliminating the need for manual feature extraction in visual tasks. Figure 2.2 illustrates how a CNN processes images. A small filter (called a "window") slides across the input image from left to right and top to bottom. At each position, the filter performs calculations on the pixels it covers, extracting important patterns like edges or textures. The results of these calculations are collected into a feature map, which represents what patterns the

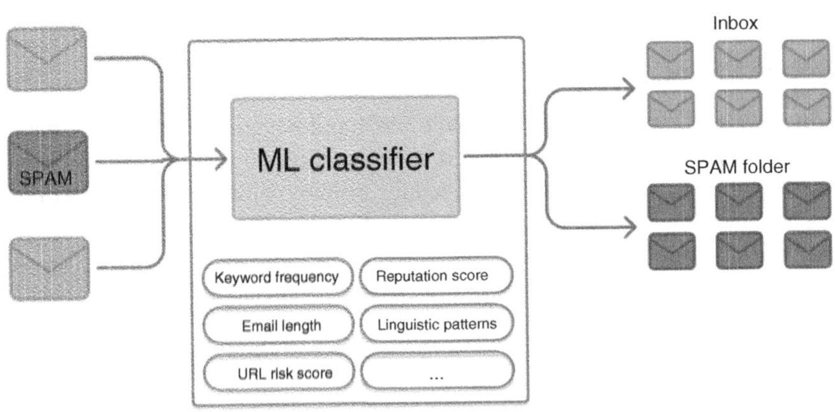

Figure 2.1 Feature engineering for an email spam classifier model.

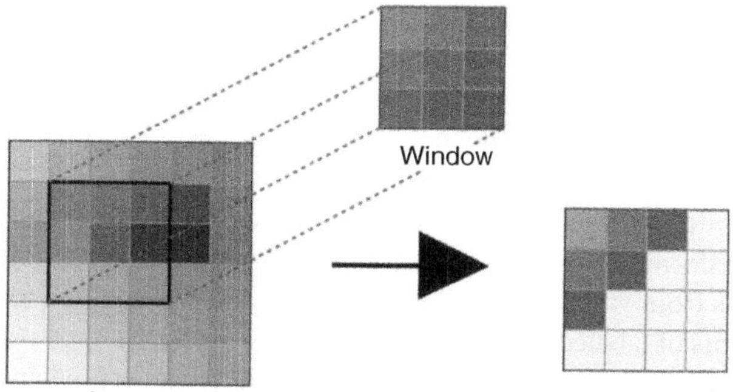

Figure 2.2 CNN: Sliding window and feature map.

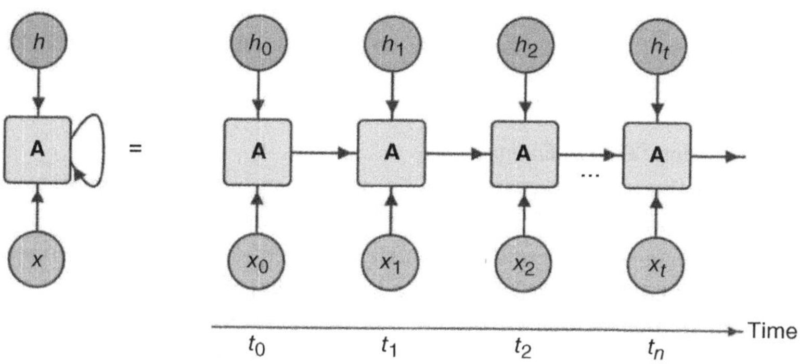

Figure 2.3 RNN architecture.

network detected in different parts of the image. This sliding window approach allows CNNs to recognize features regardless of where they appear in the image.

Simultaneously, recurrent neural networks (RNNs) [5] and their advanced variants long short-term memory (LSTM) networks [6] and gated recurrent units (GRUs) [7]—excelled at processing sequential data such as text, speech, and time series. Figure 2.3 illustrates the fundamental architecture of an RNN. On the left, we see the "folded" representation where the recurrent unit A contains a self-loop, indicating that the hidden state h feeds back into itself. The right side shows the same network "unfolded" through time, revealing how RNNs process sequential data: at each time step (t_0, t_1, t_2, ..., t_n), the network takes an input x_t and combines it with the previous hidden state to produce a new hidden state h_t. This hidden state carries information forward to the next time step, enabling RNNs to maintain memory across sequences and capture temporal dependencies in data such as text, speech, or time series.

This phase represented a crucial philosophical shift: instead of relying on human intuition to craft features, researchers focused on designing network architectures that could automatically uncover relationships and representations that human experts might never have considered.

2.1.1.1.3 Phase 3. Pretraining + Fine-tuning: Transfer Learning

The third phase introduced a game-changing concept that would lay the foundation for modern LLMs: transfer learning through pretraining and fine-tuning. The computational and data constraints of the previous era gave way to an abundance of both, enabling a new approach to model development.

Pioneering models like bidirectional encoder representations from transformers (BERT) [8] and early versions of generative pretrained transformer (GPT) [9, 10] demonstrated that models could first learn general language representations from vast, unlabeled text corpora, then be adapted to specific tasks with relatively modest amounts of labeled data.

This pretraining phase involved training models on broad, self-supervised tasks (such as predicting masked words or next tokens), allowing them to develop a deep understanding of language structure, semantics, and context. Subsequently, these pretrained models could be fine-tuned for specific applications by adjusting their parameters while preserving the rich linguistic knowledge embedded within.

The impact was transformative: transfer learning dramatically reduced data requirements for specialized tasks, democratized access to high-performance natural language processing (NLP) capabilities, and accelerated deployment across diverse domains.

2.1.1.1.4 Phase 4. Pretraining + Prompts: Prompt Engineering

The current era represents perhaps the most dramatic paradigm shift yet, as the computational and data limitations that necessitated fine-tuning have been largely overcome. We now witness the emergence of truly large-scale language models guided primarily through prompt engineering—the art and science of crafting natural language instructions to direct model behavior.

In this phase, massive pretrained models like GPT-3 [10], GPT-4 [11], and other LLMs are controlled not through traditional programming or extensive fine-tuning, but through carefully designed prompts. This approach leverages the vast knowledge and capabilities embedded within these models without requiring significant computational resources for model modification.

The breakthrough capabilities include:

- Zero-shot Learning: Models perform tasks they were never explicitly trained for by understanding natural language descriptions
- Few-shot Learning: With minimal examples in the prompt, models rapidly adapt to new contexts and requirements
- Conversational Interfaces: Users interact with artificial intelligence (AI) systems in natural language, making advanced capabilities accessible to nontechnical users

In summary, as summarized in Table 2.1, the transformation from manual feature engineering to prompt engineering represents a fundamental shift in how we conceptualize human–AI interaction. This evolution reflects not merely technological progress, but a democratization of AI capabilities, moving from highly specialized technical knowledge requirements to more intuitive, natural language interactions. Understanding this trajectory provides essential context for appreciating both the revolutionary nature of current LLMs and their potential for transforming IR and human–computer interaction.

Table 2.1 Evolution of feature engineering approaches in ML.

Phase	Era	Paradigm and key characteristics	Dominant models	Human role and limitations
Traditional ML	1990s–2000s	• Manual feature engineering • Hand-crafted features	SVMs, decision trees	• Domain expert as feature engineer • Performance ceiling limited by human intuition
Neural networks	2010s	• Architecture design • Automated feature learning	CNNs, RNNs, LSTMs, GRUs	• Researcher as architecture designer • Requires specialized deep learning expertise
Transfer learning	Late 2010s	• Pretrain + fine-tune • General-to-specific adaptation	BERT, GPT-2	• Data scientist as task adapter • Requires labeled data and fine-tuning expertise
Prompt-based learning	2020s–present	• Pretrain + prompt engineering • Natural language instruction	GPT-3, GPT-4, Claude, PaLM	• User as prompt designer • Requires prompt crafting skills and understanding

2.1.1.2 Natural Language Processing Tasks

NLP encompasses a wide range of tasks, each designed to extract or generate specific linguistic, semantic, or structural insights from text. These NLP tasks can be categorized into the following four types:

1. Classification tasks identify categories and labels:
   - Intent classification determines user goals for conversational AI.
   - Sentiment analysis detects emotional tone for social media monitoring.
   - Topic classification organizes content by subject matter.
2. Sequence labeling tasks analyze text structure:
   - Named entity recognition extracts key concepts and entities.
   - Part-of-speech tagging provides grammatical structure for downstream tasks.
3. Generation tasks create new text:
   - Next token prediction forms the foundation of language models.
   - Text generation produces coherent continuations and responses.
4. Understanding tasks extract structured meaning:
   - Query understanding transforms questions into searchable formats.
   - Language detection enables multilingual processing.
   - Text summarization condenses information while preserving meaning.

Each task contributes unique insights that collectively approach human-level language comprehension. The modular nature allows flexible combinations based on application requirements, from simple classifiers to sophisticated conversational AI systems.

As an example, we illustrate key NLP tasks using the simple yet rich example sentence: "How to extract information efficiently?" and the results are listed in Table 2.2.

Table 2.2 NLP tasks for input "How to extract information efficiently?"

NLP task	Purpose	Output example	Key insights
Intent classification	Determine user's goal/intention	How-to	User seeks instructional guidance on a process
Sentiment analysis	Identify emotional tone	Neutral	Informational text without emotional content
Topic classification	Categorize subject matter	Data science	Related to data processing and analytics
Named entity recognition	Extract entities and their types	how:ACTION	Identifies key actions, objects, and modifiers
		extract:ACTION	
		information:ENTITY	
		efficiently:MODIFIER	
Part-of-speech tagging	Grammatical role identification	how:WH-ADV	Standard grammatical structure analysis
		to:TO	
		extract:VB	
		information:NN	
		efficiently:RB	
Next token prediction	Predict likely continuation	from (38.72%)	Suggests common continuations of the phrase
		using (24.35%)	
		in (11.97%)	
Text generation	Generate relevant continuation	"...from large documents using automated NLP tools"	Completes the thought with contextually relevant content
Query understanding	Parse search/information intent	search_type:HOW_TO	Structured understanding for search systems
		domain:INFORMATION_	
		EXTRACTION	
		goal:EFFICIENCY	
Language detection	Identify text language	English	Clear language identification
Text summarization	Generate concise summary	"Methods for efficient information extraction"	Captures core meaning in fewer words

2.1.2 LLM Fundamentals

LLMs represent a major leap forward in AI and NLP. Understanding the fundamentals of LLMs is crucial as these concepts serve as the building blocks for the more advanced topics we'll cover for the rest of the book. By mastering these fundamentals, readers will have a solid understanding of the technology powering LLMs, allowing you to appreciate their role in solving real-world problems like search, recommendation systems, and data retrieval—topics that we'll explore in-depth throughout the book.

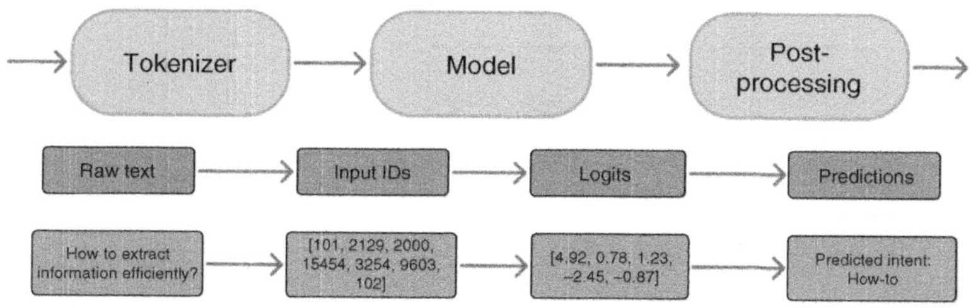

Figure 2.4 NLP workflow: Tokenizer → Model → Post-processing.

In NLP, the first essential step is data preprocessing, which involves cleaning and transforming raw text into a format that can be effectively processed by ML models. Preprocessing tasks typically include text normalization, regular expression handling, edit distance calculation, and most importantly—tokenization.

As illustrated in Figure 2.4, the NLP pipeline follows a clear sequence:

1. The tokenizer splits raw text into tokens and maps them to numerical representations (input IDs).
2. These input IDs are passed into the model, which produces logits as outputs.
3. A post-processing step transforms these logits into interpretable predictions (e.g., class labels or probabilities).

Figure 2.4 illustrates the NLP process of the example mentioned in Section 2.1.1.2, for the "intent classification" NLP task for the input text "How to extract information efficiently."

2.1.2.1 Tokenization

Tokenization is the process of breaking raw text into smaller meaningful units called tokens—which could be words, subwords, or even characters. Tokenizers serve as the primary gateway between human language and machine understanding. They reduce the complexity of text while preserving its semantic meaning and structure. A tokenizer performs several essential functions to prepare text for model processing.

2.1.2.1.1 Tokenization Process

The tokenization process follows specific steps. For example, as shown in Figure 2.5, given the input sentence: "How to extract information efficiently?":

1. Split the input text into basic units called tokens. Depending on the tokenization approach, these units may be words, subwords, or individual characters. A tokenizer might produce:
   - Word-level tokens: ["How," "to," "extract," "information," "efficiently," "?"]
   - Subword tokens (e.g., WordPiece): ["how," "to," "extra," "##ct," "information," "efficient," "##ly," "?"]
   - Character-level tokens: ["H," "o," "w," ..., "l," "y"]
2. Add special tokens that serve specific functions in NLP models:
   - [CLS] is used in classification tasks and placed at the beginning of input.
   - [SEP] separates multiple segments.
   - [MASK] marks tokens to be predicted and indicates tokens hidden for prediction tasks.
3. Map each token to an integer (token ID) using a vocabulary lookup table, creating the numerical representation required by models.

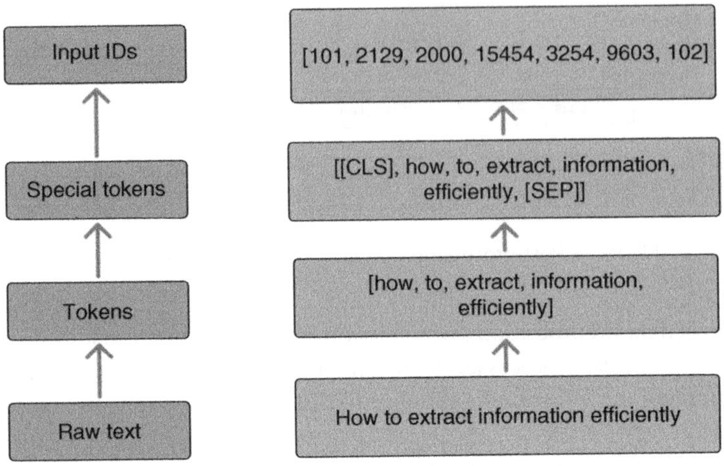

Figure 2.5 Tokenization pipeline – Raw text → Tokens → Input IDs.

Table 2.3 Comparison of tokenization methods and their characteristics.

Processing unit	Features	Limitations	Examples
Word	Splits text using whitespace or punctuation	High out-of-vocabulary (OOV) rate; poor generalization to morphology and spelling variants	NLTK, SpaCy (en), whitespace-based
Character	Extremely small vocabulary; robust to OOV, typos, and new words	Ignores word-level semantics; produces long sequences; weak at long-distance dependencies	CharRNN, some morphological models
Subword	Balances vocabulary size and expressiveness; captures common morphemes	Requires tokenization heuristics; frequent words should be kept whole, rare words should be segmented	Byte-Pair Encoding (BPE), WordPiece, Unigram LM
ML-based (subword)	Learns token boundaries from large corpora using frequency or likelihood optimization	Still relies on subword heuristics; may oversplit rare or multilingual tokens	BERT (WordPiece), RoBERTa (BPE), ALBERT, XLNet, T5 (SentencePiece)

2.1.2.1.2 Tokenization Granularities

Tokenizers can operate at different levels of granularity, each offering trade-offs between vocabulary size, flexibility, semantic richness, and model efficiency. Broadly, tokenization methods fall into three primary categories: **word-based**, **character-based**, and **subword-based**. In addition to these rule-based approaches, modern NLP models employ ML-based tokenizers, which optimize segmentation based on statistical patterns or language model objectives. Table 2.3 summarizes the differences:

By understanding the nuances of different tokenization approaches and their implementations in modern NLP models, practitioners can make informed decisions that significantly affect model performance, efficiency, and applicability across languages and domains.

Table 2.4 Cross-linguistic tokenization examples.

Language	Word-level tokens	Subword tokens (BERT tokenization)	Character tokens
English	How, to, extract, information, efficiently, ?	How, to, extra, ##ct, information, efficient, ##ly, ?	H, o, w, ..., l, y, ?
French	Comment, extraire, des, informations, efficacement, ?	Comment, extra, ##ire, des, informations, efficace, ##ment, ?	...
Chinese	如何, 高效, 地, 提取, 信息, ?	如, 何, 高, 效, 地, 提, 取, 信, 息, ?	如, 何, 高, 效, 地, 提, 取, 信, 息, ?
Korean	효율적으로 , 정보를 , 추출하는 , 방법 , ?	## 효율, 적, ## 으로, ## 정보, ## 를, ## 추출, ## 하는, 방, ## 법, ?	효 , 율 , 적 , 으 , 로 , ...

2.1.2.1.3 Language-Specific Considerations

Tokenization strategies vary significantly across languages due to structural differences. For instance:

- English uses spaces and punctuation as natural word boundaries (e.g., "extract").
- Chinese does not use spaces between words, requiring statistical or rule-based methods (e.g., "提取信息").
- Korean is morphologically rich, requiring decomposition into morphemes (e.g., "효율적으로").

The following Table 2.4 summarizes tokenization results for the phrase:

- English: "How to extract information efficiently?"
- French: "Comment extraire des informations efficacement?"
- Chinese: "如何高效地提取信息?"
- Korean: "효율적으로정보를추출하는방법?"

Here is a Python snippet that demonstrates how to perform word-, subword-, and character-level tokenization for multiple languages using standard NLP libraries:

```
from nltk.tokenize import word_tokenize
from transformers import AutoTokenizer
import jieba
from konlpy.tag import Mecab

examples = {
    'english': "How to extract information efficiently?",
    'french': "Comment extraire des informations efficacement?",
    'chinese': "如何高效地提取信息？",
    'korean': "효율적으로정보를추출하는방법?"
}

def basic_word_tokenize(text, language):
    if language == 'chinese':
        return list(jieba.cut(text))
    elif language == 'korean':
        try:
            mecab = Mecab()
            return [word for word, _ in mecab.pos(text)]
        except:
            return text.split()
    else:
        return word_tokenize(text)

```

```
def character_tokenize(text):
    return list(text)

def bert_tokenize(text, model_name="bert-base-multilingual-uncased"):
    tokenizer = AutoTokenizer.from_pretrained(model_name)
    return tokenizer.tokenize(text)

for lang, text in examples.items():
    print(f"\n--- {lang.upper()} ---")
    print("Original:", text)
    print("Word-level:", basic_word_tokenize(text, lang))
    print("Subword (BERT):", bert_tokenize(text))
    print("Character-level:", character_tokenize(text))
```

When applied to our example sentences, these tokenization methods produce distinct results that reflect both the linguistic properties of each language and the tokenization approach:

Example output for English with different tokenization methods:

```
# Word: ['How', 'to', 'extract', 'information', 'efficiently', '?']
# Character: ['H', 'o', 'w', ' ', 't', 'o', ' ', 'e', 'x', 't', 'r', 'a', 'c', 't',
    ...]
# BERT: ['how', 'to', 'extra', '##ct', 'information', 'efficient', '##ly', '?']
```

This tokenization step provides the essential bridge between human language and computational understanding, enabling subsequent model processing and ultimately leading to meaningful NLP applications.

2.1.2.2 Embeddings

NLP relies heavily on the ability to represent text in a form that ML models can interpret—namely, as numerical vectors. Once text is tokenized, each token is mapped to a high-dimensional numerical vector.

Traditional methods like one-hot encoding or bag-of-words (BoW) models suffer from sparsity and fail to capture semantic relationships. To overcome these limitations, modern NLP uses dense, low-dimensional vector representations called embeddings.

2.1.2.2.1 Word Embeddings

We briefly discussed the concept of word embeddings in Section 1.2.3.2. The embeddings are typically learned during the training process, enabling the model to capture complex linguistic patterns. Table 2.5 lists the popular word embedding models and their characteristics.

Table 2.5 Popular word embedding models.

Model	Methodology	Characteristics
Word2Vec [12]	Predictive (skip-gram, CBOW)	Learn embeddings using shallow neural networks trained on context windows
GloVe [13]	Count-based (global vectors)	Learns word vectors from word co-occurrence statistics
FastText [14]	Subword-enhanced model	Represents words as the sum of their n-gram embeddings, handling rare/misspelled words
ELMo [15]	Contextual (deep Bi-LSTM)	Generates different embeddings for the same word in different contexts

Here we demonstrate the classic "king–man + woman → queen" analogy of word embeddings using the GloVe embedding model.

```
import torch
from torchtext.vocab import GloVe
import numpy as np

# Load GloVe 100d vectors
print("Loading GloVe embeddings...")
glove = GloVe(name='6B', dim=100)
print(f"GloVe vocabulary size: {len(glove.stoi)}")

def get_vector(word):
    """Get vector for a word, returns numpy array"""
    return glove[word].numpy()  # Convert to numpy array

def find_most_similar(target_vector, vocab, exclude_words=None, top_k=5):
    """Find most similar words to target vector"""
    if exclude_words is None:
        exclude_words = []

    cos = torch.nn.CosineSimilarity(dim=1)
    similarities = {}

    # Convert target to tensor for comparison
    target_tensor = torch.from_numpy(target_vector)

    for word in vocab:
        if word in exclude_words:
            continue
        word_tensor = glove[word]
        sim = cos(target_tensor.unsqueeze(0), word_tensor.unsqueeze(0)).item()
        similarities[word] = sim

    # Return top k results
    return sorted(similarities.items(), key=lambda x: x[1], reverse=True)[:top_k]

# ==========================================
# Main Analogy Demonstration
# ==========================================
print("\n" + "=" * 50)
print("CLASSIC ANALOGY: king - man + woman = ?")
print("=" * 50)

# Compute king - man + woman
king_vec = get_vector('king')
man_vec = get_vector('man')
woman_vec = get_vector('woman')

print(f"King vector shape: {king_vec.shape}")
print(f"Man vector shape: {man_vec.shape}")
print(f"Woman vector shape: {woman_vec.shape}")

# Perform the vector arithmetic
result_vector = king_vec - man_vec + woman_vec
print(f"Result vector shape: {result_vector.shape}")

# Find most similar words (excluding the input words)
top_words = find_most_similar(result_vector, glove.stoi.keys(),
                              exclude_words=['king', 'man', 'woman'],
                              top_k=5)

print("\nTop 5 results for 'king - man + woman':")
for i, (word, score) in enumerate(top_words, 1):
    print(f"{i:2d}. {word:<15} (similarity: {score:.4f})")

# Check if 'queen' is in the results
```

```
queen_rank = next((i for i, (word, _) in enumerate(top_words, 1) if word == 'queen'),
    None)
if queen_rank:
    print(f"\n 'queen' found at rank {queen_rank}!")
else:
    print(f"\n 'queen' not in top 5 results")
```

Output of top 5 most similar words based on GloVe word embeddings and cosine similarity:

```
Loading GloVe embeddings...
GloVe vocabulary size: 400000

==================================================
CLASSIC ANALOGY: king - man + woman = ?
==================================================
King vector shape: (100,)
Man vector shape: (100,)
Woman vector shape: (100,)
Result vector shape: (100,)

Top 5 results for 'king - man + woman':
  1. queen            (similarity: 0.7834)
  2. monarch          (similarity: 0.6934)
  3. throne           (similarity: 0.6833)
  4. daughter         (similarity: 0.6809)
  5. prince           (similarity: 0.6713)

 'queen' found at rank 1!
```

Figure 2.6 illustrates the first two dimensions of the principal component analysis (PCA) [16] using the GloVe [13] word embeddings.

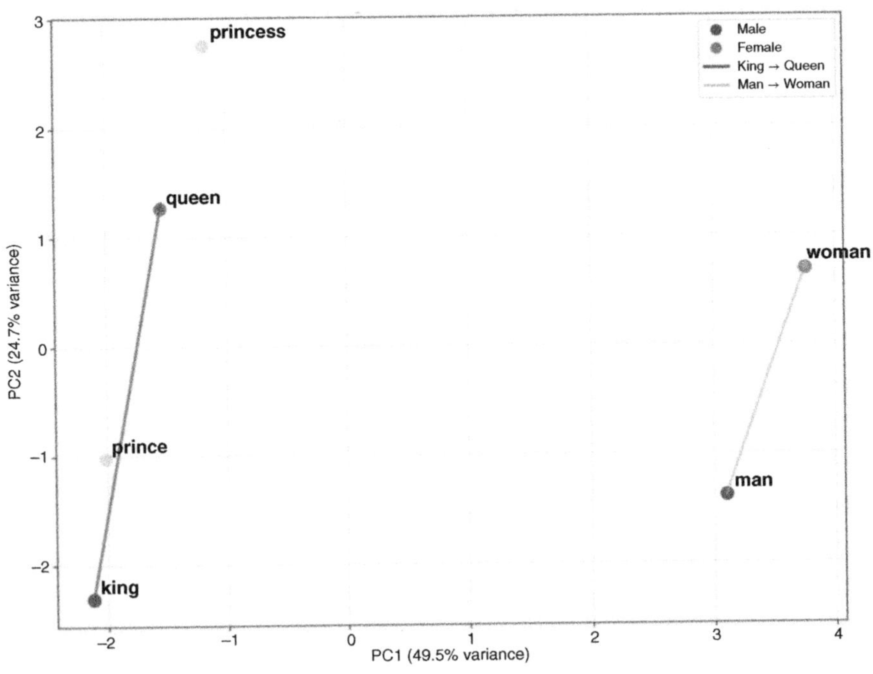

Figure 2.6 PCA projection of GloVe word embeddings (king − man + woman = queen).

While word embeddings provide powerful representations of individual words, they fall short in one critical dimension: context. A key limitation of static word embeddings—such as those produced by Word2Vec [12] or GloVe [13]—is that each word is assigned a single, fixed vector regardless of the surrounding context. For instance, the word "bank" will have the same embedding whether it appears in "river bank" or "investment bank," despite these contexts clearly invoking different meanings. To overcome the "context" limitations of word embeddings, researchers have introduced sentence embeddings.

2.1.2.2.2 Sentence Embeddings

Sentence embeddings extend the principles of word embeddings to capture the meaning of phrases, sentences, or paragraphs. By encoding higher level linguistic patterns, they enable more nuanced comparisons between text segments than word-level approaches alone.

Figure 2.7 illustrates the evolution timeline of word and sentence embedding models, from the static era (2013–2016) to the contextual era (2018) and the optimization era (2019–2020).

Simple Averaging Methods The most straightforward approach computes the arithmetic mean of word embeddings within a sentence. This method treats sentences as bags of words, where the sentence vector represents the centroid of all constituent word vectors. While computationally efficient, this approach ignores word order entirely, meaning "The dog chased the cat" and "The cat chased the dog" receive identical representations despite opposite meanings.

Python code:

```python
sentence_embedding = np.mean([embedding(word) for word in sentence], axis=0)
```

Weighted Averaging and PCA Advanced unsupervised methods, such as smooth inverse frequency (SIF) embeddings, address these limitations by downweighting frequent but less informative words (e.g., "the," "and") and removing dominant principal components.

SIF weights words inversely proportional to their corpus frequency, down-weighting common function words while emphasizing informative content words. After computing weighted averages, SIF removes the first principal component to eliminate common semantic directions that may obscure sentence-specific meanings.

This technique, introduced in "A Simple but Tough-to-Beat Baseline for Sentence Embeddings" [21], often outperforms simple averaging in semantic similarity tasks by emphasizing content-bearing terms.

Neural Architecture (RNNs, CNNs, and Transformers): RNN, including LSTMs and GRUs, process sentences sequentially to capture word order, long-range dependencies, and compositional relationships. The final hidden state serves as the sentence embedding, accumulating semantic content through sequential processing. CNNs offer an alternative approach, applying filters of various sizes to capture local n-gram patterns, then using pooling operations to create fixed-size representations.

The Transformer architecture revolutionized sentence embeddings through self-attention mechanisms that consider all words simultaneously when computing representations. BERT creates contextual embeddings for each token, often using the special [CLS] token as a sentence-level representation aggregated through self-attention.

Sentence-BERT (SBERT) [19] addresses vanilla BERT's limitations by fine-tuning using Siamese network architecture and contrastive learning. SBERT trains the model to make similar sentences have similar embeddings while maintaining BERT's contextual understanding

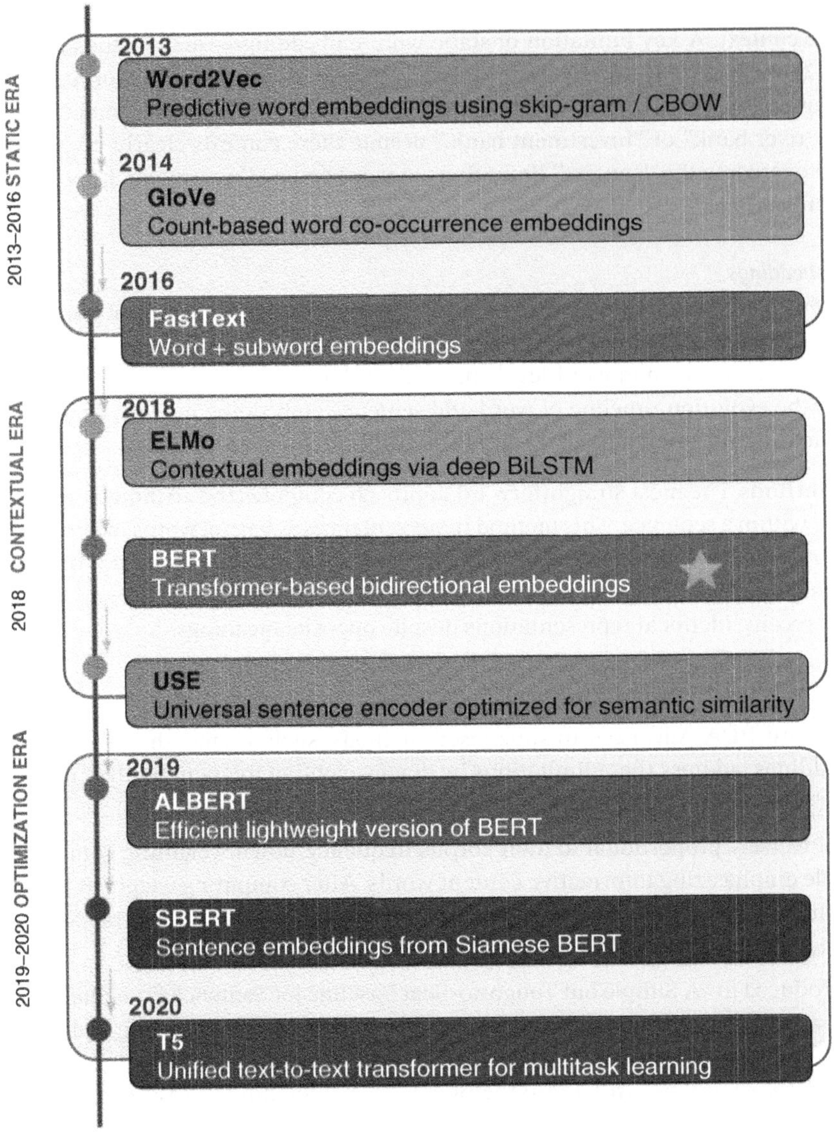

Figure 2.7 Word sentence embeddings timeline (Word2Vec [12], GloVe [13], FastText [14], ELMo [15], BERT [8], USE [17], ALBERT [18], SBERT [19], T5 [20]).

capabilities. The universal sentence encoder (USE) [17] offers another approach with two variants: a transformer-based model for accuracy and a deep averaging network for efficiency, trained on diverse tasks including skip-thought prediction and conversational response prediction.

Sentence embeddings are foundational in many applications, including semantic similarity, paraphrase detection, document classification, dense retrieval, and question answering. By translating entire sentences into dense vectors that preserve meaning, they enable models to reason beyond isolated tokens and support more human-like language understanding.

2.1.2.3 Transformer Models and Attention Mechanism

In 2017, Google's machine translation team introduced the transformer model in a groundbreaking paper titled "Attention is All You Need" [22], presented at the neural information processing systems (NeurIPS) conference. This model marked a significant shift in the field of ML and NLP, fundamentally changing how sequences are processed and setting the foundation for virtually all modern LLMs.

The transformer's revolutionary contribution lies in its ability to process sequences in parallel rather than sequentially, dramatically improving computational efficiency while maintaining or improving model performance. This parallel processing capability, combined with the self-attention mechanism, enabled the development of much larger and more capable models that form the backbone of today's LLMs.

2.1.2.3.1 Transformer Architecture

As shown in Figure 2.8, the transformer architecture is composed of two main components: an encoder and a decoder, each serving distinct but complementary roles in sequence-to-sequence tasks.

The encoder is responsible for capturing the meaning, context, and relationships within the input sequence. It processes the entire input simultaneously, creating rich representations that capture both local and long-range dependencies between tokens. The encoder consists of multiple identical layers, each containing two main sub-components: a multi-head self-attention mechanism and a position-wise feed-forward network.

The decoder leverages the encoder's understanding to generate the output sequence autoregressively. Unlike the encoder, the decoder generates tokens one at a time, using both the encoder's representations and previously generated tokens. The decoder contains an additional cross-attention layer that allows it to focus on relevant parts of the input when producing each output token.

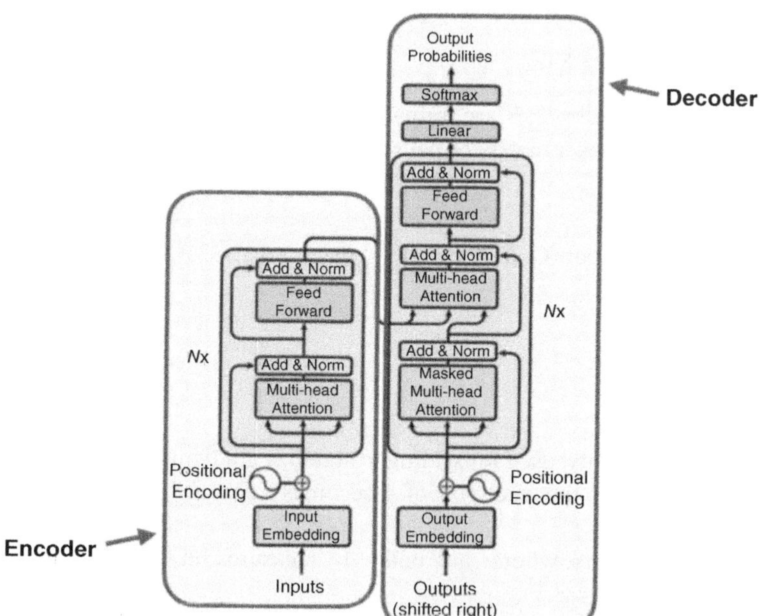

Figure 2.8 The transformer architecture [22] / Neural Information Processing Systems Foundation, Inc.

Each encoder and decoder layer is built around two core components:

- Feed-forward Neural Network: Applies a nonlinear transformation to each token's representation independently, providing the model with additional capacity for complex transformations
- Self-attention Mechanism: Allows the model to weigh the importance of different positions in the sequence when processing each token

2.1.2.3.2 *Self-attention Mechanism: The Heart of Transformers*

At the core of the transformer model lies the self-attention mechanism, a critical innovation that allows the model to understand and process sequences of words or tokens in parallel while capturing complex relationships between all positions in the sequence.

Conceptual Foundation: The self-attention mechanism addresses a fundamental limitation of previous sequential models: the ability to efficiently capture long-range dependencies. While RNNs and LSTMs process sequences step-by-step, potentially losing information over long distances, self-attention allows every position to directly attend to every other position in a single step. This mechanism operates on the principle that when processing any given token, the model should be able to look at the entire sequence and determine which other tokens are most relevant for understanding the current token's meaning and context. This global view enables the model to capture relationships that might span the entire input sequence.

Mathematical Formulation: For each input token, the self-attention mechanism generates three vectors through learned linear transformations:

- Query (Q): Represents the current token's "question" or what information it is seeking from other tokens.
- Key (K): Represents how each token can "answer" or be relevant to queries from other tokens.
- Value (V): Carries the actual content or information that each token contributes.

Formally, let $X \in \mathbb{R}^{n \times d_{\text{model}}}$ be the matrix of input embeddings for a sequence of n tokens. The model first projects X into queries, keys, and values using learned weight matrices:

$$Q = XW_Q, \qquad K = XW_K, \qquad V = XW_V$$

where $W_Q, W_K \in \mathbb{R}^{d_{\text{model}} \times d_k}$ and $W_V \in \mathbb{R}^{d_{\text{model}} \times d_v}$ are trainable parameters.

The attention mechanism then computes a matrix of pairwise relevance scores and uses it to form a weighted combination of values:

$$A = \text{softmax}\left(\frac{QK^\top}{\sqrt{d_k}}\right), \qquad \text{Attention}(Q, K, V) = AV$$

where:

- $Q = XW_Q \in \mathbb{R}^{n \times d_k}$ is the query matrix.
- $K = XW_K \in \mathbb{R}^{n \times d_k}$ is the key matrix.
- $V = XW_V \in \mathbb{R}^{n \times d_v}$ is the value matrix.
- d_k is the key (and query) dimensionality, used for scaling to stabilize gradients.
- softmax is applied row-wise to $QK^\top$ so that each row of A becomes a probability distribution over all tokens.
- $A \in \mathbb{R}^{n \times n}$ is the *attention weight matrix*, where each entry A_{ij} measures how much token i attends to token j.

Working Example: To illustrate how self-attention works in practice, let's work through a concrete example with the sentence: **"The cat sat on the mat."**

Step 1: Input Representation

First, each word is converted to an embedding vector. For simplicity, let's use four-dimensional embeddings:

$$
\begin{aligned}
\text{“The”} &\rightarrow [1.0, 0.2, 0.5, 0.1] \\
\text{“cat”} &\rightarrow [0.3, 1.0, 0.2, 0.8] \\
\text{“sat”} &\rightarrow [0.5, 0.3, 1.0, 0.4] \\
\text{“on”} &\rightarrow [0.2, 0.5, 0.3, 1.0] \\
\text{“the”} &\rightarrow [1.0, 0.2, 0.5, 0.1] \\
\text{“mat”} &\rightarrow [0.4, 0.8, 0.2, 0.9]
\end{aligned}
$$

Step 2: Generate Q, K, V Matrices

Each embedding is transformed into Query, Key, and Value vectors using learned weight matrices W_Q, W_K, and W_V. For our example, let's assume these transformations produce:

Query Matrix Q (what each token is looking for):

$$
Q = \begin{bmatrix} 0.8 & 0.1 & 0.3 \\ 0.2 & 0.9 & 0.1 \\ 0.4 & 0.2 & 0.7 \\ 0.1 & 0.6 & 0.2 \\ 0.8 & 0.1 & 0.3 \\ 0.3 & 0.5 & 0.8 \end{bmatrix}
$$

Key Matrix K (what each token can provide):

$$
K = \begin{bmatrix} 0.7 & 0.2 & 0.4 \\ 0.3 & 0.8 & 0.2 \\ 0.5 & 0.1 & 0.9 \\ 0.2 & 0.7 & 0.1 \\ 0.7 & 0.2 & 0.4 \\ 0.4 & 0.6 & 0.7 \end{bmatrix}
$$

Value Matrix V (the actual content each token contributes):

$$V = \begin{bmatrix} 0.9 & 0.3 & 0.2 \\ 0.1 & 0.8 & 0.4 \\ 0.6 & 0.2 & 0.9 \\ 0.3 & 0.7 & 0.1 \\ 0.9 & 0.3 & 0.2 \\ 0.2 & 0.5 & 0.8 \end{bmatrix}$$

Step 3: Compute Attention Scores
Now we calculate how much each token should attend to every other token. Let's focus on how the word "cat" attends to all words:
$Q_{\text{cat}} \cdot K^T$ **calculations**:

$$\begin{aligned}
\text{"cat" attending to "The"}: &\quad [0.2, 0.9, 0.1] \cdot [0.7, 0.2, 0.4] = 0.36 \\
\text{"cat" attending to "cat"}: &\quad [0.2, 0.9, 0.1] \cdot [0.3, 0.8, 0.2] = 0.80 \\
\text{"cat" attending to "sat"}: &\quad [0.2, 0.9, 0.1] \cdot [0.5, 0.1, 0.9] = 0.28 \\
\text{"cat" attending to "on"}: &\quad [0.2, 0.9, 0.1] \cdot [0.2, 0.7, 0.1] = 0.68 \\
\text{"cat" attending to "the"}: &\quad [0.2, 0.9, 0.1] \cdot [0.7, 0.2, 0.4] = 0.36 \\
\text{"cat" attending to "mat"}: &\quad [0.2, 0.9, 0.1] \cdot [0.4, 0.6, 0.7] = 0.69
\end{aligned}$$

Step 4: Apply Scaling and Softmax — Forming the Attention Weight Matrix
Scale by $\sqrt{d_k} = \sqrt{3} \approx 1.73$:

$$\text{Scaled scores}: [0.21, 0.46, 0.16, 0.39, 0.21, 0.40]$$

Apply softmax to obtain the attention weight matrix $A = \text{softmax}(QK^T/\sqrt{d_k})$:

$$\text{Attention weights (row for "cat")}: [0.12, 0.16, 0.11, 0.15, 0.12, 0.15]$$

Note: These weights sum to 1.0, representing how much "cat" should focus on each word. The full attention matrix $A \in \mathbb{R}^{6\times6}$ contains one such row for every token in the sequence, with each row specifying a probability distribution over all tokens.

The attention weights reveal interesting patterns:

- "cat" attends most strongly to itself (0.16), which is common as tokens often need to preserve their own information.
- "cat" also attends significantly to "on" (0.15) and "mat" (0.15), suggesting the model is learning relationships between the subject ("cat") and the spatial relationship it has with other objects.
- The attention to "The" tokens (0.12 each) shows the model recognizes the articles as providing contextual information.

This pattern demonstrates how self-attention captures semantic and syntactic relationships: the model learns that subjects ("cat") are related to the objects and prepositions that describe their actions and positions.

Step 5: Compute Weighted Sum
Finally, compute the output for "cat" as a weighted sum of all value vectors:

$$\begin{aligned}\text{Output}_{\text{cat}} &= 0.12 \times V_{\text{The}} + 0.16 \times V_{\text{cat}} + 0.11 \times V_{\text{sat}} \\ &\quad + 0.15 \times V_{\text{on}} + 0.12 \times V_{\text{the}} + 0.15 \times V_{\text{mat}} \\ &= [0.39, 0.47, 0.38]\end{aligned}$$

2.1.2.3.3 Computational Complexity of Self-Attention Algorithm

Algorithm 1 shows the steps of computing scaled dot-product attention.

Algorithm 1: Scaled Dot-product Attention

Require: Input embeddings $X \in \mathbb{R}^{n \times d_{\text{model}}}$; learned projection matrices $W^Q, W^K \in \mathbb{R}^{d_{\text{model}} \times d_k}$, $W^V \in \mathbb{R}^{d_{\text{model}} \times d_v}$

Ensure: Contextualized token representations $Z = \text{Attention}(Q, K, V) \in \mathbb{R}^{n \times d_v}$

$Q \leftarrow XW^Q$ ▷ Query matrix $Q \in \mathbb{R}^{n \times d_k}$

$K \leftarrow XW^K$ ▷ Key matrix $K \in \mathbb{R}^{n \times d_k}$

$V \leftarrow XW^V$ ▷ Value matrix $V \in \mathbb{R}^{n \times d_v}$

$S \leftarrow QK^\top$ ▷ Raw attention scores $S \in \mathbb{R}^{n \times n}$

$S \leftarrow S/\sqrt{d_k}$ ▷ Scale to stabilize softmax

$A \leftarrow \text{softmax}(S, \text{dim} = -1)$ ▷ Attention weights $A \in \mathbb{R}^{n \times n}$

$Z \leftarrow AV$ ▷ Contextualized output $Z \in \mathbb{R}^{n \times d_v}$

return Z

Complexity: Computing QK^T requires $O(n^2 \cdot d_k)$ operations, and the resulting attention matrix A requires $O(n^2)$ memory—the well-known quadratic bottleneck that motivates efficient attention variants.

In practice, this quadratic dependence on sequence length often dominates the cost of Transformer layers, especially when n is large. Modern implementations mitigate this cost via batching over the batch dimension B (yielding $B \times n \times n$ attention tensors) and by using highly optimized GPU kernels.

Computational Advantages: The self-attention mechanism provides several crucial advantages:

- Parallelization: Unlike RNNs, all positions can be computed simultaneously, dramatically reducing training time.
- Long-range Dependencies: Direct connections between all positions enable efficient modeling of long-distance relationships.
- Interpretability: Attention weights provide insights into which parts of the input the model considers important.
- Flexibility: The same mechanism works for various sequence lengths without architectural changes.

These advantages have made the transformer architecture the foundation for virtually all modern LLMs, from BERT and GPT to more recent models like GPT-4 and beyond. The self-attention mechanism's ability to capture complex relationships while remaining computationally efficient has proven to be the key innovation that enabled the current era of large-scale language modeling.

2.1.2.3.4 *Transformer Design Elements*

The transformer architecture also incorporates several crucial design elements:

Residual Connections: Enable training of very deep networks by allowing gradients to flow directly to earlier layers. Instead of learning a mapping $F(x)$ from an input x, each sub-layer learns a residual function and outputs

$$y = x + F(x)$$

so that even if F is poorly conditioned, gradients can still propagate along the identity path $y \approx x$ without vanishing.

Layer Normalization: Stabilizes training and improves convergence. Given a token representation $h \in \mathbb{R}^d$, layer normalization computes

$$\mu = \frac{1}{d}\sum_{j=1}^{d} h_j, \qquad \sigma^2 = \frac{1}{d}\sum_{j=1}^{d}(h_j - \mu)^2$$

and returns

$$\mathrm{LN}(h)_j = \gamma_j \frac{h_j - \mu}{\sqrt{\sigma^2 + \varepsilon}} + \beta_j$$

where $\gamma, \beta \in \mathbb{R}^d$ are learned parameters and ε is a small constant for numerical stability. Unlike batch normalization, layer normalization normalizes across features within a single token and does not depend on batch size, making it well suited for sequence models. Table 2.6 lists the comparison of batch normalization and layer normalization.

Positional Encoding: Since self-attention is inherently permutation-invariant (it would produce the same output for "The cat sat on the mat" and "mat the on sat cat The"), Transformers add positional encoding to preserve sequence order information.

Table 2.6 Comparison of batch normalization and layer normalization.

Aspect	Batch normalization (BN)	Layer normalization (LN)
Normalization axis	Normalizes across the batch dimension (and spatial dimensions in CNNs)	Normalizes across feature dimensions of a single token or timestep
Dependency on batch size	Yes—behavior changes with batch size; unstable for very small batches	No—completely independent of batch size
Typical usage	Convolutional neural networks, image models, large-batch training	RNNs, Transformers, sequence models, variable-length, or small-batch settings
Effect on training stability	Can cause training instability with small or non-i.i.d. batches	Provides consistent stabilization across layers and batch settings
Formula	Uses batch mean/variance over samples	Uses feature-wise mean and variance within each sample

Positional encodings are added to the input embeddings before the first attention layer. The original Transformer uses sinusoidal positional encoding:

$$PE_{(\text{pos},2i)} = \sin\left(\frac{\text{pos}}{10000^{2i/d_{\text{model}}}}\right)$$

$$PE_{(\text{pos},2i+1)} = \cos\left(\frac{\text{pos}}{10000^{2i/d_{\text{model}}}}\right)$$

where pos is the position and i is the dimension. This encoding allows the model to learn to attend to relative positions and understand sequence order.

Multi-head Attention: In practice, transformers use multiple attention heads, each learning different types of relationships. Some heads might focus on syntactic relationships (subject-verb connections); others might capture semantic similarities (related concepts). Still others might learn positional or structural patterns.

The outputs from all heads are concatenated and linearly transformed to produce the final attention output, allowing the model to capture a rich variety of relationships simultaneously.

2.1.2.4 Model Training and Loss Function

As introduced in Section 2.1.1.2, one of the most fundamental tasks in NLP is next-token prediction, serving as the foundation for generating coherent and contextually appropriate text. This task requires models to predict the most likely subsequent token in a sequence given all previous tokens, a capability that underpins everything from autocomplete systems to sophisticated language generation models. During the training process, models iteratively refine their predictive capabilities by adjusting internal parameters to minimize the discrepancy between their predictions and the actual next tokens observed in training data.

2.1.2.4.1 Training Paradigms for Language Models

Autoregressive Language Modeling: Traditional language models employ autoregressive training, where the model learns to predict each token based on all preceding tokens in a sequence. This left-to-right prediction strategy mirrors how humans typically process text, building understanding progressively as they read. The model processes sequences token by token, using the hidden representations of previous tokens to inform predictions about subsequent ones.

The mathematical formulation of autoregressive language modeling can be expressed as maximizing the likelihood of observing a sequence of tokens:

$$P(x_1, x_2, \dots, x_n) = \prod_{i=1}^{n} P(x_i | x_1, x_2, \dots, x_{i-1})$$

where x_i is the token at position i, and n is the sequence length.
Note: In transformer language models, the terms "autoregressive" and "causal" are often used interchangeably. Technically, "causal" refers to the dependency constraint (each position depends only on previous positions), while "autoregressive" describes the sequential generation process. In practice, causal masking in the attention mechanism enforces the autoregressive property, making these concepts functionally equivalent in this context.

This approach proves particularly effective for text generation tasks, as it naturally captures the sequential dependencies inherent in natural language. However, autoregressive training only utilizes left context, potentially missing valuable information from subsequent tokens that could inform better understanding of earlier positions.

Masked Language Modeling: Transformer models like BERT [8] revolutionized language model training through masked language modeling (MLM), a bidirectional training strategy that enables models to leverage both left and right context when making predictions. In MLM, the training process randomly masks approximately 15% of input tokens, replacing them with special [MASK] tokens, and trains the model to predict the original masked tokens based on surrounding context.

The masking strategy typically follows a specific protocol: 80% of selected tokens are replaced with [MASK], 10% are replaced with random tokens from the vocabulary, and 10% remain unchanged. This approach prevents the model from simply learning to predict [MASK] tokens and encourages more robust representations that can handle various token substitutions.

MLM's bidirectional nature allows models to develop richer contextual understanding compared to autoregressive approaches. By considering context from both directions, models can better capture phenomena like long-range dependencies, semantic relationships, and syntactic structures that require understanding of complete sentence context.

Hybrid Training Approaches: Modern language models increasingly employ hybrid training strategies that combine multiple objectives. For example, models like text-to-text transfer transformer (T5) [20] frame all tasks as text generation problems, using a unified approach that can handle both understanding and generation tasks. ELECTRA [23] introduces a "replaced token detection" objective where the model learns to identify which tokens have been replaced by a generator network, providing more efficient training than traditional MLM.

Figure 2.9 illustrates the comparisons of different training paradigms.

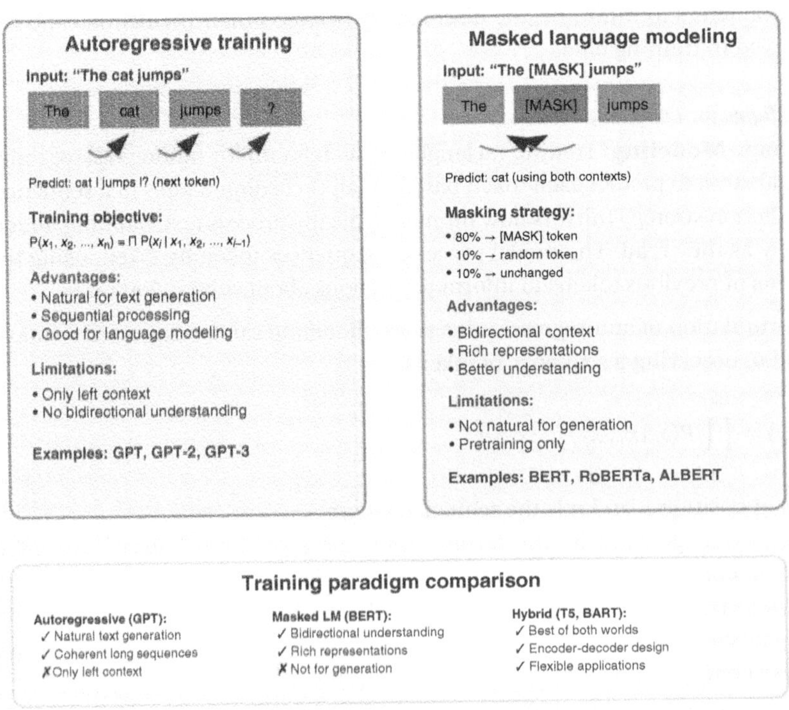

Figure 2.9 Compare autoregressive training and masked language modeling.

Here is the Python code example for MLM prediction:

```
from transformers import BertTokenizer, BertForMaskedLM
import torch
import torch.nn.functional as F

# Load pre-trained model and tokenizer
tokenizer = BertTokenizer.from_pretrained("bert-base-uncased")
model = BertForMaskedLM.from_pretrained("bert-base-uncased")

# Input sentence with a masked token
text = "The cat jumps [MASK] to the ceiling."
inputs = tokenizer(text, return_tensors="pt")
mask_index = torch.where(inputs["input_ids"] == tokenizer.mask_token_id)[1].item()

# Forward pass
with torch.no_grad():
    outputs = model(**inputs)
    logits = outputs.logits

# Get predictions at the masked token position
mask_logits = logits[0, mask_index]
probs = F.softmax(mask_logits, dim=-1)
top_probs, top_indices = torch.topk(probs, 5)
top_tokens = tokenizer.convert_ids_to_tokens(top_indices)

# Display top results
for token, prob in zip(top_tokens, top_probs):
    print(f"Token: {token:<12} | Probability: {prob.item():.4f}")
```

The output result is:

```
Token: up           | Probability: 0.7229
Token: back         | Probability: 0.0371
Token: high         | Probability: 0.0302
Token: off          | Probability: 0.0246
Token: on           | Probability: 0.0243
```

2.1.2.4.2 Loss Functions in Language Model Training

Cross-entropy Loss: Cross-entropy loss serves as the predominant loss function for language model training, measuring the divergence between predicted probability distributions and actual target distributions. For a vocabulary of size V, the model outputs a probability distribution over all possible tokens, and cross-entropy loss quantifies how well this distribution matches the true token.

The mathematical formulation of cross-entropy loss for a single prediction is:

$$L = -\sum_{i=1}^{V} y_i \log(p_i)$$

where y_i represents the true distribution (typically one-hot encoded), and p_i represents the predicted probabilities. In practice, since the target is usually a single correct token, this simplifies to the negative log probability of the correct token:

$$L = -\log(p_{\text{correct}})$$

Cross-entropy loss naturally encourages the model to assign high probability to correct tokens while penalizing confident incorrect predictions more severely than uncertain ones. If a model assigns 0.9 probability to the correct token "jumps," the loss is $-\log(0.9) \approx 0.045$. However, if it assigns only 0.1 probability to "jumps," the loss increases dramatically to $-\log(0.1) \approx 2.303$, providing strong learning signals for significant mistakes.

Advanced Loss Functions: Beyond standard cross-entropy, researchers have developed specialized loss functions to address specific training challenges. Label smoothing modifies the target distribution by assigning small probabilities to incorrect labels, preventing overconfident predictions and improving model calibration. Focal loss addresses class imbalance by focusing learning on hard examples, reducing the contribution of easy predictions to the overall loss.

Contrastive learning approaches, used in models like SimCLR [24] and sentence embedding systems, employ loss functions that encourage similar examples to have similar representations while pushing dissimilar examples apart. These approaches prove particularly effective for learning meaningful representations in self-supervised settings.

Figure 2.10 shows how cross-entropy loss decreases over training epochs for both training and validation data during MLM. It illustrates:

- A steady decline in loss as the model learns from data.
- A narrowing gap between training and validation loss, indicating good generalization.

2.1.2.4.3 *The Training Process in Detail*

As illustrated in Figure 2.11, training transformer models include the following steps:

Data Preprocessing and Tokenization: The training process begins with preprocessing massive datasets comprising billions of tokens from web pages, books, and articles. Data undergoes filtering to remove low-quality content and harmful material. Advanced tokenization schemes like byte-pair encoding (BPE) create subword vocabularies that balance word-level semantics with character-level flexibility, enabling effective handling of out-of-vocabulary words.

Embedding and Positional Encoding: Tokenized inputs convert to dense vector embeddings that capture semantic relationships between tokens. Embedding dimensions typically range from 512 to 4096 in large models. Since attention mechanisms are position-agnostic, transformer models add sinusoidal positional encodings that provide unique patterns for each sequence position and generalize beyond training sequence lengths.

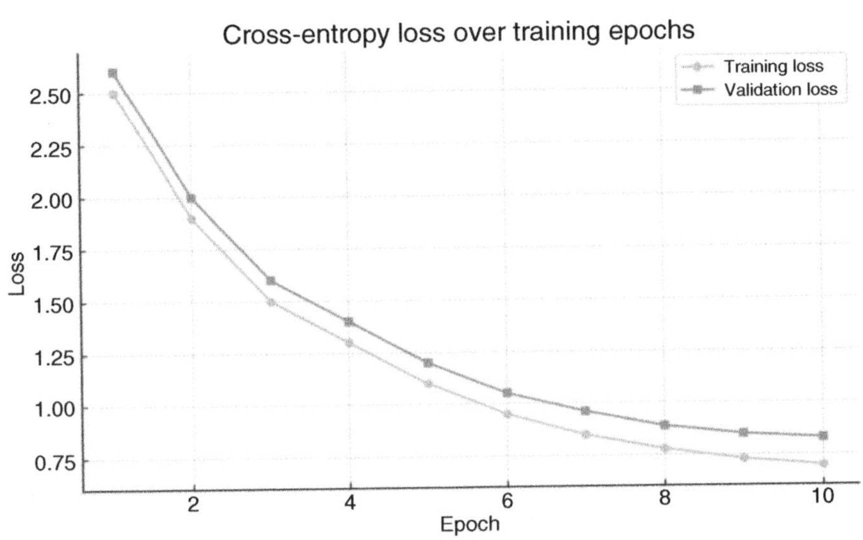

Figure 2.10 Cross-entropy loss over training epochs.

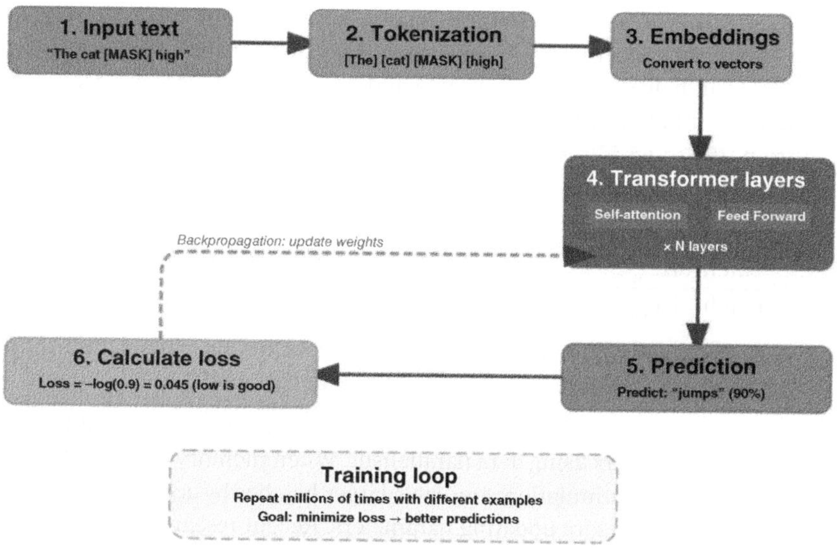

Figure 2.11 Transformer training process.

Forward Pass and Loss Calculation: Embedded inputs flow through multiple transformer layers containing self-attention and feed-forward networks. Self-attention computes relationships between all token pairs, allowing the model to focus on relevant context. After processing, the model generates probability distributions over the vocabulary for prediction positions.

The loss function compares predictions against target tokens, with cross-entropy loss providing gradients for parameter adjustment. Backpropagation propagates gradients through the network, while optimization algorithms like Adam adapt learning rates based on gradient history for stable training of large models.

2.1.2.4.4 Parameter Updates and Optimization

Training requires careful learning rate scheduling, gradient clipping, and regularization to prevent instabilities. Mixed precision training using both 16- and 32-bit arithmetic enables training larger models with limited memory while maintaining numerical stability.

Learning rate scheduling: plays a crucial role in achieving optimal convergence. Modern transformer training typically employs a linear warmup phase where the learning rate increases from zero to the target value over the first 1–10% of training steps. The closed-form expression for linear warmup is:

$$\mathrm{lr}(t) = \mathrm{lr}_{\mathrm{base}} \cdot \frac{t}{t_{\mathrm{warmup}}}, \quad \text{for } t \leq t_{\mathrm{warmup}}$$

This prevents unstable updates when gradients are large and parameters are far from optimal. Following warmup, cosine annealing has become the standard, smoothly decreasing the learning rate following a cosine curve:

$$\mathrm{lr}(t) = \mathrm{lr}_{\mathrm{min}} + \frac{1}{2}(\mathrm{lr}_{\mathrm{base}} - \mathrm{lr}_{\mathrm{min}})\left(1 + \cos\left(\frac{\pi \cdot (t - t_{\mathrm{warmup}})}{t_{\mathrm{max}} - t_{\mathrm{warmup}}}\right)\right), \quad \text{for } t > t_{\mathrm{warmup}}$$

This schedule allows aggressive early exploration followed by careful fine-tuning. Alternative schedules like inverse square root decay (used in the original transformer) or cyclical learning rates with periodic restarts can help escape local minima and improve final performance.

Batch size dynamics: fundamentally affect both training efficiency and model quality. Larger batch sizes (ranging from 256 to 65,536 tokens for transformers) reduce gradient noise and enable more efficient hardware utilization, but require proportionally scaled learning rates following the linear scaling rule:

$$\mathrm{lr}_{\mathrm{adjusted}} = \mathrm{lr}_{\mathrm{base}} \times \frac{\mathrm{batch\_size}_{\mathrm{new}}}{\mathrm{batch\_size}_{\mathrm{base}}}$$

However, this relationship breaks down for very large batches, necessitating extended warmup periods and specialized optimizers like layer-wise adaptive moments (LAMB).

Each model has a "critical batch size" that maximizes hardware efficiency—typically 1–4 million tokens for modern transformers using data parallelism. When memory constraints prevent optimal batch sizes, gradient accumulation simulates larger batches by accumulating gradients over multiple forward passes before updating parameters. Recent research suggests that dynamically increasing batch size during training—starting small for exploration and increasing for efficient convergence—can reduce total training time while maintaining model quality.

2.1.2.4.5 Training Challenges and Solutions

Computational Efficiency: Training LLMs demands enormous computational resources, with models like GPT-3 requiring thousands of GPU-years for training. Researchers have developed numerous techniques to improve training efficiency, including gradient checkpointing to reduce memory usage, model parallelism to distribute computation across multiple devices, and efficient attention mechanisms that reduce the quadratic complexity of standard attention.

Data parallelism allows training across multiple devices by distributing different data batches to different processors, while model parallelism splits the model itself across devices. Pipeline parallelism combines these approaches by dividing the model into stages that process different batches simultaneously, maximizing hardware utilization.

Stability and Convergence: Training very large models presents significant stability challenges, including gradient explosions, loss spikes, and training divergence. Several architectural and optimization techniques work together to maintain stable training dynamics:

as introduced in Section 2.1.2.3, layer normalization stabilizes activations by normalizing across features rather than across batches, making training less sensitive to parameter initialization and enabling larger learning rates. Residual connections provide direct gradient pathways that bypass intermediate layers, ensuring gradients can flow unimpeded during backpropagation even in very deep networks. Gradient clipping prevents extreme parameter updates by enforcing a maximum gradient norm:

$$\mathbf{g} \leftarrow \mathbf{g} \cdot \min\left(1, \frac{\theta}{\|\mathbf{g}\|}\right)$$

where θ is the clipping threshold (typically 1.0) and $\|\mathbf{g}\|$ is the gradient norm. Warmup learning rate schedules gradually increase learning rates during early training, preventing destabilizing updates when parameter estimates are far from optimal.

Gradient Pathologies in Practice: Vanishing vs. Exploding

The **vanishing gradient problem**—where gradients shrink exponentially as they propagate backward through many layers—was historically the dominant challenge in training deep networks, particularly RNNs and LSTMs processing long sequences [25]. Transformers largely mitigate this issue through three mechanisms: (1) residual connections that provide gradient "shortcuts" bypassing intermediate computations; (2) layer normalization that maintains stable activation magnitudes; and (3) self-attention's direct token-to-token paths that avoid the sequential gradient chains of recurrent architectures.

However, at very large scales (billions of parameters), **exploding gradients** and **loss spikes** become the primary concerns. These manifest as sudden, dramatic increases in loss during training, often caused by outlier gradient updates or numerical instabilities in attention computations. Modern training pipelines address these through aggressive gradient clipping ($\theta \leq 1.0$), extended warmup periods (up to 10% of training steps), and careful monitoring of gradient statistics throughout training.

Data Quality and Bias: The massive datasets used for language model training inevitably contain biases, misinformation, and harmful content that models can learn and reproduce. Careful data curation, filtering, and bias detection techniques help mitigate these issues, though completely eliminating bias remains an ongoing challenge. Techniques like constitutional AI and human feedback training help align model outputs with human values and preferences.

Since its introduction, the transformer architecture has had a profound impact on the field of ML. Its ability to efficiently process and understand sequences of data has led to widespread applications across various tasks. The transformer's innovative design has firmly established it as a cornerstone in the development of modern ML models, transforming how we approach problems in NLP, computer vision, and beyond.

2.1.3 LLM Families

LLM families refer to groups of models categorized by their architecture (e.g., encoder-only, decoder-only, encoder-decoder), ownership status (proprietary vs. open-source), or specialized application domains (e.g., healthcare, legal, finance).

This section dives into the key classifications of LLM families, beginning with the foundational encoder and decoder architectures that define their operational mechanics. It further examines the distinction between proprietary models, controlled by organizations for commercial or research purposes, and open-source alternatives that foster collaboration and innovation. Finally, we explore domain-specific LLMs, which are tailored to excel in specialized fields, illustrating the growing trend of customizing LLMs to meet industry-specific needs.

2.1.3.1 Encoder and Decoder

The success of the transformer architecture led to the development of various LLM families, each optimized for different tasks. Having examined the encoder and decoder components within the Transformer architecture (Section 2.1.2.3), we now classify how LLMs employ these components at the architectural level. These families can be broadly categorized into three types: encoder-only, encoder-decoder, and decoder-only models.

Encoder-only Models: Here, "Encoder-only" refers to models that utilize only the encoder stack, not to the encoder component itself. Encoder-only models, such as BERT, excel at understanding input text. They are particularly effective for tasks like classification, entity recognition, and question answering. BERT introduced a novel approach called "masking," which randomly obscures 15% of input tokens, forcing the model to learn bidirectional context by predicting these missing words. This bidirectional analysis enables a more comprehensive understanding of context, leading to significant improvements in various NLP tasks.

Encoder-decoder Models: Encoder-decoder models, exemplified by T5, are designed for tasks that involve transforming one sequence of text into another, such as translation or summarization. In these models, the encoder processes the input sequence to create a context representation, which the decoder then uses to generate the output text. T5 innovatively frames all NLP tasks as text-to-text conversions, simplifying the model's application across various tasks and reducing the need for task-specific modifications.

Decoder-only Models: Decoder-only models, represented by the GPT series, focus primarily on text generation tasks like creative writing and text completion. These models use a series of decoder blocks from the transformer architecture and are trained to predict the next word in a sequence based on the preceding words or given prompt. This approach allows GPT models to generate coherent and contextually relevant text across a wide range of applications.

Table 2.7 lists the applications of these three types of models. Over time, decoder-only models like GPT have emerged as the leading state-of-the-art performers for many NLP tasks, due to their versatility and powerful generative capabilities. Their success has led to increased interest in developing custom LLMs for specific domains and applications. Figures 2.12 and 2.13 illustrate the timeline of these LLM families:

Era 1: Balanced Ecosystem (2018–2021): This foundational period established the three major transformer architectures through intense competition between different approaches. The era began with GPT-1 vs. BERT in 2018, representing the clash between decoder-only generative models and encoder-only understanding models. Each architecture served distinct purposes—GPT models excelled at text generation, BERT dominated comprehension tasks, and BART [26] introduced encoder-decoder capabilities. The period culminated with breakthrough releases of GPT-3 and T5 in 2020, which established scaling paradigms and demonstrated that all three architectures could achieve remarkable results through increased computational resources.

Era 2: Decoder-only Dominance (2022–2025): This transformational period witnessed the consolidation around decoder-only models, beginning with encoder-only decline in 2022 as the field shifted toward generative capabilities. The ChatGPT revolution in late 2022/early 2023 demonstrated that decoder-only models could achieve human-like conversational abilities and broad task competency, triggering an explosion of similar models. By 2024–2025, decoder-only

Table 2.7 Model types and their applications.

Model type	Best for	Example use case
Encoder-only	Text classification	Sentiment analysis
Encoder-decoder	Translation/summarization	Translating legal documents
Decoder-only	Creative text generation	Writing poetry

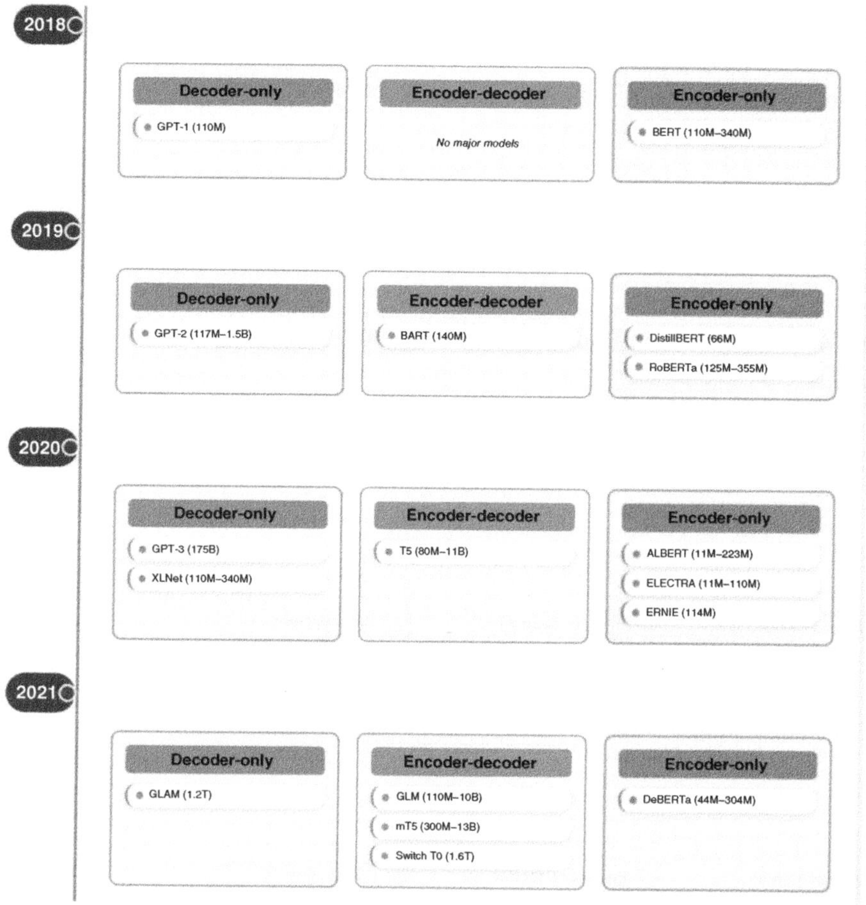

Figure 2.12 Timeline of encoder-only, encoder-decoder and decoder-only LLMs (2018–2021).

models achieved architectural dominance, with state-of-the-art models like GPT-4o (OpenAI), Claude 3.7 (Anthropic), Grok-3 (xAI), and DeepSeek R1 (DeepSeek AI) offering superior versatility across generation, reasoning, and interaction tasks. In contrast, encoder-only models became largely obsolete, and encoder-decoder models survived primarily in specialized niches—most notably in Google's Gemini series for multimodal integration.

Figure 2.13 illustrates this architectural evolution from 2022 to 2025, highlighting the decline of encoder-based models and the consolidation around decoder-only approaches.

2.1.3.2 Proprietary and Open-Source

The LLM landscape is fundamentally divided into two categories: proprietary models—developed by major technology companies—and open-weight models released by various organizations for broader community access. This distinction has become increasingly important as the field has matured, with both approaches playing complementary roles in advancing NLP and AI capabilities.

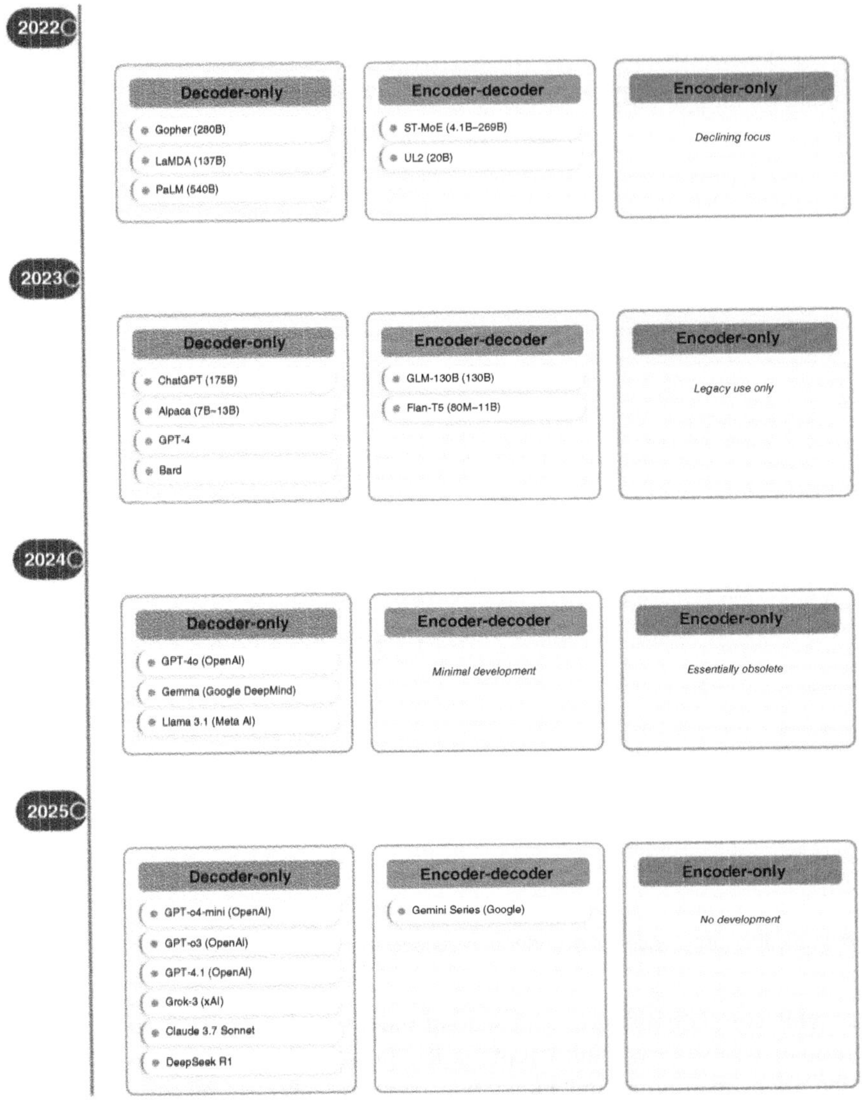

Figure 2.13 Timeline of encoder-only, encoder-decoder and decoder-only LLMs (2022–2025).

2.1.3.2.1 Proprietary LLMs

Proprietary LLMs represent the cutting edge of commercial AI development, typically created by well-funded organizations with access to massive computational resources and extensive datasets. Table 2.8 summarizes the major proprietary models and their key features.

These models are characterized by several key features:

- Scale and Performance: Proprietary models often leverage the largest training datasets and most powerful computational infrastructure, enabling them to achieve state-of-the-art performance across diverse benchmarks. They frequently pioneer new architectural innovations and training methodologies that later influence the broader field.

Table 2.8 Leading proprietary models (2024–2025).

Company	Models	Key features
OpenAI	GPT-4o, GPT-o1, GPT-o3, GPT-4.1, GPT-o4-mini	Leading reasoning capabilities, multimodal understanding, cost-effective variants
Anthropic	Claude 3.5 Sonnet, Claude 3.7 Sonnet	Safety-focused approach, strong conversational AI, and code generation
Google	Gemini 2.0 Flash, Gemini 2.5 Pro	Multimodal capabilities, integration with Google ecosystem
xAI	Grok-3	Real-time information access, X platform integration, reasoning focus
Amazon	Nova Series	Enterprise-focused, competitive pricing, AWS integration
Others	Microsoft (via OpenAI), Cohere, AI21 Labs	Specialized applications and enterprise solutions

- Controlled Access: Access is strictly managed through API endpoints, subscription services, or enterprise licensing agreements. This controlled distribution allows companies to monetize their research investments while maintaining competitive advantages.
- Closed Development: The complete development pipeline—including model architectures, training datasets, fine-tuning procedures, and safety alignments—remains proprietary. While some technical details may be disclosed in research papers, the full implementation remains confidential.
- Commercial Focus: These models are designed for production deployment with emphasis on reliability, safety, and commercial viability rather than research transparency.

2.1.3.2.2 Open-weight LLMs

In contrast, open-weight models offer transparency and flexibility, though most do not meet strict open-source definitions. While model weights, architecture, and inference code are typically public, training data, reinforcement learning human feedback (RLHF) procedures, and safety alignment details remain closed. Table 2.9 outlines what components are typically open vs. closed in these models.

Key features of open-weight models include:

- Transparency: While not completely transparent, these models provide significantly more visibility into their architecture and capabilities than proprietary alternatives.
- Customizability: Open weights enable fine-tuning, quantization, and adaptation for specific domains or hardware constraints, making them particularly valuable for research and specialized applications.
- Community-driven Development: The broader research community can analyze, improve, and build upon these models, accelerating innovation and enabling independent safety research.
- Hardware Accessibility: Many open-weight models are designed to run efficiently on consumer hardware, democratizing access to advanced AI capabilities.
- Research Enablement: Academic researchers and smaller organizations can study these models without requiring massive computational budgets for training from scratch.

Table 2.9 What is typically open vs. closed.

Component	Status	Details
Model weights and architecture	Usually open	Pretrained model parameters and architectural specifications
Basic inference code	Usually open	Code needed to run the model for inference
Model documentation	Usually open	Technical specifications, capabilities, and usage guidelines
Evaluation benchmarks	Usually open	Performance results on standard benchmarks
Complete training datasets	Often closed	Limited due to licensing, privacy, or competitive concerns
Detailed training procedures	Often closed	Hyperparameters, training infrastructure, and optimization details
RLHF data and methods	Often closed	Human feedback data and alignment procedures
Safety filtering procedures	Often closed	Content moderation and safety alignment methods
Full computational infrastructure	Often closed	Hardware specifications and distributed training setup

Table 2.10 Leading open-weight models (2024–2025).

Model series	Organization	Key strengths	Notable versions
Qwen	Alibaba	Strong multilingual capabilities, efficient performance	Qwen2.5, Qwen3
LLaMA	Meta	Widely adopted, strong performance, efficient	LLaMA 3.1
Gemma	Google	High quality, good efficiency, Google backing	Gemma 2 series
Phi	Microsoft	Strong performance with smaller sizes	Phi-3, Phi-4
DeepSeek	DeepSeek AI	Competitive reasoning capabilities	DeepSeek-R1

Based on download statistics and community adoption from Hugging Face and other platforms

Historical open-weight models like BERT [8], RoBERTa [27], and T5 [20] laid the foundation for today's ecosystem. Table 2.10 summarizes leading open-weight models in 2024–2025. Most of these can be found on Hugging Face's Open LLM Leaderboard.

2.1.3.2.3 *The Evolving Landscape*

The relationship between proprietary and open-weight models has become increasingly complex. Many organizations now pursue hybrid strategies, releasing some models openly while keeping their most advanced capabilities proprietary. This approach allows them to contribute to the research community while maintaining competitive advantages.

The open-weight movement has democratized access to powerful AI capabilities, enabling innovation in resource-constrained environments and facilitating important research into AI safety, bias, and capabilities. However, the rapid advancement of proprietary models continues to push the boundaries of what's possible, often providing the benchmarks that open-weight models strive to match.

As we move forward, the interplay between these two approaches will likely continue to shape the future of AI development, with proprietary models driving performance frontiers and open-weight models ensuring broad accessibility and enabling critical research into AI alignment and safety. The field benefits from both approaches: proprietary models provide the resources and incentives for pushing capabilities forward, while open-weight models ensure that these advances can be studied, understood, and adapted by the broader scientific community.

2.1.3.3 Domain-specific LLMs

The evolution of LLMs has reached a critical juncture where general-purpose capabilities are being refined and specialized for specific industries and domains. Domain-specific LLMs represent a paradigm shift from one-size-fits-all solutions to tailored AI systems that understand the nuanced language, regulations, and operational requirements of particular sectors. This specialization is achieved through targeted training on domain-specific datasets, fine-tuning with industry-relevant tasks, and incorporating domain knowledge into model architectures.

The development of these specialized models addresses several key limitations of general-purpose LLMs: they reduce hallucinations by grounding responses in domain-specific knowledge, improve accuracy on specialized tasks, ensure compliance with industry regulations and standards, and provide more contextually appropriate responses that align with professional practices and terminology. Table 2.11 outlines the primary approaches to domain specialization, each offering distinct advantages for different use cases and deployment scenarios.

Industry applications and transformations includes:

Software Engineering and Development: The software development industry has witnessed perhaps the most dramatic transformation through LLM integration. Code generation models like GitHub Copilot, based on OpenAI's Codex, have fundamentally changed how developers work. Table 2.12 details the transformative capabilities that code-generation LLMs have brought to software development workflows.

Table 2.11 Key approaches to domain specialization.

Approach	Description	Use cases	Benefits
Continued pretraining	Starting with general-purpose base model and continuing training on domain-specific corpora	Medical literature, legal documents, financial reports	Retains general capabilities while developing specialized knowledge
Fine-tuning and instruction tuning	Adapting pretrained models using supervised learning on domain tasks	Clinical decision support, legal document analysis	High accuracy on specific tasks with expert feedback integration
Retrieval-augmented generation (RAG)	Enhancing models with domain-specific knowledge bases queried in real time	Technical documentation, regulatory compliance	Up-to-date information with reduced hallucinations
Multimodal integration	Incorporating domain-specific data beyond text (images, charts, sensors)	Medical imaging, financial charts, code repositories	Comprehensive understanding across data types

Table 2.12 AI code generation capabilities and impact.

Capability	Description	Impact
Intelligent code completion	Context-aware suggestions understanding project structure and coding patterns	Increased developer productivity
Cross-language translation	Automated conversion between programming languages while preserving functionality	Reduced migration time and errors
Automated testing	Generation of comprehensive test suites including edge cases	Improved code quality and coverage
Documentation generation	Automatic creation of clear documentation and inline comments	Reduced documentation debt
Debugging assistance	Identification of bugs, security vulnerabilities, and performance issues	Faster issue resolution and improved code security

Table 2.13 AI applications in healthcare.

Application area	Capabilities	Regulatory considerations
Medical literature analysis	Automated summarization and synthesis of biomedical research	Must maintain citation accuracy and avoid misinterpretation
Clinical decision support	Diagnosis assistance, treatment recommendations, drug interaction analysis	Requires FDA approval for clinical use, liability concerns
EHR processing	Extraction and structuring of clinical notes and patient data	HIPAA compliance, patient privacy protection
Medical coding and billing	Automated assignment of medical codes for procedures and diagnoses	Audit trails required, financial accuracy critical
Drug discovery	Molecular property prediction, drug–target interaction modeling	Intellectual property considerations, validation requirements

Notable Models: GitHub Copilot (OpenAI Codex), Amazon CodeWhisperer, Google AlphaCode, DeepMind AlphaCode 2, StarCoder, and CodeT5.

Healthcare and Life Sciences: Healthcare represents one of the most promising yet challenging domains for LLM specialization due to the critical nature of medical decisions and strict regulatory requirements. Table 2.13 summarizes the key applications of specialized LLMs in healthcare, along with critical regulatory considerations that shape their deployment.

Key regulatory frameworks: HIPAA (US), GDPR (Europe), FDA validation requirements, clinical trial protocols

Financial Services: The financial sector has embraced LLMs for their ability to process complex regulatory requirements, analyze market data, and automate routine tasks while maintaining the precision required for financial operations. Table 2.14 illustrates how financial institutions are leveraging domain-specific LLMs across various operational areas.

Specialized models: BloombergGPT (financial data), FinGPT (financial analysis), and proprietary institutional models.

Table 2.14 AI applications in financial services.

Application	Function	Business value
Regulatory compliance	Automated analysis of regulatory changes and compliance monitoring	Reduced compliance costs, faster regulatory response
Risk assessment	Real-time market analysis, credit risk evaluation, portfolio optimization	Improved risk management, better investment decisions
Fraud Detection	Pattern recognition in transactions and communications	Reduced fraud losses, enhanced security
Customer service	Intelligent chatbots for financial queries and personalized advice	24/7 availability, improved customer satisfaction
Algorithmic trading	NLP of news, earnings calls, and market sentiment	Enhanced trading strategies, faster market response

Beyond software development, healthcare, and financial services, LLMs are increasingly being adopted across a wide range of other industries. In the legal domain, they assist with contract analysis, legal research, and summarizing case law. In education, LLMs support personalized tutoring, automated grading, and curriculum generation. In manufacturing and logistics, they enhance operational planning and enable real-time decision support through natural language interfaces. The adaptability of LLMs across domains highlights their transformative potential in driving efficiency, improving decision-making, and enabling intelligent automation at scale.

2.1.4 LLM Paradigms

As discussed in Section 2.1.1, the historical evolution of ML provides essential context for understanding contemporary approaches. Modern LLM development revolves around three fundamental paradigms that collectively define how these sophisticated models learn, adapt, and interact with users. These paradigms represent the methodological framework that has enabled LLMs to achieve remarkable capabilities in language understanding and generation.

Pretraining—Building the Foundation: Pretraining serves as the foundational phase where models acquire broad language understanding from massive, diverse text corpora using self-supervised learning objectives such as MLM or next-token prediction. The goal is to develop general-purpose language representations that capture grammar, semantics, factual knowledge, and reasoning patterns without requiring labeled data. This phase enables the model to serve as a versatile foundation for downstream applications.

Post-training—Specialization and Enhancement: Post-training encompasses the sophisticated suite of techniques applied after initial pretraining to transform raw language models into capable, aligned, and specialized systems. This critical phase includes supervised fine-tuning (SFT), RLHF, alignment procedures, reasoning enhancement, and efficient adaptation methods. These processes collectively adapt base models for specific domains, improve safety and helpfulness, and refine behavior to align with human expectations and ethical guidelines.

Prompt Engineering—Task Adaptation via Natural Language: Prompt engineering represents the application phase where models are guided to perform specific tasks through carefully crafted natural language instructions. This paradigm eliminates the need for parameter updates, making it highly efficient and accessible. Techniques such as zero-shot and few-shot prompting, chain-of-thought (CoT) reasoning, and in-context learning (ICL) allow users to leverage LLM capabilities using only textual cues, democratizing access to powerful AI functionality.

These paradigms often work synergistically—a model might be pretrained on general data, undergo post-training for medical applications, then be guided by prompts for specific clinical queries. We will explore each paradigm in greater technical depth, including implementation challenges and optimization strategies.

2.1.4.1 Pretraining: Foundation Building

Pretraining represents the foundational phase in LLM development, where models undergo initial training on vast amounts of unlabeled text data. This process equips models with a comprehensive understanding of human language, enabling them to grasp syntax, semantics, and nuanced relationships between words and phrases across diverse contexts.

2.1.4.1.1 Core Objectives and Methodology

The primary objective during pretraining is for models to learn general language patterns and structures from diverse sources, including Wikipedia, news articles, books, and web content, without focusing on specific downstream tasks. This broad exposure enables models to develop robust internal representations that capture multiple facets of language understanding.

Different model architectures employ distinct pretraining objectives optimized for their intended use cases. BERT utilizes MLM and next sentence prediction to develop bidirectional understanding, while GPT models focus on autoregressive next-token prediction for generation capabilities. These foundational training objectives shape the model's subsequent capabilities and determine optimal application domains.

2.1.4.1.2 Pretraining Task Variants

Modern pretraining employs various sophisticated objectives beyond basic language modeling:

- MLM randomly masks tokens and trains models to predict them using bidirectional context, enabling rich contextual understanding crucial for comprehension tasks [8].
- Permuted language modeling shuffles token sequences to force models to learn context-independent relationships, enhancing robustness to varied input structures, particularly beneficial for code generation tasks [28].
- Contrastive learning trains models to distinguish relevant pairs from irrelevant ones through positive and negative sampling, improving representation quality and semantic understanding [29].
- Span corruption masks continuous token spans rather than individual tokens, encouraging models to understand longer-range dependencies and coherent text reconstruction [20].

The selection of pretraining tasks depends on factors including data characteristics, target applications, and computational constraints. Models focused on code generation may prioritize permuted language modeling to handle scrambled syntax, while conversational systems might emphasize contrastive learning to differentiate relevant responses from noise.

2.1.4.1.3 Computational Challenges and Considerations

Developing LLMs from scratch represents a significant undertaking requiring substantial financial and technical resources, often costing millions of dollars due to massive computational requirements [30]. This process demands deep understanding of complex neural architectures and sustained research commitment, typically restricting custom LLM development to organizations with substantial resources and specialized AI expertise. Only a handful of institutions (e.g., OpenAI, Anthropic, Google, Meta) possess the resources and expertise to conduct such training end-to-end.

While training on more recent data can significantly extend knowledge bases, this approach faces considerable challenges including enormous computational and data requirements, alongside the demanding task of maintaining data quality and relevance across diverse sources and domains.

2.1.4.2 Post-training: Specialization and Alignment

Post-training has evolved into a sophisticated multistage process that transforms pretrained foundation models into capable, aligned, and specialized systems ready for real-world deployment. This critical phase encompasses several interconnected approaches that collectively enhance model capabilities while ensuring safe and beneficial behavior.

2.1.4.2.1 Supervised Fine-tuning

SFT represents the initial post-training stage where pretrained models adapt to specific tasks or domains through training on curated, labeled datasets. This transfer learning approach updates model parameters to better align with target task distributions while preserving broad language understanding acquired during pretraining.

The SFT process involves several critical steps: task specification and learning paradigm selection (supervised, weakly supervised, or reinforcement learning), pretrained model selection and weight initialization, preparation of task-relevant training data and labels, hyperparameter optimization including learning rate scheduling and regularization, and iterative evaluation on validation sets.

As shown in Figure 2.14, pretraining and fine-tuning typically employ a two-stage approach: first, they learn about human language by leveraging large-scale text datasets such as Wikipedia, news articles, or web content to complete the pretraining of the model. Then, according to the specific task objective, the pretrained model is fine-tuned in order to achieve better performance on this task.

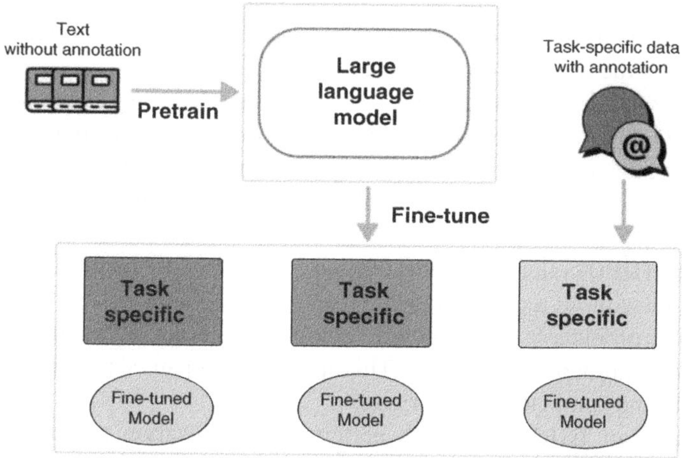

Figure 2.14 Pretraining and fine-tune LLMs.

Instruction fine-tuning represents a specialized SFT variant where models learn to follow diverse instructions through training on instruction-response pairs [31, 32]. This approach teaches models to understand and execute various types of commands, questions, and requests, forming the foundation for general-purpose AI assistants.

However, SFT faces inherent challenges. The objective mismatch between pretraining (e.g., next-token prediction) and fine-tuning (task-specific optimization) can create training instabilities. Additionally, overfitting risks are significant due to typically smaller fine-tuning datasets compared to pretraining corpora, requiring careful learning rate adjustment and regularization techniques such as dropout or data augmentation.

2.1.4.2.2 Efficient Adaptation with LoRA

Low-rank adaptation (LoRA) [33] has revolutionized fine-tuning efficiency by enabling parameter-efficient adaptation of LLMs. Instead of updating all model parameters, LoRA introduces trainable low-rank decomposition matrices that approximate full parameter updates with significantly reduced computational requirements.

LoRA works by decomposing weight updates into two smaller matrices: $W = W_0 + BA$, where W_0 represents frozen pretrained weights, and B and A are trainable low-rank matrices. This approach typically requires updating only 0.1–1% of original parameters while maintaining comparable performance to full fine-tuning.

The advantages of LoRA include dramatic reduction in memory requirements and training time, ability to store multiple task-specific adaptations efficiently, and preservation of original model capabilities through frozen base weights. This efficiency enables organizations with limited computational resources to adapt large models for specific applications.

2.1.4.2.3 Alignment and Human Feedback

Modern LLMs require sophisticated alignment techniques to ensure outputs align with human values, preferences, and safety requirements. This process has evolved beyond simple supervised learning to incorporate human judgment and reinforcement learning principles.

RLHF represents a breakthrough approach that optimizes models based on human preference data rather than traditional loss functions [34, 35]. The RLHF process involves three stages: SFT on high-quality examples, training a reward model on human preference comparisons, and reinforcement learning optimization using the learned reward model.

Proximal policy optimization (PPO) serves as the primary reinforcement learning algorithm in RLHF implementations [36]. PPO maintains training stability by constraining policy updates within a trusted region, preventing catastrophic performance degradation during optimization. This approach enables models to learn complex preference patterns while maintaining coherent language generation capabilities.

Direct preference optimization (DPO) offers an innovative alternative to PPO-based RLHF by directly optimizing preference data without requiring explicit reward model training [37]. DPO reformulates the preference learning problem as direct policy optimization, reducing computational complexity while maintaining alignment effectiveness. This approach has gained popularity due to its implementation simplicity and competitive performance.

Group relative policy optimization (GRPO) extends preference optimization to handle group-based comparisons, enabling more nuanced preference learning from complex human feedback patterns [38]. GRPO considers relative preferences within groups of responses, capturing subtle preference distinctions that traditional pairwise methods might miss.

2.1.4.2.4 Constitutional AI and Safety Training

Constitutional AI represents an advanced alignment approach that trains models to follow a set of principles or "constitution" governing their behavior [39]. This method combines supervised learning on constitutional examples with reinforcement learning from AI feedback (RLAIF), where models critique and revise their own outputs according to constitutional principles.

This approach enables models to develop intrinsic understanding of ethical guidelines and safety considerations, reducing reliance on extensive human oversight during deployment while maintaining alignment with human values.

2.1.4.2.5 Reasoning Enhancement

Post-training increasingly focuses on enhancing model reasoning capabilities through specialized training techniques:

- CoT [32] Training explicitly teaches models to break down complex problems into step-by-step reasoning processes, significantly improving performance on mathematical, logical, and analytical tasks.
- Self-consistency training [40] encourages models to generate multiple reasoning paths and select the most consistent solution, improving reliability and accuracy in complex reasoning scenarios.
- Tool-use training [41] enables models to learn when and how to utilize external tools, calculators, databases, or APIs, extending their capabilities beyond text generation to include interactive problem-solving.

2.1.4.3 Prompt Engineering

Prompt engineering is a crucial technique for harnessing the full power of LLMs. By carefully crafting input prompts, users can guide these models to perform specific tasks without the need for extensive task-specific training. This technique leverages the vast knowledge embedded in pretrained models, making them adaptable and versatile for a wide range of applications. In this section, we will explore key aspects of prompt engineering, including ICL, instruction following, and CoT reasoning.

2.1.4.3.1 In-context Learning

ICL allows LLMs to adapt their responses based on examples or context provided within the prompt [10]. This ability is particularly useful in scenarios where minimal or no task-specific data is available. However, ICL's effectiveness hinges on the model's pretraining data. For example, translating rare languages may fail if the model lacks exposure to them during training. ICL is ideal for tasks like customer support (e.g., "Classify this query as billing/technical") where minimal training data exists.

ICL is divided into two primary techniques:

Zero-shot Learning: In zero-shot learning, LLMs tackle tasks without any prior examples. For instance, if tasked with translating a sentence from English to French without any training on translation tasks, the model might still succeed by leveraging its pretrained knowledge.

For example, given the prompt: the model might generate:

> **user** Translate the following sentence to French: 'The weather is nice today.'
>
> **assistant**
> Le temps est agréable aujourd'hui.

This demonstrates the model's ability to generalize its pretrained knowledge to new tasks without needing explicit examples.

Few-shot Learning: Few-shot learning takes ICL a step further by providing the model with a small number of examples within the prompt. These examples help the model understand the task better, leading to more accurate results.

For instance, if we wanted the model to identify the sentiment of a sentence, we might provide a few examples: The model, using the context of previous examples, would likely respond with:

> **user** Sentence: 'I love this product!' Sentiment: Positive
> Sentence: 'This is the worst experience ever.' Sentiment: Negative
> Sentence: 'The movie was just okay.' Sentiment: Neutral
> Sentence: 'The service was fantastic!' Sentiment: [Fill in the blank]
>
> **assistant**
> Positive

This demonstrates the model's ability to learn from minimal data and apply that learning to new, similar tasks.

2.1.4.3.2 Instruction Following

Instruction following is an ability that enables LLMs to comprehend and execute tasks described in natural language instructions [35]. This capability allows LLMs to perform tasks they were not explicitly trained for, simply by following the instructions given in the prompt.

Direct Instruction: Suppose you ask the model to summarize a lengthy article with a prompt like: "Summarize the following article in two sentences":

followed by the article text. The model can effectively generate a concise summary, showcasing its ability to understand and follow instructions. A clear instruction like "Summarize in one sentence" yields better results than ambiguous prompts like "Make this shorter."
For example:

> **user** Summarize the following article in two sentences:
> Artificial Intelligence is transforming the tech industry. It is being used in various applications such as natural language processing, image recognition, and autonomous driving.
>
> **assistant**
> AI is revolutionizing the tech industry with applications in NLP, image recognition, and autonomous driving.

Task Adaptation: Instruction fine-tuning has been shown to significantly enhance model performance, especially for larger models. Studies have demonstrated that LLMs with around 100 billion parameters, like GPT-3, can adapt to perform new tasks simply by receiving instructions, without needing additional task-specific training.
For example:

user Write a Python function that checks if a number is prime.

assistant

```
def is_prime(n):
  if n <= 1:
    return False
  for i in range(2, int(n**0.5) + 1):
    if n % i == 0:
      return False
  return True
```

This example illustrates the model's ability to understand a programming task and generate the appropriate code to accomplish it.

2.1.4.3.3 *Chain-of-thought Reasoning*

CoT reasoning is an advanced prompting technique that guides LLMs to articulate intermediate steps or thoughts before arriving at the final answer. This approach mirrors human problem-solving processes, where complex tasks are broken down into smaller, more manageable parts. CoT reasoning significantly improves performance on complex tasks and enhances the interpretability and reliability of model outputs. Let's take a look at an example of using CoT reasoning for math problem-solving.

Consider a math problem where the model needs to solve a word problem:

user If a train travels 60 miles per hour for 2 hours and then 40 miles per hour for 3 hours, how far did the train travel in total?

First, calculate the distance traveled at 60 miles per hour. Then, calculate the distance traveled at 40 miles per hour. Finally, add the two distances together to find the total distance.

assistant

Distance at 60 mph: $60 * 2 = 120$ miles.
Distance at 40 mph: $40 * 3 = 120$ miles.
Total distance $= 120 + 120 = 240$ miles.

This example shows how CoT reasoning helps the model break down the problem into logical steps, leading to a correct and well-structured solution.

2.1.4.4 From Prompt Engineering to Context Engineering

As LLM applications matured from demos to production systems, it became clear that achieving consistent, reliable performance required far more sophisticated approaches. The term "prompt engineering" itself has proven inadequate to describe the full scope of work required to optimize

LLM systems. What started as writing better prompts has evolved into **context engineering**—a comprehensive discipline focused on architecting the entire context window and information flow of LLM applications.

2.1.4.4.1 *Why Context Engineering*

Prompt engineering has evolved from a simple technique of crafting input prompts to a sophisticated discipline now increasingly referred to as context engineering. This evolution reflects the growing complexity of LLM applications and the recognition that optimizing model performance requires more than just well-written prompts—it demands careful orchestration of the entire context window.

The shift from prompt to context engineering reflects several fundamental realizations:

1. **Context Is Everything**
   Modern LLMs don't just need instructions; they need the right information at the right time in the right format. Consider a customer support bot—it needs not only just the user's question but also their purchase history, previous interactions, current system status, and company policies. This isn't prompt engineering; it's context orchestration.

2. **Beyond Static Prompts**
   Production LLM systems are dynamic. They must handle:
   - Changing user states and conversation history
   - Real-time data integration (current time, live metrics, API responses)
   - Variable document lengths and formats
   - Multistep reasoning chains
   - Tool interactions and function calls

3. **The Hidden Complexity**
   What appears as a simple chat interaction often involves:
   `User Input → Query Augmentation → Retrieval → Ranking → Context Assembly → Prompt Construction → LLM Call → Output Parsing → Validation → Response`
   Each step requires careful engineering to ensure the LLM receives optimal context.

2.1.4.4.2 *Core Components of Context Engineering*

Context engineering encompasses several interconnected elements:

Prompt engineering remains the foundation-crafting clear instructions, using appropriate delimiters, and structuring few-shot examples. However, it's now just one component of a larger system. Effective prompt design involves understanding how to frame tasks, provide examples that guide model behavior, and structure information for optimal comprehension.

RAG brings dynamic knowledge into the context, transforming static prompts into living systems. This involves critical decisions about what information to retrieve and when, how to chunk and index large documents effectively, and strategies for ranking and filtering results based on relevance. Perhaps most importantly, RAG requires seamless integration of retrieved content into the prompt without disrupting the natural flow of information. We are going to dive into more details of RAG in Chapter 3.

Memory management operates on two temporal scales. Short-term memory maintains conversation coherence across turns and manages the working context within token limits. Long-term memory involves persistent knowledge storage, intelligent caching of frequently accessed information, and strategic decisions about what information to retain vs. discard. This component ensures that systems can handle extended interactions while remaining performant and cost-effective.

Structured I/O forms the reliability backbone of production systems. This encompasses input validation and parsing to handle diverse user inputs gracefully, output formatting using JSON schemas or other structured formats to ensure consistent responses, and comprehensive error handling with fallback strategies. Type safety across the entire pipeline prevents cascading failures and enables robust system behavior even when individual components encounter issues.

State and history tracking enables sophisticated multistep workflows and learning from past interactions. This includes managing complex reasoning processes that may require multiple iterations, implementing revision and self-correction loops based on intermediate results, maintaining audit trails for debugging and compliance, and handling user session management across potentially disconnected interactions. These capabilities transform simple question-answering systems into intelligent agents capable of complex, multi-turn problem-solving.

Each component influences and depends on the others, creating a web of interactions that must be carefully orchestrated for optimal performance.

Context engineering is rapidly evolving, with new patterns and best practices emerging. As LLMs become more capable, the differentiator won't be the models themselves but how effectively we engineer their context. The future belongs to those who master not just prompt writing, but the full art and science of context engineering.

2.2 LLM Use Case in Information Retrieval

Having explored the foundational concepts of IR in Chapter 1, we now turn to examine how LLMs are redefining this field. LLMs bring advanced natural language understanding capabilities that address many traditional IR challenges, from interpreting complex user queries to capturing semantic relationships between documents and search intent. Their ability to process and generate human-like text enables more sophisticated matching between what users actually want and what systems can deliver.

This section explores key applications where LLMs are transforming IR systems, focusing on semantic search and personalized recommendations. These use cases demonstrate how LLMs move beyond simple keyword matching to enable truly intelligent information discovery that understands context, intent, and user preferences.

2.2.1 Use LLM for Search Systems

Traditional semantic similarity search, using sparse and dense retrieval, confronts significant challenges, such as an overwhelming list of similar results due to large corpuses, the need for up-to-date content, and the oversight of crucial details relevant to a search. These issues underscore the limitations of traditional semantic search methods that often struggle to balance relevance, recency, and the completeness of information effectively.

This highlights the necessity for incorporating LLMs, which enhance key search modules including query understanding, document understanding, and retrieval and ranking through the following approaches:

- Query Understanding: LLMs excel at interpreting nuanced user intent by capturing the semantics, context, and variations in natural language queries. This enables more accurate matching even when queries are vague, complex, or under-specified.

- Document Representation: By leveraging their deep language modeling capabilities, LLMs can extract key concepts, summarize content, and synthesize features from unstructured documents. This improves the quality of indexing and relevance estimation.
- Retrieval: LLMs enhance retrieval by refining semantic representations of queries and documents—often through contrastive learning techniques. Moreover, they can be used to automatically label relevance by identifying salient information, which enables the training of domain-adapted retrieval models.
- Ranking: In many cases, LLMs can serve directly as off-the-shelf zero-shot or few-shot rankers, delivering strong performance without requiring task-specific fine-tuning. However, to further optimize ranking systems—especially in domain-specific applications—LLMs can be used for training data augmentation. This includes generating synthetic query-document pairs, rewriting queries to simulate different intents, or creating diverse negative samples. These augmented datasets improve generalization and robustness in learned ranking models. Table 2.15 lists the comparisons of using LLMs for direct inference vs. data augmentation.

By supporting both direct inference and indirect supervision, LLMs provide flexible pathways to enhance end-to-end IR pipelines across diverse settings. As shown in Figure 2.15, besides enhancing traditional IR components, LLMs can also function as agents (such as acting as conversational agents to guide users through iterative search refinement) and perform tasks to improve search experience.

2.2.1.1 LLM-enhanced Query Processing

Building upon the traditional foundations introduced in Section 1.3.1.2, modern LLM-based query processing takes search understanding to the next level. Instead of relying solely on static heuristics, LLMs dynamically infer intent, rewrite queries, and manage complex inputs through decomposition and routing. This section explores how LLMs augment the three major pillars of query processing: intent recognition (understanding what users want), query rewriting (refining the user input), and routing with decomposition (managing complexity), delivering deeper understanding, richer contextualization, and higher precision.

Table 2.15 LLM usage comparison: Inference vs. augmentation.

Aspect	LLMs for direct inference	LLMs for data augmentation
Purpose	Use LLMs as ready-to-deploy retrieval/retrieval components	Generate synthetic data to train or improve traditional models
Fine-tuning Required	Not required (zero-shot or few-shot)	Optional (used to fine-tune downstream models with augmented data)
Use cases	Zero-shot reranking, prompt-based document scoring	Synthetic query generation, Relevance labeling, Query rewriting
Efficiency	Ideal for low-latency or small-scale settings	Scalable for batch training and large-scale optimization
Customization	Limited to prompt engineering or few-shot templates	Highly customizable to specific domains or tasks
Dependency	Depends on API or pre-deployed LLMs	Can use LLMs during preprocessing; no runtime cost
Examples	Use ChatGPT to rerank top-k retrieved docs	Use GPT-4 to create new query-document pairs for contrastive training
Benefits	Fast deployment, no training cost	Improves model robustness and performance in domain-specific tasks

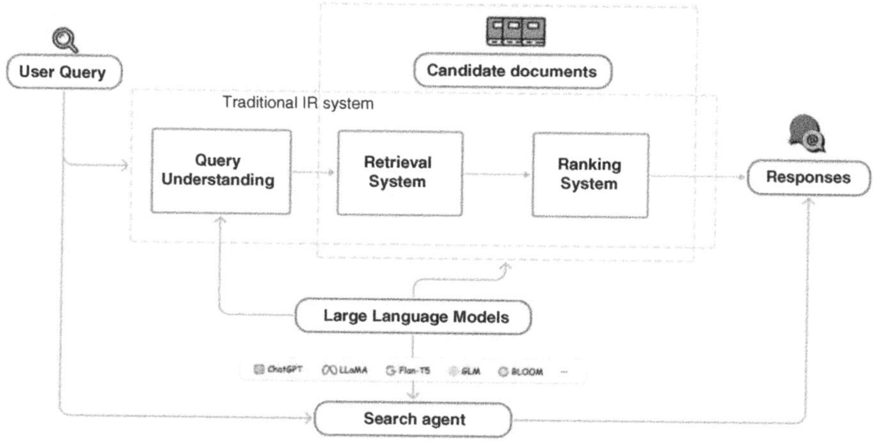

Figure 2.15 Overview of applying LLMs into information retrieval systems.

2.2.1.1.1 Intent Recognition

As discussed in Section 1.3.1.2, search intents typically fall into three broad categories: *informational*, *navigational*, and *transactional*.

Unlike traditional classifiers, LLMs can contextualize intent even when queries are ambiguous or underspecified. Table 2.16 demonstrates how LLMs like gpt-3.5-turbo can interpret intent dynamically by incorporating contextual cues.

user	You are an intelligent assistant that understands search query intentions. Analyze the following search query and provide the intent category (Informational, Instructional, or Transactional) and an explanation. Consider the context if provided.

Handling Long-tail Queries: LLMs are particularly effective at handling long-tail queries—uncommon, specific searches that are challenging for traditional systems:
- They can understand and process complex, conversational queries.
- LLMs can infer relationships between concepts to find relevant results even when exact matches are scarce.

Cross-lingual Search: LLMs trained on multiple languages can facilitate cross-lingual search:
- They can understand queries in one language and find relevant documents in another.
- This expands the scope of searchable content for users.

2.2.1.1.2 Query Rewriting

After intent classification, LLMs perform sophisticated query rewriting, a multistep process encompassing normalization, compression, expansion, and enrichment—each introduced in Section 1.3.1.2, but now enhanced with LLM-specific strategies.

Query Normalization: LLMs can recognize variations in user queries, such as typos or synonyms, and normalize them. This capability leverages LLMs' contextual understanding rather than relying solely on dictionary-based corrections. For example, if a user types "Azuer Devpos" instead of "Azure DevOps," an LLM can correct the misspelling and provide accurate results. Here is an example input prompt and output from an LLM for query normalization:

Table 2.16 Use LLM to generate intent classification results.

Input query	Input context	Output category from LLM	Output explanations from LLM
"What is the capital of France?"	None	Informational	The search query "What is the capital of France?" is seeking factual information. The user is looking for the answer to a specific question, indicating an informational intent to acquire knowledge about the capital city of France. The goal is not to perform a transaction or to receive step-by-step instructions but to gain general knowledge and information on the topic
"Buy plane tickets to New York."	None	Transactional	The search query "Buy plane tickets to New York" indicates a transactional intent, as the user is looking to make a purchase The user is specifically seeking to buy plane tickets to New York, showing an intent to complete a transaction related to air travel
"pizza"	None	Informational	The search query "pizza" suggests that the user is looking for information about pizza, such as its ingredients, different types of pizza, recipes, or pizza places near them. There is no specific action requested in the query, indicating an informational intent to gather knowledge or details about pizza
"pizza"	"I am a chef but don't know how to make pizzas."	Instructional	The search query "pizza" combined with the context provided indicates the user is looking for instructions on how to make pizzas as they mentioned being a chef but not knowing how to make them. This falls under the instructional intent category as the user is seeking guidance or step-by-step instructions on how to make pizzas
"pizza"	"I am hungry and it's dinner time."	Transactional	Based on the context provided ("I am hungry and it's dinner time"), the search query "pizza" suggests a transactional intent. The user is likely looking to order or find a nearby pizza restaurant to satisfy their hunger and have dinner, indicating a desire to make a transaction such as ordering food
"bat"	"I love playing baseball."	Informational	Based on the search query "bat" and the context provided, the user is most likely seeking information or general knowledge about bats related to baseball (such as types of bats, features, uses in the game, etc.) rather than looking to buy or engage in a specific transaction related to bats. Therefore, the intent falls under the Informational category

user Normalize the following query by correcting typos and understanding intent:
Azuer Devpos tutorials

assistant
Azure DevOps tutorials

Query Compression: Query compression involves reducing redundant information in chat histories or queries while preserving critical context. This technique involves summarizing or condensing the chat history to focus on the most relevant parts, improving the accuracy and efficiency of the retrieval process by eliminating irrelevant historical messages (e.g., unrelated questions from prior sessions); compression reduces computational overhead while maintaining query focus. For example, compressing the chat history to the last question ensures relevant context is preserved without overwhelming the system. Here is an example input and output from an LLM for query compression:

> **user** User: Hi, I've been doing some research on budget travel options in Europe.
> User: Can you tell me about Eurail passes and how they work?
> User: Oh, and by the way, I'm also interested in learning about museum discounts in Paris.
> User: Sorry, forgot to mention——I'm mostly traveling in France and Italy.
> User: What are some good apps for public transport in Rome?
>
> **assistant**
> What are the best public transport apps for Rome?

In this example, the LLM compresses the multi-turn history by dropping unrelated context (e.g., Eurail, museum discounts) and isolating the final user intent. This preserves essential information ("Rome" + "public transport") while reducing noise, improving both retrieval accuracy and system efficiency.

Query Expansion: Query expansion is a powerful technique that enhances search accuracy by broadening or refining the original query. LLMs excel at this task by analyzing the query's context and intent, then suggesting additional terms or rephrasing the query for improved results. Here is an example input prompt and output from an LLM for query expansion:

> **user** Expand the following query to a list of related queries to better understand the intent of the search query
> Best laptops 2024
>
> **assistant**
> 1. Top–rated laptops for professionals in 2024
> 2. Affordable laptops with high performance in 2024
> 3. Best gaming laptops for 2024
> 4. Sleek and lightweight laptops for students in 2024
> 5. Laptops with longest battery life in 2024

Context enrichment plays a crucial role in query expansion, adding relevant information to improve retrieval accuracy. If we add some context to the query expansion prompt, the output from the LLM model can change based on different contexts.

> **user** I want to read reviews of products
> Expand the following query to a list of related queries to better understand the intent of the search query
> Query: "Best laptops 2024"

assistant
1. Best laptops reviews 2024
2. Top–rated laptops 2024
3. Latest laptop models 2024
4. Laptop rankings 2024
5. Laptop comparisons 2024

Here are two notable frameworks for query rewriting.

1. **Hypothetical Document Embeddings**
   Hypothetical document embedding (HyDE)generates a single hypothetical document to answer a query, enhancing retrieval by aligning the query with relevant documents [42]. HyDE operates zero-shot, making it versatile for ad-hoc IR tasks. The following prompt and result illustrate HyDE for e-commerce search query:

user Generate a hypothetical passage to answer the query.
Query: What are the best features of wireless headphones in 2024?

assistant
In 2024, top wireless headphones offer advanced noise cancellation, up to 30 hours of battery life, and high–fidelity audio with Bluetooth 5.3 for seamless connectivity. They often include touch controls and compatibility with voice assistants like Siri or Google Assistant, ideal for both casual and professional use.

This passage enriches the query, enabling better matching with WANDS product descriptions.

2. **Query2Doc**
   Query2Doc, introduced in 2023, generates pseudo-documents (synthetic texts expanding query context) to improve sparse and dense retrieval [43]. For e-commerce search queries, Query2Doc adds related terms and context, boosting search engine performance. An example using LLM to generate the pseudo-documents:

user Generate a pseudo–document to expand the query.
Query: Best laptops for gaming 2024

assistant
In 2024, the best gaming laptops feature high–refresh–rate displays (e.g., 144Hz), powerful GPUs like NVIDIA RTX 4080, and at least 16GB of RAM for smooth performance. They include efficient cooling systems and customizable RGB keyboards, catering to gamers seeking immersive experiences.

This pseudo-document enriches the query with specific attributes, aiding retrieval of relevant WANDS product listings.

2.2.1.1.3 Routing and Decomposition

Beyond understanding and rewriting queries, LLMs enhance query routing and query decomposition by dynamically analyzing query content and context to determine the appropriate execution path or to simplify complex requests. These capabilities are essential in distributed environments where information spans multiple domains or retrieval systems.

Query routing involves identifying the relevant tool, database, or retrieval module that best serves the user's intent. LLMs can assist routing by classifying queries into backend task categories or outputting the correct module name. Here is an example of routing classification:

user You are an intelligent assistant that classifies queries into appropriate data sources. Given a user query, return the most relevant system: [financial_db, media_search, web_index, product_catalog].

Query: "Find photos of the Eiffel Tower"

assistant
media_search

Here is an example of tool selection:

user Classify the following queries to the correct tool:
1. "How much did I spend last month?"
2. "Pictures of cats wearing hats"
3. "Buy iPhone 15 in San Francisco"
4. "What is the capital of Japan?"

Options: [product_catalog, knowledge_base, media_search, financial_db]

assistant
1. financial_db
2. media_search
3. product_catalog
4. knowledge_base

Query decomposition is the process of splitting complex or multi-intent queries into atomic sub-questions that can be independently processed. Here is an example of LLM-based query decomposition:

user Decompose the following query into multiple focused sub–questions:
"What are the causes, effects, and mitigation strategies for climate change?"

assistant
1. What are the primary causes of climate change?
2. What are the major effects of climate change?
3. What strategies are used to mitigate climate change?

2.2.1.2 Document Representation

In modern search systems, representing documents in a way that captures their true semantic meaning is fundamental for achieving high retrieval relevance. Traditional approaches—such as BoW and term frequency-inverse document frequency (TF-IDF)—represent documents based on surface-level word frequency statistics, which often miss deeper semantic connections. In contrast, LLMs have revolutionized document representation by generating rich, contextual embeddings that encapsulate both syntactic structure and semantic nuance.

These embeddings position documents within a high-dimensional vector space, where semantically similar documents are located closer to one another. This allows search systems to move beyond simple keyword overlap and perform comparisons based on meaning, intent, and contextual relevance.

LLMs like Sentence-BERT [19] or MiniLM[44] encode documents into dense vector representations using their deep neural architectures. These embeddings reflect not only the lexical features of a document but also its contextual meaning across sentences or paragraphs. This shift from symbolic to semantic representations enables more accurate document ranking and retrieval.

The following Python example illustrates how to generate document and query embeddings using a lightweight sentence embedding model (sentence-transformers/all-MiniLM-L6-v2) and compute their semantic similarity for retrieval.

```
from transformers import AutoTokenizer, AutoModel
import torch
import numpy as np

class DocumentEmbedder:
    def __init__(self, model_name='sentence-transformers/all-MiniLM-L6-v2'):
        self.tokenizer = AutoTokenizer.from_pretrained(model_name)
        self.model = AutoModel.from_pretrained(model_name)

    def embed_document(self, document_text):
        """Generate semantic embedding for a document"""
        # Tokenize and encode
        inputs = self.tokenizer(document_text,
                                return_tensors='pt',
                                max_length=512,
                                truncation=True,
                                padding=True)

        # Generate embeddings
        with torch.no_grad():
            outputs = self.model(**inputs)
            # Mean pooling of token embeddings
            embeddings = outputs.last_hidden_state.mean(dim=1)

        return embeddings.numpy().flatten()

    def compute_similarity(self, query_embedding, doc_embeddings):
        """Compute cosine similarity between query and documents"""
        similarities = []
        for doc_emb in doc_embeddings:
            similarity = np.dot(query_embedding, doc_emb) / (
                    np.linalg.norm(query_embedding) * np.linalg.norm(doc_emb)
            )
            similarities.append(similarity)
        return similarities

# Example usage
embedder = DocumentEmbedder()

documents = [
    "Machine learning algorithms can automatically identify patterns in data.",
    "Artificial intelligence systems are transforming healthcare diagnostics.",
    "Deep learning models require large amounts of training data."
]

# Generate embeddings for documents
doc_embeddings = [embedder.embed_document(doc) for doc in documents]

```

```
# Query example
query = "AI in medical applications"
query_embedding = embedder.embed_document(query)

# Find most similar documents
similarities = embedder.compute_similarity(query_embedding, doc_embeddings)
ranked_docs = sorted(zip(documents, similarities), key=lambda x: x[1], reverse=True)

for doc, score in ranked_docs:
    print(f"Score: {score:.3f} | Document: {doc}")
```

Given a search query "AI in medical applications" and a list of candidate documents, the previous python code computes the similarity of query and each candidate document using the query embedding and document embeddings. The final output is ranked based on the similarity score, where larger values indicate the document is more likely to be relevant to the search query.

Results:

```
Score: 0.624 | Document: Artificial intelligence systems are transforming healthcare
    diagnostics.
Score: 0.396 | Document: Machine learning algorithms can automatically identify
    patterns in data.
Score: 0.237 | Document: Deep learning models require large amounts of training data.
```

Note how the top-ranked document shares no exact keywords with the query "AI in medical applications" yet achieves the highest similarity score due to semantic alignment between "AI/artificial intelligence" and "medical/healthcare diagnostics."

2.2.1.3 Relevance Ranking

Beyond retrieval, LLMs also elevate the precision of search result ranking. In traditional search engines, ranking often depends on statistical features like TF-IDF scores or PageRank. However, LLMs offer a richer toolkit for relevance estimation.

2.2.1.3.1 Multifactor Ranking

LLMs improve ranking by assessing semantic alignment between query and documents, analyzing document structure and coherence, and integrating contextual factors such as document recency or source authority. These signals can be combined using a weighted scoring framework to compute a final relevance score. Figure 2.16 lists the workflow diagram of the multifactor ranking.

The following example demonstrates how to compute a composite ranking score using three signals: semantic similarity, recency, and authority:

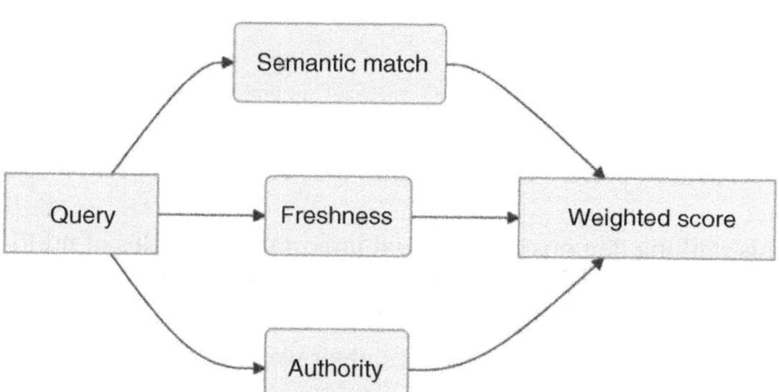

Figure 2.16 Multifactor ranking.

```
from transformers import AutoTokenizer, AutoModel
import torch
import pandas as pd

# Load pre-trained sentence embedding model
tokenizer = AutoTokenizer.from_pretrained("sentence-transformers/all-MiniLM-L6-v2")
model = AutoModel.from_pretrained("sentence-transformers/all-MiniLM-L6-v2")

# Example query and document corpus
query = "What are the recent advancements in NLP?"
documents = [
    {"text": "Transformers have revolutionized natural language processing.", "recency
    ": 0.9, "authority": 0.8},
    {"text": "Machine learning models are used to predict housing prices.", "recency":
     0.6, "authority": 0.7},
    {"text": "The stock market showed significant volatility last year.", "recency":
    0.7, "authority": 0.6}
]

# Tokenize the query and document texts
texts = [query] + [doc["text"] for doc in documents]
inputs = tokenizer(texts, padding=True, truncation=True, return_tensors="pt")

# Compute sentence embeddings
with torch.no_grad():
    outputs = model(**inputs)
    embeddings = outputs.last_hidden_state
    mask = inputs['attention_mask'].unsqueeze(-1).expand(embeddings.size()).float()
    sentence_embeddings = torch.sum(embeddings * mask, 1) / torch.clamp(mask.sum(1),
     min=1e-9)

# Compute semantic similarity
query_embedding = sentence_embeddings[0]
doc_embeddings = sentence_embeddings[1:]
cos_sim = torch.nn.functional.cosine_similarity(query_embedding.unsqueeze(0),
    doc_embeddings)

# Incorporate semantic similarity, document recency, and authority into final ranking
results = []
for i, doc in enumerate(documents):
    similarity = cos_sim[i].item()
    recency = doc["recency"]
    authority = doc["authority"]
    final_score = 0.6 * similarity + 0.2 * recency + 0.2 * authority
    results.append({
        "Document": doc["text"],
        "Semantic Similarity": round(similarity, 4),
        "Recency": recency,
        "Authority": authority,
        "Final Score": round(final_score, 4)
    })

# Sort and display results
df = pd.DataFrame(sorted(results, key=lambda x: x["Final Score"], reverse=True))
print(df)
```

The multifactor ranking system produces more nuanced results that better reflect real-world information needs:

As summarized in Table 2.17, this ranking demonstrates several important principles of multifactor scoring. The top-ranked document achieves the highest final score (0.6419) through a combination of strong semantic relevance (0.5032) and excellent metadata scores (recency: 0.9, authority: 0.8). While the other documents have reasonable metadata scores, their weak semantic similarity to the NLP query results in significantly lower final rankings.

Table 2.17 Multifactor ranking example output.

Document	Semantic similarity	Recency	Authority	Final score
Transformers have revolutionized natural language processing.	0.5032	0.9	0.8	0.6419
Machine learning models are used to predict housing prices.	0.1323	0.6	0.7	0.3394
The stock market showed significant volatility last year.	0.0941	0.7	0.6	0.3165

2.2.1.3.2 Data Augmentation

In addition to improving runtime ranking, LLMs can also help train better ranking models through **data augmentation**. In low-resource settings where labeled query-document pairs are limited, LLMs can generate paraphrased queries or hypothetical documents to enrich the training corpus.

The following code demonstrates how to use GPT-2 to paraphrase queries and generate synthetic training data for ranking tasks:

```
from transformers import GPT2LMHeadModel, GPT2Tokenizer
import random
import torch

class RankingDataAugmenter:
    def __init__(self):
        self.tokenizer = GPT2Tokenizer.from_pretrained('gpt2')
        self.model = GPT2LMHeadModel.from_pretrained('gpt2')
        self.tokenizer.pad_token = self.tokenizer.eos_token

    def generate_query_variations(self, original_query, num_variations=5):
        """Generate paraphrased versions of a query"""
        variations = []

        prompts = [
            f"Rephrase this search query: '{original_query}' -> ",
            f"Another way to ask '{original_query}' is: ",
            f"Synonymous query for '{original_query}': "
        ]

        for _ in range(num_variations):
            prompt = random.choice(prompts)
            inputs = self.tokenizer.encode(prompt, return_tensors='pt')

            with torch.no_grad():
                outputs = self.model.generate(
                    inputs,
                    max_length=inputs.shape[1] + 20,
                    num_return_sequences=1,
                    temperature=0.8,
                    pad_token_id=self.tokenizer.eos_token_id
                )

            generated_text = self.tokenizer.decode(outputs[0], skip_special_tokens=True)
            variation = generated_text.replace(prompt, "").strip()

            if variation and len(variation.split()) <= 10:  # Keep reasonable length
                variations.append(variation)

        return list(set(variations))  # Remove duplicates

    def create_synthetic_pairs(self, queries, documents, labels):
```

```
        """Create augmented query-document pairs for training"""
        augmented_data = []

        for query, doc, label in zip(queries, documents, labels):
            # Original pair
            augmented_data.append((query, doc, label))

            # Generate query variations
            query_variations = self.generate_query_variations(query)

            for variation in query_variations:
                # Add some noise to labels for variations
                if label == 1:  # Relevant
                    new_label = random.choice([1, 1, 1, 0])  # Mostly relevant
                else:  # Not relevant
                    new_label = random.choice([0, 0, 0, 1])  # Mostly not relevant

                augmented_data.append((variation, doc, new_label))

        return augmented_data

# Example usage
augmenter = RankingDataAugmenter()

# Original training data
original_queries = ["machine learning healthcare", "AI medical diagnosis"]
original_docs = [
    "ML algorithms detect cancer in medical scans",
    "Weather forecast using satellite data"
]
original_labels = [1, 0]  # 1 = relevant, 0 = not relevant

# Generate augmented dataset
augmented_dataset = augmenter.create_synthetic_pairs(
    original_queries, original_docs, original_labels
)

print("Augmented Dataset Examples:")
for query, doc, label in augmented_dataset[:5]:
    print(f"Query: {query}\nDocument: {doc}\nLabel: {label}\n" + "-" * 30)
```

This approach demonstrates how LLMs can help generate high-quality synthetic examples that mirror natural query variations. Such augmentations improve generalization, reduce overfitting, and are especially valuable when labeled data is scarce.

In summary, these techniques—query understanding, query processing, document semantic embedding, relevance ranking, and data augmentation—illustrate how LLMs serve as both run-time inference engines and training-time data generators, enhancing every stage of the retrieval and ranking pipeline in search systems.

2.2.2 Use LLM for Recommendation Systems

LLMs have redefined the landscape of modern recommender systems. Beyond traditional collaborative filtering or content-based techniques, LLMs bring language understanding, reasoning, and abstraction capabilities to user modeling and recommendation logic. These models allow systems to interpret not just what users do, but why they do it—based on how they express preferences in natural language and how they behave over time and context.

2.2.2.1 User Profiling

User profiling is foundational to personalized recommendations. With the help of LLMs, user profiles can now go beyond static lists of interests or click histories. LLMs enable dynamic, rich, and multidimensional user profiles that combine explicit feedback, behavioral patterns, sentiment analysis, temporal activity, and even inferred context or emotion.

2.2.2.1.1 *Explicit Preferences*

LLMs can extract fine-grained product or service attributes (known as aspects from user reviews, and analyze sentiment toward each of them. For example, if a user writes, "Great product with excellent build quality, but the price is quite high," an LLM-based profiler can identify "quality" and "price" as relevant aspects, and assign separate sentiment scores to each. This allows systems to recognize that a user values quality highly, but is price-sensitive. In our profiling results, users showed strong positive preferences for aspects like quality, performance, and delivery, with high average sentiment and ratings.

2.2.2.1.2 *Implicit Preferences*

By analyzing user behavior such as views, clicks, purchases, and dwell time, LLM-based systems can infer implicit preferences, including:

- Category Affinity: Certain categories (e.g., electronics, books) receive disproportionately higher interaction time.
- Brand Loyalty: Repeated actions favoring specific brands like Apple or Samsung reflect brand preference.
- Price Sensitivity: LLMs can estimate how likely users are to engage with higher-priced items based on weighted interactions, helping personalize results based on willingness to pay.

2.2.2.1.3 *Sentiment Patterns*

LLMs perform sentiment analysis across user reviews to detect both emotional tone and consistency over time. This includes:

- An overall sentiment trend (positive or negative drift).
- Sentiment by rating or category (e.g., electronics vs. home).
- Emotional intensity (how strongly users react).
- Sentiment consistency (variability in reactions).

Additionally, LLMs extract themes from text—such as performance, service, or value—which can be tracked as positive or negative across reviews. This provides deeper insight into what users care about emotionally, not just behaviorally.

2.2.2.1.4 *Temporal Patterns*

Temporal insights from user activity logs—such as timestamped clicks or purchases—reveal when users are most active and what they prefer across seasons or days. LLM-based profiling extracts:

- Hourly activity patterns (e.g., peak times like 9AM, 2PM).
- Preferred weekdays (e.g., weekends vs. weekdays).
- Seasonal preferences (e.g., buying more books in winter).
- Activity consistency (how stable user habits are over time).

This allows the recommendation engine to not only match content to the user's interests, but to deliver it at the right time.

For instance, one user profile generated from LLM-based analysis revealed strong interest in electronics, high engagement with the Apple brand, and a consistent positive sentiment trend across

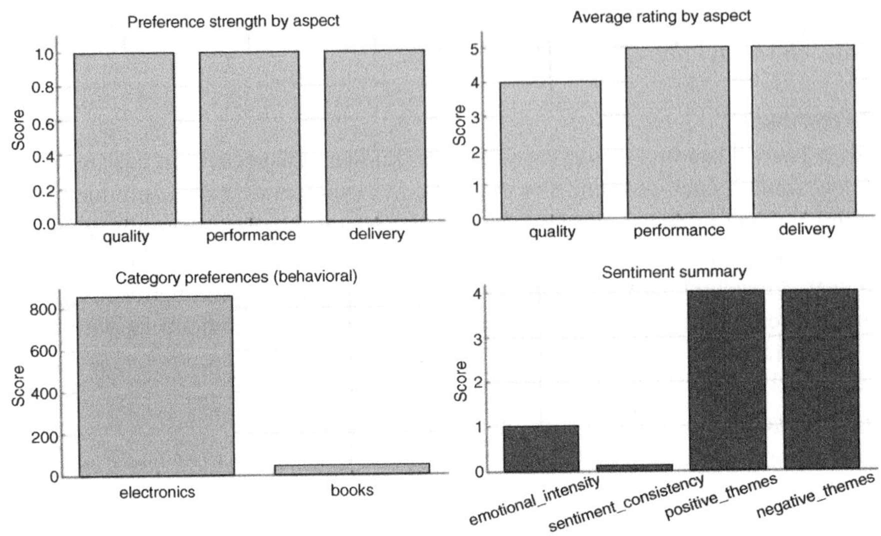

Figure 2.17 LLM-based user profile visualization.

several interactions (Figure 2.17). Another profile captured a more emotionally volatile pattern, with strong preferences but also a notable negative experience, suggesting the need for more personalized attention or recovery strategies.

```
# Simplified example of LLM user profiling output
user_profile = {
    "explicit_preferences": {
        "quality": {"preference_strength": 0.999, "average_rating": 4.0},
        "delivery": {"preference_strength": 0.999, "average_rating": 5.0}
    },
    "implicit_preferences": {
        "category_preferences": {"electronics": 855.0, "books": 45.0},
        "price_sensitivity": 0.79,
        "brand_loyalty": {"Apple": 5, "Samsung": 3}
    },
    "sentiment_patterns": {
        "emotional_intensity": 0.999,
        "sentiment_consistency": 0.999,
        "positive_themes": ["positive_quality", "positive_performance"]
    },
    "temporal_patterns": {
        "peak_activity_hours": [14, 19],
        "seasonal_preferences": {"winter": ["electronics", "books"]}
    }
}
```

Together, these capabilities show that LLMs enable a new era of dynamic and explainable user modeling. Rather than relying solely on historical purchases or similarity metrics, systems can now understand users as evolving entities—shaped by emotion, time, context, and language.

In the following sections, we'll explore how these profiles translate into item representations, context-aware recommendations, and explainability features within modern LLM-powered recommender systems.

2.2.2.2 Item Representation

Just as understanding user preferences is essential for effective recommendations, a rich and contextualized representation of items is equally critical. Traditional recommender systems often rely on structured metadata—such as category, price, or brand—and basic keyword-based descriptions. However, such shallow features limit personalization, especially for cold-start items or when dealing with nuanced user intents.

LLMs introduce a powerful paradigm shift by enabling semantically rich and multidimensional representations of items. These representations are generated from diverse information sources, including item descriptions, specifications, reviews, and contextual tags. The result is a deeper understanding of an item's features, user perception, and intended usage context.

In our code example, a pair of headphones was described by its title, description, and specifications. The resulting semantic vector enables fine-grained comparisons not just by category but by actual functional similarities (e.g., noise cancellation, battery life).

Input item data:

```
item_data = {
    'title': 'Wireless Bluetooth Headphones',
    'description': 'High-quality wireless headphones with noise cancellation',
    'price': 150,
    'brand': 'Sony',
    'category': 'Electronics/Audio',
    'specifications': {
        'battery_life': '30 hours',
        'connectivity': 'Bluetooth 5.0',
        'noise_cancellation': 'Active'
    },
    'tags': ['wireless', 'bluetooth', 'noise-cancelling'],
    'reviews': [
        {'text': 'Excellent sound quality and great battery life', 'rating': 5},
        {'text': 'Good headphones but a bit expensive', 'rating': 4}
    ]
}
```

Example of LLM item profile output:

```
Item Profile:
Semantic embedding shape: (384,)
Features: {
  "price_tier": "mid_range",
  "brand_tier": "premium",
  "feature_density": 0.6,
  "category_specificity": 0.5,
  "text_richness": 0.6528695652173913,
  "availability_score": 0.8
}
User perception: {
  "overall_sentiment": 0.13741767406463623,
  "key_strengths": [
    "quality"
  ],
  "key_weaknesses": [
    "value"
  ],
  "common_use_cases": [],
  "sentiment_distribution": {
    "positive": 1,
    "negative": 1,
    "neutral": 0
  },
  "average_rating": 4.5,
  "review_count": 2
}
```

2.2.2.2.1 *Semantic Embedding of Items*

LLMs can encode item information into high-dimensional vectors that capture underlying semantics. For example, the model `sentence-transformers/all-MiniLM-L6-v2` produces 384-dimensional embeddings from textual data like titles, descriptions, specifications, and category labels. These embeddings allow the model to represent similar items in nearby regions of semantic space, facilitating improved similarity matching and ranking.

2.2.2.2.2 *Feature Extraction and Structured Profiling*

Beyond embeddings, LLM-powered systems can also extract structured features such as:

- Price tier (e.g., budget, mid-range, premium).
- Brand tier (e.g., standard, premium, luxury).
- Feature density, which reflects the richness of the item's attribute set.
- Text richness, derived from vocabulary diversity and completeness of descriptions.
- Category specificity, indicating how niche or general an item is.
- Availability score, estimating the item's current stock status.

These features allow recommendation systems to reason beyond embeddings and incorporate practical constraints like stock status or affordability.

2.2.2.2.3 *User Perception and Sentiment-driven Insights*

By analyzing review content using sentiment analysis pipelines, LLMs can further profile items based on how users perceive them. For instance, a set of headphone reviews might reveal:

```
Key Strengths: "quality" or "sound"
Key Weaknesses: "value" (e.g., expensive)
Common Use Cases: "work", "travel"
Sentiment Distribution: 2 positive, 1 negative
Average Rating: 4.67
```

These insights contribute to a holistic understanding of the item from the user's perspective, allowing for more emotionally resonant and context-aware recommendations.

2.2.2.2.4 *Contextual Attributes and Use Scenarios*

LLMs can also extract implicit contextual information from item descriptions and tags, identifying:

- Suitable Environments: Professional, casual
- Target Demographics: Adults, youth, seniors
- Usage Scenarios: Entertainment, commuting, work

Such contextual cues support more adaptive recommendation strategies—e.g., recommending professional-grade headphones for work contexts, or casual use headphones for travel.

In short, LLMs unlock a powerful new level of item representation by blending deep semantic understanding, structured feature extraction, and real-world user sentiment. These capabilities are foundational for building recommender systems that are not only accurate but also adaptable to nuanced user needs and dynamic environments.

2.2.2.3 Explanation Generation

One of the most significant challenges in modern recommendation systems is the "black box" problem—users receive recommendations without understanding why those particular items were suggested. This lack of transparency can lead to reduced user trust, lower engagement, and missed opportunities for users to refine their preferences. LLM-powered explanation generation transforms opaque recommendation systems into transparent, trustworthy advisory platforms that not only suggest items but also articulate the reasoning behind each recommendation in natural, personalized language.

Traditional recommendation systems, even sophisticated ones, typically provide minimal explanation beyond simple statements like "users who bought this also bought that" or "based on your purchase history." LLM-powered explanation systems create rich, multifaceted explanations that help users understand not just what is being recommended, but why it makes sense for their specific situation, preferences, and current context.

2.2.2.3.1 Multifaceted Explanations

Explanations generated by LLMs are often multidimensional. Here are some examples:

1. Given a context-aware recommendation for a "Smart Coffee Maker," the model may produce a narrative like:

```
assistant
    I recommend 'Smart Coffee Maker' because it provides a good balance of quality
    and price. It also is perfect for your morning routine.
    Furthermore, it offers premium features not found in basic alternatives.
```

2. Given a context-aware recommendation for a "Wireless Noise-Cancelling Headphones," the model may produce a narrative like:

```
assistant
    I recommend 'Wireless Noise–Cancelling Headphones' because it strongly
    matches your interest in electronics and comes from Sony, a brand you've liked
    before. It also is perfect for your morning routine.
    Furthermore, it offers premium features not found in basic alternatives.
```

These explanations weave together several components:

- Relevance Explanation: Matches the user's interest in product category (e.g., electronics) or brand preferences (e.g., Apple, Sony).
- Context Explanation: Incorporates contextual cues such as time of day, location (e.g., during commute), device type (e.g., mobile), and activity (e.g., exercising).
- Comparison Explanation: Highlights why this item stands out vs. alternatives by citing unique features (e.g., noise cancellation, premium build).
- Timing Explanation: Aligns the recommendation with current user activities or schedule (e.g., morning routine, rainy weather).
- Personalization Explanation: Leverages the user's implicit behavioral data (e.g., price sensitivity, brand loyalty, past purchases).

2.2.2.3.2 Template-based and Generative Approaches
Explanations can be produced in two ways:

- Template-Based: These are rule-driven and rely on predefined patterns. They are deterministic, interpretable, and efficient but limited in flexibility.
- Generative LLM-Based: These leverage models like GPT or T5 to produce fluid, personalized explanations. They can adapt tone, wording, and depth depending on the situation.

This level of linguistic nuance makes recommendations feel more trustworthy and customized, thus improving user experience and decision confidence.

2.2.2.3.3 Implementation Considerations
Generating these explanations requires integrating LLMs with multiple data sources:

- User Profile: Captures behavioral patterns, brand and category preferences, sentiment trends, and temporal activity.
- Context: Includes real-time variables such as current activity, location, weather, and device.
- Item Profile: Provides metadata like category, tags, features, brand, availability, and contextual suitability.

An LLM-based explanation module dynamically pulls from these profiles to compose justifications that are both rational and contextually meaningful.

In conclusion, LLMs have significantly enhanced both semantic search and recommender systems, offering more intelligent, context-aware, and personalized IR. However, their implementation requires careful consideration of computational resources, data quality, privacy, and ethical implications. As LLM technology continues to evolve, we can expect further improvements in these critical areas of IR.

2.3 Challenges of LLMs for Information Retrieval

While LLMs have demonstrated remarkable capabilities across various IR tasks, they face significant limitations that can impact their reliability and effectiveness in real-world applications. Understanding these challenges is crucial for developing robust AI systems that can meet enterprise and user requirements.

2.3.1 Core Limitations: Hallucinations and Knowledge Staleness

2.3.1.1 The Knowledge Cutoff Problem
The most fundamental limitation of LLMs stems from their training methodology. LLMs are trained on static datasets up to a specific point in time, creating what's known as the knowledge cutoff problem. This temporal boundary creates several critical issues:

- Outdated Information: LLMs may provide responses based on information that is no longer current, particularly problematic for queries about recent events, evolving technologies, or dynamic business environments.
- Real-time Data Gap: Traditional LLMs cannot incorporate live data feeds, breaking news, or recent developments into their responses without complete retraining.
- Misinformation Risk: In rapidly changing situations, LLMs may inadvertently spread outdated or incorrect information by relying on their fixed knowledge base.

2.3.1.1.1 *Practical Example: The Twitter CEO Question*

Consider a common scenario where knowledge cutoff dates significantly impact response accuracy. Let's examine how different models with varying knowledge cutoffs handle this question. Using the following example code of asking LLMs this question:

> **user** Who is the current CEO of Twitter?

```python
import openai
from openai import OpenAI

print(f"OpenAI version: {openai.__version__}")

# Initialize the OpenAI client
client = OpenAI(api_key=api_key)

content = "Who is Twitter's CEO"

response = client.chat.completions.create(
    model="gpt-4o", # can change to other models
    messages=[
        {"role": "user", "content": content}
    ]
)

print(response.choices[0].message.content)
```

Model with 2022 knowledge cutoff will have these results:

> **assistant**
>
> As of my last training data update, Jack Dorsey was the CEO of Twitter. However, please note that my knowledge cutoff is in 2022, and there may have been changes since then. I recommend checking a current and reliable source for the most up–to–date information on Twitter's CEO.

Model with 2023 knowledge cutoff will have these results:

> **assistant**
>
> As of my last update in September 2023, Twitter does not have a named CEO. The company has been without a permanent CEO since Parag Agrawal was removed from his position on October 27, 2022, following changes in ownership after Elon Musk's acquisition of the company. Since then, Elon Musk has been involved in making decisions for the company, but he does not hold the official title of CEO. For the most current information, you may want to check the latest news sources.

This example clearly illustrates how knowledge cutoff dates can lead to potentially outdated or incorrect information, especially for queries about recent organizational changes.

2.3.1.1.2 *Possible Solutions and Their Limitations*

Several approaches have been proposed to address the knowledge cutoff problem, each with distinct trade-offs:

- Continuous Pretraining: Regularly updating LLMs with recent data can reduce the knowledge gap. However, this approach is resource-intensive, requiring significant computational power and time for frequent retraining cycles.
- Fine-tuning: Adapting LLMs to newer contexts through fine-tuning on recent datasets allows models to incorporate current information without complete retraining. While more efficient than full retraining, this method still requires substantial resources and technical expertise.
- Static External Integration: Some systems attempt to supplement LLM knowledge with preloaded external databases. However, these approaches often lack the dynamic retrieval capabilities needed for truly current information.

2.3.2 The RAG Solution: Dynamic Knowledge Integration

RAG emerges as a compelling solution to address these fundamental limitations. RAG systems enhance LLMs by providing real-time access to external knowledge sources, creating a hybrid approach that combines the language understanding capabilities of LLMs with the currency and accuracy of external data sources.

2.3.2.1 RAG vs. Alternative Approaches

As a comparison, Table 2.18 lists the strengths and weakness of RAG vs. the alternative approaches.

2.3.2.1.1 *RAG vs. Prompt Engineering*

Prompt engineering, which we covered in Section 2.1.4.3 can help steer a model to respond in the way we want. However, it has fundamental limitations:

- Knowledge Gaps: Prompt engineering cannot address situations where the LLM simply lacks required information. No amount of clever prompting can make a model aware of events that occurred after its training cutoff.
- Consistency Challenges: Prompt-based approaches may produce inconsistent results across different queries or user sessions, as they rely on the LLM's internal knowledge representation.
- Limited Scope: Prompts cannot provide access to proprietary or domain-specific information that wasn't included in the model's training data.

RAG addresses these limitations by providing external knowledge access, ensuring consistent and current information regardless of the query complexity.

Table 2.18 Comparisons of RAG vs. prompt engineering vs. fine-tuning.

Technique	Strengths	Weaknesses
Prompt engineering	Fast, simple to apply	Cannot access missing knowledge or domain-specific context
Fine-tuning	High task accuracy	Costly, slow, requires expertise and periodic retraining
RAG	Up-to-date, explainable, domain-adaptable	Requires access to reliable retrieval systems

2.3.2.1.2 RAG vs. Pretraining or Fine-tuning

Pretraining and fine-tuning LLMs, which we discussed in Section 2.1.4, can improve performance on specific tasks but presents several challenges:

- Resource Requirements: Fine-tuning large models requires significant computational resources, making it expensive and inaccessible for many organizations.
- Technical Expertise: Successful fine-tuning requires deep NLP expertise and careful hyperparameter optimization, creating barriers for smaller teams.
- Time Investment: The fine-tuning process is time-intensive, involving dataset preparation, training setup, and iterative refinement.
- Static Knowledge: Even fine-tuned models become outdated over time and require periodic retraining to maintain currency.
- Limited Flexibility: Fine-tuned models are optimized for specific tasks and may not generalize well to related but different use cases.

2.3.2.2 RAG Implementation Example

Let's demonstrate how RAG can solve the Twitter CEO problem by integrating real-time search capabilities. Here is the example Python code for the implementations:

```
import openai
from openai import OpenAI
import requests
import json

# Initialize OpenAI client
client = OpenAI(api_key=api_key) # replace with your own api key

def search_web(query):
    """
    Search the web for current information using SERP API
    """
    api_key = serp_api_key  # Replace with your SERP API key
    params = {
        "q": query,
        "api_key": api_key,
        "engine": "google",
        "num": 20,
        "output": "json",
    }

    response = requests.get("https://serpapi.com/search", params=params)
    data = response.json()

    results = data.get("organic_results", [])
    if not results:
        return "No search results found."

    # Combine multiple snippets for comprehensive context
    snippets = [result.get("snippet", "") for result in results if "snippet" in result
    ]
    combined_snippets = "\n\n".join(snippets)[:2000]  # Limit to 2000 characters
    return combined_snippets

# Define the search function tool
search_function = {
    "name": "search_web",
    "description": "Search the web for current information",
    "parameters": {
        "type": "object",
        "properties": {
            "query": {
                "type": "string",
```

```
                "description": "The search query"
            }
        },
        "required": ["query"]
    }
}

# Step 1: Initial LLM call with tool capability
content = "Who is Twitter's CEO"

response = client.chat.completions.create(
    model="gpt-4o",
    messages=[
        {"role": "user", "content": content}
    ],
    tools=[{"type": "function", "function": search_function}],
    tool_choice="auto"
)

# Step 2: Extract and execute the search query
tool_call = response.choices[0].message.tool_calls[0]
search_query = json.loads(tool_call.function.arguments)["query"]

# Execute the web search
search_result = search_web(search_query)
print(f"=== Search Results ===")
print(search_result)

# Step 3: Provide search results back to LLM for final response
response2 = client.chat.completions.create(
    model="gpt-4",
    messages=[
        {"role": "user", "content": content},
        {
            "role": "assistant",
            "tool_calls": [tool_call]
        },
        {
            "role": "tool",
            "tool_call_id": tool_call.id,
            "name": "search_web",
            "content": search_result
        }
    ]
)

print(f"=== Final LLM Response ===")
print(response2.choices[0].message.content)
```

Before July 2025

The results if running the script before July 2025:

- Search Results Output:

```
I am excited to welcome Linda Yaccarino as the new CEO of Twitter! @LindaYacc will
      focus primarily on business operations, while I focus on...

Elon Musk on Friday named longtime media executive Linda Yaccarino as the new CEO
     of Twitter, months after he promised to step back from the...

Linda Yaccarino succeeded Musk as CEO on June 5, 2023, with Musk remaining as the
     chairman and the chief technology officer...

Former NBCUniversal advertising chief Linda Yaccarino will become Twitter's new
     CEO, current chief executive Elon Musk says in a tweet...
```

- LLM Response:

> **assistant**
> The CEO of Twitter is Linda Yaccarino.

In/After July 2025

The results of running the script in/after July 2025:

- Search Results Output:

```
Linda Yaccarino succeeded Musk as CEO on June 5, 2023, with Musk remaining as the
    chairman and the chief technology officer. In July 2023, Musk announced that
    ...

Linda Yaccarino announced Wednesday that she is stepping down as CEO at X, the
    social media site formerly known as Twitter.

We still don't know why Linda Yaccarino stepped down as CEO of X on Tuesday. Her
    sudden abdication came off the back of a legendarily ...

Chaos continues at Elon Musk's X with departure of CEO Linda Yaccarino ... In the
    nearly three years since Musk acquired Twitter, the social-media ...
```

- LLM Response:

> **assistant**
> As of July 2025, Linda Yaccarino–––who had been CEO of X (formerly Twitter) since June 5, 2023–––stepped down from the role. Since her departure, no public successor has been officially named.

This example demonstrates how RAG successfully provides accurate, current information by combining LLM capabilities with real-time data retrieval.

2.3.2.3 Key Motivations for RAG Adoption

This section outlines the core motivations behind the growing adoption of RAG systems in real-world applications.

2.3.2.3.1 Dynamic Knowledge Updates

Unlike traditional LLMs that require expensive retraining to incorporate new information, RAG systems can access updated knowledge sources in real time. This capability is particularly valuable for:

- Breaking news and current events.
- Market data and financial information.
- Technical documentation and product updates.
- Regulatory changes and policy updates.

2.3.2.3.2 Enhanced Data Privacy and Security

RAG offers significant advantages for enterprise applications concerned with data privacy:

- Local Data Control: Sensitive business data can remain within enterprise boundaries while still benefiting from LLM capabilities. The RAG system retrieves relevant internal documents without exposing proprietary information to external services.

- Granular Access Control: Organizations can implement sophisticated permission systems that control which users can access specific knowledge sources, ensuring information security at multiple levels.
- Audit Trails: RAG systems provide transparency by showing exactly which sources were used to generate responses, enabling better compliance and accountability.

2.3.2.3.3 Domain-specific Customization

RAG systems excel at integrating specialized, domain-relevant data sources:

- Industry-specific Knowledge: Legal firms can integrate case law databases, healthcare organizations can access medical literature, and financial institutions can incorporate regulatory documents.
- Enterprise Knowledge Bases: Companies can leverage their internal documentation, best practices, and institutional knowledge to enhance LLM responses.
- Real-time Data Integration: RAG can incorporate live data feeds, sensor readings, and dynamic databases to provide contextually relevant responses.

The challenges faced by traditional LLMs—particularly hallucinations and knowledge staleness—represent significant barriers to their deployment in critical applications. RAG emerges as a transformative approach that addresses these limitations while providing additional benefits in terms of security, transparency, and customization. By combining the natural language understanding capabilities of LLMs with dynamic knowledge retrieval, RAG systems offer a path toward more reliable, current, and trustworthy AI applications.

As we move forward in this book, we'll explore the technical implementation details of RAG systems, examine different architectural approaches, and provide practical guidance for building production-ready RAG solutions that meet real-world requirements.

2.3.3 Practice: LLM for Search and Recommendation Tasks

This section demonstrates how static pretrained LLMs perform on search and recommendation tasks without retrieval, establishing a baseline for comparison with retrieval-augmented results in Section 2.3.4.

The objective of this case study is to examine the performance of an LLM in basic search and recommendation tasks. We evaluate two scenarios: one centered on IR (search), and the other on personalized suggestions (recommendation). In both cases, the retrieval source is based on the model's default pretrained knowledge. The setup of the LLM server and the configuration of the web interface are detailed in Appendix A.3. The case studies presented in the remainder of this chapter will be built upon this foundation.

2.3.3.1 Search Scenario

To illustrate the LLM's capabilities in an informational search context—commonly encountered in search-based systems—we present a query regarding the 2024 U.S. Presidential Election.

user Who are the candidates in the final round of the 2024 U.S. Presidential Election? Please provide a brief answer of fewer than 100 words.

assistant

As of now, the 2024 U.S. presidential election's final–round candidates are not yet confirmed. Incumbent President Joe Biden (Democrat) has announced his

> re–election bid, while former President Donald Trump (Republican) is the leading GOP contender. Other potential candidates include Vice President Kamala Harris (if Biden steps down) and Republicans like Nikki Haley or Ron DeSantis. The general election will feature the major party nominees after primary contests conclude. This information is subject to change as the election approaches.

This response highlights a key limitation of LLMs: their reliance on static, pretraining knowledge. As such, they cannot provide real-time updates or current event details, which restricts their utility in time-sensitive search applications.

2.3.3.2 Recommendation Scenario

We next examine the LLM's ability to generate recommendations by asking for movie suggestions tailored to specific preferences.

> **user** Sandie has watched *Garfield Gets Real* and *Garfield's Pet Force*. What movie should she watch next? Please provide a brief answer of fewer than 100 words.
>
> **assistant**
> Sandie should watch *Garfield: The Movie* (2004), the first theatrical film, to experience Garfield's origin story before the sequels. If she prefers chronological order, *Garfield 2* (2006) follows next.

While the recommendation is logically sound and contextually relevant, it also illustrates the model's limitations in capturing nuanced user preferences or providing deeply personalized suggestions. Enhancing the system prompt or incorporating additional contextual cues could significantly improve the relevance and quality of the recommendations. See Section 2.3.4 for improved results when the same queries are augmented with retrieved context.

2.3.4 Practice: Contextual LLM Responses

This section's objective is to explore how integrating external context into prompt engineering can refine the quality of responses generated by LLMs. Building upon the earlier examples—where we examined the LLM's performance in answering search queries about the 2024 U.S. Presidential Election and providing movie recommendations for "Sandie"—we now demonstrate how supplementing prompts with relevant background information can significantly improve both the accuracy and relevance of model outputs. To enrich the model's understanding, we explicitly provide external data within the prompt. The following two use cases illustrate this approach.

2.3.4.1 Use Case 1: Search Query (2024 U.S. Presidential Election)

In this example, we enhance the model's knowledge of the 2024 U.S. Presidential Election by incorporating contextual information from a reliable source. Specifically, we extract leading paragraphs from the Wikipedia page on the 2024 election[1] and include them directly in the prompt to guide the response generation.

---

1 https://en.wikipedia.org/wiki/2024_United_States_presidential_election

user Given the information extracted from Wikipedia:

""""

Presidential elections were held in the United States on November 5, 2024. The Republican Party's ticket––Donald Trump, who served as the 45th president of the United States from 2017 to 2021, and JD Vance, a U.S. senator from Ohio––defeated the Democratic Party's ticket––Kamala Harris, the incumbent U.S. vice president, and Tim Walz, the incumbent governor of Minnesota.

...

Trump won the Electoral College with 312 electoral votes to Harris' 226. Trump won every swing state, including the first win of Nevada by Republicans since 2004. Trump won the national popular vote with a plurality of 49.8%, making him the first Republican to win the popular vote since George W. Bush in 2004. Trump became the first person since Democrat Grover Cleveland in 1892 (and the second person overall) to be elected to a nonconsecutive second term as president of the United States. Analysts attributed the outcome to the 2021–2023 inflation surge, a global anti–incumbent wave, the unpopularity of the Biden administration, and Trump's gains with the working class and minority voters.

""""

Answer this question:
Who are the candidates in the final round of the 2024 U.S. Presidential Election?
Please provide a brief answer of fewer than 100 words.

assistant

The 2024 U.S. presidential election featured Republican Donald Trump (with running mate JD Vance) and Democrat Kamala Harris (with Tim Walz). Trump, the former president, defeated Harris, the incumbent vice president, winning 312 electoral votes to her 226. This followed Joe Biden's withdrawal after a debated performance, making Harris the first non–primary nominee since 1968.

By embedding factual, up-to-date context directly into the prompt, the LLM delivers a more accurate and informed response than it would relying solely on internal knowledge.

2.3.4.2 Use Case 2: Recommendation Query (Movie Recommendation for Sandie)

Next, we examine how providing additional context about user preferences can improve recommendation quality. Here, we incorporate movie metadata from a Netflix dataset available on Kaggle,[2] which includes genres, descriptions, and other attributes. This enables the model to generate tailored suggestions based on detailed historical viewing patterns.

user Given the following records extracted from the movie database:

""""

show_id,title,director,actors,country,date_added,release_year,rating,duration,genres, description

...

s3058,Garfield Gets Real,"Mark A.Z. Dippé, Kyung Ho Lee","Frank Welker, Wally Wingert, Gregg Berger, Jennifer Darling, Pat Fraley, Jason Marsden, Neil Ross, Audrey Wasilewski, Stephen Stanton, Greg Eagles",United States,"January 1, 2020",2007,

2 https://www.kaggle.com/datasets/shivamb/netflix-shows

TV–Y,78 min,"Children & Family Movies, Comedies","After trying to live in the ""real world"" and concluding it isn't for him, Garfield must return to his comic–strip home before getting trapped forever."
s3059,Garfield's Fun Fest,"Mark A.Z. Dippé, Eondeok Han","Frank Welker, Tim Conway, Gregg Berger, Wally Wingert, Audrey Wasilewski, Fred Tatasciore, Stephen Stanton, Jennifer Darling, Neil Ross, Greg Eagles","United States, South Korea"," January 1, 2020",2008,TV–Y7,79 min,"Children & Family Movies, Comedies","Garfield has long been the champ at a yearly competition for the funniest comic strip. But this time, Garfield must battle tall, handsome Ramone."
s3060,Garfield's Pet Force,Mark A.Z. Dippé,"Frank Welker, Vanessa Marshall, Gregg Berger, Wally Wingert, Audrey Wasilewski, Jason Marsden, Fred Tatasciore, Stephen Stanton, Jennifer Darling, Neil Ross, Greg Eagles, Cathy Cavadini, David Michie, Greg Finley","South Korea, United States","January 1, 2020",2009,G,78 min,"Children & Family Movies, Comedies","When an alien menace and her army of undead minions threaten to enslave the world, Garfield heads a team of comic–strip superheroes to save the day."
...
""""

Answer this question:
Sandie has watched *Garfield Gets Real* and *Garfield's Pet Force*. What movie should she watch next? Please provide a brief answer of fewer than 100 words.

assistant

Sandie should watch *Garfield's Fun Fest* next. It shares the same Garfield animated universe, has a similar comedic tone, and features recurring voice actors like Frank Welker. The movie's plot about Garfield competing in a comic–strip contest aligns with her interest in lighthearted, family–friendly stories.

This enhanced prompt allows the LLM to analyze specific movie characteristics and user history, resulting in a more personalized and contextually appropriate recommendation.

2.3.4.3 Contextual Influence on LLM Output

This section investigates how the integration of external context significantly enhances the performance of LLMs in generating accurate, relevant, and detailed responses. By comparing two scenarios—one where an LLM operates without additional contextual information and another where it is supplemented with structured background data—we demonstrate the transformative effect of context on model output quality.

2.3.4.3.1 Case 1: Knowledge Query

When asked, "Who are the candidates in the final round of the 2024 U.S. Presidential Election?" without any supplementary information, the LLM responded with a general acknowledgment that the election had not yet occurred from its knowledge cutoff date. It listed potential contenders such as President Joe Biden (Democrat) and former President Donald Trump (Republican), while noting other possible alternatives like Vice President Kamala Harris or Governor Nikki Haley. However, since the model lacked access to real time or updated political developments, it could not provide definitive or current candidate names. Its response remained speculative and limited to pretraining knowledge.

In contrast, when the same question was posed alongside a contextual excerpt from the Wikipedia page describing the actual outcome of the November 5, 2024 election, the LLM delivered a precise and factually grounded response. It correctly identified the final-round candidates as Republican Donald Trump (with running mate JD Vance) and Democrat Kamala Harris (with Tim Walz). The model also explained how Harris became the Democratic nominee after Biden withdrew due to concerns about his health and debate performance, illustrating how external context enables the LLM to produce timely, accurate, and actionable answers.

2.3.4.3.2 Case 2: Movie Recommendation

In the second scenario, the LLM was initially asked for a movie recommendation based solely on Sandie's viewing history of "Garfield Gets Real" and "Garfield's Pet Force," without any additional background. The model offered general suggestions based on genre similarity, recommending movies like "Garfield: The Movie" and "Garfield 2." While these were plausible options, they did not fully consider nuances such as narrative continuity, voice actors, or thematic consistency.

However, when the model was provided with enriched contextual data—specifically, metadata extracted from a Netflix dataset detailing Sandie's viewing history, including genres, descriptions, and production details—it generated a more personalized suggestion. The LLM recommended "Garfield's Fun Fest," noting that it shared the same animated universe, featured recurring voice actors like Frank Welker, and maintained a consistent comedic tone. This enhanced recommendation demonstrated a deeper understanding of user preferences, highlighting how contextual data can refine and personalize AI-generated outputs.

2.3.4.3.3 Key Takeaways: Implications of Context Integration in LLM Applications

These comparative cases clearly illustrate the critical role of context in improving LLM performance:

- Specificity and Accuracy: In both cases, the LLM's responses were notably more specific and accurate when context was provided. Without context, the model defaulted to general knowledge, which, although informative, lacked the precision expected in time-sensitive or personalized applications.
- Relevance: Context allows the model to align its responses with the user's underlying needs or situational requirements. For instance, in the movie recommendation case, context enabled the model to offer a suggestion that closely matched Sandie's likely tastes, rather than simply matching genre elements.
- Confidence and Precision: With access to contextual data, the LLM could confidently provide direct answers—such as naming exact candidates or recommending a specific movie—rather than hedging or offering vague possibilities. This increased confidence leads to higher user trust and satisfaction.

These findings underscore the value of RAG techniques in enhancing LLM capabilities by supplementing internal knowledge with relevant, up-to-date, and domain-specific context. When applied effectively, contextual augmentation transforms generic model outputs into highly relevant, detailed, and tailored responses, significantly improving the utility of AI-driven systems in search, recommendation, and decision-support applications.

2.4 Summary

This chapter provides a foundational overview of LLMs and their application in IR systems. It begins with a detailed explanation of the core concepts that underpin LLMs, particularly the transformer models and attention mechanisms that have revolutionized NLP. Various types of LLMs are discussed, including encoder-decoder architectures, proprietary and open-source models, as well as domain-specific variants.

The chapter also introduces key learning paradigms for LLMs, such as pretraining, fine-tuning, and prompt engineering, which are critical for adapting these models to specific tasks. This leads into a discussion of the emerging role of LLMs in IR.

One highlighted use case involves using the open source model within the OpenAI compatible API server for search and recommendation tasks. Step-by-step instructions for setting up and running the model, integrating it with Jupyter Notebooks, and performing both search and recommendation queries are provided.

The chapter concludes by exploring challenges faced by LLMs in IR, including issues like hallucinations and knowledge staleness. It also introduces RAG as a potential solution, comparing it to other techniques like prompt engineering and fine-tuning. A series of use cases are presented to demonstrate how LLMs can be enhanced with external contextual information, with examples ranging from search queries related to the U.S. Presidential Election to movie recommendations. The impact of providing additional context on the quality of LLM responses is also analyzed.

In Chapter 3, we are going to dive into more details of how RAG can improve the IR domain. We'll discuss the major components of the RAG system, major RAG approaches and frameworks.

References

1 Corinna Cortes and Vladimir Vapnik. Support-vector networks. *Machine Learning*, 20:273–297, 1995.

2 J Ross Quinlan. Induction of decision trees. *Machine Learning*, 1:81–106, 1986.

3 David D Lewis. Naive (bayes) at forty: The independence assumption in information retrieval. In *European Conference on Machine Learning*, pages 4–15. Springer, 1998.

4 Yann LeCun, Léon Bottou, Yoshua Bengio, and Patrick Haffner. Gradient-based learning applied to document recognition. *Proceedings of the IEEE*, 86(11):2278–2324, 2002.

5 David E Rumelhart, Geoffrey E Hinton, and Ronald J Williams. Learning representations by back-propagating errors. *Nature*, 323(6088):533–536, 1986.

6 Sepp Hochreiter and Jürgen Schmidhuber. Long short-term memory. *Neural Computation*, 9(8):1735–1780, 1997.

7 Kyunghyun Cho, Bart Van Merriënboer, Caglar Gulcehre, Dzmitry Bahdanau, Fethi Bougares, Holger Schwenk, and Yoshua Bengio. Learning phrase representations using RNN encoder-decoder for statistical machine translation. *arXiv preprint arXiv:1406.1078*, 2014.

8 Jacob Devlin, Ming-Wei Chang, Kenton Lee, and Kristina Toutanova. Bert: Pre-training of deep bidirectional transformers for language understanding. In *Proceedings of the 2019 Conference of the North American Chapter of the Association for Computational Linguistics: Human Language Technologies*, volume 1 *(Long and Short Papers)*, pages 4171–4186, 2019.

9 Alec Radford, Karthik Narasimhan, Tim Salimans, and Ilya Sutskever. Improving language understanding by generative pre-training. *OpenAI Blog*, 2018.

10 Tom Brown, Benjamin Mann, Nick Ryder, Melanie Subbiah, Jared D Kaplan, Prafulla Dhariwal, Arvind Neelakantan, Pranav Shyam, Girish Sastry, Amanda Askell, et al. Language models are few-shot learners. *Advances in Neural Information Processing Systems*, 33:1877–1901, 2020.

11 Josh Achiam, Steven Adler, Sandhini Agarwal, Lama Ahmad, Ilge Akkaya, Florencia Leoni Aleman, Diogo Almeida, Janko Altenschmidt, Sam Altman, Shyamal Anadkat, et al. GPT-4 technical report. *arXiv preprint arXiv:2303.08774*, 2023.

12 Tomas Mikolov, Kai Chen, Greg Corrado, and Jeffrey Dean. Efficient estimation of word representations in vector space. *arXiv preprint arXiv:1301.3781*, 2013.

13 Jeffrey Pennington, Richard Socher, and Christopher D Manning. Glove: Global vectors for word representation. In *Proceedings of the 2014 Conference on Empirical Methods in Natural Language Processing (EMNLP)*, pages 1532–1543, 2014.

14 Piotr Bojanowski, Edouard Grave, Armand Joulin, and Tomas Mikolov. Enriching word vectors with subword information. *Transactions of the Association for Computational linguistics*, 5:135–146, 2017.

15 Matthew E Peters, Mark Neumann, Mohit Iyyer, Matt Gardner, Christopher Clark, Kenton Lee, and Luke Zettlemoyer. Deep contextualized word representations. In *Proceedings of the 2018 Conference of the North American Chapter of the Association for Computational Linguistics: Human Language Technologies,* volume 1 *(Long Papers)*, pages 2227–2237, 2018.

16 Ian T Jolliffe. *Principal Component Analysis for Special Types of Data*. Springer, 2002.

17 Daniel Cer, Yinfei Yang, Sheng-yi Kong, Nan Hua, Nicole Limtiaco, Rhomni St John, Noah Constant, Mario Guajardo-Cespedes, Steve Yuan, Chris Tar, et al. Universal sentence encoder. *arXiv preprint arXiv:1803.11175*, 2018.

18 Zhenzhong Lan, Mingda Chen, Sebastian Goodman, Kevin Gimpel, Piyush Sharma, and Radu Soricut. Albert: A lite bert for self-supervised learning of language representations. In *International Conference on Learning Representations*, 2020.

19 Nils Reimers and Iryna Gurevych. Sentence-bert: Sentence embeddings using siamese bert-networks. In *Proceedings of the 2019 Conference on Empirical Methods in Natural Language Processing and the 9th International Joint Conference on Natural Language Processing (EMNLP-IJCNLP)*, pages 3982–3992, 2019.

20 Colin Raffel, Noam Shazeer, Adam Roberts, Katherine Lee, Sharan Narang, Michael Matena, Yanqi Zhou, Wei Li, and Peter J Liu. Exploring the limits of transfer learning with a unified text-to-text transformer. *The Journal of Machine Learning Research*, 21(1):5485–5551, 2020.

21 Sanjeev Arora, Yingyu Liang, and Tengyu Ma. A simple but tough-to-beat baseline for sentence embeddings. In *International Conference on Learning Representations*, 2017.

22 Ashish Vaswani, Noam Shazeer, Niki Parmar, Jakob Uszkoreit, Llion Jones, Aidan N Gomez, Łukasz Kaiser, and Illia Polosukhin. Attention is all you need. *Advances in Neural Information Processing Systems*, 30:5998–6008, 2017.

23 Kevin Clark, Minh-Thang Luong, Quoc V Le, and Christopher D Manning. Electra: Pre-training text encoders as discriminators rather than generators. *arXiv preprint arXiv:2003.10555*, 2020.

24 Ting Chen, Simon Kornblith, Mohammad Norouzi, and Geoffrey Hinton. A simple framework for contrastive learning of visual representations. In *International Conference on Machine Learning*, pages 1597–1607. PmLR, 2020.

25 Sepp Hochreiter. The vanishing gradient problem during learning recurrent neural nets and problem solutions. *International Journal of Uncertainty, Fuzziness and Knowledge-Based Systems*, 6(02):107–116, 1998.

26 Mike Lewis, Yinhan Liu, Naman Goyal, Marjan Ghazvininejad, Abdelrahman Mohamed, Omer Levy, Ves Stoyanov, and Luke Zettlemoyer. BART: Denoising sequence-to-sequence pre-training for natural language generation, translation, and comprehension. *arXiv preprint arXiv:1910.13461*, 2019.

27 Yinhan Liu, Myle Ott, Naman Goyal, Jingfei Du, Mandar Joshi, Danqi Chen, Omer Levy, Mike Lewis, Luke Zettlemoyer, and Veselin Stoyanov. RoBERTa: A robustly optimized BERT pretraining approach. *arXiv preprint arXiv:1907.11692*, 2019.

28 Zhilin Yang, Zihang Dai, Yiming Yang, Jaime Carbonell, Russ R Salakhutdinov, and Quoc V Le. XLNet: Generalized autoregressive pretraining for language understanding. *Advances in Neural Information Processing Systems*, 32:5753–5763, 2019.

29 Tianyu Gao, Xingcheng Yao, and Danqi Chen. SimCSE: Simple contrastive learning of sentence embeddings. *arXiv preprint arXiv:2104.08821*, 2021.

30 Emma Strubell, Ananya Ganesh, and Andrew McCallum. Energy and policy considerations for modern deep learning research. In *Proceedings of the AAAI Conference on Artificial Intelligence*, volume 34, pages 13693–13696, 2020.

31 Jason Wei, Maarten Bosma, Vincent Y Zhao, Kelvin Guu, Adams Wei Yu, Brian Lester, Nan Du, Andrew M Dai, and Quoc V Le. Finetuned language models are zero-shot learners. *arXiv preprint arXiv:2109.01652*, 2021.

32 Jason Wei, Xuezhi Wang, Dale Schuurmans, Maarten Bosma, Fei Xia, Ed Chi, Quoc V Le, and Denny Zhou. Chain-of-thought prompting elicits reasoning in large language models. *Advances in Neural Information Processing Systems*, 35:24824–24837, 2022.

33 Edward J Hu, Yelong Shen, Phillip Wallis, Zeyuan Allen-Zhu, Yuanzhi Li, Shean Wang, Lu Wang, and Weizhu Chen. LoRA: Low-rank adaptation of large language models. *ICLR*, 1(2):3, 2022.

34 Paul F Christiano, Jan Leike, Tom Brown, Miljan Martic, Shane Legg, and Dario Amodei. Deep reinforcement learning from human preferences. *Advances in Neural Information Processing Systems*, 30:4302–4310, 2017.

35 Long Ouyang, Jeffrey Wu, Xu Jiang, Diogo Almeida, Carroll Wainwright, Pamela Mishkin, Chong Zhang, Sandhini Agarwal, Katarina Slama, Alex Ray, et al. Training language models to follow instructions with human feedback. *Advances in Neural Information Processing Systems*, 35:27730–27744, 2022.

36 John Schulman, Filip Wolski, Prafulla Dhariwal, Alec Radford, and Oleg Klimov. Proximal policy optimization algorithms. *arXiv preprint arXiv:1707.06347*, 2017.

37 Rafael Rafailov, Archit Sharma, Eric Mitchell, Christopher D Manning, Stefano Ermon, and Chelsea Finn. Direct preference optimization: Your language model is secretly a reward model. *Advances in Neural Information Processing Systems*, 36:53728–53741, 2023.

38 Zhihong Shao, Peiyi Wang, Qihao Zhu, Runxin Xu, Junxiao Song, Xiao Bi, Haowei Zhang, Mingchuan Zhang, YK Li, Y Wu, et al. DeepSeekMath: Pushing the limits of mathematical reasoning in open language models. *arXiv preprint arXiv:2402.03300*, 2024.

39 Yuntao Bai, Saurav Kadavath, Sandipan Kundu, Amanda Askell, Jackson Kernion, Andy Jones, Anna Chen, Anna Goldie, Azalia Mirhoseini, Cameron McKinnon, et al. Constitutional AI: Harmlessness from AI feedback. *arXiv preprint arXiv:2212.08073*, 2022.

40 Xuezhi Wang, Jason Wei, Dale Schuurmans, Quoc Le, Ed Chi, Sharan Narang, Aakanksha Chowdhery, and Denny Zhou. Self-consistency improves chain of thought reasoning in language models. *arXiv preprint arXiv:2203.11171*, 2022.

41 Timo Schick, Jane Dwivedi-Yu, Roberto Dessì, Roberta Raileanu, Maria Lomeli, Eric Hambro, Luke Zettlemoyer, Nicola Cancedda, and Thomas Scialom. Toolformer: Language models can teach themselves to use tools. *Advances in Neural Information Processing Systems*, 36:68539–68551, 2023.

42 Luyu Gao, Xueguang Ma, Jimmy Lin, and Jamie Callan. Precise zero-shot dense retrieval without relevance labels. In *Proceedings of the 61st Annual Meeting of the Association for Computational Linguistics (volume 1: Long Papers)*, pages 1762–1777, Toronto, Canada: Association for Computational Linguistics, July 2023. https://aclanthology.org/2023.acl-long.99.

43 Liang Wang, Nan Yang, and Furu Wei. Query2doc: Query expansion with large language models. *arXiv preprint arXiv:2303.07678*, March 2023. https://arxiv.org/abs/2303.07678.

44 Wenhui Wang, Furu Wei, Li Dong, Hangbo Bao, Nan Yang, and Ming Zhou. MiniLM: Deep self-attention distillation for task-agnostic compression of pre-trained transformers. In *Proceedings of the 34th International Conference on Neural Information Processing Systems*, Advances in Neural Information Processing Systems 33. Curran Associates, Inc., 2020.

3

Retrieval-Augmented Generation

As large language models (LLMs) continue to evolve, their ability to generate human-like text and handle complex queries has opened up many possibilities. However, traditional LLMs often struggle with two main challenges: staying current with up-to-date information and avoiding inaccuracies (like hallucinations) in their responses. This is where retrieval-augmented generation (RAG) comes in.

RAG systems combine the power of LLMs with external knowledge sources, such as databases or documents, to provide more accurate, up-to-date, and contextually relevant answers. Instead of relying solely on the data the model was trained on, a RAG system uses a retrieval mechanism to pull in relevant information from an external knowledge base. This information is then used to improve the LLM's responses, making them more precise and grounded in reality.

In this chapter, we'll dive into how RAG works, exploring its key components—knowledge indexing, retrieval, and the language generation process. We'll look at how RAG systems handle data processing, indexing, retrieval methods, and generation models, providing a practical framework for building systems that blend LLMs with real-world knowledge to enhance their output.

3.1 RAG Overview

A RAG system comprises three core components: a retrieval mechanism, an external knowledge base, and a language model for generation. The retrieval mechanism accesses the external knowledge base (which includes documents, databases, and knowledge graphs [KGs]) to fetch relevant information beyond the model's training data. The retrieved information is either concatenated with the user's query or directly fed into the LLM's input layer, thereby guiding the generation process toward data-backed responses. This seamless integration of retrieval, external knowledge, and language generation empowers RAG systems to provide responses by dynamically accessing and incorporating diverse data sources.

3.1.1 RAG Components

The primary components of RAG systems are presented in Figure 3.1, which illustrates how the retrieval mechanism interacts with the knowledge base and feeds information into the language model to form a cohesive response pipeline.

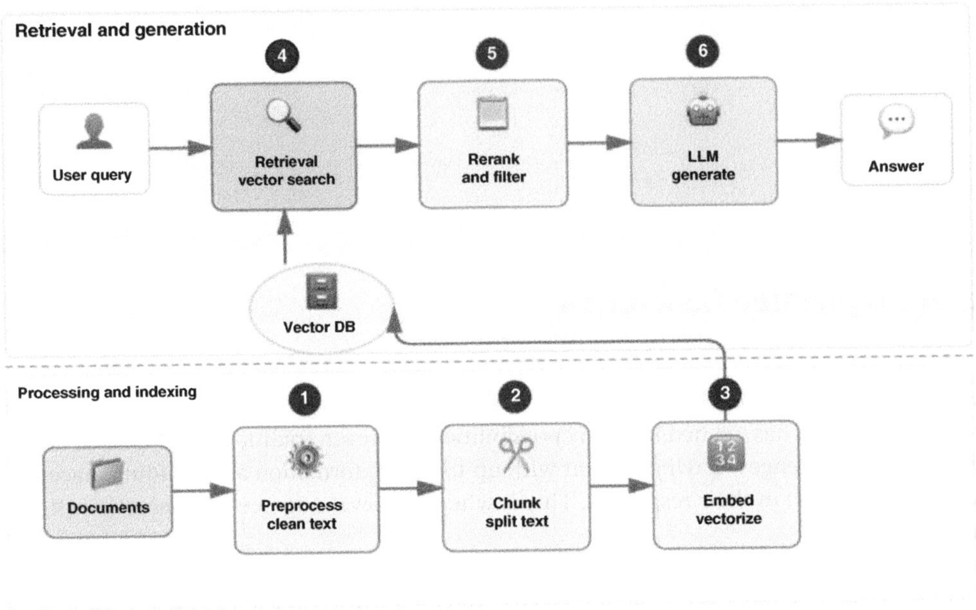

Figure 3.1 Naive RAG workflow overview.

As shown in Figure 3.1, the "naive" RAG workflow includes several steps, which we can roughly categorize into three major components: data preparation and indexing, retrieval and generation:

- **Data Preparation and Indexing**
  1. **Data Preprocessing:** Raw documents are loaded, cleaned, and parsed.
  2. **Text Splitting:** Documents are chunked into manageable segments with overlap.
  3. **Embedding Generation:** Text chunks are converted to vector embeddings and stored in a vector database for efficient retrieval.
- **Retrieval**
  4. **Query-document Matching:** User query triggers retrieval of relevant documents from vector database (DB).
  5. **Post-retrieval:** Reranking and filtering retrieved results.
- **Generation**
  6. **Response Generation:** LLM generates answers using retrieved context.

We are going to dive into more details of each component in the next few sections.

3.1.2 From Tokens to Passage Embeddings

A common point of confusion concerns how text becomes a vector suitable for similarity search. The process involves multiple stages, as illustrated in the following:

1. **Tokenization:** The input text is segmented into subword tokens using the encoder's tokenizer (e.g., WordPiece, BPE). For example, "The capital of France is Paris" might become `[CLS] The capital of France is Paris [SEP]`.
2. **Token Embedding:** Each token is mapped to an initial embedding vector via a learned embedding table. At this stage, each token has its own vector (e.g., 768-dimensional for BERT-base).

3. **Contextual Encoding:** The token embeddings pass through the transformer encoder layers, producing contextualized representations where each token's vector incorporates information from the entire sequence.
4. **Pooling:** The sequence of contextualized token vectors is aggregated into a single passage-level embedding. Common strategies include:
   - **[CLS] Pooling:** Use the final hidden state of the `[CLS]` token.
   - **Mean Pooling:** Average all token embeddings (often excluding special tokens).
   - **Max Pooling:** Take the element-wise maximum across token embeddings.
5. **Vector Storage:** The resulting single vector (e.g., 768 or 1,024 dimensions) is stored in a vector database alongside metadata pointing to the original text chunk.

Token-to-passage Embedding Pipeline

Step 1 – Tokenization:
`"The capital of France"`$\mapsto$ `[CLS], The, cap, ##ital, of, France, [SEP]`
Step 2 – Token Embedding:
Each token $\mapsto$ vector $\mathbf{e}_i \in \mathbb{R}^d$ (Result: n vectors)
Step 3 – Transformer Encoding:
$[\mathbf{e}_1, \dots, \mathbf{e}_n] \mapsto [\mathbf{h}_1, \dots, \mathbf{h}_n]$ (Contextualized representations)
Step 4 – Pooling:
$[\mathbf{h}_1, \dots, \mathbf{h}_n] \mapsto \mathbf{p} = \frac{1}{n}\sum_{i=1}^{n} \mathbf{h}_i$ (Single vector)
Step 5 – Storage:
$\mathbf{p} \in \mathbb{R}^d$ stored in vector database with metadata

Note: The same pipeline applies to queries at retrieval time, producing a query vector for similarity matching against stored passage vectors.

This pipeline clarifies why RAG systems use vector databases: passage-level embeddings enable efficient approximate nearest neighbor (ANN) search across millions of documents. The query undergoes the identical tokenization-embedding-encoding-pooling process, producing a query vector that can be compared against all stored passage vectors using cosine similarity or inner product.

3.1.3 RAG as a Retrieval-agnostic Architecture

While the original RAG formulation [1] employed dense retrieval as its retrieval backbone, the RAG paradigm is fundamentally *retrieval-agnostic*—it can incorporate any method that maps queries to relevant passages. Modern RAG implementations leverage these broad categories of retrieval:

- **Sparse Retrieval:** Lexical matching methods such as TF–IDF and BM25; fast, interpretable, and strong for exact-term queries.
- **Dense Retrieval:** Neural embeddings (e.g., DPR [2], BGE [3]) capturing semantic similarity; effective for paraphrase and concept-level queries.
- **Graph-structured Retrieval:** Retrieval over graph-based or graph-enhanced indexes, including KGs, HNSW-style vector graphs, and multi-hop reasoning via neighborhood or path traversal.
- **Hybrid Retrieval:** Any combination of heterogeneous retrieval signals—sparse, dense, graph-based, hierarchical, or multimodal—fused through score combination, learned weighting, or reranking.

This flexibility allows practitioners to select retrieval strategies based on their specific requirements—whether prioritizing speed, semantic understanding, or domain-specific terminology. The subsequent sections detail each retrieval paradigm and their integration into RAG pipelines.

3.2 Data Preparation and Indexing

Data preparation and indexing is the foundational step in any RAG pipeline. Its goal is to transform raw, heterogeneous data into an indexed and queryable format that the retriever can efficiently use. The quality, structure, and accessibility of prepared data directly impact the system's ability to retrieve relevant information and generate accurate responses.

The data preparation process follows a systematic pipeline that ensures data quality and retrieval efficiency: Raw Data Sources → Data Ingestion→ Preprocessing→ Chunking→ Embedding→ Indexing→ Knowledge Base.

Each stage in this pipeline serves a specific purpose in optimizing data for retrieval and generation tasks, with careful attention to maintaining information integrity while maximizing searchability and relevance.

3.2.1 Understanding Data Source Types

3.2.1.1 Structured vs. Unstructured Data

The foundation of any RAG system begins with understanding the diverse types of data sources it must accommodate. Data sources generally fall into two primary categories, each requiring distinct preparation approaches: structured and unstructured data.

3.2.1.1.1 Structured Data

Structured data refers to information highly organized in a predefined format, typically found in:

- Relational Databases: Customer records, transaction logs, inventory systems, and business intelligence data stored in SQL databases like PostgreSQL, MySQL, or Oracle.
- NoSQL Databases: Document stores (MongoDB), key-value pairs (Redis), columnar databases (Cassandra), and graph databases (Neo4j) that maintain consistent schema structures.
- Spreadsheets and CSV Files: Financial reports, data exports, research datasets, and analytical summaries with defined columns and data types.
- API Responses: JSON or XML formatted data from web services, providing structured access to real-time information like stock prices, weather data, or social media metrics. Example: `SELECT * FROM company_profiles WHERE industry = 'pharmaceuticals'`

3.2.1.1.2 Unstructured Data

Unstructured data encompasses information without a predefined organizational structure, including:

- Text Documents: Research papers, reports, manuals, emails, legal documents, and articles that contain rich narrative content and context.
- Web Content: Blog posts, news articles, forum discussions, and social media content that provide diverse perspectives and current information.
- Multimedia Content: While primarily text-focused, RAG systems increasingly need to handle metadata and transcriptions from images, videos, and audio files.

- Communication Records: Chat logs, meeting transcripts, support tickets, and collaborative documents that capture organizational knowledge and decision-making processes.

3.2.1.1.3 Programmatic Data Access

Modern RAG systems increasingly rely on programmatic data access through APIs and web services, which bridges the gap between structured and unstructured data sources.

This programmatic source must often be queried on-the-fly and transformed into intermediate chunks for retrieval. There are different types of APIs and the most popular ones are: REST, GraphQL, Streaming, and WebSockets. Table 3.1 lists the comparisons of various types of APIs.

3.2.1.1.4 Choosing the Right Data Type

Table 3.2 lists the comparisons of three data types.

The choice between structured, unstructured, and programmatic data depends on the specific requirements of the RAG application:

Structured data excels in scenarios demanding precision, such as retrieving exact product specifications or financial figures. Its rigid schema ensures consistency but limits adaptability to new or evolving information.

Unstructured data is indispensable when depth and context matter, such as answering complex medical questions or summarizing legal precedents. However, it requires significant preprocessing to be usable.

Programmatic data bridges the gap, offering real-time access to semi-structured information. APIs are ideal for integrating live updates—like weather forecasts or stock prices—but introduce dependencies on external services and rate limits.

In practice, most sophisticated RAG systems employ a hybrid approach (illustrated in Figure 3.2), combining structured databases for foundational knowledge, APIs for real-time updates, and unstructured documents for comprehensive context.

Table 3.1 API comparisons.

API aspect	REST APIs	GraphQL APIs	Streaming APIs	WebSocket APIs
Data structure	Fixed endpoint responses	Flexible query responses	Continuous data streams	Bidirectional real-time data
RAG integration	Request-response chunking	Query-specific chunking	Stream segmentation	Real-time context updates
Rate limiting	HTTP rate limits	Query complexity limits	Throughput throttling	Connection limits
Caching strategy	Response caching	Query result caching	Stream buffering	State synchronization
Error handling	HTTP status codes	GraphQL error responses	Stream interruption handling	Connection drop recovery
Authentication	API keys, OAuth tokens	Similar to REST	Token-based streaming	WebSocket authentication
Versioning	URL or header versioning	Schema evolution	Stream format versioning	Protocol versioning
Best for RAG	Standard data retrieval	Complex, nested queries	Real-time data feeds	Interactive applications

Table 3.2 Structured vs API vs unstructured data: comprehensive comparison tables.

Aspect	Structured data	API data	Unstructured data
Definition	Data organized in predefined formats with clear schema	Semi-structured data accessed through programmatic interfaces	Data without predefined organizational structure
Schema	Fixed, well-defined schema	Semiflexible schema, versioned APIs	No fixed schema, flexible format
Format	Tables, rows, columns, fixed fields	JSON, XML, REST responses, GraphQL	Free-form text, documents, multimedia
Predictability	Highly predictable structure	Moderately predictable (API versioning)	Unpredictable structure
Storage	Relational databases, spreadsheets	API endpoints, microservices, cloud services	File systems, document stores, data lakes
Preprocessing complexity	Low—minimal cleaning required	Moderate—JSON parsing, rate limiting	High—extensive cleaning and normalization
Examples	SQL Databases, Spreadsheets, CSV Files	REST APIs, GraphQL, Streaming APIs, WebSockets	Text documents, Web contents, Emails, Research papers

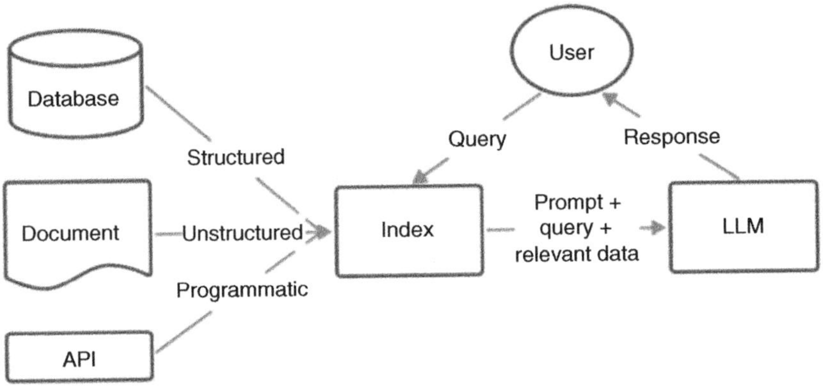

Figure 3.2 Hybrid approaches of data sources.

For example:

- A "customer support" RAG might:
  - Query a structured database for order history (structured).
  - Check an API for current shipping status (programmatic).
  - Search unstructured FAQs for troubleshooting advice (unstructured).
- A "legal assistant" RAG might:
  - Get client data from a CRM (structured).
  - Find current regulations from government APIs (programmatic).
  - Draw precedent cases from a document store (unstructured).

This multisource strategy ensures responses are not only accurate and up-to-date but also rich in explanatory detail—key to delivering value in real-world applications. By weaving these data sources together, a RAG system can ensure its responses are grounded, current, and deeply contextual. By understanding these data types and their trade-offs, developers can design RAG systems that leverage the strengths of each while mitigating their respective limitations.

3.2.1.2 Multimodal Data Considerations

Traditional RAG systems focus primarily on textual information. As RAG systems evolve, they increasingly need to handle multimodal data sources.

As LLMs evolve toward multimodality, RAG systems are also expanding to incorporate information from various modalities beyond text. This could include:

Images: Where visual content (e.g., diagrams, charts, photographs) needs to be described or understood. This often involves converting visual information into textual embeddings or descriptions. The techniques being utilized involve: image recognition and description, chart and graph interpretation, and visual document analysis.

Audio/Video: Requiring transcription into text or embedding extraction to make the content searchable and retrievable. Some example techniques include speech-to-text conversion, podcast transcriptions, scene description, subtitle extraction, and visual content summarization.

Structured Tables Within Documents: Often requiring specialized parsing to maintain their tabular structure and relationships. Integrating multimodal data introduces additional complexity in preprocessing and retrieval techniques.

Specialized models (e.g., Flamingo [4], GPT-4V [5]) can handle this type of input with appropriate embedding pipelines.

3.2.2 Data Preprocessing

Data preprocessing is the critical bridge between raw data sources and a functional RAG system. This stage transforms heterogeneous, often messy data into clean, standardized formats suitable for embedding and retrieval. The quality of preprocessing directly impacts retrieval accuracy and generation relevance.

3.2.2.1 Core Preprocessing Pipeline

The preprocessing pipeline follows a systematic approach tailored to each data type:

```
Raw Data → Data Validation → Cleaning → Normalization → Format Standardization → Feature Extraction → Quality Assurance
```

Each stage serves a specific purpose in preparing data for optimal retrieval performance.

3.2.2.2 Text Data Preprocessing

Text preprocessing forms the backbone of most RAG systems. The process involves multiple stages, each addressing specific data quality issues that could interfere with semantic understanding and retrieval accuracy.

3.2.2.2.1 Cleaning and Normalization

Text cleaning removes noise while preserving semantic meaning. This involves eliminating HTML tags, fixing encoding issues, normalizing whitespace, and removing special characters that don't contribute to meaning.

Unicode normalization ensures consistent character representation across different encoding systems. This process converts characters to canonical forms and handles special cases like smart quotes, em dashes, and various apostrophe styles. Such normalization is crucial for matching queries with content that may have been authored in different text editors or systems.

3.2.2.2.2 Language-specific Processing

Different languages require tailored preprocessing approaches. English text benefits from contraction expansion (converting "don't" to "do not") and careful case normalization that preserves proper nouns. Asian languages like Chinese and Japanese require special segmentation algorithms since they lack explicit word boundaries. Right-to-left languages such as Arabic and Hebrew need directional handling, while German text may require compound word splitting to improve searchability.

3.2.2.3 Structured Data Preprocessing

Structured data requires different preprocessing strategies to preserve relationships while making content searchable. The challenge lies in transforming rigid schemas into natural language representations without losing the inherent structure that makes the data valuable.

3.2.2.3.1 Database Record Transformation

Database records must be converted into searchable text while maintaining their semantic relationships.

For example, a customer database record containing fields like company name, industry, revenue, and employee count would be transformed into a natural language representation: "Customer Profile: Acme Corp, Industry: Manufacturing, Annual Revenue: $50,000,000, Company Size: 250 employees, Established: 2010." This transformation may also include derived information, such as classifying the company as a "Large Enterprise" based on revenue thresholds.

3.2.2.3.2 Table Processing

Tables within documents require special handling to preserve their structural relationships.

For example, a sales performance table might be processed using row-based representation, where each row becomes a sentence: "Product Widget A: Q1 sales $100K, Q2 sales $150K, growth +50%." Alternatively, column-based representation might be used for queries focusing on specific metrics across all products. The choice depends on the expected query patterns and the nature of the data.

3.2.2.4 Multimodal Data Preprocessing

As RAG systems evolve to handle diverse data types, preprocessing must adapt to extract searchable information from non-textual sources.

3.2.2.4.1 Image Processing Pipeline

Image preprocessing involves multiple extraction layers. First, metadata such as filename, format, and dimensions provides basic searchable attributes. Vision models then generate textual descriptions of the visual content, while OCR extracts any text present in the image. These elements combine into a comprehensive textual representation that captures both the visual and textual content of the image.

3.2.2.4.2 Audio/Video Processing

Audio and video content undergoes transcription to convert speech to text, with temporal markers added to maintain chronological context. Speaker identification enhances the transcript by attributing statements to specific individuals, crucial for meeting recordings or interviews. Video processing additionally includes scene descriptions and visual element extraction, creating a multilayered textual representation of the multimedia content.

3.2.2.5 Data Quality Assurance

Preprocessing must include validation to ensure data quality meets system requirements. Quality checks verify minimum and maximum length constraints, detect the language of the content, validate encoding, identify excessive repetition, and assess readability scores. These checks prevent low-quality or corrupted data from entering the system, which could degrade retrieval performance.

Preprocessing metrics help monitor the effectiveness of the transformation process. Key metrics include size reduction ratios, token count changes, information density calculations, and semantic similarity scores between original and processed content. These metrics ensure that preprocessing achieves its goals of noise reduction without significant information loss.

3.2.2.6 Preprocessing Best Practices

Effective preprocessing balances several competing objectives:

Preserve Semantic Information: Semantic information must be preserved, which means maintaining punctuation that affects meaning, preserving document structure and section headers, and retaining crucial metadata like dates and sources. Overaggressive cleaning that strips essential context is as problematic as under-processing that leaves noise in the data.

Handle Edge Cases: Edge cases require special attention in preprocessing pipelines. Scientific notation needs readable conversion while maintaining precision. Currency symbols should be standardized to their three-letter codes. Mathematical expressions require careful handling to remain both searchable and meaningful. These special formats often appear in technical or financial documents and can significantly impact retrieval accuracy if not handled properly.

Optimize for Retrieval: Optimization for retrieval involves adding contextual markers that aid in chunking, including section summaries for long documents, and maintaining cross-references between related content. These additions improve the retrieval system's ability to find relevant passages and understand document structure.

Scale Considerations: Scale considerations become crucial when processing large document collections. Batch processing strategies, parallel execution, and efficient memory management ensure that preprocessing remains performant even with millions of documents. The preprocessing pipeline must be designed to handle both individual document updates and bulk ingestion scenarios.

In summary, effective data preprocessing is not a one-size-fits-all process. It requires careful consideration of data types, use cases, and downstream requirements. By implementing robust preprocessing pipelines that balance information preservation with noise reduction, RAG systems can achieve superior retrieval accuracy and generation quality. The key is to maintain semantic richness while ensuring data is clean, consistent, and optimized for embedding and retrieval operations. Success in preprocessing directly translates to success in the overall RAG system performance.

3.2.3 Chunking Techniques

In RAG systems, chunking is a critical preprocessing step that transforms raw documents into manageable text segments for embedding and retrieval. The goal of chunking is to divide content in a way that preserves semantic coherence while aligning with model constraints, such as token limits imposed by embedding models and downstream LLMs.

Chunking strategies are not one-size-fits-all. The art of chunking requires balancing competing demands: maintaining sufficient context for meaningful understanding while creating focused segments that enable precise retrieval. Too large, and chunks may exceed model token limits or dilute retrieval precision; too small, and critical context gets fragmented across multiple segments, potentially losing coherent meaning. A well-designed chunking strategy improves retrieval quality, enhances answer accuracy, and ensures efficient processing in RAG pipelines.

Why Chunking Remains Essential Despite Long Context Windows

Modern LLMs support dramatically expanded context windows—GPT-4 Turbo (128K tokens), Claude 3 (200K), and Gemini 1.5 Pro (1M). Yet chunking remains essential for three reasons:

1. Cost and Latency: Long-context inference is expensive: API pricing scales with input tokens, and latency grows superlinearly due to $O(n^2)$ attention. Processing 100K tokens takes 30–60 seconds; retrieving 5 relevant 500-token chunks completes in under 2 seconds at 10–50 × lower cost [6].

2. The "Lost in the Middle" Effect: Liu et al. [7] showed that LLM accuracy drops 20–30% when key information appears mid-context versus at the beginning or end. Retrieval places relevant chunks at the prompt's start, ensuring consistent attention.

3. Production System Requirements: Modern systems use chunking for embedding reuse, incremental index updates, and fine-grained source attribution—benefits unavailable with full-context approaches.

Long context windows complement rather than replace chunking: they expand post-retrieval capabilities, but chunked retrieval remains optimal for locating relevant information in large collections.

3.2.3.1 Understanding Chunking Fundamentals

There are several factors that play a role in determining the best chunking strategy, depending on the use case.

The nature of the content is fundamental. For instance, whether dealing with lengthy documents like articles or books, or shorter forms such as tweets or instant messages, requires the suitable model and chunking approach. The choice of embedding model and its optimal block size also plays a crucial role. User query expectations, whether short and specific or long and complex, influence the chunking method to ensure a better match between embedded queries and content chunks. Additionally, the application of retrieved results, be it for semantic search, summarization, or other purposes. Any limitations it imposes, such as token limits for subsequent LLM processing, must be considered to tailor chunk size and strategy accordingly.

3.2.3.1.1 The Context-precision Trade-off

The size of each chunk is a central design parameter. Smaller chunks typically improve retrievability and precision, particularly for short or specific queries. However, they may fragment meaning—especially when context is split across multiple sentences. Conversely, larger chunks preserve more context but risk exceeding token limits, leading to truncation or inefficiencies in embedding and generation.

For example, OpenAI's text-embedding-3-small model supports up to 8,191 tokens, but chunking to that maximum length may hinder precision in retrieval. In contrast, traditional transformer-based encoders such as BERT or Sentence-BERT are constrained to around 512 tokens. Balancing these constraints is essential: the ideal chunk size should be long enough to preserve meaningful context but short enough to fit within token limits and maintain relevance.

Table 3.3 Comparisons of chunking strategies for different content types.

Content type	Optimal chunk strategy	Example scenario
Long-form articles	Fixed-size (500–800 tokens)	Academic papers requiring context
Short tweets	Intent-based (50–100 tokens)	Sentiment analysis of social data
Legal documents	Recursive + entity-aware	Extracting clauses with named laws

3.2.3.1.2 Content-aware Segmentation

Effective chunking strategies must consider the inherent structure and nature of different content types. Academic papers benefit from different segmentation approaches than legal documents, which in turn require different handling than conversational transcripts or technical manuals (Table 3.3).

Long-form content such as research articles, reports, and books typically contains hierarchical information structures with sections, subsections, and supporting details. These documents benefit from semantic-aware chunking that respects logical boundaries while maintaining sufficient context for understanding.

Short-form content, including social media posts, news briefs, and messaging data, often requires intent-based approaches that preserve complete thoughts or interactions within individual chunks. The brevity of the source material changes the chunking calculus entirely.

Structured documents like legal contracts, technical specifications, and regulatory texts contain formal organizational patterns that should guide segmentation decisions. These documents often benefit from entity-aware chunking that maintains references and relationships between defined terms and their applications.

3.2.3.2 Core Chunking Methodologies

Table 3.4 lists the comparisons of these different chunking strategies.

Here, we use an example of an input document listed in Figure 3.3, and will show results of chunking/splitting using each chunking method.

3.2.3.2.1 Fixed-size Chunking

Fixed-size chunking is the most straightforward technique, where text is split into uniform blocks of tokens or characters. It's efficient and easy to implement, especially when combined with overlapping windows to prevent information loss at chunk boundaries.

This approach splits text based on token count while respecting tokenization constraints. The overlap parameter proves crucial for maintaining semantic continuity across chunk boundaries. A 200-token overlap ensures that concepts spanning chunk boundaries remain accessible, though this redundancy increases storage requirements and computational overhead. Figure 3.4 demonstrates the chunking results of using the TokenTextSplitter in LlamaIndex.

Fixed-size chunking is especially effective for embedding large corpora quickly, but it can break sentence structure, potentially degrading the quality of downstream retrieval.

3.2.3.2.2 Sentence-based Chunking

Sentence-aware chunking addresses fixed-size chunking's main limitation by respecting sentence boundaries during segmentation. This method prevents mid-sentence cuts that can create confusing or meaningless fragments, improving both retrieval quality and generation coherence.

Table 3.4 Chunking strategy comparison.

Strategy	Best use cases	Advantages	Limitations	Chunk size control
TokenTextSplitter	Model compatibility, large-scale processing	Predictable sizes, efficient processing, overlap control	May break sentences, less semantic coherence	Precise token count control
SentenceSplitter	General documents, articles, reports	Natural boundaries, readable chunks, preserves grammar	Variable sizes, may lose cross-sentence context	Sentence-based with token limits
SentenceWindow	Q&A systems, precise fact retrieval with context	Precise retrieval, context preservation, flexible	Complex metadata management, storage overhead	Single sentences with context metadata
RecursiveCharacter Splitter	Mixed content, structured documents, balanced approach	Hierarchical splitting, natural boundaries, size control	Complex configuration, may still break context	Target size with hierarchy respect
SemanticSplitter	Research papers, complex documents, topic clustering	Coherent topics, intelligent breakpoints, adaptive sizing	Computationally intensive, requires embedding model	Content-driven, variable sizes

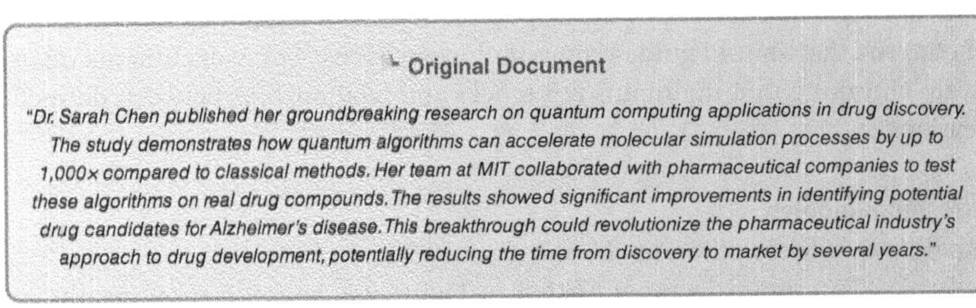

Figure 3.3 Input document to the text splitter.

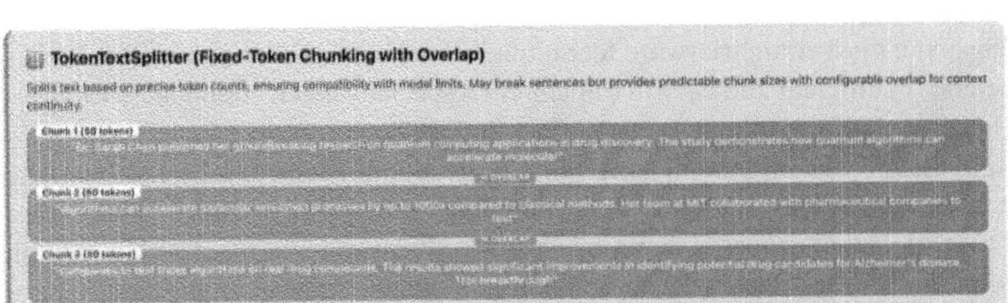

Figure 3.4 Chunking results of TokenTextSplitter.

This approach splits documents at sentence level using NLP-based parsing. It can be combined with overlapping windows of adjacent sentences, which adds neighboring sentences as metadata for added context during retrieval. Figures 3.5 and 3.6 demonstrate the chunking results of using the SentenceSplitter and SentenceWindowNodeParser in LlamaIndex.

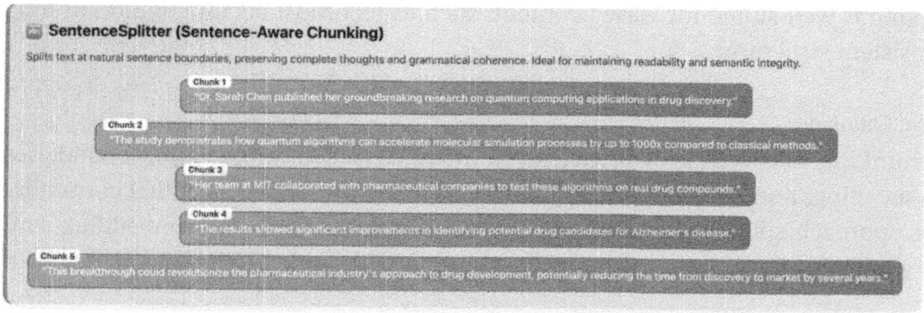

Figure 3.5 Chunking results of SentenceSplitter.

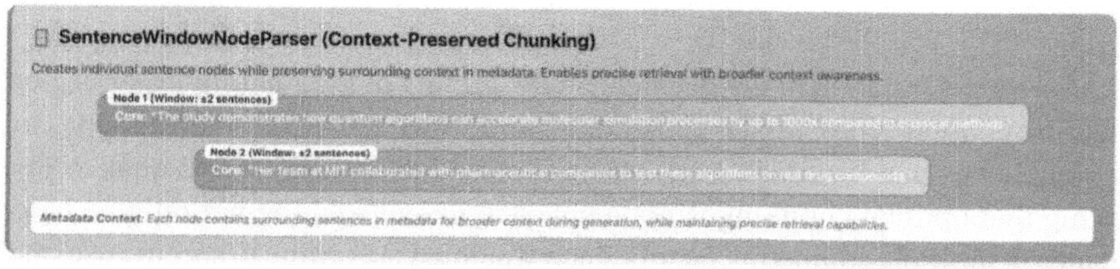

Figure 3.6 Chunking results of SentenceWindowNodeParser.

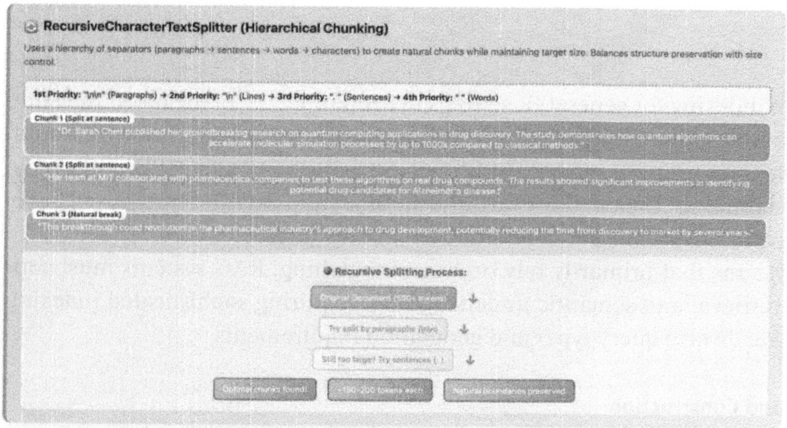

Figure 3.7 Chunking results of RecursiveCharacterTextSplitter.

This strategy is ideal for summarization, question answering, and other tasks where grammatical integrity matters.

3.2.3.2.3 *Recursive Chunking*

Recursive chunking is a flexible approach that splits text hierarchically using delimiters like paragraphs, sentences, or punctuation. If the text still exceeds the target chunk size, it recursively divides it further.

This approach allows fine-tuned control over chunk sizes while preserving semantic structure. This method attempts to split first by paragraph, then sentence, and finally character if necessary, balancing chunk size and readability. Figure 3.7 demonstrates the chunking results of using the RecursiveCharacterTextSplitter in LlamaIndex.

Recursive chunking is well-suited for varied content, such as technical documentation or long articles with inconsistent structure.

3.2.3.2.4 Semantic Chunking

Semantic-aware chunking uses vector- or model-based methods to determine chunk boundaries based on shifts in meaning, not just length or formatting. It helps preserve conceptual continuity across topics. This approach splits text at semantically meaningful points using embedding similarity. This method adapts chunk sizes dynamically, aligning with topic boundaries rather than syntactic markers.

Figure 3.8 demonstrates the chunking results of using the SemanticSplitterNodeParser. In LlamaIndex.

Semantic chunking is especially effective for applications like legal document review or topic-driven summarization.

The success of RAG systems fundamentally depends on the quality of their chunking strategies. By carefully considering content types, model constraints, and usage patterns, organizations can develop chunking approaches that maximize both retrieval precision and generation quality. The key lies in recognizing that optimal chunking is not a one-size-fits-all solution but rather a carefully calibrated balance of multiple competing factors, each weighted according to specific use case requirements and performance objectives. We are going to demonstrate the code details of using LlamaIndex for chunking in Chapter 4.

3.2.4 Indexing

We've discussed document indexing for general IR settings in Section 1.2.2.2 of the book. Indexing is a foundational step in both traditional IR and RAG. It transforms raw content into structured, searchable data that enables fast, accurate, and scalable query responses. Although the underlying goal remains the same—facilitating efficient retrieval—the methods and data structures used differ significantly between classical IR systems and modern RAG architectures.

Unlike traditional IR systems that primarily rely on lexical matching, RAG systems must support both precise factual retrieval and semantic understanding, requiring sophisticated indexing strategies that accommodate diverse query types and generation requirements.

3.2.4.1 RAG Index Types and Construction

RAG systems typically employ multiple index types simultaneously, each optimized for different retrieval scenarios and content characteristics.

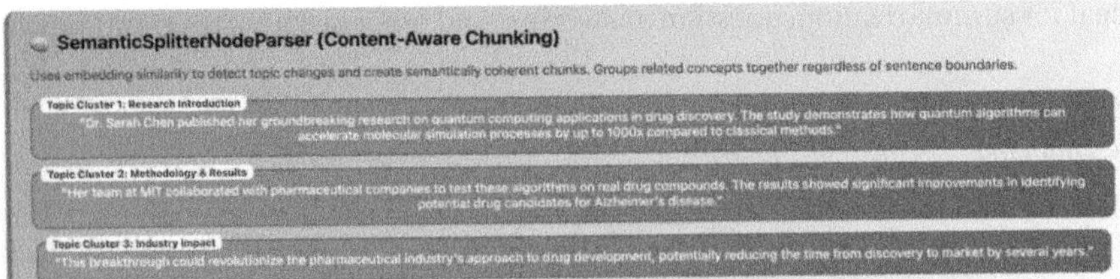

Figure 3.8 Chunking results of SentenceSplitterNodeParser.

Summary Index: Creates hierarchical abstractions of document collections, enabling efficient retrieval of high-level concepts and overviews. This index type proves particularly valuable for generating comprehensive responses that require synthesis across multiple sources.

```
from llama_index.core import SummaryIndex

# Create summary index for high-level concept retrieval
summary_index = SummaryIndex.from_documents(
    documents=processed_documents,
    show_progress=True
)

# Optimized for broad, conceptual queries
response = summary_index.as_query_engine().query(
    "What are the main themes in climate research?"
)
```

Vector Store Index: Represents the core of semantic retrieval in RAG systems, storing dense embeddings that capture semantic relationships between content segments. This index enables retrieval based on meaning rather than exact-term matches.

```
from llama_index.core import VectorStoreIndex
from llama_index.embeddings.openai import OpenAIEmbedding

# Configure embedding model
embed_model = OpenAIEmbedding(model="text-embedding-3-large")

# Create vector index for semantic retrieval
vector_index = VectorStoreIndex.from_documents(
    documents=chunked_documents,
    embed_model=embed_model,
    show_progress=True
)
```

Tree Index: Organizes information hierarchically, enabling efficient retrieval of related concepts at different levels of abstraction. This structure proves particularly effective for documents with clear hierarchical organization, such as technical manuals or academic papers.

```
from llama_index.core import TreeIndex
from llama_index.embeddings.openai import OpenAIEmbedding

# Configure embedding model
embed_model = OpenAIEmbedding(model="text-embedding-3-large")

# Create vector index for semantic retrieval
tree_index = TreeIndex.from_documents(
    documents=chunked_documents,
    embed_model=embed_model,
    show_progress=True
)
```

Keyword Table Index: Maintains traditional keyword-based retrieval capabilities, ensuring precise matching for specific terms, entities, and exact phrases. This index complements semantic retrieval by providing fallback precision for factual queries.

```
from llama_index.core import KeywordTableIndex
from llama_index.embeddings.openai import OpenAIEmbedding

# Configure embedding model
embed_model = OpenAIEmbedding(model="text-embedding-3-large")

# Create vector index for semantic retrieval
keyword_index = KeywordTableIndex.from_documents(
    documents=chunked_documents,
    embed_model=embed_model,
    show_progress=True
)
```

Property Graph Index: Represents knowledge as interconnected entities and relationships, enabling retrieval that considers semantic connections between concepts, entities, and facts.

```
from llama_index.core import PropertyGraphIndex
from llama_index.llms.openai import OpenAI

# Create knowledge graph index
property_graph_index = PropertyGraphIndex.from_documents(
    documents=documents,
    llm=OpenAI(model="gpt-4"),
    embed_model=embed_model,
    show_progress=True
)
```

3.2.4.2 Storage Component Architecture

RAG systems require sophisticated storage architectures that can handle the diverse data types and access patterns inherent in knowledge-augmented generation. As illustrated in Figure 3.9, there are multiple storage architectures:

Document Stores: Manage ingested documents as Node objects, preserving original content structure, metadata, and relationships. These stores maintain the authoritative source content that serves as the foundation for all retrieval operations.

Index Stores: Handle lightweight metadata associated with different index types, including configuration parameters, statistics, and structural information necessary for query processing and index maintenance.

Vector Stores: Store high-dimensional embedding vectors along with associated metadata, optimized for similarity search operations. Modern vector stores like Pinecone, Weaviate, and Chroma provide both storage and retrieval capabilities specifically designed for semantic search.

Property Graph Stores: Manage KGs with entities, relationships, and properties, enabling complex traversal queries that consider semantic connections between concepts.

Chat Stores: Organize conversational context and history, enabling RAG systems to maintain coherent multi-turn interactions while preserving relevant context for subsequent queries.

Table 3.5 lists the comparisons of these storage types.

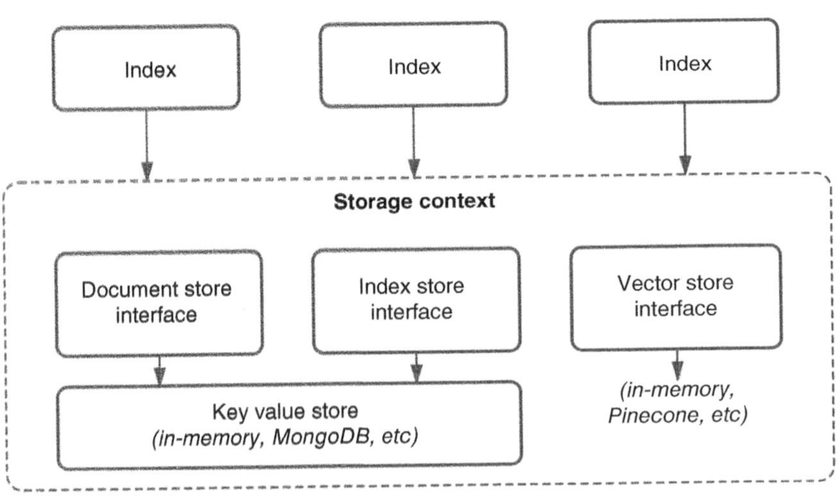

Figure 3.9 RAG index storage context [8] / LlamaIndex.

Table 3.5 RAG index storage types.

Store type	Contents	Implementation examples
Document store	Original node objects	MongoDB, PostgreSQL
Index store	Index metadata	SQLite, Redis
Vector store	Embedding vectors	FAISS, Pinecone, Weaviate
Property graph store	Knowledge graph triples	Neo4j, Amazon Neptune
Chat store	Conversation history	Firebase, Custom JSON

3.2.4.3 Advanced RAG Indexing Considerations

3.2.4.3.1 Multimodal Index Integration

Modern RAG systems increasingly need to handle diverse content types simultaneously, requiring integrated indexing strategies that can seamlessly combine textual, visual, and structured information.

Unified Retrieval Architecture: Combines multiple index types through intelligent routing and fusion mechanisms, selecting optimal retrieval strategies based on query characteristics and content types.

```
from llama_index.core import ComposableGraph

# Create composite index with multiple retrieval strategies
graph = ComposableGraph.from_indices(
    SummaryIndex,
    [vector_index, keyword_index, property_graph_index],
    index_summaries=["Vector semantic search", "Keyword exact match", "Knowledge graph
      traversal"],
    max_keywords_per_chunk=10
)
```

3.2.4.3.2 Dynamic Index Optimization

Adaptive Indexing: RAG systems implement continuous optimization based on query patterns, retrieval performance, and generation quality feedback. This includes automatic parameter tuning, index restructuring, and performance monitoring.

Incremental Updates: Unlike traditional batch indexing, RAG systems must support real-time content updates while maintaining query performance, requiring sophisticated incremental indexing strategies.

3.3 Retrieval Approaches

The retrieval component represents the intelligence layer of RAG systems, responsible for identifying and extracting the most relevant information from external knowledge sources to enhance generation quality. Building upon the foundational retrieval models discussed in Section 1.2.3, RAG systems implement specialized retrieval approaches that address the unique challenges of knowledge-augmented generation.

3.3.1 The Distinctive Nature of RAG Retrieval

Unlike traditional information retrieval systems that serve end users directly, RAG retrieval operates as an intermediate step in a larger generation pipeline. This fundamental difference creates unique requirements: retrieved content must not only be relevant to user queries but also optimally structured for language model consumption and response synthesis.

The primary goal extends beyond simple relevance matching to encompass four critical dimensions. **Relevance** ensures retrieved information directly addresses both the user's query and the language model's generation needs. **Speed** maintains responsive user experiences even when searching vast knowledge bases. **Completeness** provides sufficient context without overwhelming the model with extraneous information. **Scalability** enables the system to grow with expanding knowledge repositories while maintaining performance.

3.3.1.1 Context Window Optimization

RAG systems face a unique constraint absent in traditional retrieval: the language model's context window. While conventional search engines can return extensive result lists for human review, RAG must carefully curate content to maximize information density within token limits. This constraint drives innovative approaches to chunking and context assembly.

Intelligent chunking extends beyond mechanical document splitting. Technical documentation might benefit from larger chunks that preserve procedural integrity—a troubleshooting guide loses value if steps are separated. Conversely, factual queries often work well with focused segments that directly answer specific questions. The optimal chunk size varies not just by content type but also by the intended use case and model architecture.

Dynamic context assembly represents another crucial innovation. Rather than simply concatenating retrieved chunks, sophisticated RAG systems organize information to maintain logical flow and topical coherence. Consider a query about climate change impacts: the system might retrieve data on temperature trends, economic effects, and mitigation strategies from different sources, then assemble them in a narrative structure that facilitates comprehensive response generation.

3.3.1.2 Generation-aware Retrieval

The most sophisticated RAG systems implement retrieval strategies that anticipate generation needs. This involves understanding not just what information answers a query, but what information enables high-quality response synthesis. A query about "explaining quantum computing to beginners" requires different retrieval strategies than "quantum computing implementation challenges"—the former benefits from analogies and simplified explanations, while the latter needs technical specifications and research findings.

Multi-intent queries present particular challenges. A request to "compare renewable energy policies in European countries and their economic impact" spans policy databases, economic statistics, and analytical reports. Effective retrieval must coordinate across these domains while maintaining coherence in the assembled context.

3.3.2 Core Retrieval Methods in RAG

As illustrated in Figure 3.10, this section explores the various retrieval methods employed in RAG, each leveraging distinct techniques to identify and retrieve content based on the structure and nature of the information being sought.

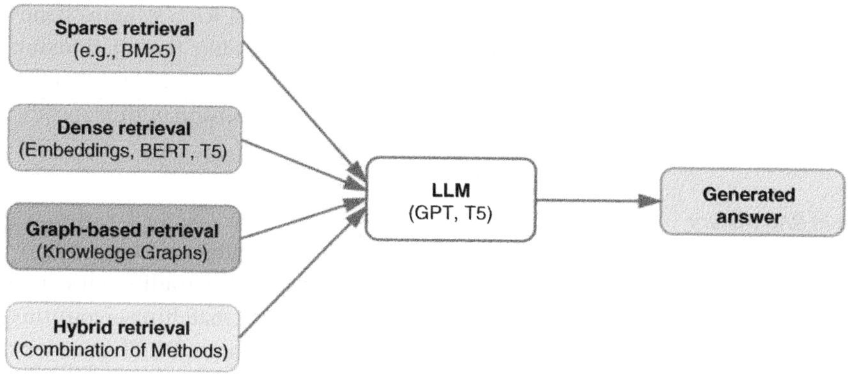

Figure 3.10 Retrieval methods in RAG.

3.3.2.1 Sparse Retrieval: Precision Through Exact Matching

Sparse retrieval methods, exemplified by BM25, remain fundamental to RAG systems despite the rise of neural approaches. Their strength lies in precision—when users need specific facts, dates, or technical specifications, exact-term matching proves invaluable.

Technical domains particularly benefit from sparse retrieval's precision. Consider a customer support system handling error messages: when a user reports "Error E2319: Database connection timeout," sparse retrieval efficiently locates the exact error documentation. This precision extends to regulatory compliance, where legal references require exact statute numbers and policy identifiers.

The efficiency of sparse methods also makes them ideal for initial filtering in large-scale systems. By quickly narrowing millions of documents to thousands of candidates based on keyword matches, sparse retrieval creates a manageable subset for more computationally intensive semantic analysis.

3.3.2.2 Dense Retrieval: Understanding Through Semantics

Dense retrieval transforms both queries and documents into high-dimensional vector representations that capture semantic meaning beyond surface-level word matching. This approach, powered by models like BERT, Sentence-BERT, or specialized embeddings from providers like OpenAI and Cohere [9], enables RAG systems to understand conceptual relationships and retrieve relevant information even when vocabulary differs.

The power of dense retrieval becomes evident in handling natural language queries. A question about "environmentally friendly transportation" successfully retrieves content about electric vehicles, public transit efficiency, and urban planning initiatives—even when these exact phrases don't appear in the query. This semantic understanding proves crucial for open-domain question answering where users express needs in varied ways.

Multilingual capabilities represent another significant advantage. Dense embeddings trained on multilingual corpora enable cross-language retrieval, allowing English queries to access relevant content from research papers in Chinese or technical documentation in German. This dramatically expands the effective knowledge base without requiring translation of all content.

Domain adaptation further enhances dense retrieval effectiveness. Medical RAG systems benefit from embeddings fine-tuned on biomedical literature, capturing nuances like the relationship between symptoms and conditions. Financial systems use embeddings that understand market terminology and regulatory language. This specialization improves retrieval precision within specific domains while maintaining the benefits of semantic understanding.

3.3.2.2.1 DPR's Role in RAG History

Dense passage retrieval (DPR) [2] marked a turning point in open-domain question answering by demonstrating that learned dense representations could significantly outperform traditional sparse methods like BM25. Prior to DPR, retrieval systems relied primarily on lexical matching—counting shared terms between queries and documents. DPR instead trained two separate BERT encoders: one for queries and one for passages. By optimizing these encoders with contrastive learning on question-passage pairs, DPR learned to map semantically related texts to nearby points in embedding space, enabling retrieval based on meaning rather than surface-level word overlap.

The RAG framework [1] built directly upon DPR's dense retrieval capability. Lewis et al. demonstrated that by combining a DPR retriever with a sequence-to-sequence generator, the system could dynamically access external knowledge during generation—effectively giving the language model a "long-term memory" beyond its fixed parameters. This retrieve-then-generate paradigm proved remarkably effective: RAG achieved state-of-the-art results on multiple knowledge-intensive benchmarks while using fewer parameters than purely parametric alternatives.

DPR's influence extends beyond the original RAG paper. Its core innovations—dual-encoder architecture, in-batch negatives, and hard negative mining—have been refined in subsequent models:

- ANCE [10]: Uses asynchronously updated ANN indices to mine harder negatives during training.
- ColBERT [11]: Retains token-level embeddings for late interaction, improving fine-grained matching while maintaining efficiency.
- Contriever [12]: Removes the need for labeled data through unsupervised contrastive pretraining.
- BGE [3]: A family of general-purpose embedding models from BAAI, with BGE-M3 [13] supporting multilingual, multi-functionality (dense, sparse, and multi-vector retrieval), and multi-granularity capabilities.

Understanding DPR's contributions provides essential context for the retrieval techniques discussed in Section 3.3.2.2 and helps explain why dense embeddings have become the default choice in production RAG systems.

3.3.2.3 Graph-based Retrieval: Navigating Relationships

Unlike traditional sparse or dense retrieval models that rely on independent document representations, graph-based methods explicitly capture interdependencies, hierarchies, and semantic relationships among data entities or passages. It enables systems to extract and navigate relevant trees/subgraphs based on the user's information needs. This retriever surfaces not just isolated documents but also contextual connections, supporting more accurate and holistic reasoning in downstream language models.

We explore three key variants within graph-based RAG, as summarized in Table 3.6

Table 3.6 RAG graph types comparison.

Type	Node type	Edge type	Example	Best for
Tree graph RAG	Summarized clusters	Semantic hierarchy	RAPTOR	Long documents, multi-scale summarization
Passage graph RAG	Passages	Lexical and semantic similarity	KGP for MD-QA	Multi-document reasoning, structured QA
Knowledge graph RAG	Entities	Factual relations	Wikidata-based RAG systems	Fact-based QA, recommender systems

3.3.2.3.1 Tree Graph RAG

Tree-structured retrieval introduces a hierarchical abstraction over large corpora by recursively clustering and summarizing documents into a tree, where each node represents increasingly abstracted or summarized views of the underlying content.

A recent state-of-the-art example of this approach is recursive abstractive processing for tree-organized retrieval (RAPTOR) [14]. RAPTOR constructs a multilayered tree by embedding text chunks, grouping them using soft clustering via Gaussian mixture models, and recursively summarizing the clusters using language models. This process produces a semantically organized tree, where leaf nodes contain detailed content and parent nodes summarize higher-level concepts.

RAPTOR supports two main retrieval strategies: tree traversal, which hierarchically navigates the tree layer by layer, and collapsed tree, which flattens the tree into a single layer to select nodes from all abstraction levels based on query relevance. The collapsed tree approach was shown to yield superior performance on complex multi-hop QA tasks when combined with GPT-4. This flexible retrieval enables the system to dynamically access granular or thematic information, depending on the query's intent, which is often critical for reasoning-intensive tasks.

The tree-based graph in RAPTOR embodies both semantic structure and abstraction-aware compression, two properties that are essential for scalable and intelligent retrieval. Unlike traditional RAG methods that retrieve flat, contiguous text spans, tree graph RAG facilitates multiscale reasoning by integrating evidence across the hierarchy. We are going to discuss more details of the RAPTOR method in Section 10.1.1 of the book.

3.3.2.3.2 Passage Graph RAG

Passage graph RAGs model passages (e.g., paragraphs, table rows, or structured sections) as nodes in a graph, with edges denoting semantic or lexical similarity. This form of retrieval emphasizes local coherence and logical flow, enabling multi-hop reasoning across adjacent or related passages. This method is particularly suited for multi-document question answering (MD-QA), where reasoning often involves synthesizing information scattered across multiple sources.

The knowledge graph prompting (KGP) method [15] exemplifies passage graph RAG by constructing a dynamic KG over multiple documents (illustrated in Figure 3.11), where:

- Node Creation: Each passage becomes a node with its textual content as its feature.
- Edge Creation: Links are formed based on lexical overlaps (e.g., shared entities or keywords) and semantic similarity (e.g., paraphrased meanings or shared relations).

To query this graph, KGP introduces an LLM-based graph traversal agent, which navigates the graph by progressively collecting relevant passages while staying aligned with the query intent. The

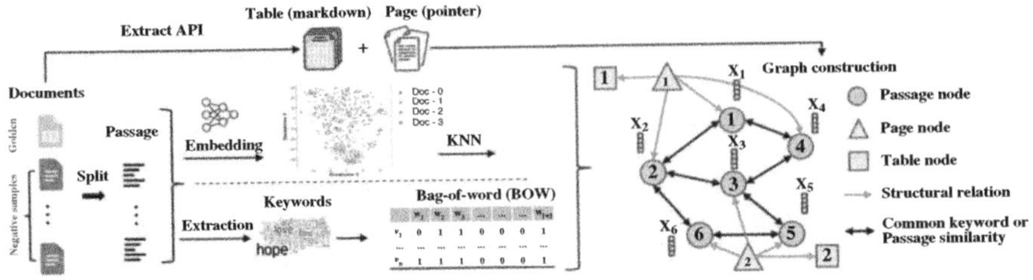

Figure 3.11 Passage graph construction [15] / Association for the Advancement of Artificial Intelligence.

graph acts as a global planner, ensuring retrieval coherence across documents, while the traversal agent serves as a local navigator, ensuring retrieval quality.

This two-level strategy—graph-level regulation and LLM-level traversal—enables more intelligent and efficient retrieval. Experiments showed that KGP improves retrieval quality and answer accuracy on multi-document QA tasks, highlighting the strength of passage graph modeling in contexts where structured documents (e.g., legal filings, financial reports) must be interpreted holistically.

3.3.2.3.3 *Knowledge Graph RAG*

A KG organizes information into entities (nodes) and relationships (edges) between those entities. These graphs can represent various types of relationships, such as hierarchies, associations, or dependencies. For example, in healthcare, relationships might include `prescribes`, `has_side_effect`, or `belongs_to_class_of_drugs`. In legal contexts, edges might represent `references_law`, `overrules_precedent`, or `applies_to_state`.

By querying a KG, RAG systems can retrieve not only relevant documents but also gain insight into how different pieces of information are related, enhancing the overall contextual understanding.

For example, consider a user query: "What are the treatments for COVID-19 and their known side effects?". By using the KG illustrated in Figure 3.12, a graph-based retriever can surface not only textual passages on treatments but also the relational connections among:

```
COVID-19 --> treated_by --> Remdesivir
Remdesivir --> referenced_in --> NEJM Study 2020
Remdesivir --> has_side_effect --> Liver toxicity
Liver toxicity --> monitored_by --> ALT/AST blood tests
```

The RAG model, armed with this retrieved subgraph, can generate a response that includes not only names of drugs but also clinical trial references, safety profiles, and diagnostic links—all deeply grounded in medical knowledge.

Graphs provide a rich source of context, enabling the system to retrieve not just isolated pieces of information but a web of related knowledge. This is particularly useful for answering complex, multifaceted queries where understanding the relationships between entities is crucial. By grounding the generated output in a well-structured knowledge base, the RAG system can produce more accurate and reliable responses.

We will explore the advanced implementation of graph-based RAG, including graph embedding strategies, graph traversal algorithms, and graph-to-text generation techniques, in Part II and Part III of this book.

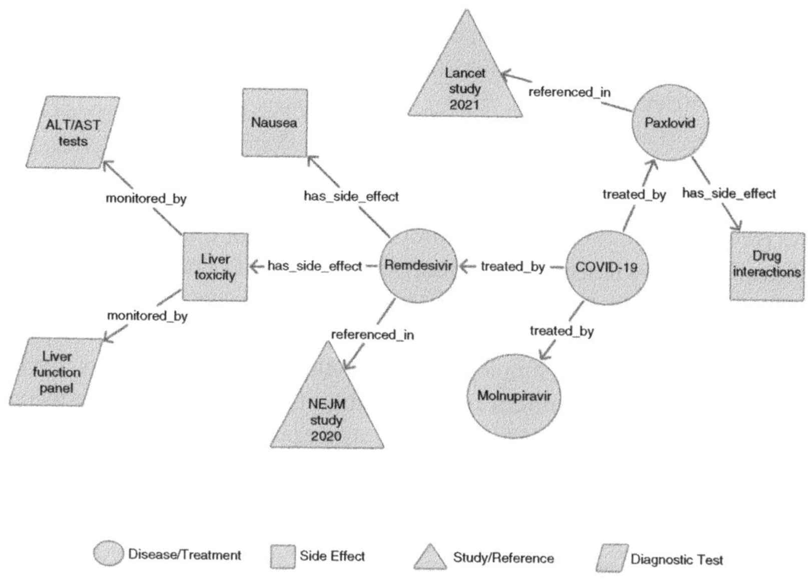

Figure 3.12 COVID-19 treatment knowledge graph.

3.3.2.4 Hybrid Approaches: Combining Strengths

Real-world RAG systems increasingly adopt hybrid strategies that leverage complementary strengths of different retrieval methods. Rather than choosing between precision and recall, hybrid systems achieve both through intelligent combination strategies.

Late fusion represents the most straightforward approach, combining relevance scores from different retrievers after independent operation. A typical implementation might weight BM25 scores at 0.3 and dense retrieval scores at 0.7, though optimal weights vary by domain and query type. More sophisticated approaches use learned fusion models that dynamically adjust weights based on query characteristics.

Query routing takes a different approach by selecting appropriate retrievers based on query analysis. Queries with specific entity mentions route to sparse retrieval for precision, while conceptual questions leverage dense retrieval for semantic understanding. This routing can be rule-based for predictable patterns or learned through query classification models.

Cascaded retrieval implements multistage strategies where each stage refines results from the previous one. Initial sparse retrieval might identify relevant documents, followed by dense retrieval for semantic ranking, and finally graph-based exploration for relationship discovery. This progression from broad to specific ensures comprehensive coverage while maintaining precision.

3.3.3 Advanced Retrieval Patterns

3.3.3.1 Hierarchical Index Retrieval

Large-scale RAG systems often implement hierarchical indexing to manage retrieval efficiency across massive document collections. This approach creates multiple index levels: document-level summaries for initial filtering and chunk-level indices for detailed retrieval.

The hierarchy enables efficient navigation through scale. A query about "machine learning in healthcare" first identifies relevant documents through summary matching, then retrieves specific chunks about applications, challenges, or case studies. This two-stage approach reduces computational load while maintaining retrieval quality.

3.3.3.2 Temporal and Source-aware Retrieval

Information currency and source credibility significantly impact RAG system effectiveness. Temporal awareness ensures recent information receives appropriate priority for time-sensitive queries while maintaining historical context when needed. A query about "current COVID-19 treatments" prioritizes 2024 research over 2020 studies, while "history of mRNA vaccine development" benefits from chronological coverage.

Source credibility integration goes beyond simple authority ranking. Academic papers, government publications, and established news sources receive credibility weights, but these weights adjust based on domain and query context. Medical queries prioritize peer-reviewed research, while technology queries might value recent blog posts from recognized experts over older academic papers.

3.3.3.3 Interactive and Conversational Patterns

Multi-turn conversations require sophisticated retrieval strategies that maintain context across interactions. Each query builds upon previous exchanges, requiring retrieval systems to understand not just the current question but its relationship to the conversation history. A discussion about renewable energy might progress from general concepts to specific technologies to implementation challenges, with each retrieval stage informed by the established context.

Clarification-driven retrieval identifies ambiguous queries and guides users toward more specific information needs. When users ask about "Python performance," the system might retrieve information about execution speed, memory usage, and optimization techniques while recognizing the query's ambiguity. This approach improves user experience by anticipating multiple interpretations.

3.3.4 Retrieval Performance Optimization

3.3.4.1 Latency Optimization Strategies

Real-time performance demands careful optimization across the retrieval pipeline. Precomputed embeddings eliminate runtime encoding costs for documents, requiring only query encoding during retrieval. Sophisticated caching strategies store not just final results but intermediate computations, enabling rapid response for similar queries.

ANN algorithms like HNSW or IVF trade minimal accuracy loss for substantial speed improvements in dense retrieval. These algorithms prove essential for production systems handling millions of documents with millisecond latency requirements.

3.3.4.2 Quality Assurance Mechanisms

Continuous monitoring ensures retrieval quality maintenance over time. Automated relevance scoring tracks retrieval performance across query types, while user feedback provides ground truth for system optimization. This monitoring enables identification of retrieval failures, bias detection, and performance degradation.

Bias mitigation requires particular attention in RAG systems, where retrieval biases can propagate to generated responses. Systems must ensure fair representation across perspectives, sources, and viewpoints. This involves not just technical solutions but also careful curation of knowledge sources and regular auditing of retrieval patterns.

3.3.5 Reranking and Filtering

As retrieval in RAG systems evolves from simple document fetching to nuanced information orchestration, **reranking and filtering** emerge as essential post-retrieval steps. These operations refine raw retrieval outputs into an optimal context set that maximizes utility for downstream generation. While initial retrieval prioritizes coarse relevance, reranking and filtering optimize for *context quality, complementarity, diversity, and coherence*—all of which significantly impact the final generation.

In complex settings, retrieval and reranking may be applied in multiple rounds. An initial retrieval stage casts a wide net, followed by reranking and filtering. A second retrieval stage may then refine or expand results based on underrepresented query intents. This **iterative refinement** enables dynamic coverage expansion and improved factual grounding.

3.3.5.1 Index Fusion

RAG systems often maintain multiple indices, each optimized for different sources (e.g., product databases, user manuals, scientific papers) or retrieval paradigms (e.g., sparse, dense, graph-based). Combining results from these indices creates a richer and more complete retrieval set.

Index fusion techniques range in complexity:

- **Simple Union:** Merges the top-k results from each index.
- **Weighted Score Fusion:** Combines scores using linear weights based on retriever trust or domain priority.
- **Reciprocal Rank Fusion (RRF):** A rank-based fusion method that avoids score normalization [16]. Each document's fused score is computed as the reciprocal of its rank across different retrievals:

$$\mathrm{RRF}(d) = \sum_{i=1}^{n} \frac{1}{k + \mathrm{rank}_i(d)}$$

where $\mathrm{rank}_i(d)$ is the rank of document d in the ith retriever's results, and k is a smoothing constant (commonly $k = 60$) [16].

Figure 3.13 demonstrates RRF combining outputs from a BM25 index, a BERT-based dense index, and a graph-based index. Despite differing scoring scales, RRF yields a unified top-k list by emphasizing agreement in rankings. Here is a diagram illustrating the RRF process. It shows how multiple retrieval systems (BM25, dense, graph-based) assign different ranks to the same set of documents. This fusion technique aggregates rankings—not raw scores—making it robust to different scoring scales across systems.

3.3.5.2 Reranking for Generation Utility

Once an initial candidate set is retrieved, **reranking** models reorder them based on deeper semantic evaluation. Unlike bi-encoders, **cross-encoder rerankers** jointly encode query-document pairs, allowing rich query–passage interactions. Popular models include MonoBERT [17], ColBERTv2 [18], and Cohere Rerank [19]

These models provide fine-grained relevance signals and are typically applied to the top-k retrieved results due to their computational cost.

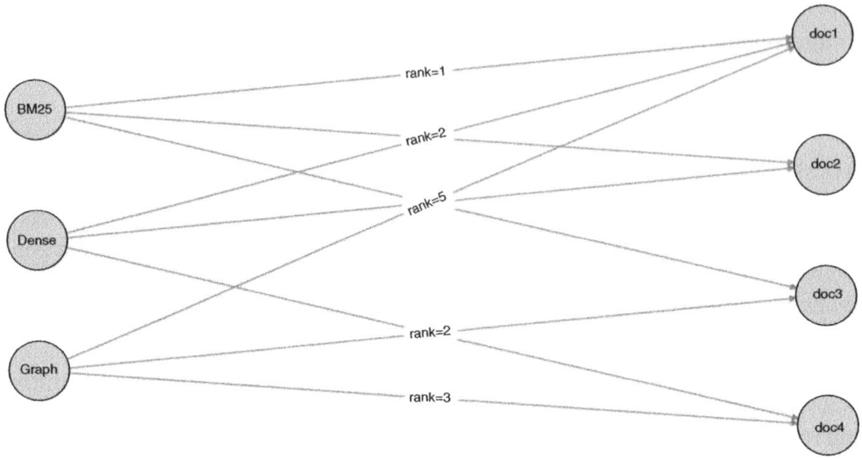

Figure 3.13 Reciprocal rank fusion: Document rankings from multiple retrievers.

3.3.5.3 Generation-aware Filtering

Beyond surface-level relevance, not all retrieved documents are equally useful for generation. Some may be redundant, off-topic, or exceed token limits. **Generation-aware reranking and filtering** refine the context window to include:

- Complementary Information: Prioritize documents with non-overlapping insights.
- Deduplication: Remove near-duplicate content.
- Diversity: Ensure topical breadth for comparative or open-ended queries.
- Structural Coherence: Select passages that maintain logical and narrative flow.

For example, to answer "Compare renewable energy strategies of Germany, France, and Spain," generation-aware filtering selects concise, high-quality passages from each country—avoiding redundancy and improving answer completeness.

3.3.5.3.1 Summary

Reranking and filtering bridge the gap between retrieved relevance and generative usefulness. By combining cross-encoder reranking, index fusion, and filtering heuristics, RAG systems deliver more accurate, comprehensive, and coherent outputs.

3.4 Generation

The generation phase represents the culmination of the RAG pipeline, where retrieved knowledge is synthesized with language model capabilities to produce coherent, contextually relevant, and factually accurate responses. This critical stage transforms raw retrieved information into polished outputs that directly address user queries while maintaining both accuracy and fluency.

Success in the generation phase requires careful orchestration of multiple components: strategic model selection, precise parameter tuning, efficient inference optimization, and robust accuracy mechanisms. Each decision in this phase directly impacts the quality, reliability, and usability of the final system output.

3.4.1 Generation Pipeline

The generation process follows a systematic pipeline that ensures retrieved knowledge is effectively integrated into high-quality responses.

- Step 1: Document integration retrieved documents from the RAG retrieval stage are processed and formatted as contextual input. This involves selecting the most relevant passages, organizing them logically, and structuring them for optimal model consumption.
- Step 2: Context formatting retrieved texts are formatted into prompts using established patterns such as "Based on the following sources...." This formatting provides clear context boundaries and establishes the authoritative nature of the retrieved information.
- Step 3: Parameter configuration generation parameters are set according to task requirements. For factual accuracy, typical settings include `temperature=0.2` for reduced randomness, `max_length=150` for concise responses, and `top-p=0.9` for focused token selection.
- Step 4: Model execution the language model processes the formatted prompt and retrieved context to generate candidate responses. This stage involves the core neural computation where language understanding meets retrieved knowledge.
- Step 5: Post-processing validation generated responses undergo validation checks including fact-checking filters, coherence evaluation, and citation verification to ensure quality and accuracy before delivery.
- Step 6: Response finalization the validated response is formatted with appropriate citations to source documents, providing transparency and enabling users to verify information independently.

Here is an example generation pipeline implementation:

```
def generate_rag_response(query, retrieved_docs, model, parameters):
    # Step 1-2: Format context with retrieved documents
    context = "Based on the following sources:\n"
    for i, doc in enumerate(retrieved_docs[:3]):  # Use top 3 documents
        context += f"Source {i+1}: {doc.content}\n"

    prompt = f"{context}\nQuestion: {query}\nAnswer:"

    # Step 3: Configure parameters for factual accuracy
    generation_config = {
        'temperature': 0.2,          # Low randomness for accuracy
        'max_length': 150,           # Concise responses
        'top_p': 0.9,               # Focused token selection
        'do_sample': True,           # Enable controlled sampling
        'pad_token_id': model.config.eos_token_id
    }

    # Step 4: Model execution and generate response
    response = model.generate(prompt, **generation_config) # Specify model selection

    # Step 5: Post-processing validation
    validated_response = apply_fact_checking(response, retrieved_docs)

    # Step 6: Add citations
    final_response = add_source_citations(validated_response, retrieved_docs)

    return final_response
```

3.4.2 Model Selection Framework

Choosing the appropriate generation model requires systematic evaluation across multiple dimensions that directly impact system performance and user experience. There are several major factors that affect the model selection decisions: *task type*, *response length* and *latency and resource constraints*.

Different **task types** lead to different model selections. For creative tasks (e.g., story writing, content generation, brainstorming, etc.), models such as GPT-4 and Claude are good candidates with their flexibility and creative capabilities. For domain-specific factual tasks (e.g., legal advice, medical information, etc.) require models with strong grounding capabilities. Domain-specific models such as Legal-BERT [20] for legal applications or BioBERT [21] for medical contexts, which often outperform general models by providing specialized knowledge and terminology understanding.

Different **response length** requirements also lead to different model selections. For short responses (<100 tokens), lightweight models like DistillBERT [22] or smaller GPT variants provide adequate quality with faster inference. These models suit FAQ systems, quick lookups, and brief explanations. For large responses (500+ tokens), larger models like GPT-4 or T5-Large handle extended outputs better, maintaining coherence across long passages while managing complex multipart responses effectively.

Different **latency and resource** constraints lead to different model selections. For real-time applications such as chatbots, live support, and interactive systems, they require models optimized for speed. Models such as DistilBERT [22], TinyBERT [23] and optimized GPT variants are usually good candidates. For batch processing such as report generation, offline analysis, and content creation, they can utilize larger models such as GPT-4, Claude, etc.

With these model selection strategy considerations, Table 3.7 lists some examples of recommended models for various task categories and use cases.

3.4.3 Parameter Tuning

Language models offer multiple parameters to steer generation: quality, diversity, and precision. Effective parameter tuning requires understanding how each parameter affects output characteristics and systematically optimizing for specific use cases.

Some of the major generation parameters include the following:

Table 3.7 Practical model selection examples.

Task category	Recommended models	Key characteristics	Use case examples
Creative tasks	GPT-4, GPT-3.5, Claude	High creativity, diverse vocabulary	Story writing, marketing copy, brainstorming
Factual tasks	GPT-4, Llama-2-Chat, PaLM	Strong reasoning, accuracy focus	Legal advice, medical info, technical docs
Analytical tasks	T5-Large, GPT-4, FLAN-T5	Structured thinking, summarization	Data analysis, report generation, comparisons
Conversational	ChatGPT, Claude, Llama-2-Chat	Natural dialog, context awareness	Customer support, tutoring, assistance
Domain-specific	BioBERT, Legal-BERT, FinBERT	Specialized knowledge, terminology	Medical diagnosis, legal research, finance

Temperature Control (0.0–1.0)

Temperature governs the randomness of token selection during generation:

- Temperature = 0.0: Completely deterministic output, selecting the highest probability token at each step. Ideal for factual queries requiring consistent answers.
- Temperature = 0.2–0.3: Low randomness suitable for technical documentation, legal advice, or medical information where accuracy is paramount.
- Temperature = 0.5–0.7: Moderate creativity appropriate for educational content, explanations, or business communications.
- Temperature = 0.8–1.0: High creativity for storytelling, marketing copy, or brainstorming applications.

Top-p Nucleus Sampling (0.1–0.9)

Top-*p* sampling controls vocabulary diversity by limiting token selection to the smallest set whose cumulative probability exceeds the threshold:

- Top-*p* = 0.1: Extremely focused, selecting only the most probable tokens. Suitable for precise factual responses.
- Top-*p* = 0.3–0.5: Moderate focus for balanced accuracy and variety in professional writing.
- Top-*p* = 0.7–0.9: Broad exploration allowing creative expression while maintaining coherence.

Repetition Penalties

- Frequency Penalty (0.0–2.0): Reduces likelihood of repeating tokens proportional to their frequency.
- Presence Penalty (0.0–2.0): Discourages any token repetition regardless of frequency.

Length Controls

- Max Tokens: Hard limit on response length.
- Stop Sequences: Specific phrases that terminate generation.
- Length Penalty: Bias toward longer or shorter responses.

3.4.4 Inference Speed and Optimization Techniques

Efficient inference is critical for providing responsive user experiences while managing computational costs effectively. Real-time RAG applications must deliver fast, low-latency outputs without sacrificing generation quality.

To meet the dual goals of low-latency inference and cost-effective scalability, a variety of optimization techniques can be applied to the generation phase. Each of these methods balances trade-offs in speed, memory, accuracy, and implementation complexity.

Table 3.8 lists a comparison of inference optimization techniques.

Key-value Caching: Key-value caching is an optimization strategy particularly effective for transformer-based autoregressive models (e.g., GPT). During generation, transformers compute key and value vectors at each layer for all previous tokens in the sequence. In naive implementations, these computations are repeated at each step. With caching, these vectors are stored once and reused in subsequent decoding steps, drastically reducing redundant computation.

- Benefit: Significant reduction in compute time per token (especially noticeable in long outputs).

Table 3.8 Inference optimization techniques.

Technique	Purpose	Trade-offs
Key-value caching	Reuses transformer states for faster decoding	Increases memory usage
Model quantization	Converts models to lower precision (e.g., FP32 → INT8)	Slight accuracy degradation, but ~4x speedup
Parallel decoding	Uses beam search to generate multiple tokens concurrently	Can increase complexity and GPU demand
Distillation	Trains smaller student model from larger teacher	Reduced latency, some loss in language quality
MoE (mixture of experts) models	Activates only subsets of expert modules	Efficient at scale, more complex to train

- Trade-off: Increased memory consumption due to storage of intermediate activations.
- Use Case: Essential for streaming chat interfaces or long-form summarization.

Model Quantization: Model quantization reduces the size of a model by converting its parameters from high-precision formats (e.g., 32-bit floating point) to low-precision representations (e.g., 8-bit integers). This reduces memory usage, speeds up matrix multiplication, and enables better hardware utilization, especially on CPUs or edge devices.

- Benefit: Up to 4x speedups with 75.
- Trade-off: Potential minor degradation in generation quality, especially for nuanced text.
- Use Case: Ideal for deployment in resource-constrained environments (mobile, browser, edge devices).

Parallel Decoding: Parallel decoding attempts to speed up generation by predicting multiple tokens at once. One common approach is beam search, where multiple candidate sequences are expanded in parallel, and the most promising one is selected at the end.
Newer approaches like speculative decoding or token-level parallelism (e.g., in FlashAttention2) also aim to speed up inference by batching and speculative execution.

- Benefit: Reduced wall-clock time, especially in long-generation tasks.
- Trade-off: Increased GPU demand and more complex control flow.
- Use Case: Summarization, document generation, and reranking where latency matters.

Knowledge Distillation: Knowledge distillation is a training-time optimization that creates a smaller "student" model by transferring knowledge from a larger, high-performance "teacher" model. The student attempts to mimic the output distributions or hidden states of the teacher, achieving similar performance at a fraction of the size.

- Benefit: Dramatically faster inference and smaller deployment footprint.
- Trade-off: May sacrifice some generalization ability or fluency on complex tasks.
- Use Case: Chatbots, virtual assistants, and recommendation systems with hard latency budgets.

MoE: MoE models are a form of sparse activation architecture. Instead of activating the full model for every input, only a subset ("experts") are dynamically selected based on the input query. This allows extremely large models to operate efficiently, as only a few components run per input.

- Benefit: Scales to hundreds of billions of parameters with manageable compute.
- Trade-off: Complex training and serving infrastructure; sensitive to expert routing quality.
- Use Case: Large-scale RAG systems in enterprise or research with high accuracy demands.

3.5 Summary

RAG systems represent a sophisticated synthesis of information retrieval and language generation capabilities. This chapter has systematically unpacked the core components that enable RAG architectures to deliver accurate, context-rich responses across diverse applications.

The foundation lies in thoughtful data preparation, where the distinction between structured databases, unstructured content, and real-time programmatic sources shapes preprocessing strategies. Multimodal considerations further expand the system's ability to handle visual, auditory, and tabular data alongside traditional text. Effective chunking methodologies emerge as critical for balancing contextual continuity with retrieval precision, whether through fixed-size segmentation or advanced semantic-aware partitioning.

At the heart of RAG systems, retrieval components demonstrate specialized adaptations beyond conventional search paradigms. Sparse retrieval excels in precision-critical scenarios requiring exact terminology matches, while dense vector approaches capture nuanced semantic relationships. Graph-based methods enable sophisticated multi-hop reasoning through explicit relationship modeling, and hybrid strategies combine these strengths for robust performance. Innovations like hierarchical retrieval and temporal weighting address generation-specific challenges around context management and information recency. The generation phase synthesizes retrieved knowledge through careful model selection and parameter optimization. Performance tuning focuses on aligning inference speed with quality requirements, ensuring responses maintain both factual grounding and conversational flow. Throughout the pipeline, techniques like dynamic indexing and retrieval quality monitoring maintain system responsiveness and reliability.

Three cross-cutting principles define effective RAG implementation:

1. Retrieval-for-Generation Design: Every retrieval decision must consider how results will fuel coherent generation.
2. Progressive Refinement: Layered processing from data ingestion to final output ensures quality at each stage.
3. Adaptive Architectures: The best systems fluidly combine approaches based on query context and domain needs.

These concepts form the blueprint for building RAG systems that transform raw information into actionable knowledge—a capability increasingly vital in our data-rich landscape. The following sections in the upcoming chapters will explore how these technical foundations power real-world applications across industries.

References

1 Patrick Lewis, Ethan Perez, Aleksandra Piktus, Fabio Petroni, Vladimir Karpukhin, Naman Goyal, Heinrich Küttler, Mike Lewis, Wen-tau Yih, Tim Rocktäschel, et al. Retrieval-augmented generation for knowledge-intensive NLP tasks. *Advances in Neural Information Processing Systems*, 33:9459–9474, 2020.

2 Vladimir Karpukhin, Barlas Oguz, Sewon Min, Patrick SH Lewis, Ledell Wu, Sergey Edunov, Danqi Chen, and Wen-tau Yih. Dense passage retrieval for open-domain question answering. In *EMNLP (1)*, pages 6769–6781, 2020.

3 Shitao Xiao, Zheng Liu, Peitian Zhang, and Niklas Muennighoff. C-pack: Packaged resources to advance general Chinese embedding. *arXiv preprint arXiv:2309.07597*, 2023.

4 Jean-Baptiste Alayrac, Jeff Donahue, Pauline Luc, Antoine Miech, Iain Barr, Yana Hasson, Karel Lenc, Arthur Mensch, Katherine Millican, Malcolm Reynolds, et al. Flamingo: A visual language model for few-shot learning. *Advances in Neural Information Processing Systems*, 35:23716–23736, 2022.

5 OpenAI. GPT-4V(ision) system card, September 2023. https://openai.com/research/gpt-4v-system-card. Accessed: 2024-03-04.

6 Zihan Liu, Wei Ping, Rajarshi Roy, Peng Xu, Chankyu Lee, Mohammad Shoeybi, and Bryan Catanzaro. ChatQA: Surpassing GPT-4 on conversational QA and RAG. *arXiv preprint arXiv:2401.10225*, 2024.

7 Nelson F Liu, Kevin Lin, John Hewitt, Ashwin Paranjape, Michele Bevilacqua, Fabio Petroni, and Percy Liang. Lost in the middle: How language models use long contexts. *Transactions of the Association for Computational Linguistics*, 12:157–173, 2024.

8 LlamaIndex. Storing—LlamaIndex documentation, 2025. https://docs.llamaindex.ai/en/stable/module_guides/storing/#concept. Accessed: 2025-06-21.

9 Cohere. Cohere platform documentation, 2024. https://docs.cohere.com. Accessed: 2025-07-05.

10 Lee Xiong, Chenyan Xiong, Ye Li, Kwok-Fung Tang, Jialin Liu, Paul Bennett, Junaid Ahmed, and Arnold Overwijk. Approximate nearest neighbor negative contrastive learning for dense text retrieval. In *International Conference on Learning Representations*, 2021.

11 Omar Khattab and Matei Zaharia. ColBERT: Efficient and effective passage search via contextualized late interaction over BERT. In *Proceedings of the 43rd International ACM SIGIR Conference on Research and Development in Information Retrieval*, pages 39–48, 2020.

12 Gautier Izacard, Mathilde Caron, Lucas Hosseini, Sebastian Riedel, Piotr Bojanowski, Armand Joulin, and Edouard Grave. Unsupervised dense information retrieval with contrastive learning. *Transactions on Machine Learning Research. arXiv preprint arXiv:2112.09118*, 2022.

13 Jianlv Chen, Shitao Xiao, Peitian Zhang, Kun Luo, Defu Lian, and Zheng Liu. BGE M3-embedding: Multi-lingual, multi-functionality, multi-granularity text embeddings through self-knowledge distillation, In *Findings of the association for computational linguistics: ACL*, pages 2318–2335, 2024.

14 Parth Sarthi, Salman Abdullah, Aditi Tuli, Shubh Khanna, Anna Goldie, and Christopher D Manning. Raptor: Recursive abstractive processing for tree-organized retrieval. In *The Twelfth International Conference on Learning Representations*, 2024.

15 Yu Wang, Nedim Lipka, Ryan A Rossi, Alexa Siu, Ruiyi Zhang, and Tyler Derr. Knowledge graph prompting for multi-document question answering. In *Proceedings of the AAAI Conference on Artificial Intelligence*, volume 38, pages 19206–19214, 2024.

16 Gordon V Cormack, Charles LA Clarke, and Stephan Buettcher. Reciprocal rank fusion outperforms condorcet and individual rank learning methods. In *Proceedings of the 32nd International ACM SIGIR Conference on Research and Development in Information Retrieval*, pages 758–759. ACM, 2009.

17 Rodrigo Nogueira, Wei Yang, Kyunghyun Cho, and Jimmy Lin. Multi-stage document ranking with BERT. *arXiv preprint arXiv:1910.14424*, 2019. https://arxiv.org/abs/1910.14424.

18 Omar Khattab, Keshav Santhanam, Xiang Lisa Li, David Hall, Percy Liang, Christopher Potts, and Christopher D Manning. ColBERTv2: Effective and efficient retrieval via lightweight late interaction. *arXiv preprint arXiv:2112.01488*, Dec 2022. https://arxiv.org/abs/2112.01488.

19 Thibault Formal, Kriti Popat, Amir Veyseh, Nils Reimers, Sylvie Shi, Lucas Fayoux, Elliott Choi. Cohere rerank: Scalable retrieval with large language models. Cohere Blog, 2023. https://cohere.com/blog/rerank. Accessed: 2026-03-04.

20 Ilias Chalkidis, Manos Fergadiotis, Prodromos Malakasiotis, Nikolaos Aletras, and Ion Androutsopoulos. Legal-BERT: The muppets straight out of law school. *arXiv preprint arXiv:2010.02559*, Oct 2020. https://arxiv.org/abs/2010.02559.

21 Jinhyuk Lee, Wonjin Yoon, Sungdong Kim, Donghyeon Kim, Sunkyu Kim, Chan Ho So, and Jaewoo Kang. BioBERT: A pre-trained biomedical language representation model for biomedical text mining. *Bioinformatics*, 36(4):1234–1240, Feb 2020. https://doi.org/10.1093/bioinformatics/btz682.

22 Victor Sanh, Lysandre Debut, Julien Chaumond, and Thomas Wolf. DistilBERT, a distilled version of BERT: Smaller, faster, cheaper and lighter. *arXiv preprint arXiv:1910.01108*, Oct 2019. https://arxiv.org/abs/1910.01108.

23 Xiaoqi Jiao, Yichun Yin, Lifeng Shang, Xin Jiang, Xiao Chen, Linlin Li, Fang Wang, and Qun Liu. TinyBERT: Distilling BERT for natural language understanding. *arXiv preprint arXiv:1909.10351*, Sep 2020. https://arxiv.org/abs/1909.10351.

4

Practice: RAG Implementation

Having established the theoretical foundations and architectural principles of retrieval-augmented generation (RAG) systems in Chapter 3, we now turn to the practical implementation of these concepts. While understanding the interplay between data preparation, retrieval mechanisms, and generation strategies is essential, translating these principles into working systems requires robust software frameworks that can efficiently orchestrate the complex RAG pipeline.

This chapter examines two leading frameworks that have emerged to address this challenge: **LangChain/LangGraph** and **LlamaIndex**. Each framework embodies different design philosophies—LangChain prioritizes flexibility and extensibility for general-purpose large language model (LLM) applications, while LlamaIndex focuses on streamlined, RAG-specific workflows. Through detailed exploration of LlamaIndex's implementation stages, from offline indexing to online querying, we will demonstrate how theoretical RAG concepts translate into practical code.

Later sections explore advanced implementations like the WANDS dataset for product search and agentic RAG designs for tool-augmented reasoning. Together, these examples show how RAG theory translates into real-world applications.

4.1 Overview of LangChain and LlamaIndex

A comparison of the LangChain and LlamaIndex frameworks is presented in Table 4.1.

4.1.1 LangChain and LangGraph

Both are general-purpose LLM frameworks:

- LangChain [1], first released in October 2022, is a modular framework for developing LLM-powered applications, including RAG systems. Its core strength lies in its flexibility, enabling developers to construct complex workflows by combining components such as chains (sequential task pipelines), agents (autonomous decision-making modules), and memory modules (for context persistence in conversational applications). LangChain supports integration with external tools, including vector databases (e.g., FAISS [2], Pinecone [3]), APIs, and custom logic, but requires explicit configuration of data pipelines and retrieval strategies (e.g., BM25, semantic search). This granularity makes it suitable for diverse applications, such as chatbots, code generation, and task automation, beyond RAG.

Table 4.1 Comparison of LangChain/LangGraph and LlamaIndex for RAG workflows.

Aspect	LangChain/LangGraph	LlamaIndex
Core purpose	General-purpose LLM applications (e.g., RAG, chatbots, automation)	Specialized RAG workflows (e.g., document QA, knowledge bases)
Architecture	Modular components with explicit configuration	High-level abstractions with automated indexing and retrieval
Data handling	Utilities for vector stores and retrieval, but requires manual setup	Built-in indexes, auto-chunking, hybrid retrieval (vector + keyword)
Use cases	Broad, including RAG, conversational agents, and task automation	Document-centric RAG, with support for agentic workflows
Flexibility	Highly customizable, suitable for complex workflows	Streamlined for RAG; less flexible for non-RAG tasks
Community and ecosystem	Large community, integrations with OpenAI, Hugging Face, etc.	Growing community, strong focus on document processing tools
Trade-offs	Overkill for simple RAG tasks; ideal for complex workflows	Limited for non-RAG applications; excels in rapid prototyping

- LangGraph [4], a companion library to LangChain, focuses on graph-based workflows for modeling complex relationships and multistep reasoning. It represents tasks (e.g., data retrieval, LLM inference) as nodes in a directed acyclic graph (DAG), with edges defining dependencies. This structure supports conditional logic (e.g., retrieving specific data based on user input) and iterative processes (e.g., query refinement). LangGraph is particularly effective for applications requiring precise control over execution flow, such as multistage decision-making or stateful systems. However, its advanced features demand deeper technical expertise compared to LangChain's core functionality.

LangChain and LangGraph offer remarkable flexibility for customized LLM applications, but their modular design introduces complexity. Developers must manually configure RAG components, such as vector store connections and query logic, which can be cumbersome for simple document-based tasks. Supported by a large community and integrations with tools like Hugging Face and OpenAI, LangChain/LangGraph excels in scenarios requiring dynamic APIs, persistent memory, or graph-based reasoning. However, for projects focused solely on document retrieval and integration, their extensive feature set may introduce unnecessary overhead.

4.1.2 LlamaIndex

LlamaIndex [5], released in late 2022, is a specialized framework designed for streamlined RAG workflows. It emphasizes ease of use and efficiency, providing built-in indexes (e.g., vector, tree, knowledge graph) that automate data organization, text chunking, and retrieval. LlamaIndex supports hybrid retrieval strategies, combining vector similarity with keyword matching, and advanced techniques like reranking or metadata filtering. Its high-level abstractions simplify integration with diverse data sources (e.g., PDFs, CSVs, APIs) and LLMs, minimizing boilerplate code. These features enable rapid development of document-centric applications, such as legal research tools, customer support systems, or internal knowledge bases. While primarily focused on RAG, LlamaIndex also supports agentic workflows and query transformations, broadening its applicability. Its growing community and integrations with document processing tools make it a strong choice for prototyping, though it offers less flexibility for non-RAG tasks compared to LangChain.

4.1.3 Comparison Summary

LangChain/LangGraph provides a versatile platform for building complex LLM applications, including RAG, but its configuration overhead may deter developers working on straightforward document-based tasks. LlamaIndex, with its automated indexing and retrieval pipelines, offers a more accessible approach for RAG, particularly for applications requiring fast query-response cycles over static or semi-static knowledge bases. Both frameworks are viable for RAG, with the choice depending on project requirements: LangChain/LangGraph for highly customized, multi-step workflows, and LlamaIndex for rapid prototyping and document-centric use cases. For most of the experiments and code demonstrations in this book, we adopt LlamaIndex due to its simplicity and strong alignment with typical RAG use cases. Nonetheless, we also recognize the strengths of LangChain for advanced applications and include one illustrative case study in Section 8.2 to showcase its capabilities.

4.2 Implementing RAG Pipelines with LlamaIndex

The objective of this section is to demonstrate how to build a basic RAG pipeline using LlamaIndex. LlamaIndex provides a comprehensive framework for building advanced AI applications through its robust RAG pipeline API, agent API, and auxiliary tools that streamline data processing and querying workflows. The RAG implementation in LlamaIndex follows a two-phase architecture:

1. Offline Phase (two stages):
   - Loading: Data ingestion and preprocessing.
   - Indexing: Structuring data into searchable formats.
2. Online Phase (one stage):
   - Querying: Real-time interaction with the indexed data.

Before diving into implementation details, we first configure the environment by connecting to local LLM endpoints. With the local OpenAI API-compatible servers prepared in Appendix A, the following code establishes a global configuration for LlamaIndex to use these services:

```
from llama_index.core import Settings
from llama_index.llms.openai_like import OpenAILike

Settings.llm = OpenAILike(
    model="o1-mini",
    api_base=api_base,                      # [1]
    api_key="sk-fake-api-key-abcdefghijklmnopqrstuvwxyz-0123456789",
    context_window=32768,
    is_chat_model=True,                     # [1]
    is_function_calling_model=True,         # [1]
    temperature=0.6,                        # [2]
    timeout=timeout,                        # [3]
)
```

[1] Local model hosting with chat and function-calling capabilities enabled.
[2] Model-specific temperature override—following model card guidelines.
[3] Extended timeout recommended when running on less powerful hardware.

In addition to the language model, we configure the embedding model that will be used for document retrieval from vector stores:

```
from llama_index.embeddings.openai_like import OpenAILikeEmbedding

Settings.embed_model = OpenAILikeEmbedding(
    model_name="text-embedding-ada-002",
    api_base=api_base,
    embed_batch_size=2048,
    timeout=timeout,
)
```

4.2.1 Offline Stages: Loading and Indexing

The goal of offline stages is to build an index ready for retrieval. We will first load documents into memory, next calculate the embeddings for the vector store, then insert the embeddings together with documents into vectors, and finally persist the vector stores.

4.2.1.1 Loading

The loading stage in LlamaIndex takes various data sources as input, such as files, web pages, and databases. It processes these sources and produces a list of `Document` objects as output. These `Document` objects serve as the source for further processing in the LlamaIndex pipeline.

4.2.1.1.1 Document

A document is a container for any data source (e.g., PDF, API output, described in Section 3.2.1.1), storing text, metadata, and relationships. It can be created manually or via data loaders, with beta support for multimodal content like images.

Directory Reader: The `SimpleDirectoryReader` in LlamaIndex offers a flexible and user-friendly interface for loading files from a directory. It automatically detects supported file extensions (e.g., `.epub`, `.docx`, `.pdf`, `.mp4`, `.jpg`, `.ipynb`, and more) and uses the appropriate reader to extract content from each file. The following example demonstrates how to load Markdown files from the `/usr/share/doc/gnome-user-docs/` directory—on Ubuntu 24.04, this directory contains the user documentation for the GNOME desktop environment:

```
from llama_index.core import SimpleDirectoryReader
from llama_index.core.readers.file.base import default_file_metadata_func
import os

documents_directory = SimpleDirectoryReader(
    "/usr/share/doc/gnome-user-docs/",
    required_exts=[".md"],
    file_metadata=lambda file_path: {
        **default_file_metadata_func(file_path),
        "extension": os.path.splitext(file_path)[1],
    },
).load_data()
```

This call returns a list of documents, each enriched with metadata. The metadata is stored as a Python dictionary that includes both default fields (such as file path, size, and dates) and any custom fields added via the `file_metadata` function. An example of the resulting metadata dictionary is shown below:

```
{
  "file_path": "/usr/share/doc/gnome-user-docs/README.md",
  "file_name": "README.md",
  "file_type": "text/markdown",
```

```
    "file_size": 537,
    "creation_date": "2024-11-02",
    "last_modified_date": "2023-09-17",
    "extension": ".md"
}
```

Web Page Reader: The `SimpleWebPageReader` in LlamaIndex retrieves and returns the raw HTML content of web pages by default. In addition to the page content, we enrich each document with custom metadata—including the source URL and file extension—so that downstream components such as node parsers can appropriately handle HTML files. The following example demonstrates how to load two web pages while attaching useful metadata to each document:

```
from llama_index.readers.web import SimpleWebPageReader

documents_web = SimpleWebPageReader(
    metadata_fn=lambda url: {
        "webpage_url": url,
        "extension": ".html",
    }
).load_data(
    [
        "https://example.com/",
        "https://www.whitehouse.gov/administration/melania-trump/",
    ]
)
```

4.2.1.1.2 Transformation

A transformation operates as a function that accepts a list of `Nodes` as input and produces a modified list of `Nodes` as output. These transformations can be chained together to execute a sequence of operations on the `Nodes`. Since `Documents` are a specific type of `Node`, the transformation chain can begin with a set of `Documents`. `Nodes`, as the foundational building blocks for constructing an index, allow the chain to conclude with the creation of an index.

Node: `Node` is a "chunk" of a `Document` (e.g., text, image), inheriting the `Document`'s metadata and relationships. `Nodes` are essential in LlamaIndex and can be parsed from `Documents` using `NodeParser` classes.

File-based Node Parsers: As described in Section 3.2.2.2, for HTML, JSON, and Markdown files—identified by their file extensions in the `Document` metadata—the following code employs the corresponding parser to split the `Document` into `Nodes`:

```
from llama_index.core.node_parser import SimpleFileNodeParser
from llama_index.core.node_parser import HTMLNodeParser

sfnp = SimpleFileNodeParser()
htmlnp = HTMLNodeParser(tags=["p", "h1"])

nodes_file = sfnp.get_nodes_from_documents(
    documents_directory,
) + htmlnp.get_nodes_from_documents(
    documents_web,
)
```

This example demonstrates two types of node parsers. The `SimpleFileNodeParser` automatically selects the appropriate parser based on the file extension. For HTML files, the `HTMLNodeParser` is used to extract content from specified HTML tags (e.g., <p> and <h1> in this case), enabling structured parsing of web page content.

Text Splitters: Splitting text using fixed-size blocking (described in Section 3.2.3.2) with a preference for complete sentences (described in Section 3.2.3.2) ensures that chunks are less likely to end with incomplete phrases or fragmented sentences. This approach is particularly important for embedding models, which often have strict context-length limitations. Unlike file-based splitters, which cannot guarantee that chunk lengths align with these constraints, text splitters are designed to produce chunks that remain within the valid input range for embedding models, ensuring compatibility and optimal performance.

```
from llama_index.core.node_parser import SentenceSplitter

ss = SentenceSplitter(
    chunk_size=512,
    chunk_overlap=64,
    paragraph_separator="\n",
)
nodes_sentence = ss(nodes_file)
```

Splitting text by intent (described in Section 3.2.3.2) can be achieved using an embedding model, which calculates the embedding distance between neighboring sentences. When this distance exceeds a default threshold of 0.95, the model identifies a significant shift in meaning or topic and splits the text accordingly. This method ensures that chunks align with semantically coherent boundaries, improving the relevance and usability of the resulting segments for downstream tasks such as classification or clustering.

```
from llama_index.core.node_parser import SemanticSplitterNodeParser

ssnp = SemanticSplitterNodeParser(
    embed_model=Settings.embed_model,
    sentence_splitter=nltk.sent_tokenize,
)
nodes_semantics = ssnp(nodes_sentence)
```

Relation-based Node Parsers: As described in Section 3.2.3.2, `HierarchicalNodeParser` chunks nodes into hierarchical nodes. This means a single input will be chunked into several hierarchies of chunk sizes, with each node containing a reference to its parent node.

```
from llama_index.core.node_parser import HierarchicalNodeParser

nodes_hierarchy = HierarchicalNodeParser.from_defaults(
    chunk_sizes=[2048, 512, 128],
).get_nodes_from_documents(nodes_file)
```

This example shows a three-level parser by `SentenceSplitter`. The split nodes form a forest data structure, in which each tree is organized by four types of links: `parent_node`, `child_nodes`, `prev_node`, `next_node`.

Metadata: As demonstrated in Section 4.2.1.1, key-value metadata can be added to documents during the loading phase. This metadata is retained and carried through the entire document transformation pipeline—including node parsing and chunking stages—enabling downstream components to make use of contextual information. In addition to manually defined metadata, LLMs can be used to automatically enrich documents with semantic information. For example, LLMs can extract summaries, infer topics, or generate descriptive titles. Below, we demonstrate two commonly used metadata extractors in LlamaIndex: `SummaryExtractor` and `TitleExtractor`.

The `SummaryExtractor` operates at the node level, enriching each document fragment with a semantic summary. It appends a `section_summary` key to the node's metadata dictionary, providing valuable context for downstream tasks such as querying or analysis. Here's an example of how it can be configured and applied:

```
from llama_index.core.extractors import SummaryExtractor

se = SummaryExtractor(
    prompt_template="""Summarize the main topics and key entities from the following
     section in a single sentence:

{context_str}"""
)
nodes_metadata = se(nodes_semantics)
```

In this configuration, the LLM is prompted to generate a concise, one-sentence summary of the content within each node. This summary is then stored in the node's metadata under the `section_summary` field. Below is an illustrative example of the input provided to the LLM and the resulting generated summary:

> **user** Summarize the main topics and key entities from the following section in a single sentence:
>
> extension: .html
> url: https://www.whitehouse.gov/administration/melania–trump/
> tag: p
>
> In addition to BE BEST, Mrs. Trump has a deep appreciation for the military and their families. She traveled to military bases domestically and overseas, including Iraq in 2018, and regularly incorporated military families in public engagements. In her time away from the White House, Mrs. Trump established Fostering the Future, a BE BEST initiative, which provides college–level scholarships to those aging out of the foster care system. The scholarship funds are provided by a portion of the revenue generated from her Web 2 and Web 3 platforms. Her platforms utilize blockchain technology, marking her as a pioneer within the industry while empowering the next generation of young Americans with the resources needed to excel.
>
> **assistant**
> Melania Trump championed military outreach through domestic and international engagements, including a 2018 Iraq visit, established the BE BEST initiative's Fostering the Future scholarship program for foster care youth, and leveraged blockchain–powered Web 2 and Web 3 platforms to fund these efforts, emerging as a tech industry pioneer.

The `TitleExtractor` operates at the document level, generating a `document_title` key in the metadata. It uses a two-step process: first extracting candidate titles from individual nodes, then combining them into a final, comprehensive title. Below is an example configuration and usage:

```
from llama_index.core.extractors import TitleExtractor

te = TitleExtractor(
    nodes=3,
    node_template="""Generate a concise and relevant title for the following content.
     Output only the title text-do not include explanations or formatting:

{context_str}""",
    combine_template="""Based on the candidate titles and the content below, generate
     the most accurate and comprehensive title for the document. Output only the title
     text-no explanations or formatting:

{context_str}""",
)
nodes_metadata = te(nodes_semantics)
```

This two-phase approach ensures that the final title accurately reflects the overall content by first generating local titles from up to three leading nodes, then synthesizing them into a unified document title. The LLM generates a title for each selected node based on its content. Here's an example input and output for one node:

> **user** Generate a concise and relevant title for the following content. Output only the title text—do not include explanations or formatting:
>
> # gnome–user–docs
>
> This package contains documents which are packaged together and shipped as gnome–user–docs in the core GNOME distribution.The documents are targeted for end–users and system administrators of GNOME and have general GNOME applicability.
>
> **assistant**
> GNOME User and System Administrator Documentation (gnome–user–docs)

After collecting titles from multiple nodes, the LLM combines them into a single, representative title for the entire document. Here's an example of that stage:

> **user** Based on the candidate titles and the content below, generate the most accurate and comprehensive title for the document. Output only the title text—no explanations or formatting:
>
> GNOME User and System Administrator Documentation (gnome–user–docs),
>
> GNOME Documentation Project (GDP) Homepage,
>
> Report Bugs in GNOME User Docs via GitLab
>
> **assistant**
> GNOME User and System Administrator Documentation (gnome–user–docs), GNOME Documentation Project (GDP) Homepage, and Report Bugs in GNOME User Docs via GitLab

Ingestion Pipeline: To streamline the transformation process, we can chain the previously defined node transformations into a single pipeline using LlamaIndex's `IngestionPipeline`. This allows for efficient and consistent processing of documents through multiple stages. The following code demonstrates how to construct and execute such a pipeline:

```
from llama_index.core.ingestion import IngestionPipeline

pipeline = IngestionPipeline(transformations=[ss, ssnp, se, te])
nodes = pipeline.run(
    documents=nodes_file,
    num_workers=16,
)
```

In this example, the transformations `ss`, `ssnp`, `se`, and `te` refer to objects initialized in earlier sections.

4.2.1.2 Vector Store Index

An index is a core data structure in LlamaIndex that enables efficient retrieval of relevant context for user queries. It serves as the foundation for RAG, allowing models to access external knowledge beyond their training data. Indexes are built from `Document` objects, which are processed into `Node` instances representing semantically meaningful chunks of the original content. These nodes are stored in a structured format, and the index provides a `Retriever` interface to automate the querying process. The most widely used index type in LlamaIndex is the `VectorStoreIndex`. This index leverages embeddings—dense vector representations of text—to enable semantic search, where results are based on meaning rather than exact keyword matches. Internally, the `VectorStoreIndex` relies on three primary components for data storage:

- `SimpleDocumentStore`: Stores raw document content.
- `SimpleIndexStore`: Manages metadata associated with the index.
- `SimpleVectorStore`: Stores precomputed embeddings for fast similarity search.

Constructing an index from a set of nodes is straightforward:

```
from llama_index.core import VectorStoreIndex

index = VectorStoreIndex(nodes)
```

Behind the scenes, the `VectorStoreIndex` processes each node by converting its text into numerical embeddings using an LLM API (e.g., OpenAI or Hugging Face). This step—commonly referred to as "embedding your text"—enables the system to perform semantic similarity searches efficiently. However, this embedding process can be computationally expensive, especially when dealing with large volumes of text. Since it typically involves sequential API calls, generating embeddings for many text chunks may result in significant processing time and resource usage. Therefore, it is important to consider both performance and cost when working with extensive datasets.

4.2.1.2.1 Storing

To improve efficiency and avoid recomputing expensive embeddings on each run, indexes can be persisted to disk (described in Section 3.2.4.2). This allows for offline embedding generation and enables fast reloading for repeated querying. The following line saves the index and all its components to the default directory (`./storage`):

```
index.storage_context.persist()
```

By default, this creates a folder named `./storage` containing several JSON files that represent the different components of the index:

```
$ tree --charset=ascii storage
./storage
|-- default__vector_store.json
|-- docstore.json
|-- graph_store.json
|-- image__vector_store.json
`-- index_store.json

1 directory, 5 files
```

To reload a previously saved index, use the following code:

```
from llama_index.core import load_index_from_storage
from llama_index.core.storage.storage_context import DEFAULT_PERSIST_DIR
from llama_index.core import StorageContext

index = load_index_from_storage(
    StorageContext.from_defaults(persist_dir=DEFAULT_PERSIST_DIR)
)
```

This reconstructs the index from disk, restoring all stored metadata and embeddings—allowing you to immediately begin querying without reprocessing documents or regenerating embeddings.

4.2.1.2.2 Chroma Vector Store

For scalable or distributed workflows, external vector stores such as Chroma can be used in place of local storage. Unlike file-based persistence, Chroma runs as a long-lived service, maintaining data availability across sessions without requiring manual reloading. To begin using Chroma with LlamaIndex, first initialize the Chroma client. This client can be treated as a singleton throughout your application, ensuring consistent access to the vector store for all components:

```
import chromadb

chroma_client = chromadb.HttpClient()
chroma_collection = chroma_client.get_or_create_collection("demo_collection")
```

Once the collection is created, it can be used to instantiate a `ChromaVectorStore`, which integrates with LlamaIndex's indexing pipeline:

```
from llama_index.core import StorageContext, VectorStoreIndex
from llama_index.vector_stores.chroma import ChromaVectorStore

index_chroma = VectorStoreIndex(
    nodes,
    storage_context=StorageContext.from_defaults(
        vector_store=ChromaVectorStore(chroma_collection=chroma_collection)
    ),
)
```

This integration allows you to leverage Chroma's efficient similarity search capabilities while benefiting from LlamaIndex's high-level abstractions for document processing and retrieval.

4.2.2 Online Stage: Querying

During the online stage, the system retrieves relevant information and generates meaningful responses based on user queries. The primary input is the index, and the final output is a natural language response generated by the LLM. The querying process typically involves three main steps: retrieve, post-process, and synthesize.

4.2.2.1 Retriever

A retriever fetches relevant nodes from the index or node list based on a given query string. LlamaIndex supports multiple retrieval strategies, including keyword-based, vector-based, and hybrid approaches.

The following example uses the BM25 algorithm (described in Section 3.3.2.1) for keyword-based retrieval. It selects the top two most relevant nodes based on lexical similarity:

```
from llama_index.retrievers.bm25 import BM25Retriever

retriever_bm25 = BM25Retriever.from_defaults(
    nodes=nodes,
    similarity_top_k=2,
```

```
)
retrieved_nodes = retriever_bm25.retrieve("What was the focus of the BE BEST campaign
    ?")
```

This example demonstrates a vector-based retriever (described in Section 3.3.2.2) that applies metadata filtering to narrow down results. Here, we filter nodes whose `section_summary` field contains any of the specified keywords: "Melania," "BE BEST," or "Trump."

```
from llama_index.core.vector_stores.types import (
    MetadataFilter,
    MetadataFilters,
    FilterOperator,
)

retriever_vector_store = index.as_retriever(
    similarity_top_k=2,
    filters=MetadataFilters(
        filters=[
            MetadataFilter(
                key="section_summary",
                value=["Melania", "BE BEST", "Trump"],
                operator=FilterOperator.ANY,
            )
        ],
    ),
)
retrieved_nodes = retriever_vector_store.retrieve(
    "What was the focus of the BE BEST campaign?"
)
```

You can combine multiple retrievers (described in Sections 3.3.2.4 and 3.3.5.1) to improve recall and relevance. The following example fuses the results from a BM25 and a vector-based retriever using simple fusion without additional query generation:

```
from llama_index.core.retrievers import QueryFusionRetriever

retriever_fusion = QueryFusionRetriever(
    [retriever_vector_store, retriever_bm25],
    similarity_top_k=2,
    num_queries=1,  # set this to 1 to disable query generation
    use_async=True,
)
retrieved_nodes = retriever_fusion.retrieve(
    "What was the focus of the BE BEST campaign?"
)
```

Each strategy has its strengths—BM25 excels at matching exact terms, vector retrievers capture semantic meaning, and fused retrieval combines both for more robust results.

4.2.2.2 Node Postprocessors

Node postprocessors are used to refine or filter retrieved nodes based on relevance, keywords, or LLM-based ranking.

The following example uses the `LLMRerank` postprocessor (described in Section 3.3.5.2) to reorder retrieved nodes by semantic relevance using an LLM:

```
from llama_index.core.postprocessor import LLMRerank

reranker = LLMRerank(top_n=2)
reranked_nodes = reranker.postprocess_nodes(
    retrieved_nodes,
    query_str="What was the focus of the BE BEST campaign?",
)
```

This reranking step ensures that only the most relevant documents are passed to the synthesizer for response generation. Below is an example of how the LLM evaluates and ranks the retrieved content:

user A list of documents is shown below. Each document has a number next to it along with a summary of the document. A question is also provided.
Respond with the numbers of the documents you should consult to answer the question, in order of relevance, as well
as the relevance score. The relevance score is a number from 1–10 based on how relevant you think the document is to the question.
Do not include any documents that are not relevant to the question.
Example format:
Document 1:
<summary of document 1>

Document 2:
<summary of document 2>

...

Document 10:
<summary of document 10>

Question: <question>
Answer:
Doc: 9, Relevance: 7
Doc: 3, Relevance: 4
Doc: 7, Relevance: 3

Let's try this now:

Document 1:
[Excerpt from document]
extension: .html
url: https://www.whitehouse.gov/administration/melania–trump/
tag: p
section_summary: Melania Trump, a former model and naturalized U.S. citizen, leveraged her philanthropy, business ventures (e.g., her jewelry line), and roles as First Lady (2017–2021) to champion children's well–being through the BE BEST initiative and support military families, while maintaining a legacy of community engagement with organizations like the American Red Cross and American Heart Association.
document_title:

Melania Trump: First Lady of the United States, Wife of the 45th and 47th Presidents, Second Foreign–Born First Lady, Model, Philanthropist, and Advocate for Military Families and Foster Youth via Blockchain–Powered Scholarships
Excerpt:
–––––

At a young age, she began what would soon become a highly successful modeling career, appearing in many high–profile ad campaigns and working with some of the best photographers in the fashion industry.In 1996, Mrs. Trump moved to New York, and 10 years later, she proudly became a United States Citizen, making her the only First Lady to become a naturalized citizen.Mrs. Trump has always been an active member of her community.In 2005, she was the Honorary Chairwoman of the Martha Graham Dance Company.That same year, she was awarded Goodwill Ambassador by the American Red Cross––a role she served for 4 years.Mrs. Trump served 5 years as Honorary Chairwoman for the Boys' Club of New York and was named Woman of the Year in 2006 by the Police Athletic League.In 2010, Mrs. Trump was the Chairwoman of the American Heart Association.In addition to her philanthropic work and modeling and fashion career, Mrs. Trump is an accomplished businesswoman.In 2010, she launched her own successful jewelry collection, Melania Trump Timepieces and Jewelry.While Melania Trump became a household name, she is, first and foremost, a mother and wife, and in 2017, she made the White House and Washington home for her family.In her role as First Lady from 2017 to 2021, Mrs. Trump focused on the many issues affecting children across the Nation.This passion led to the establishment of BE BEST in 2018; Mrs. Trump's awareness campaign focused on the well–being of children and highlighted the people and programs dedicated to ensuring a better future for the next generation.Mrs. Trump promoted BE BEST throughout the country and internationally.In addition to BE BEST, Mrs. Trump has a deep appreciation for the military and their families.She traveled to military bases domestically and overseas, including Iraq in 2018, and regularly incorporated military families in public engagements.
–––––

Document 2:
[Excerpt from document]
extension: .html
url: https://www.whitehouse.gov/administration/melania–trump/
tag: p
section_summary: Melania Trump, as First Lady, promoted the BE BEST initiative, supported military families through domestic and international engagements including Iraq in 2018, and launched the Fostering the Future scholarship program funded by her blockchain–based Web 2 and Web 3 platforms to aid foster care youth.
document_title:

Melania Trump: First Lady of the United States, Wife of the 45th and 47th Presidents, Second Foreign–Born First Lady, Model, Philanthropist, and Advocate for Military Families and Foster Youth via Blockchain–Powered Scholarships
Excerpt:
–––––
In addition to BE BEST, Mrs. Trump has a deep appreciation for the military and their families. She traveled to military bases domestically and overseas, including Iraq in 2018, and regularly incorporated military families in public engagements.In her time away from the White House, Mrs. Trump established Fostering the Future, a BE BEST initiative, which provides college–level scholarships to those aging out of the foster care system.The scholarship funds are provided by a portion of the revenue generated from her Web 2 and Web 3 platforms.Her platforms utilize blockchain technology, marking her as a pioneer within the industry while empowering the next generation of young Americans with the resources needed to excel.

Question: What was the focus of the BE BEST campaign?
Answer:

assistant
Answer:
Doc: 1, Relevance: 9
Doc: 2, Relevance: 7

Explanation:
– **Document 1** explicitly states that the BE BEST initiative focused on "the well–being of children across the Nation," emphasizing efforts to highlight programs and people working toward a better future for the next generation. This directly answers the question.
– **Document 2** reinforces the focus on children's well–being by mentioning BE BEST in the context of supporting foster youth through scholarships, but it adds less detail about the campaign's overarching goals compared to Document 1.

You can also filter retrieved nodes (described in Section 3.3.5.3) using required and excluded keywords—this functionality is implemented using spaCy's `PhraseMatcher`. It is particularly useful for removing irrelevant content or focusing on specific topics such as "BE BEST" while excluding unrelated terms.

```
from llama_index.core.postprocessor import KeywordNodePostprocessor

filtered_nodes = KeywordNodePostprocessor(
    required_keywords=["BE BEST"],
    exclude_keywords=["Kamala Harris", "Harris", "Kamala"],
).postprocess_nodes(retrieved_nodes)
```

This ensures that only nodes containing the desired keywords—and none matching the exclusion list—are included in the final results.

4.2.2.3 Response Synthesizer

The response synthesizer (described in Section 3.4.1) is responsible for generating the final answer by invoking an LLM using the user's query and a set of relevant document nodes. It produces a structured `Response` object that includes the synthesized answer, along with additional metadata such as source nodes. LlamaIndex's synthesizer supports advanced features like structured answer filtering and adaptive prompt selection based on the chosen `ResponseMode`.

```
from llama_index.core import get_response_synthesizer
from llama_index.core.response_synthesizers import ResponseMode

synthesizer = get_response_synthesizer(
    response_mode=ResponseMode.COMPACT,
    structured_answer_filtering=True,
)
response = synthesizer.synthesize(
    query="What was the focus of the BE BEST campaign?",
    nodes=reranked_nodes,
)
```

The response synthesizer incorporates two critical features to enhance its functionality. First, when enabled via `structured_answer_filtering=True`, this feature automatically filters out

Table 4.2 RAG response synthesis modes.

Mode name	Description	Use case
REFINE	Iteratively refines responses across nodes. Initial answer → N-1 refine passes	Best for precision when context must be carefully integrated
COMPACT	Combines nodes into larger chunks first, then refines. Fewer API calls	Optimal balance of speed and accuracy for moderate-sized datasets
SIMPLE_SUMMARIZE	Merges all chunks into a single context block before LLM call	Risk of context overflow; use only for small/short documents
TREE_SUMMARIZE	Builds a hierarchical tree index, summarizing top-down in a tree response	Ideal for complex, multilevel document structures requiring deep analysis
GENERATION	Ignores provided context entirely; relies solely on LLM creativity	Use for hypothetical scenarios or when external context isn't required
NO_TEXT	Returns raw retrieved nodes without synthesis	Useful for debugging or when manual interpretation is preferred
CONTEXT_ONLY	Returns concatenated text chunks without LLM involvement	Quick access to raw context data
ACCUMULATE	Synthesizes individual responses per node, then combines results	Preserves granular insights but increases API costs
COMPACT_ACCUMULATE	Combines nodes first, then accumulates results across consolidated chunks	Faster alternative to ACCUMULATE with reduced API calls

irrelevant or low-confidence context nodes before generating the final response. This is especially valuable when using function-calling LLMs (e.g., OpenAI models), where excessive or noisy context can lead to hallucinations or parsing errors. Second, the synthesizer dynamically selects an appropriate prompt based on the specified `ResponseMode`. Each mode—such as `COMPACT`, `SIMPLE_SUMMARIZE`, or `ACCUMULATE`—is associated with a tailored prompt designed to optimize the synthesis strategy for that use case. A full list of these modes and their corresponding prompts is provided in Table 4.2.

When selecting a response mode, consider the following trade-offs. For speed vs. accuracy, prioritize `COMPACT` or `COMPACT_ACCUMULATE` for efficiency, while opting for `REFINE` or `ACCUMULATE` when higher precision is needed, albeit with increased API usage. Regarding context limitations, avoid `SIMPLE_SUMMARIZE` for large documents to prevent exceeding the LLM's context window; instead, use `TREE_SUMMARIZE` for hierarchical processing of extensive data. In scenarios requiring no context, `GENERATION` enables creative answers, whereas `NO_TEXT` and `CONTEXT_ONLY` allow direct access to raw data for hybrid approaches.

The following is an example of an LLM dialog used to generate the final response:

system

You are an expert Q&A system that is trusted around the world.
Always answer the query using the provided context information, and not prior knowledge.

Some rules to follow:
1. Never directly reference the given context in your answer.
2. Avoid statements like 'Based on the context, ...' or 'The context information ...' or anything along those lines.

user Context information is below.

[Excerpt from document]
extension: .html
url: https://www.whitehouse.gov/administration/melania–trump/
tag: p
section_summary: Melania Trump, a former model and naturalized U.S. citizen, leveraged her philanthropy, business ventures (e.g., her jewelry line), and roles as First Lady (2017–2021) to champion children's well–being through the BE BEST initiative and support military families, while maintaining a legacy of community engagement with organizations like the American Red Cross and American Heart Association.
document_title:

Melania Trump: First Lady of the United States, Wife of the 45th and 47th Presidents, Second Foreign–Born First Lady, Model, Philanthropist, and Advocate for Military Families and Foster Youth via Blockchain–Powered Scholarships
Excerpt:

At a young age, she began what would soon become a highly successful modeling career, appearing in many high–profile ad campaigns and working with some of the best photographers in the fashion industry.In 1996, Mrs. Trump moved to New York, and 10 years later, she proudly became a United States Citizen, making her the only First Lady to become a naturalized citizen.Mrs. Trump has always been an active member of her community.In 2005, she was the Honorary Chairwoman of the Martha Graham Dance Company.That same year, she was awarded Goodwill Ambassador by the American Red Cross––a role she served for 4 years.Mrs. Trump served 5 years as Honorary Chairwoman for the Boys' Club of New York and was named Woman of the Year in 2006 by the Police Athletic League.In 2010, Mrs. Trump was the Chairwoman of the American Heart Association.In addition to her philanthropic work and modeling and fashion career, Mrs. Trump is an accomplished businesswoman.In 2010, she launched her own successful jewelry collection, Melania Trump Timepieces and Jewelry.While Melania Trump became a household name, she is, first and foremost, a mother and wife, and in 2017, she made the White House and Washington home for her family.In her role as First Lady from 2017 to 2021, Mrs. Trump focused on the many issues affecting children across the Nation.This passion led to the establishment of BE BEST in 2018; Mrs. Trump's awareness campaign focused on the well–being of children and highlighted the people and programs dedicated to ensuring a better future for the next generation.Mrs. Trump promoted BE BEST throughout the country and internationally.In addition to BE BEST, Mrs. Trump has a deep appreciation for the military and their families.She traveled to military bases domestically and overseas, including Iraq in 2018, and regularly incorporated military families in public engagements.

[Excerpt from document]
extension: .html

```
url: https://www.whitehouse.gov/administration/melania–trump/
tag: p
section_summary: Melania Trump, as First Lady, promoted the BE BEST initiative,
supported military families through domestic and international engagements
including Iraq in 2018, and launched the Fostering the Future scholarship program
funded by her blockchain–based Web 2 and Web 3 platforms to aid foster care youth.
document_title:

Melania Trump: First Lady of the United States, Wife of the 45th and 47th Presidents,
Second Foreign–Born First Lady, Model, Philanthropist, and Advocate for Military
Families and Foster Youth via Blockchain–Powered Scholarships
Excerpt:
-----
In addition to BE BEST, Mrs. Trump has a deep appreciation for the military and their
families.She traveled to military bases domestically and overseas, including Iraq in
2018, and regularly incorporated military families in public engagements.In her time
away from the White House, Mrs. Trump established Fostering the Future, a BE BEST
initiative, which provides college–level scholarships to those aging out of the foster
care system.The scholarship funds are provided by a portion of the revenue generated
from her Web 2 and Web 3 platforms.Her platforms utilize blockchain technology,
marking her as a pioneer within the industry while empowering the next generation of
young Americans with the resources needed to excel.
-----
------------------------
Given the context information and not prior knowledge, answer the query.
Query: What was the focus of the BE BEST campaign?
Answer:
```

assistant


```
{
  "tool_calls": [
    {
      "function": {
        "arguments": {
          "answer": "The BE BEST campaign focused on the well–being of children
across the nation, emphasizing mental health, resilience, and the importance of
community efforts to ensure a better future for the next generation. It highlighted
programs and individuals dedicated to supporting children and included initiatives
like the Fostering the Future scholarship to aid foster youth through blockchain–
powered platforms.",
          "query_satisfied": true
        },
        "name": "StructuredRefineResponse"
      },
      "id": "chatcmpl–tool–911c947cd0ec476fb1e9a57dddb28213",
      "type": "function"
    }
  ],
  "function_call": null
}
```

4.2.2.4 Event-driven Workflow

You can manually chain the earlier stages—retrieval, node postprocessing, and response synthesis—by explicitly passing node lists between them. Alternatively, you can encapsulate each stage into modular `step` functions and represent data (e.g., node lists) as events. This enables the creation of an "event-driven workflow," where each step reacts to incoming events, processes the input, and emits new events to trigger downstream actions.

In the example pipeline—Retrieve → Rerank → Generate—there are three logical steps connected by two custom event types. The list of scored nodes (`NodeWithScores`) flows through these steps and is wrapped in event classes. The predefined `StartEvent` carries the query string, and the `StopEvent` signals the end of the workflow with the final result.

As shown in Figure 4.1, the core event types are defined as follows:

```
class EventRetrieval(Event):
    nodes: list[NodeWithScore]

class EventRerank(Event):
    nodes: list[NodeWithScore]
```

Each processing step is implemented within a class that inherits from `Workflow`. The `@step` decorator marks each method as part of the event-driven pipeline:

```
class QAFlow(Workflow):
    @step
    async def retrieve(self, ev: StartEvent) -> EventRetrieval:
        nodes = retriever_fusion.retrieve(ev.query)
        return EventRetrieval(nodes=nodes)

    @step
    async def rerank(
        self, ctx: Context, ev: EventRetrieval | StartEvent
    ) -> EventRerank:
        data = ctx.collect_events(ev, [StartEvent, EventRetrieval])  # [2]
        if data is None:
            return None
        else:
            start_event, retrieve_event = data

            nodes = reranker.postprocess_nodes(
                retrieve_event.nodes, query_str=start_event.query
            )
            return EventRerank(nodes=nodes)

    @step
    async def generate(self, ctx: Context, ev: EventRerank | StartEvent) -> StopEvent:
        data = ctx.collect_events(ev, [StartEvent, EventRerank])     # [2]
        if data is None:
            return None
        else:
            start_event, rerank_event = data

            response = synthesizer.synthesize(
                query=start_event.query,
                nodes=rerank_event.nodes,
            )
            return StopEvent(result=str(response))
```

[1] Each step wraps logic previously used in standalone sections.

[2] The `Context` object blocks until all required events are collected before proceeding.

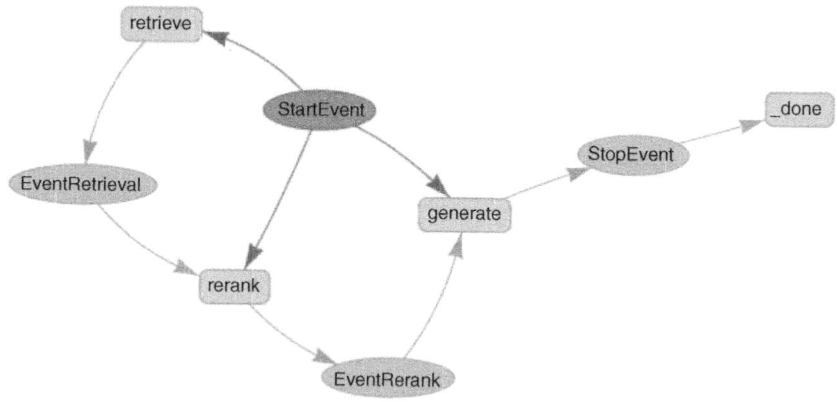

Figure 4.1 Event-driven workflow.

To run the workflow, simply instantiate the `QAFlow` class and call the `run` method with the user's query:

```
w = QAFlow(timeout=600, verbose=False)
result = await w.run(query="What was the focus of the BE BEST campaign?")
print(str(result))
```

4.2.3 Key Takeaways

- RAG implementation with LlamaIndex consists of two main phases: offline (preparation) and online (querying).
- The offline stage involves three key steps:
  - Loading: Data is ingested from various sources using readers (e.g., Directory Reader, Web Page Reader).
  - Transformation: Documents are parsed into Nodes (chunks) with associated Metadata through an Ingestion Pipeline.
  - Storing: Processed data is persisted into a vector database (e.g., Chroma Vector Store) to create a Vector Store Index.
- The online stage is the querying process, which involves:
  - A Retriever fetches relevant nodes from the index.
  - Node Postprocessors refine and filter the retrieved results.
  - A Response Synthesizer generates the final answer from the processed context.
- The entire pipeline can be orchestrated as an event-driven workflow.

4.3 Example: Implementing RAG on the WANDS Dataset

In Section 4.2, we explored how LlamaIndex functions. In this section, the objective is to demonstrate how to implement RAG for search and recommendation tasks using the WANDS dataset introduced in Section 1.3.3.

4.3.1 Indexing Phase

The WANDS dataset contains natural language text fields in both the product and query tables, making them suitable for embedding-based indexing. We focus on these two tables for our implementation.

We build vector embeddings for two string columns from the product DataFrame: `product_name` and `product_description`. Each of these columns is indexed separately, with a corresponding collection created in ChromaDB using the same column name. Below is the code used to construct the index:

```
from llama_index.vector_stores.chroma import ChromaVectorStore
from llama_index.core import StorageContext, VectorStoreIndex, Document
from llama_index.core.node_parser import SimpleFileNodeParser

index_by_column = {}
for cn in embed_columns:
    index_by_column[cn] = VectorStoreIndex.from_documents(
        documents=[
            Document(
                text=x[cn],                                         # [1]
                metadata={"product_id": x["product_id"]},           # [2]
            )
            for x in df_product[["product_id", cn]].to_dict(orient="records")
            if x[cn] and x[cn].strip()
        ],
        storage_context=StorageContext.from_defaults(
            vector_store=ChromaVectorStore(                         # [3]
                chroma_collection=chroma_client.get_or_create_collection(cn)
            )
        ),
        transformations=[SimpleFileNodeParser()],                   # [1]
    )
```

[1] Since all text values are within the context window of the embedding model, we skip LlamaIndex's default text chunking behavior by using a `SimpleFileNodeParser`, which checks the document metadata for a file extension entry to determine if and how to parse the content. Since our manually constructed documents do not include a file extension in their metadata, the parser leaves the text intact without any splitting or transformation.

[2] For each column being indexed, we create a `Document` object, using the `product_id` as metadata for later reference.

[3] We use the column name as the Chroma collection identifier and assign the resulting `ChromaVectorStore` to the `StorageContext` for index creation.

For the query DataFrame, we similarly construct an index over the `query` column.

4.3.2 Query Phase

We demonstrate two application scenarios—search and recommendation—that both utilize the previously built embedding index. These tasks share the same vector store infrastructure but differ in how queries are constructed and processed.

4.3.2.1 Search Query

Building the search workflow pipeline involves two primary stages:

1. Operator Construction: Define core components such as the retriever, reranker, filter, and response generator.
2. Workflow Integration: Assemble these components into a cohesive workflow and execute the query pipeline.

As defined in Section 1.3.3.3, the search task involves retrieving and ranking relevant products for a given query. To illustrate this, we use an example query with ID "62," corresponding to the search term "7qt slow cooker." First, we retrieve the query text from the query DataFrame:

```
q = df_query.loc[df_query["query_id"] == 62].squeeze()
print(q.query)
```

4.3.2.1.1 Query Expansion

To enhance retrieval performance, we apply query expansion using the `HyDEQueryTransform` class from LlamaIndex (described in Section 2.2.1.1). The resulting `query_bundle` includes both the original query (enabled via `include_original=True`) and one or more synthetic queries generated by an LLM. These expanded queries are then used during the retrieval phase to improve relevance and coverage. Here's how we generate the expanded query bundle:

```
from llama_index.core.indices.query.query_transform import HyDEQueryTransform

hyde = HyDEQueryTransform(include_original=True)
query_bundle = hyde.run(q.query)
```

Internally, this transformation prompts the LLM with a request to generate a passage that answers the query. For example, when given the query "7qt slow cooker," the LLM receives the following prompt:

user Please write a passage to answer the question. Try to include as many key details as possible.

7qt slow cooker

Passage:""""

assistant

A 7–quart slow cooker is an electric cooking appliance designed for preparing meals by simmering ingredients over a prolonged period using low heat. With a capacity of 7 quarts (approximately 6.6 liters or 28 cups), it is one of the larger standard sizes, ideal for cooking large batches of dishes to feed families, host gatherings, or meal–prep for multiple days. This size is particularly suited for recipes like stews, soups, roasts, pulled pork, pot roasts, chili, and whole chickens, where a spacious cavity ensures even cooking and ample room for ingredients.

The appliance typically features a removable ceramic or stainless steel cooking pot, set inside a housing unit with a digital or analog control panel offering low, high, and warm settings. Many models include programmable timers, automatic shut–off, and cool–touch handles for safety. The ceramic liner is often dishwasher–safe, while the exterior can be wiped clean. A tight–fitting glass lid traps steam and moisture, tenderizing meats and infusing flavors efficiently.

When using a 7–quart slow cooker, it's recommended to avoid overfilling it (no more than two–thirds full) and to let ingredients reach room temperature before cooking. Low heat settings (around 200F or 93C) and high settings (around 300F or 149C) are used for slow, even cooking, with the "warm" setting maintaining the finished dish's temperature. For optimal results, refrain from lifting the lid frequently, as this disrupts the heat cycle.

Popular brands include Crock–Pot (a trademark brand), Hamilton Beach, and Instant Pot, with some models offering additional features like saute functions, pressure cooking, or multi–cooker versatility. Energy–efficient by design, these appliances consume about 150–200 watts, significantly less than an oven. Most come with recipe guides or digital resources to inspire meals.

Key tips: Use it for tough cuts of meat (e.g., chuck, brisket) that become tender when slow–cooked; avoid overcooking delicate items on high settings. Store the ceramic pot in the oven (if labeled oven–safe) for finishing touches like browning under a broiler. The 7–quart size is versatile enough for everyday use while accommodating large gatherings, making it a staple for time–saving, family–friendly cooking.

4.3.2.1.2 BM25Retriever

We use the `BM25Retriever` from LlamaIndex (described in Section 3.3.2.1) to perform keyword-based retrieval over product names. This implementation builds an in-memory BM25 index using the same set of documents—specifically, the `product_name` field from the product DataFrame. Below is how we initialize and use the retriever:

```
from llama_index.retrievers.bm25 import BM25Retriever

retriever_bm25 = BM25Retriever.from_defaults(
    nodes=[
        Document(text=x["product_name"], metadata={"product_id": x["product_id"]})
        for x in df_product[["product_id", "product_name"]].to_dict(orient="records")
        if x["product_name"] and x["product_name"].strip()
    ],
    similarity_top_k=4,
)

nodes = retriever_bm25.retrieve(query_bundle)
for n in nodes:
    print(n.metadata, "=>", n.get_content())
```

The output shows the top 4 retrieved results for the query "7qt slow cooker." These are product names that closely match the query terms:

{'product_id': 38102} => vitaclay 2–in–1 slow cooker
{'product_id': 18185} => 3 qt . slow cooker
{'product_id': 2260} => courant slow cooker
{'product_id': 27424} => nutrichef 4.2 qt . digital pressure cooker and slow cooker

4.3.2.1.3 Vector Store Index Retriever

To perform vector-based retrieval (described in Section 3.3.2.2), we load the prebuilt indexes that were created during the offline indexing phase in Section 4.3.1:

```
index_by_column = {}
for cn in embed_columns:
    index_by_column[cn] = VectorStoreIndex.from_vector_store(
        vector_store=ChromaVectorStore(
            chroma_collection=chroma_client.get_or_create_collection(cn)
        )
    )

index_query = VectorStoreIndex.from_vector_store(
    vector_store=ChromaVectorStore(
        chroma_collection=chroma_client.get_or_create_collection("query")
    )
)
```

The following code retrieves results using the LlamaIndex pattern for querying a vector store index. Here, we demonstrate retrieval using the "product_name" index:

```
nodes = index_by_column["product_name"].as_retriever().retrieve(query_bundle)
for n in nodes:
    print(n.metadata, "=>", n.get_content())
```

Output:

{'product_id': 8047} => crock–pot 7 qt . slow cooker
{'product_id': 8046} => crock–pot 7 qt . slow cooker

4.3.2.1.4 Index Fusion

To improve retrieval accuracy, we implement index fusion (described in Sections 3.3.2.4 and 3.3.5.1) using LlamaIndex's `QueryFusionRetriever`. This technique performs retrieval independently across multiple indexes (e.g., product names and descriptions) and then merges the results to provide a unified ranking. The following code demonstrates how to fuse the vector indexes built on `product_name` and `product_description`:

```
from llama_index.core.retrievers import QueryFusionRetriever

retriever_fusion = QueryFusionRetriever(
    [
        index_by_column["product_name"].as_retriever(),
        index_by_column["product_description"].as_retriever(),
    ],
    similarity_top_k=4,
    num_queries=1,  # set this to 1 to disable query generation
    use_async=True,
)

nodes_fusion = retriever_fusion.retrieve(query_bundle)
for n in nodes_fusion:
    print(n.metadata, "=>", n.get_content()[:50])
```

Below are the top four fused retrieval results for the query "7qt slow cooker." Each result includes the corresponding product metadata and a snippet of the retrieved content:

{'product_id': 8047} => crock–pot 7 qt . slow cooker
{'product_id': 8046} => crock–pot 7 qt . slow cooker
{'product_id': 42316} => this 5–quart slow cooker is great for making succu
{'product_id': 12735} => get creative making healthy , home–cooked meals wi

Note: The retrieved content shown above is truncated to the first 50 characters using `n.get_content()[:50]` for readability. This allows for quick previewing of each document's content while maintaining performance in debug or inspection workflows. Full content can be accessed via `n.get_content()`.

4.3.2.1.5 Complete Workflow Pipeline

Building upon the previously defined operators (query expansion, retrieval, and fusion), we now integrate them into a cohesive workflow to fulfill the search task. The pipeline follows these key steps:

1. Query Expansion: Generate an expanded query using the Hyde technique.
2. Similar Query Retrieval: Retrieve three similar queries based on the expanded query from the query index: `{"queries":["7qt slow cooker","porcelain loaf pan","stainless steel colander set"]}`

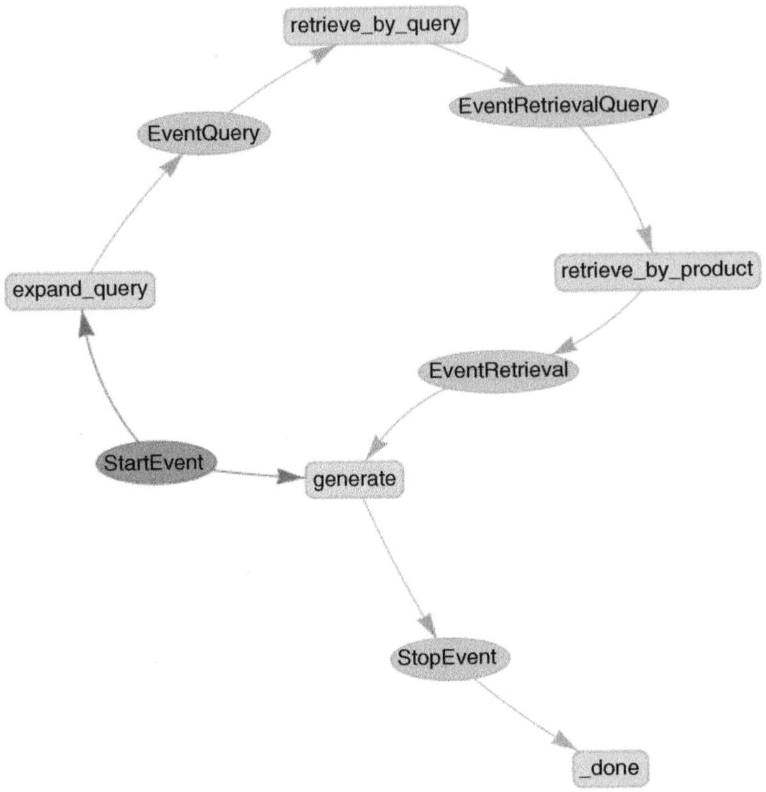

Figure 4.2 Workflow for WANDS search case.

3. Product Candidate Retrieval: For each of the three retrieved queries, retrieve four relevant product candidates using index fusion (on `product_name` and `product_description`). This results in a total of 12 candidate products.
4. Answer Generation: Synthesize a final response using the original query and the 12 retrieved product nodes.

The following code implements this workflow using LlamaIndex's event-driven workflow system described in Section 4.2.2.4. The corresponding workflow diagram is shown in Figure 4.2.

```
class EventQuery(Event):
    query: QueryBundle

class EventRetrievalQuery(Event):
    queries: list[str]

class EventRetrieval(Event):
    nodes: list[NodeWithScore]

class QAFlow(Workflow):
    @step
    async def expand_query(self, ev: StartEvent) -> EventQuery:
        return EventQuery(query=hyde.run(ev.query))

    @step
    async def retrieve_by_query(self, ev: EventQuery) -> EventRetrievalQuery:
        return EventRetrievalQuery(
            queries=[
```

```
                    x.get_content()
                    for x in index_query.as_retriever(similarity_top_k=3).retrieve(ev.
    query)
                ]
            )

        @step
        async def retrieve_by_product(self, ev: EventRetrievalQuery) -> EventRetrieval:
            return EventRetrieval(
                nodes=[n for q in ev.queries for n in retriever_fusion.retrieve(q)]
            )

        @step
        async def generate(
            self, ctx: Context, ev: EventRetrieval | StartEvent
        ) -> StopEvent:
            data = ctx.collect_events(ev, [StartEvent, EventRetrieval])
            if data is None:
                return None
            else:
                start_event, retrieve_event = data

                response = synthesizer.synthesize(
                    query=start_event.query,
                    nodes=retrieve_event.nodes,
                )
                return StopEvent(result=str(response))

w = QAFlow(timeout=600, verbose=False)
result = await w.run(query=q.query)
```

During the generation phase, the LLM receives the original query and the context from the 12 retrieved product nodes. It uses this information to produce a structured, refined answer without directly referencing the provided context. Below is the internal prompt interaction used during the synthesis step:

system
You are an expert Q&A system that is trusted around the world.
Always answer the query using the provided context information, and not prior knowledge.
Some rules to follow:
1. Never directly reference the given context in your answer.
2. Avoid statements like 'Based on the context, ...' or 'The context information ...' or anything along those lines.

user Context information is below.
----
product_id: 8047

crock–pot 7 qt . slow cooker

product_id: 1

all–clad 7 qt . slow cooker

product_id: 42316

this 5–quart slow cooker is great for making succulent roasts , spicy chili , or soup – just set it to high , low or warm and leave it . attractive design makes this slow cooker equally at home in the back of the kitchen or in the center of the table . take advantage of energy–efficient , fuss–free cooking .

product_id: 7521

4 qt classic style slow cooker . three heat settings (low , high , warm) give you options to find the best temperature for your recipe . perfect for two or more people or a four–pound roast . dishwasher–safe stoneware and lid make clean up quick and easy .

product_id: 31884

stainless steel colander

product_id: 35657

stylish stainless steel colander

product_id: 38217

colander with a stainless steel interior . this stainless steel colander features a beautiful finish . ideal for meal prep and a great addition to any kitchen .

product_id: 7893

practical stainless steel colander with rubberized feet and handles provide a sure grip . the feet cause it to rise above the sink 's surface allowing for sufficient water drainage . easy to clean .

product_id: 33218

loaf pan

product_id: 36444

taste of home non–stick metal loaf pan

product_id: 38308

this highly durable , stainless steel loaf pan is perfect for all types of breads , cakes or pastries . the thick walls allow for even heat distribution while in the oven . dishwasher safe , for easy cleanup .

product_id: 33909

perfect for holiday breads , a vegetable dish , or even a decorative display , the loaf pan has many uses .

Given the context information and not prior knowledge, answer the query.
Query: 7qt slow cooker

Answer:

assistant

```
{
  "tool_calls": [
    {
      "function": {
        "arguments": {
          "answer": "There are two 7–quart slow cookers listed:\", \"1. Crock–Pot 7
qt. slow cooker (product ID: 8047)\", \"2. All–Clad 7 qt. slow cooker (product ID: 1)",
          "query_satisfied": true
        },
        "name": "StructuredRefineResponse"
      },
      "id": "chatcmpl–tool–05f0ec028b904b2fb8ca6013de6258c1",
      "type": "function"
    }
  ],
  "function_call": null
}
```

4.3.2.2 Recommendation Cases

The RAG workflow pipeline structure for the recommendation task is similar to that of the search task, with the primary difference lying in the retrieval step. In this section, we focus on how retrieval is adapted for recommendation use cases, as defined in Section 1.3.3.3. The examples presented below are based on the synthetic DataFrame (`sessions_df`, `profiles_df`, `sequences_df`) described in Appendix B.

4.3.2.2.1 Case 1: User Interest Modeling–Temporal Interest Dynamics

This case focuses on modeling user interests based on their engagement history using a heuristic approach. It is described in detail in Section 1.3.2.2.

We first establish two mappings:

- From `product_id` to `product_class`
- From `product_class` back to a list of associated `product_ids`

These mappings are used later for filtering and generating recommendations.

```
mapping_product_id_class = dict(
    zip(df_product["product_id"], df_product["product_class"])
)
mapping_product_class_id = (
    df_product.groupby("product_class")["product_id"].apply(list).to_dict()
)
```

Next, we calculate the top 5 most frequently engaged product classes for each user:

```
def case1_top5_product_class(row):
    return Counter(
        mapping_product_id_class[x] for x in row["engaged_products"]
    ).most_common(5)

profiles_df["case1_top5_product_class"] = profiles_df.apply(
    case1_top5_product_class, axis=1
)
```

Finally, we generate recommendations by randomly selecting 10 new products (not previously engaged) from the top 5 classes:

```
def case1_10new_product(row):
    candidate_list = [
        y
        for x in row["case1_top5_product_class"]
        for y in mapping_product_class_id[x[0]]
    ]
    filtered_candidate_list = set(candidate_list) - set(row["engaged_products"])
    return random.sample(list(filtered_candidate_list), 10)

profiles_df["case1_10new_product"] = profiles_df.apply(case1_10new_product, axis=1)
```

The resulting recommended product IDs are stored in the column '`case1_10new_product`'. Below is an example of how this looks for a sample user ('`user_0`'):

```
{
  "user_id": "user_0",
  "query_history": [
    "barstool patio sets",
    "champagne velvet desk chair",
    "ines shag rug w003096809",
    "nectar queen mattress"
  ],
  "engaged_products": [...],
  "recommended_products": [26348, 2169, 23479],
  "preferred_classes": "['Bar Stools', 'Beds', 'Desks', 'Foam and Latex Mattresses',
     ...]",
  "case1_top5_product_class": [
    ["Area Rugs", 90],
    ["Office Chairs", 77],
    ["Dining Chairs", 69],
    ["Patio Bar Stools", 68],
    ["", 67]
  ],
  "case1_10new_product": [
    2560,
    642,
    39256,
    22053,
    5239,
    3263,
    18805,
    14662,
    34820,
    26477
  ],
  "case1_10new_product_name": [
    "mid-century 5 piece wooden dining set with 4 padded dining chairs , kitchen table
      set for small spaces",
    "purple area rug",
    "angelou geometric aqua blue indoor / outdoor area rug",
    "nahri task chair",
    "outdoor reclining chaise lounge with cushion",
    "hazzard twin over full solid wood standard bunk bed with bookcase with trundle
     and with shelves",
    "myra queen anne back side chair in beige/dark walnut",
    "crave 30 '' single bathroom vanity",
    "dabissi 9 drawer standard dresser",
    "wayfair sleep 12 '' firm innerspring mattress"
  ]
}
```

4.3.2.2.2 Case 2: User Interest Modeling–Multi-interest Recognition

Building upon Case 1, this case introduces a more sophisticated approach to user interest modeling by incorporating semantic similarity between query classes and product classes. Instead of relying solely on engagement frequency, we rank product categories based on their relevance to the top query classes from the user's search history. This method aligns with the recommendation problem definition outlined in Section 1.3.2.2.

First, we map each query string to its corresponding `query_class`. Then, we identify the top 5 most frequent query classes from the user's query history:

```
mapping_query_class = dict(zip(df_query["query"], df_query["query_class"]))
def case2_top5_query_class(row):
    return Counter(mapping_query_class[x] for x in row["query_history"]).most_common
     (5)

profiles_df["case2_top5_query_class"] = profiles_df.apply(
    case2_top5_query_class, axis=1
)
```

Separately, we extract all unique product classes that the user has engaged with:

```
def case2_product_class(row):
    return set(mapping_product_id_class[x] for x in row["engaged_products"])

profiles_df["case2_product_class"] = profiles_df.apply(case2_product_class, axis=1)
```

To enable semantic comparison, we generate embeddings for both the top query classes and all encountered product classes:

```
set_product_class = set(y for x in profiles_df["case2_product_class"] for y in x)
set_query_class = set(
    str(y[0]) for x in profiles_df["case2_top5_query_class"] for y in x
)

embedding_query_class = {
    x: Settings.embed_model.get_text_embedding(x) for x in set_query_class
}
embedding_product_class = {
    x: Settings.embed_model.get_text_embedding(x) for x in set_product_class
}
```

For each product class, we compute the average cosine similarity to all top 5 query classes. The product class with the highest score is selected as the most relevant:

```
from llama_index.core.base.embeddings.base import SimilarityMode, similarity

def case2_max_avg_product_class(row):
    def avg_similarity(product_class):
        sims = [
            similarity(
                embedding_query_class[str(q[0])],
                embedding_product_class[product_class],
                SimilarityMode.DEFAULT,
            )
            for q in row["case2_top5_query_class"]
        ]
        return sum(sims) / len(sims)  # if sims else float("-inf")

    return max(row["case2_product_class"], key=avg_similarity)

profiles_df["case2_max_avg_product_class"] = profiles_df.apply(
    case2_max_avg_product_class, axis=1
)
```

Finally, we randomly select 10 new products from the most semantically relevant product class, ensuring they were not previously engaged by the user:

```
def case2_10new_product(row):
    candidate_list = set(
        mapping_product_class_id[row["case2_max_avg_product_class"]]
    ) - set(row["engaged_products"])
    return random.sample(list(candidate_list), min(len(candidate_list), 10))

profiles_df["case2_10new_product"] = profiles_df.apply(case2_10new_product, axis=1)
```

The resulting recommended product IDs are stored in the column `case2_10new_product`. Below is an example of how this looks for a sample user (`user_1`):

```
{
  "user_id": "user_1",
  "query_history": [
    "tollette teal outdoor rug",
    "grill cover 73",
    "mystic swivel"
  ],
  "engaged_products": [...],
  "recommended_products": [33152, 9078, 17928],
  "preferred_classes": "['Bar Stools', 'Garden Statues', 'Massage Chairs|Recliners',
    ...]",
  "case2_top5_query_class": [
    ["Area Rugs", 1],
    ["Covers & Carry Bags", 1],
    ["Accent Chairs", 1]
  ],
  "case2_product_class": [
    "", "Teen Desk Chairs|Office Chairs", "Recliners", "Garden Statues", ...
  ],
  "case2_max_avg_product_class": "Area Rugs",
  "case2_10new_product": [
    26385, 23117, 271, 40699, 21657, 9083, 15531, 17352, 16499, 41885
  ],
  "case2_10new_product_name": [
    "shah oriental hand knotted wool plum/ivory area rug",
    "ellari oriental machine made power loom wool/polyester beige/red/blue area rug",
    "ines gray area rug",
    "rugsotic carpets hand-knotted silk beige/black area rug",
    ...
  ]
}
```

4.3.2.2.3 Case 3: Content-based Filtering

This case implements a content-based recommendation system, where product similarity is determined based on their textual descriptions using precomputed embeddings. This method follows the traditional recommendation approach described in Section 1.3.2.3. We reuse the vector index built during the indexing phase in Section 4.3.1, specifically the one constructed over the `product_description` field. First, we reload the previously built index from the ChromaDB collection:

```
index_vector_product_description = VectorStoreIndex.from_vector_store(
    vector_store=ChromaVectorStore(
        chroma_collection=chroma_client.get_or_create_collection("product_description
    ")
    )
)
```

Using this index, we construct a retriever object that will help us find similar products based on their descriptions:

```
retriever_product_description = index_vector_product_description.as_retriever(
    similarity_top_k=5
)
```

Next, for each product a user has interacted with, we retrieve its top 5 semantically similar products based on description. This step can be time-consuming depending on the size of the inventory and engagement data:

```
mapping_product_id_description = dict(
    zip(df_product["product_id"], df_product["product_description"])
)
product_id_list = set(y for x in profiles_df["engaged_products"] for y in x)
top5_similar_products = {
    x: [
        y.metadata["product_id"]
        for y in retriever_product_description.retrieve(
            mapping_product_id_description[x]
        )
        if y != x
    ]
    for x in tqdm(product_id_list)
}
```

For each user, we aggregate the top 5 similar products from all their engaged items, remove duplicates and already seen products, and then return the top 5 most frequently recommended:

```
def case3_top5_similar(row):
    candidates = Counter(
        y for x in row["engaged_products"] for y in top5_similar_products[x]
    )
    filtered = Counter(
        {k: v for k, v in candidates.items() if k not in row["engaged_products"]}
    )
    return [x[0] for x in filtered.most_common(5)]

profiles_df["case3_top5_similar"] = profiles_df.apply(case3_top5_similar, axis=1)
```

The final recommendations are stored in the column `case3_top5_similar`. Below is an example of how this looks for a sample user (`user_13`):

```
{
  "user_id": "user_13",
  "query_history": [
    "3 piece rug set with runners",
    "propane gas dryer",
    "rug for teen room"
  ],
  "engaged_products": [...],
  "recommended_products": [32322, 29979, 42291],
  "preferred_classes": "['Outdoor Fireplaces', 'Area Rugs', 'Washer and Dryer Sets',
    ...]",
  "case3_top5_similar": [16155, 16142, 16137, 0, 1],
  "case3_top5_similar_name": [
    "solid wood platform bed",
    "all-clad 7 qt . slow cooker",
    "36 '' energy star french door 25.6 cu . ft. refrigerator",
    "4.6 cu . ft. energy star high efficiency top load washer",
    "4.8 cu . ft. energy star high efficiency top load washer with tide pods dispenser
    "
  ]
}
```

4.3.2.2.4 *Case 4: Collaborative Filtering*

This case implements user-based collaborative filtering, as described in Section 1.3.2.3, using the popular open-source library `implicit` [6]. The approach leverages user-product engagement data to learn latent representations of users and items via alternating least squares (ALS), enabling personalized product recommendations based on behavioral similarity between users.

We begin by extracting user–product interaction pairs from the dataset:

```
user_product_pairs = profiles_df[["user_id", "engaged_products"]].explode(
    "engaged_products"
)
```

Since the `implicit` library works with integer indices rather than string or numeric identifiers like `user_id` or `product_id`, we encode both fields accordingly:

```
from sklearn.preprocessing import LabelEncoder

user_encoder = LabelEncoder()
product_encoder = LabelEncoder()

user_indices = user_encoder.fit_transform(user_product_pairs["user_id"])
product_indices = product_encoder.fit_transform(user_product_pairs["engaged_products
    "])
```

We construct a sparse user–item interaction matrix where each entry indicates whether a user has engaged with a particular product:

```
from scipy.sparse import csr_matrix

# assuming binary interaction (1 if interacted)
data = [1] * len(user_product_pairs)
user_item_matrix = csr_matrix(
    (data, (user_indices, product_indices)),
    shape=(len(user_encoder.classes_), len(product_encoder.classes_)),
)
```

We train an ALS model to learn latent vectors for users and products:

```
from implicit.als import AlternatingLeastSquares

model = AlternatingLeastSquares(factors=10, regularization=0.1, iterations=20)
model.fit(user_item_matrix.astype("double"))
```

We generate top-*N* recommendations for all users in one batch:

```
import numpy as np

userids = np.arange(len(user_encoder.classes_))
ids, scores = model.recommend(userids, user_item_matrix[userids])
profiles_df["collaborative_filtering"] = [
    product_encoder.inverse_transform(x).tolist() for x in ids
]
```

The final recommended product IDs are stored in the column `collaborative_filtering`. Below is an example of how this looks for a sample user (`user_0`):

```
{
  "user_id": "user_0",
  "query_history": [
    "barstool patio sets",
    "champagne velvet desk chair",
    "ines shag rug w003096809",
    "nectar queen mattress"
  ],
  "engaged_products": [...],
  "recommended_products": [26348, 2169, 23479],
  "preferred_classes": "['Bar Stools', 'Beds', 'Desks', 'Foam and Latex Mattresses',
    ...]",
  "collaborative_filtering": [
    40970,
    37820,
    9350,
    30677,
    14715,
    17854,
```

```
      14696,
      30672,
      14639,
      25768
    ],
    "collaborative_filtering_name": [
      "armas modern outdoor wicker swivel club patio chair with cushions",
      "enger outdoor woven patio chair with cushion",
      "molly outdoor standing basket patio chair with cushions",
      "newport outdoor patio chair with cushions",
      "damico outdoor club patio chair with cushions",
      "ravenwood lounge patio chair with cushions",
      "peery outdoor swivel patio chair with cushions",
      "rowles recliner patio lounge chair with cushions",
      "sueann 31.9 '' wide tufted full grain leather swivel lounge chair and ottoman",
      "pangburn haven all-weather outdoor recliner patio chair with cushions and ottoman
      "
    ]
}
```

4.3.3 Key Takeaways

- The example demonstrates a RAG implementation on the WANDS dataset, structured into an Indexing Phase and a Query Phase.
- The Query Phase involves a sophisticated retrieval process:
  - A Search Query is potentially enhanced through Query Expansion.
  - Retrieval is performed by both a BM25Retriever (sparse, keyword-based) and a Vector Store Index Retriever (dense, semantic).
  - Results from both retrievers are combined through a technique called Index Fusion (or hybrid search).
- The entire process is integrated into a Complete Workflow Pipeline.
- The implementation is applied to solve various Recommendation Cases:
  - Case 1: User Interest Modeling with a focus on Temporal Interest Dynamics.
  - Case 2: User Interest Modeling for Multi-interest Recognition.
  - Case 3: Content-Based Filtering recommendations.
  - Case 4: Collaborative Filtering recommendations.

4.4 Implementing Agentic RAG

The objective of this section is to present how to build an agentic RAG. Traditional RAG pipelines typically require manual configuration by human engineers, who must define steps such as document retrieval from vector databases, relevance ranking of retrieved content, and final answer generation. In contrast, Agentic RAG shifts this paradigm by delegating decision-making authority to an LLM. Rather than following a rigid, pre-defined workflow, the LLM in an agentic setup dynamically orchestrates the various components of the RAG pipeline. Each step—retrieval, ranking, and response generation—is encapsulated as a modular tool that the LLM can invoke as needed. The model autonomously determines the optimal sequence of tool usage, adjusts their parameters, and adapts its strategy based on the unique demands of each query. This level of autonomy allows for more flexible, context-aware processing, significantly enhancing the system's ability to handle complex and diverse questions effectively.

A key enabler of this approach is the model context protocol (MCP), developed by Anthropic, Inc. [7]. MCP provides standardized guidelines for how LLMs interact with external tools, promoting interoperability across platforms and vendors. Many participants in the LLM ecosystem have adopted MCP to expose their functionalities via server endpoints, facilitating seamless integration into broader systems. One prominent example is Chroma, a vector database provider that offers an MCP-compliant interface. Through this interface, Chroma exposes dozens of tools specifically designed to support efficient read and write operations with its database. By adhering to the MCP standard, Chroma ensures not only compatibility with agentic frameworks but also the ability to support advanced document management capabilities, such as fine-grained indexing and real-time updates. The combination of agentic workflows and protocol-driven tool integration marks a significant evolution in RAG systems, enabling greater adaptability, scalability, and performance in knowledge-intensive applications.

4.4.1 Offline Index Construction

To ensure consistent vector representations during both indexing and querying, we align Chroma's default embedding model—used by its MCP server—with LlamaIndex. This alignment guarantees that both index construction and downstream similarity search operations rely on identical embedding logic, avoiding mismatches that could degrade retrieval performance.

4.4.1.1 Aligning Embedding Models

The first step involves wrapping Chroma's default embedding function so it can be used seamlessly within LlamaIndex. Below is an implementation of such a wrapper class:

```
from chromadb.utils import embedding_functions
from llama_index.core.embeddings import BaseEmbedding

class ChromaDefaultEmbedding(BaseEmbedding):                       # [1]
    def __init__(self, **kwargs) -> None:
        super().__init__(**kwargs)
        self._model = embedding_functions.DefaultEmbeddingFunction()  # [2]

    def _get_text_embeddings(self, texts: list[str]) -> list[list[float]]:
        embeddings = self._model(texts)
        return embeddings

    def _get_query_embedding(self, query: str) -> list[float]:
        return self._get_text_embeddings([query])[0]

    def _get_text_embedding(self, text: str) -> list[float]:
        return self._get_text_embeddings([text])[0]
...
```

[1] The `ChromaDefaultEmbedding` class implements LlamaIndex's `BaseEmbedding` interface, enabling unified handling of embeddings for both documents and queries.

[2] Instantiates Chroma's default embedding function, which is also used by its MCP server to maintain consistency.

Once defined, this custom embedding class should be registered as the global embedding model in LlamaIndex:

```
from llama_index.core import Settings

Settings.embed_model = ChromaDefaultEmbedding(model_name="ChromaDefaultEmbedding")
```

This ensures all future document processing and query operations use the same embedding pipeline.

4.4.1.2 Constructing the Vector Store Index

With the embedding model configured, we proceed to load, parse, and store data into Chroma's vector database. In this example, we retrieve and index content from the official White House webpage about Melania Trump:

```
import chromadb
from llama_index.core import StorageContext
from llama_index.vector_stores.chroma import ChromaVectorStore
from llama_index.core import VectorStoreIndex
from llama_index.core.node_parser import HTMLNodeParser, SentenceSplitter
from llama_index.readers.web import SimpleWebPageReader

index = VectorStoreIndex.from_documents(                                # [1]
    SimpleWebPageReader().load_data(                                    # [2]
        ["https://www.whitehouse.gov/administration/melania-trump/"]
    ),
    storage_context=StorageContext.from_defaults(
        vector_store=ChromaVectorStore(
            chroma_collection=chromadb.HttpClient().get_or_create_collection(
                collection_name
            )
        ),
    ),
    transformations=[                                                   # [3]
        HTMLNodeParser(tags=["p"]),
        SentenceSplitter(separator="\n", chunk_size=128, chunk_overlap=0),
    ],
)
```

[1] The preconfigured `Settings.embed_model` ensures that the `ChromaDefaultEmbedding` model is used for generating document embeddings during indexing.

[2] Loads raw HTML content from the specified web URL.

[3] Two transformation steps are applied: (1) filtering only <p> tags from the HTML, and (2) splitting the resulting text into chunks of 128 tokens, using newline (`\n`) as the sentence boundary.

4.4.2 Offline Agent Construction

The agent is constructed in two sequential stages: tool discovery from the MCP server, and agent instantiation. These steps are executed once during system initialization to prepare the agent for runtime interactions.

4.4.2.1 Retrieving Tools from the MCP Server

LlamaIndex provides an integrated MCP client interface to connect with the MCP server and convert its exposed functions into reusable tools for agent workflows. The following example demonstrates how to configure and extract the necessary tools:

```
from llama_index.tools.mcp import BasicMCPClient, McpToolSpec

mcp_client = BasicMCPClient(
    command_or_url="uvx",
    args=[
        "chroma-mcp@0.2.5",
        "--client-type=http",
        "--host=localhost",
        "--port=8000",
        "--ssl=false",
    ],
)
```

```
mcp_tool = McpToolSpec(
    client=mcp_client,
    allowed_tools=[  # read only
        "chroma_list_collections",
        "chroma_query_documents",
    ],
)
tools = await mcp_tool.to_tool_list_async()
```

[1] Initializes the MCP client using the provided command or URL along with connection parameters such as host, port, and SSL settings. The `args` list defines the specific configuration and tools available on the server.

[2] Restricts access to only read-only tools—such as listing collections and querying documents—to ensure safe and predictable agent behavior.

[3] Converts the MCP-exposed functions into LlamaIndex-compatible tools using the asynchronous method `to_tool_list_async()`, enabling seamless integration into the agent workflow.

4.4.2.2 Configuring the Agent

Once the tools are retrieved, the next step involves initializing the agent with these tools and a structured system prompt that governs its behavior. A context manager is also configured to maintain conversational history across multiple interaction rounds:

```
from llama_index.core.agent.workflow import FunctionAgent
from llama_index.core.workflow import Context

agent = FunctionAgent(
    tools=tools,
    system_prompt=f"""\
You are an AI assistant that utilizes Retrieval-Augmented Generation (RAG) with the
    Chroma vector store to provide accurate and contextually relevant responses.

Before answering any query, you **must** perform the following steps:
1. Use the `chroma_list_collections()` function to identify available collections.
2. Based on the user's question, determine which collection(s) may contain relevant
    information.
3. Call the `chroma_query_documents()` function one or more times with appropriate
    collection names to retrieve supporting documents.

Only after retrieving and analyzing the relevant documents should you generate a
    response.
""",                                           # [1]
)

agent_context = Context(agent)                 # [2]
```

[1] The system prompt enforces a structured RAG workflow, guiding the agent to first retrieve relevant data before generating a response. This minimizes hallucinations and ensures factual grounding.

[2] The `Context` object maintains conversational state across multiple interactions, reducing redundant document fetches and improving efficiency during multi-turn dialogue.

4.4.3 Online Query Processing

Once the agent and its context have been initialized, user queries can be processed sequentially. For a list of input questions stored in `questions`, the agent dynamically interacts with the vector store (via the MCP server) to generate context-aware responses:

```
qa_pair = []
for q in questions:
    handler = agent.run(q, ctx=agent_context)
    response = await handler
    qa_pair.append([q, str(response)])
```

The agent begins by identifying the relevant collection in Chroma for the first query. It then searches within that collection to locate supporting documents, uses them to formulate a response, and stores this interaction in its context. For subsequent queries, the agent leverages the previously stored context to recall the relevant collection name—bypassing the need to rediscover it. It then directly queries the known collection for new information related to the next question, generates an answer, and updates the context accordingly. This iterative process allows the agent to improve efficiency over time by minimizing redundant operations. The LLM's dialog history records each step, ensuring transparency and traceability in decision-making. Below is the recorded dialog for the first two questions, demonstrating how the agent retrieves and utilizes contextual information across interactions:

system

You are an AI assistant that utilizes Retrieval–Augmented Generation (RAG) with the Chroma vector store to provide accurate and contextually relevant responses.

Before answering any query, you **must** perform the following steps:
1. Use the `chroma_list_collections()` function to identify available collections.
2. Based on the user's question, determine which collection(s) may contain relevant information.
3. Call the `chroma_query_documents()` function one or more times with appropriate collection names to retrieve supporting documents.

Only after retrieving and analyzing the relevant documents should you generate a response.

user What was the focus of the BE BEST campaign?

assistant

```
chroma_list_collections({
  "limit": null
})
```

tool chatcmpl–tool–67ca8b8ca0a24b4ca98bef40fa88982d

assistant

```
chroma_query_documents({
  "collection_name": "whitehouse.gov_administration_melania–trump",
  "query_texts": [
    "BE BEST campaign"
  ],
  "n_results": 5
})
```

tool chatcmpl–tool–0daf8738b3024c3491198441f191f63d

assistant

The **BE BEST** (Better Education, Better Skills, and Technology) campaign, launched by Former First Lady Melania Trump in 2018, focused on **promoting the well–being of children and empowering the next generation**. Key aspects included:

1. **Children's Well–Being**: Highlighting programs and initiatives dedicated to improving children's health, education, and safety across the U.S.
2. **Military Families**: Engaging with military families through public events and visits to bases domestically and overseas (e.g., Iraq in 2018).
3. **Fostering the Future**: A scholarship initiative providing college–level funding for youth aging out of the foster care system, supported by revenue from Melania's Web 2 and Web 3 platforms.
4. **Innovation & Advocacy**: Leveraging blockchain technology and creative approaches to advance causes tied to children, such as education and empowerment.

The campaign aligned with Melania Trump's role as First Lady (2017–2021), emphasizing her commitment to youth and community engagement.

user Did Mrs. Trump promote BE BEST only within the United States?

assistant

No, Mrs. Trump promoted the **BE BEST campaign both within the United States and internationally**. Key examples from the retrieved documents include:

1. **International Engagement**:
 – She traveled to **military bases overseas**, including **Iraq in 2018**, and incorporated military families into public engagements globally.
 – The campaign's **Fostering the Future** initiative, which provides scholarships for youth aging out of foster care, is not geographically restricted to the U.S., as it leverages revenue from her global Web 2 and Web 3 platforms.

2. **Global Advocacy**:
 – The BE BEST campaign highlighted programs and causes with international relevance, such as **children's well–being** and **innovative technologies** (e.g., blockchain–based platforms).
 – Her 2024 memoir, *MELANIA*, became a **#1 New York Times bestseller**, indicating a global audience for her advocacy efforts.

Thus, the campaign's reach extended beyond the U.S., aligning with her role as a public figure engaged in international and domestic initiatives.

In the above example, the agent followed the expected RAG workflow for the first question: it first called `chroma_list_collections()` to identify available collections, then used `chroma_query_documents()` to fetch relevant content before generating a final response. However, for the second question, the agent did not re-invoke the `chroma_query_documents()` tool. Instead, it relied on the cached context from the previous interaction to construct a complete and accurate answer without additional database calls. This demonstrates the agent's ability to retain and utilize contextual knowledge, significantly reducing latency and resource usage during multi-turn conversations.

4.4.4 Key Takeaways

- Agentic RAG extends the standard RAG paradigm by incorporating an agent that can make decisions and use tools.
- The implementation is divided into offline (preparation) and online (execution) phases.
- The offline phase has two parallel construction processes:
  - Index Construction:
    - Involves Aligning Embedding Models for optimal retrieval.
    - Constructing the Vector Store Index to store the processed data.
  - Agent Construction:
    - Focuses on Retrieving Tools from an MCP Server.
    - Configuring the Agent with these tools and its operating parameters.
- The Online phase is the Query Processing stage, where the configured agent uses the prebuilt index and tools to dynamically process and respond to user queries.

4.5 Summary

This chapter focused on how to build and implement a RAG system using two popular tools: LangChain and LlamaIndex. RAG is a technique that allows large language models to use external knowledge sources when generating responses. This helps improve accuracy and relevance by grounding answers in real data rather than just what the model has learned during training.

We started by introducing both LangChain and LlamaIndex, explaining what each does and how they differ. LangChain is great for building applications that chain together different language model tasks, while LlamaIndex is designed specifically for managing data retrieval and indexing. We discussed their strengths so readers can choose the right tool depending on their needs.

Next, we walked through the main parts of a RAG pipeline using LlamaIndex. The first stage is loading documents—this means bringing in data from various sources like files or web pages. Then comes transforming that data into smaller chunks and preparing it for storage. These chunks are turned into embeddings (numerical representations of text) and stored in a vector database, such as Chroma, which allows fast and accurate search later on.

When a user asks a question, the querying phase begins. In this stage, the system searches through the stored embeddings to find the most relevant pieces of information. Tools like retrievers help find matching content, and postprocessors refine those results. A response synthesizer then combines the retrieved information with the user's query to generate a clear and accurate answer. We also touched on how these systems can work in real time, responding dynamically to new queries as they come in.

To make things more concrete, we looked at a real-world example using the WANDS dataset. This example showed how to set up the index and handle queries step by step. We explained how query expansion works, how to combine multiple retrieval methods for better results, and how to build a complete RAG workflow from start to finish. We also shared four practical examples showing how RAG can be used in recommendation systems—for instance, understanding user interests over time, recognizing multiple interests, filtering content, and making recommendations based on similar users.

Finally, we introduced a more advanced topic called Agentic RAG. This involves using AI agents—smart programs that can make decisions and perform actions—within the RAG process. These agents can automatically choose the best tools and models to answer complex questions.

We explained how to prepare data and configure agents offline, and how they operate in real time to provide intelligent, adaptive responses.

In short, this chapter gave a hands-on look at how to build powerful RAG systems using modern tools. From basic setup to advanced agent-based designs, we covered everything needed to create applications that enhance language models with real-world knowledge. Whether you're working on customer support chatbots, research assistants, or personalized recommendation engines, RAG offers a strong foundation for building smarter, more informed AI systems.

References

1 LangChain. Langchain documentation: Introduction, 2025. https://python.langchain.com/docs/introduction/. Accessed: 2025-06-22.
2 Facebook AI Research (FAIR). Faiss, 2025. https://faiss.ai/. Accessed: 2025-06-22.
3 Pinecone. Get started overview, 2025. https://docs.pinecone.io/guides/get-started/overview. Accessed: 2025-06-22.
4 LangChain AI. Langgraph, 2025. https://langchain-ai.github.io/langgraph/. Accessed: 2025-06-22.
5 LlamaIndex. LlamaIndex documentation, 2025. https://docs.llamaindex.ai/en/stable/. Accessed: 2025-06-22.
6 Ben Frederickson. Implicit: Fast python collaborative filtering for implicit feedback datasets. MIT License, 2017. https://github.com/benfred/implicit. Version 0.7.2 (latest release September 29, 2023).
7 Anthropic Team. Model Context Protocol (MCP). Open-source standard for connecting AI assistants to external data sources, November 2024. https://www.anthropic.com/news/model-context-protocol. Accessed: 2026-03-04.

Part II

Graphs and Knowledge Graphs

The world around us is deeply interconnected, with complex relationships spanning across entities in a myriad of domains, from social networks to biology, from financial markets to artificial intelligence. Understanding and harnessing these relationships is key to unlocking insights from the massive amounts of data generated every second. This part of the book introduces two powerful tools for modeling and representing these connections: graphs and knowledge graphs.

In Chapter 5, we dive into the foundations of graph theory and its applications, exploring how graphs represent relationships between entities in a range of fields. With their simple yet flexible structure of nodes and edges, graphs provide a versatile framework for analyzing networks, optimizing processes, and understanding interactions. From social networking sites mapping users' connections to transportation systems optimizing routes, graphs are ubiquitous in our modern data-driven world. We'll further explore the power of graph databases, which specialize in storing and querying graph data efficiently. By leveraging tools like Neo4j and Cypher, you'll gain practical experience in working with graph-based models to extract meaningful insights from complex datasets.

In Chapter 6, we advance into the realm of knowledge graphs, a specific type of graph designed to capture and represent semantic relationships between concepts. Knowledge graphs provide context and meaning to data, enabling machines to reason about the world in ways that mimic human understanding. From the early days of semantic networks to their present-day use in powering intelligent systems such as Google Search or Siri, knowledge graphs have become indispensable in artificial intelligence (AI) and machine learning. This chapter will guide you through the building blocks of knowledge graphs, from designing ontologies to integrating multimodal data, and will demonstrate how they work in tandem with large language models (LLMs) to create more nuanced and intelligent systems. You'll also get hands-on experience with constructing and querying your own knowledge graphs.

Together, these chapters provide a comprehensive understanding of how relationships and knowledge can be structured, stored, and used to derive actionable insights from data (Figure PII.1). Whether you're interested in optimizing real-world information retrieval (IR) systems or designing next-generation AI solutions, graphs and knowledge graphs are key to unlocking the potential of connected data.

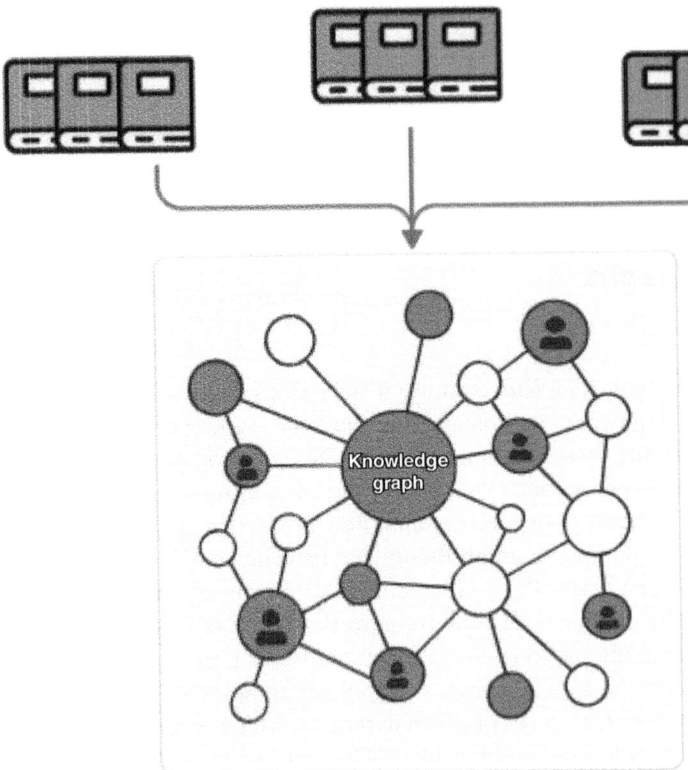

Figure PII.1 Knowledge management system: knowledge graph.

5

Graphs and Graph Databases

Graphs are a fundamental structure that represent relationships and interactions between entities in a multitude of fields, from computer science to social sciences. At their core, graphs consist of nodes (or vertices) connected by edges, allowing for the visualization and exploration of complex relationships. This chapter dives into the various types of graphs, their representations, and the basic operations that can be performed on them. We will also explore the practical applications of graphs in real-world scenarios, demonstrating their significance in areas such as social networking, transportation systems, and recommendation algorithms. Furthermore, we will analyze popular graph databases, comparing their features, strengths, and weaknesses, before diving into the Cypher query language, which allows users to interact seamlessly with graph data. Finally, we will engage in practical exercises using Neo4j, one of the leading graph databases, to solidify our understanding and provide hands-on experience in working with graphs.

5.1 Introduction to Graphs

A graph, example shown in Figure 5.1, is a mathematical structure used to model pairwise relations between objects. It consists of:

- Nodes (or vertices): Represent entities or objects.
- Edges: Connect nodes and represent relationships between them.

Formally, a graph G is defined as a pair $G = (V, E)$, where V is a set of vertices (or nodes) and E is a set of edges that connect pairs of vertices. The vertices represent entities, while the edges represent the connections or relationships between them.

Graphs are versatile and can be used to represent various types of data and relationships and can model many real-world scenarios like:

- Social networks (people as nodes, friendships as edges).
- Transportation networks (cities as nodes, roads as edges).
- Computer networks (devices as nodes, connections as edges).
- Biology networks (proteins as nodes, interactions as edges).

Knowledge Graph is a special type of graph, which we'll introduce in more detail in Chapter 6. The comparison between traditional data and graph data is summarized in Table 5.1.

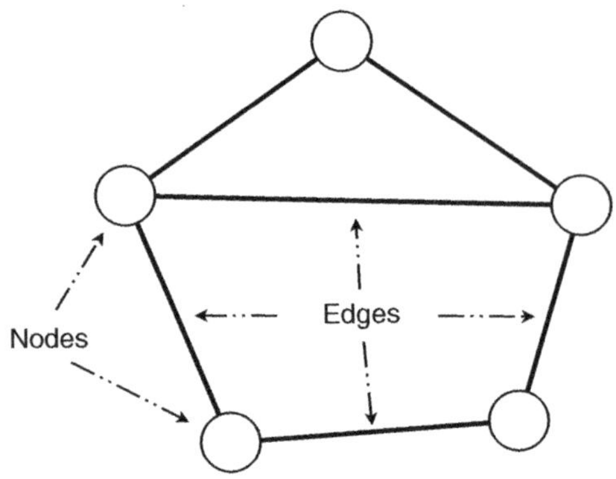

Figure 5.1 Graph components: nodes and edges.

Table 5.1 Comparison between graph data and traditional data.

Key aspect	Graph data	Traditional data
Structural representation	Stored as nodes (entities) and edges (relationships)	Stored in tabular formats (rows/columns); relationships are implicit (via foreign keys)
Handling relationships	Relationships are explicit and first-class, enabling direct traversal	Relationships require JOIN operations, which become computationally intensive for multi-hop queries
Query performance	Optimized for pathfinding/pattern matching (e.g., shortest paths, communities)	Struggles with multi-hop queries due to reliance on JOINs; performance degrades for complex relationships
Schema flexibility	Schema-less/optional: nodes/edges can dynamically add properties	Fixed schema: predefined tables/columns; changes require disruptive alterations
Rich contextual data	Edges/nodes can store metadata (e.g., weights, timestamps)	Metadata must fit into predefined columns; ad hoc data requires new columns/tables
Query languages	Uses domain-specific languages (Cypher/Gremlin) for pattern matching	Relies on SQL, limited to set-based operations and basic JOINs
Visualization	Natively visualizable as network diagrams for intuitive exploration	Relationships are abstracted; visualization requires ER diagrams or external tools

5.1.1 Types of Graphs

Graphs can be classified along multiple dimensions that reflect their structural properties, expressive power, and temporal behavior [1]. This section organizes graph types into five major categories, providing a foundation for choosing appropriate models in domain-specific applications.

5.1.1.1 Fundamental Structural Properties

Graphs can be categorized based on their structural properties, such as edge directionality and edge multiplicity.

5.1.1.1.1 *Edge Directionality*

The most fundamental distinctions among graphs arise from the way edges define relationships between vertices. In terms of directionality, a graph may be either directed or undirected. **Directed graphs** (or digraphs) use ordered pairs (u, v) to encode one-way relationships, such as in citation networks, where one paper cites another. **Undirected graphs**, by contrast, use unordered pairs $\{u, v\}$ to represent mutual or bidirectional relationships—typical of social networks or symmetric molecular bonds.

Figure 5.2 illustrates these two types of graphs. The left one is a directed graph showing one-way relationships like citations or web links; the right one is an undirected graph showing mutual relationships like friendships or molecular bonds.

5.1.1.1.2 *Edge Multiplicity*

Multiplicity refers to how many edges may exist between vertex pairs and whether self-loops are allowed. **Simple graphs** prohibit both multiple edges between any two vertices and self-loops, which is suitable for structures like family trees. **Multigraphs** permit multiple edges but no self-loops, as seen in parallel railway systems between stations. **Pseudographs** relax both constraints, allowing multiple edges and self-references, such as in server communication networks where a node may send requests to itself.

Figure 5.3 illustrates these three types of graphs. The left graph is a simple graph with no multiple edges or self-loops; the center graph is a multigraph where multiple edges are allowed between vertices; the right graph is a pseudograph with multiple edges and self-loops allowed.

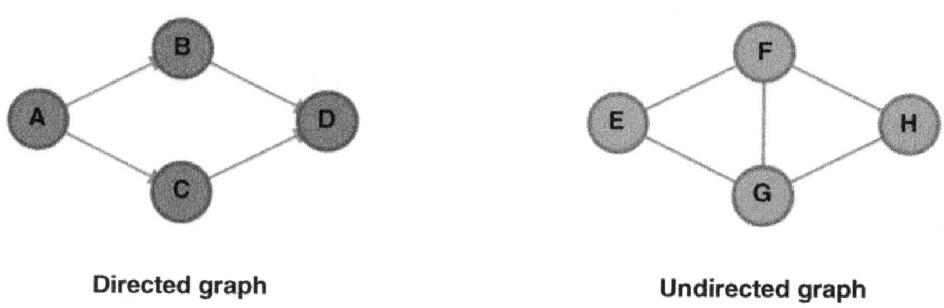

Figure 5.2 Edge directionality: directed vs. undirected graphs.

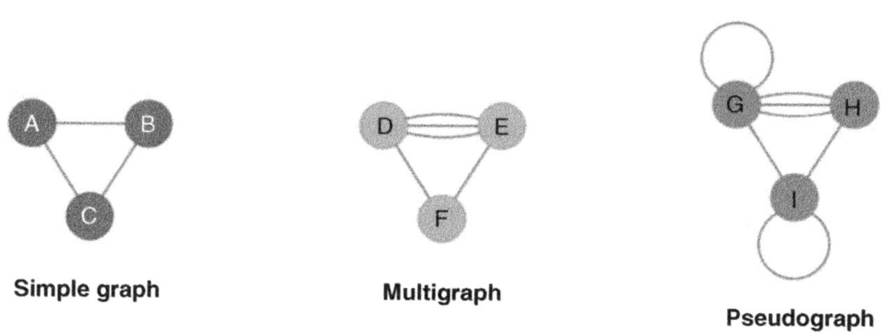

Figure 5.3 Edge multiplicity: simple, multigraph, and pseudograph.

5.1.1.2 Enriched Graph Models

Beyond their structural skeleton, graphs gain expressiveness through additional data associated with vertices and edges. These enrichments transform abstract mathematical objects into powerful tools for real-world modeling.

5.1.1.2.1 Weighted Graphs

Beyond basic structure, graphs gain expressiveness by attaching metadata or semantics to vertices and edges. Weighted graphs assign numerical values to edges, representing costs, capacities, or distances. In transportation networks, edge weights often correspond to geographical distances, while in neural networks, they quantify synaptic strength.

5.1.1.2.2 Labeled Graphs

Labeled graphs use categorical or symbolic tags rather than numerical weights. For example, biological networks label protein interactions with terms like "activation" or "inhibition," and semantic networks in linguistics link words with syntactic roles, such as "dog" → [noun] → "bark."

Figure 5.4 illustrates examples of a weighted graph and a labeled graph. The left is a weighted graph with numerical edge values (e.g., distances); the right is a labeled graph with categorical edge labels (e.g., relationship types).

5.1.1.2.3 Property Graphs

Property or attributed graphs take this further by associating key-value metadata with nodes and edges. A social network might model users as vertices with attributes such as age or location, while edges (e.g., friend relationships) include properties like connection timestamps. Knowledge graphs follow this pattern, representing entities such as "Einstein" with properties like "profession = physicist." Property graphs form the basis of many graph databases, such as Neo4j, enabling expressive and efficient querying.

5.1.1.3 Specialized Graph Structures

Certain applications impose structural constraints that yield specialized graph classes with unique properties and tailored algorithms.

5.1.1.3.1 Topologically Constrained Graphs

Planar graphs occupy a special position in both theory and practice. A graph is planar if it can be drawn on a plane with no edge crossings—a property with profound implications. Circuit designers must ensure their layouts are planar to manufacture on two-dimensional silicon wafers. Geographic information systems naturally work with planar graphs, as political boundaries and transportation networks embedded in physical space rarely cross without intersection.

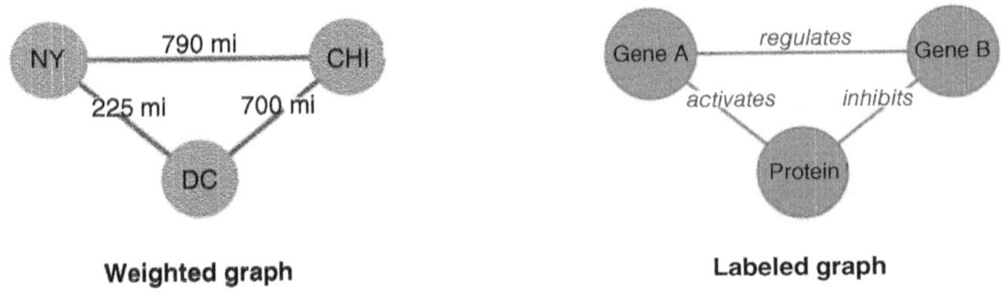

Figure 5.4 Enriched models: weighted and labeled graphs.

Kuratowski's theorem [2] provides an elegant characterization: a graph is planar if and only if it contains no subdivision of K_5 (the complete graph on five vertices) or $K_{3,3}$ (the complete bipartite graph with three vertices in each partition). This theoretical result translates into practical algorithms for planarity testing and planar embedding, crucial for applications from VLSI design to cartographic visualization.

Bipartite graphs partition vertices into two disjoint sets with edges only between sets, never within. This structure naturally models relationships between different entity types: students and courses, users and products, or documents and terms. Recommendation systems leverage bipartite graphs extensively, where one partition represents users and the other items, with edges encoding ratings or interactions. The bipartite structure enables specialized algorithms like maximum bipartite matching for task assignment or the Hungarian algorithm for optimal pairing [3].

Figure 5.5 illustrates examples of a planar graph and a bipartite graph. The left one is a planar graph (can be drawn without edge crossings), and the right one is a bipartite graph (vertices in two disjoint sets with edges only between sets).

Trees and directed acyclic graphs (DAGs) represent hierarchical relationships without cycles. Every tree with n vertices contains exactly $n - 1$ edges, making it the sparsest possible connected graph. File systems, organizational charts, and decision trees all exhibit tree structure. DAGs generalize trees by allowing vertices to have multiple parents while maintaining acyclicity—a property essential for dependency resolution in package managers, task scheduling in distributed systems, and causal inference in statistics.

5.1.1.3.2 Higher-order Structures

Traditional graphs model pairwise relationships, but many phenomena involve interactions among multiple entities simultaneously. **Hypergraphs** address this limitation by generalizing edges to hyperedges that connect arbitrary vertex subsets [4, 5]. A research collaboration hypergraph might use hyperedges to represent paper authorship, where each hyperedge connects all co-authors of a publication. Unlike traditional graphs where an edge connects exactly two vertices, a hyperedge might link three, five, or even hundreds of vertices.

Figure 5.6 illustrates an example of a hypergraph, showing hyperedges that connect multiple vertices simultaneously (e.g., research collaborations, chemical reactions).

Chemical reaction networks exemplify hypergraph applications: a reaction like $2H_2 + O_2 \rightarrow 2H_2O$ involves three reactant molecules producing two product molecules—naturally represented as a hyperedge connecting five vertices. Database systems use hypergraphs to model complex

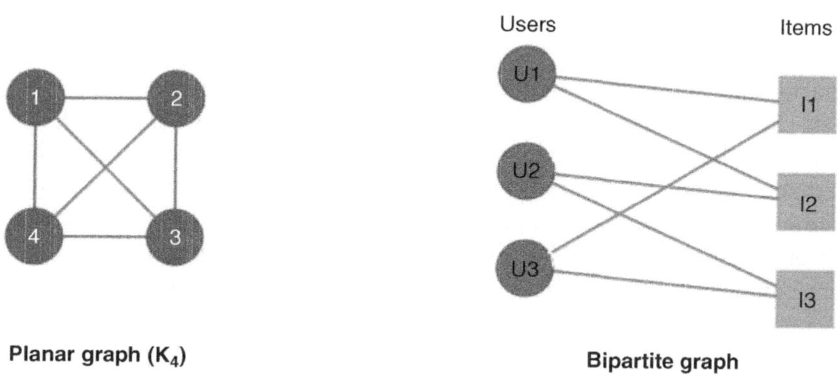

Figure 5.5 Topologically constrained: planar and bipartite graphs.

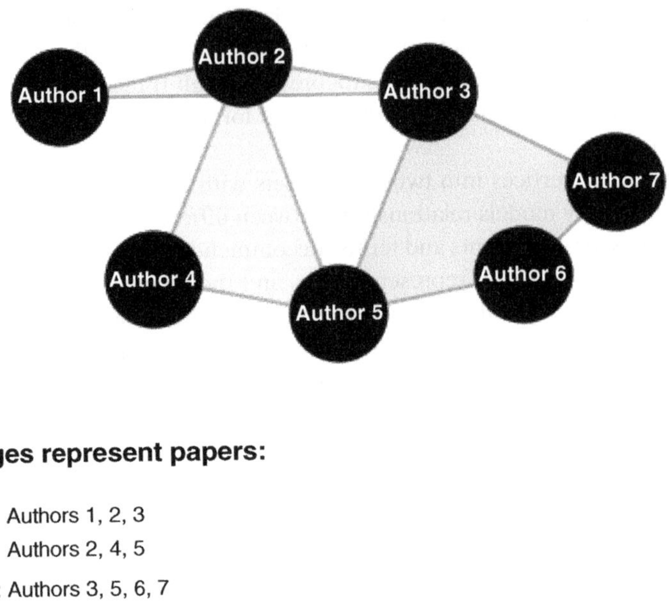

Figure 5.6 Hypergraph example.

queries joining multiple tables, where each hyperedge represents a join operation connecting all participating relations.

Simplicial complexes extend graphs into higher dimensions by including simplices: edges (1-simplices), triangles (2-simplices), tetrahedra (3-simplices), and beyond [6]. While graphs only capture pairwise connections, simplicial complexes can represent three-way relationships (triangles) or four-way relationships (tetrahedra). Topological data analysis employs simplicial complexes to detect holes, voids, and other topological features in high-dimensional data, revealing structure invisible to traditional graph analysis.

5.1.1.4 Temporal and Dynamic Graphs

Static graphs capture relationships at a single moment, but real-world networks evolve continuously. Temporal graphs address this limitation through various models of time-varying structure [7, 8].

Discrete-time dynamic graphs represent networks as sequences of static snapshots. Each snapshot $G_t = (V_t, E_t)$ captures the graph state at time t, suitable for systems with natural temporal boundaries. Daily social network states, hourly traffic patterns, or annual corporate board networks fit this model. Analysts can track metrics across snapshots: How does the clustering coefficient evolve? When do communities merge or split? Which vertices maintain central positions over time?

Continuous-time temporal graphs offer finer granularity by associating validity intervals with edges. Each edge exists only during specific time periods, encoded as intervals $[t_{\text{start}}, t_{\text{end}}]$. Professional networks model employment relationships this way, where edges between people and companies exist only during employment periods. Disease transmission networks use continuous-time models to track contact duration and timing, crucial for understanding outbreak dynamics.

Figure 5.7 illustrates an example of temporal graph, showing edge evolution over time. Solid edges are currently active, and dotted edges are inactive.

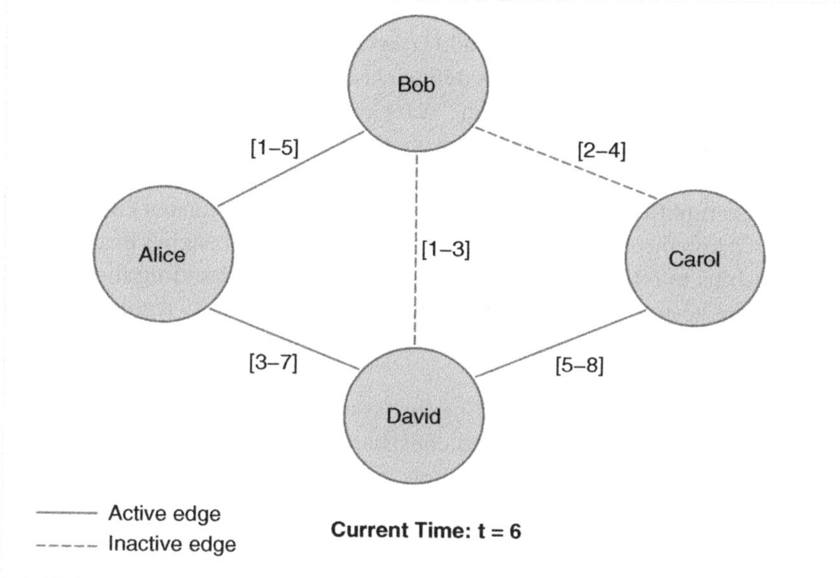

Figure 5.7 Temporal graph.

These temporal models enable sophisticated queries impossible on static graphs. Finding time-respecting paths requires edges along the path to have compatible timing: each edge must exist when traversed, and arrival at a vertex must precede departure. Temporal motifs—small recurring patterns with specific timing relationships—reveal communication patterns, information cascades, and coordination mechanisms in complex systems.

5.1.1.5 Knowledge and Semantic Graphs

The need to represent meaning, context, and inference has driven the development of specialized graph models for knowledge representation.

5.1.1.5.1 RDF Graphs

The resource description framework (RDF) structures knowledge as subject-predicate-object triples, forming directed labeled graphs where vertices represent entities and edges represent relationships [9]. Each triple makes an atomic assertion:

```
<http://dbpedia.org/resource/Marie_Curie>
<http://dbpedia.org/property/birthPlace>
<http://dbpedia.org/resource/Warsaw>
```

RDF's power lies in its simplicity and standardization. Uniform resource identifiers (URIs) provide global unique identifiers, enabling distributed knowledge bases to reference shared concepts. SPARQL, the query language for RDF, supports federated queries across multiple endpoints, allowing seamless integration of diverse data sources. The semantic web vision builds on RDF to create a web of data as interconnected as the current web of documents.

5.1.1.5.2 Knowledge Graphs

While RDF provides the foundation, knowledge graphs add layers of sophistication through ontologies, inference rules, and entity resolution. Ontologies define class hierarchies and constraints: if "Warsaw" is a "City" and every "City" is a "Place," then "Warsaw" is a "Place" through transitive inference. Rules encode domain knowledge: if someone won a Nobel Prize in Physics, they must be a physicist.

Google's knowledge graph exemplifies this approach at scale, containing billions of entities and relationships that power search enhancements [10]. When users search for "Einstein's birthplace," the system traverses relationships to return "Ulm, Germany" directly, understanding the query's intent rather than just matching keywords. Knowledge graphs also support entity resolution, recognizing that "Marie Curie," "Maria Skłodowska-Curie," and "Madame Curie" refer to the same person despite surface differences.

Modern knowledge graphs integrate machine learning for tasks like link prediction (inferring missing relationships), entity extraction from text, and consistency checking. They serve as the backbone for question-answering systems, virtual assistants, and recommendation engines that require deep understanding of entity relationships and properties.

We are going to dive into more details of knowledge graphs in Chapter 6.

5.1.1.6 Composite and Hybrid Graphs

In practice, many graph-based systems combine multiple features. A financial knowledge graph, for example, may be directed, weighted, temporal, and attributed: nodes represent organizations, edges encode acquisition events with associated transaction amounts and timestamps, and metadata describes industries, valuations, and regulatory statuses.

Similarly, recommendation systems often use bipartite, attributed graphs in which users and items are nodes, edges reflect interactions (e.g., purchases or views), and metadata captures user preferences, item categories, and interaction timestamps.

Such hybrid representations guide algorithmic design. Weighted graphs support shortest path and flow algorithms, while attributed graphs enable node classification and link prediction. By understanding the interplay among graph features, system designers can select appropriate models and techniques for their specific applications.

5.1.1.7 Selecting Appropriate Graph Types

The diversity of graph types reflects the complexity of real-world modeling challenges. Selecting the appropriate type requires understanding both the data characteristics and analytical goals. Table 5.2 lists examples of graph type selections based on data characteristics.

For example, a social media platform might model its network as directed (follower relationships), weighted (interaction frequency), temporal (relationship evolution), and attributed (user profiles and post metadata). This layered approach combines multiple graph types to capture different aspects of the same underlying system.

The art of graph modeling lies not in selecting a single type but in combining multiple types to create representations that balance expressiveness with computational tractability. As graph applications continue to expand across domains, new hybrid types emerge to address specific modeling challenges, demonstrating that our taxonomy of graphs remains a living, evolving framework.

Table 5.2 Graph type selection based on data characteristics.

Data characteristics	Suggested graph type	Typical applications
Asymmetric relationships	Directed graphs	Email networks, citation networks
Relationship strengths	Weighted graphs	Transportation systems, neural networks
Multiple relationship types	Property graphs	Social platforms, enterprise knowledge graphs
Embedding or layout constraints	Planar graphs	VLSI circuit design, cartographic maps
Entity type separation	Bipartite graphs	Recommender systems, task assignments
Hierarchical dependencies	Trees or DAGs	File systems, build pipelines, task scheduling
Multi-way interactions	Hypergraphs	Scientific collaborations, chemical reactions
Time-evolving relationships	Temporal graphs	Epidemiology, messaging platforms
Semantic and ontological meaning	Knowledge graphs	Search engines, question-answering systems

5.1.2 Graph Representation

Graphs can be represented in various ways, each tailored to different computational needs and structural characteristics. The choice of representation often depends on factors such as the density of edges, the size of the graph, and the operations required (e.g., checking edge existence, iterating over neighbors). Three common methods include adjacency matrices, adjacency lists, and edge lists, each with distinct advantages and use cases.

5.1.2.1 Edge List

The simplest form of graph representation is an edge list, which is merely a collection of all edges in the graph. Each edge is stored as a pair (or triplet for weighted edges) indicating its endpoints. For example, a graph representing flight routes between cities might have an edge list like ("New York," "London"), ("London," "Paris"), ("Tokyo," "San Francisco").

This method is highly space-efficient for sparse graphs, requiring only $O(m)$ space, and it's easy to construct from raw data. However, checking whether two vertices are connected requires scanning the entire list, resulting in $O(m)$ time in the worst case. Edge lists are thus ideal for scenarios where the graph is static and only basic queries are needed, such as input/output operations or initial setup in algorithms.

The edge lists for the example graphs in Figure 5.8 are:

(a) Undirected Edges:

$$\mathrm{A} - \mathrm{B},$$

$$\mathrm{A} - \mathrm{C},$$

$$\mathrm{B} - \mathrm{C},$$

$$\mathrm{C} - \mathrm{D}$$

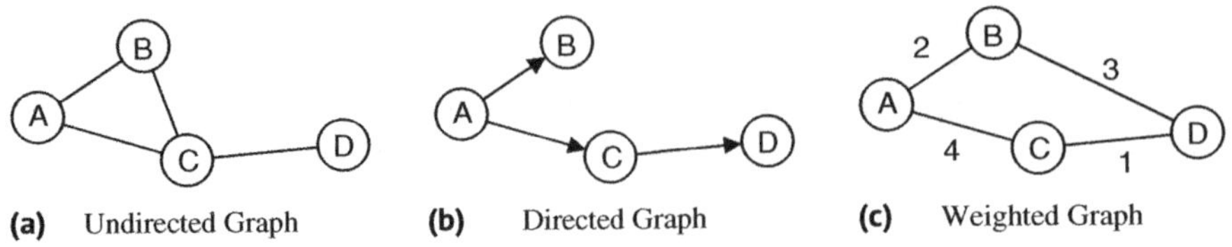

Figure 5.8 Graph types and examples.

(b) Directed Edges:

$$A \rightarrow B,$$

$$A \rightarrow C,$$

$$C \rightarrow D$$

(c) Edges with Weights:

$$A — B\,(2),$$

$$A — C\,(4),$$

$$B — D\,(3),$$

$$C — D\,(1)$$

5.1.2.2 Adjacency Matrix

An adjacency matrix is a two-dimensional array where rows and columns correspond to vertices in the graph. The value at position $A[i][j]$ indicates whether there is an edge from vertex i to vertex j. In unweighted graphs, entries are typically binary (0 for no edge, 1 for an edge), while weighted graphs store the weight of the connection.

For example, consider a graph with three vertices labeled A, B, and C, where edges exist between $A - B$, $A - C$, and $B - C$. The adjacency matrix would have ones at positions $A[0][1]$, $A[0][2]$, $A[1][0]$, $A[1][2]$, $A[2][0]$, and $A[2][1]$, forming a symmetric matrix for an undirected graph.

This structure allows for constant-time ($O(1)$) edge lookups but requires $O(n^2)$ space, making it inefficient for large, sparse graphs.

The adjacency matrices for the example graphs in Figure 5.8 are:

(a) Undirected Edges:

	A	B	C	D
A	0	1	1	0
B	1	0	1	0
C	1	1	0	1
D	0	0	1	0

(b) Directed Edges:

	A	B	C	D
A	0	1	1	0
B	0	0	0	0
C	0	0	0	1
D	0	0	0	0

(c) Edges with Weights:

	A	B	C	D
A	0	2	4	0
B	2	0	0	3
C	4	0	0	1
D	0	3	1	0

5.1.2.3 Adjacency List

Adjacency lists provide a more flexible and space-efficient alternative for sparse graphs. Instead of a matrix, they use an array of linked lists or dynamic arrays, where each index corresponds to a vertex. The list at each index contains references to the vertices adjacent to it.

For instance, imagine modeling a social network where each person (vertex) has relatively few connections compared to the total number of users. An adjacency list would store each user's friends in a list, avoiding the storage overhead of an adjacency matrix. Suppose user X follows Y and Z, while Y follows only X; the adjacency lists for X, Y, and Z would hold $[Y, Z]$, $[X]$, and $[]$, respectively.

This representation reduces space complexity to $O(n + m)$, where m is the number of edges, and it efficiently handles operations like iterating through a vertex's neighbors in $O(d)$ time, where d is the degree of the vertex.

The adjacency lists for the example graphs in Figure 5.8 are:

(a) Undirected Edges:

$$\text{A} : \{\text{B, C}\}$$

$$\text{B} : \{\text{A, C}\}$$

$$\text{C} : \{\text{A, B, D}\}$$

$$\text{D} : \{\text{C}\}$$

(b) Directed Edges:

$$A : \{B, C\}$$

$$B : \{\}$$

$$C : \{D\}$$

$$D : \{\}$$

(c) Edges with Weights:

$$A : \{(B, 2), (C, 4)\}$$

$$B : \{(A, 2), (D, 3)\}$$

$$C : \{(A, 4), (D, 1)\}$$

$$D : \{(B, 3), (C, 1)\}$$

Each representation balances trade-offs between space usage and query performance. Adjacency matrices excel in dense graphs and frequent edge existence checks but suffer in memory for sparse structures. Adjacency lists are versatile for most practical applications, especially when iterating over neighbors is common, while edge lists serve as a lightweight option for minimal requirements. The optimal choice depends on the specific context, such as whether the graph is static or dynamic, the proportion of edges to vertices, and the computational tasks involved.

5.1.3 Graph Operations

Graphs offer a rich mathematical framework to represent and manipulate complex relationships among entities. Fundamental graph operations such as traversal, pathfinding, and structural analysis underpin a wide range of applications in search, recommendation, network routing, and knowledge representation. In this section, we review essential graph algorithms, emphasizing their principles, time complexities, and practical use cases.

5.1.3.1 Graph Traversal

Graph traversal is the backbone of graph analysis—it systematically explores nodes and edges to uncover structural properties, answer connectivity queries, and serve as a foundation for more advanced algorithms.

5.1.3.1.1 Random Walk

A random walk is a stochastic traversal where, starting from a node, the next step is selected uniformly or probabilistically from its neighbors. Formally, for a graph $G = (V,E)$, a random walk is a Markov chain over V, where the transition probability from node u to node v depends on the edge weight $w(u,v)$.

We will discuss more details of random walk variations in the Graph-RAG context in Section 7.2.2.2 of the book.

5.1.3.1.2 BFS and DFS

Breadth-first search (BFS) explores a graph level-by-level from a given source node. It guarantees the shortest path in unweighted graphs and is optimal for minimal-hop exploration.

Depth-first search (DFS) dives deep along a branch before backtracking, making it suitable for tasks such as cycle detection, path enumeration, and connectivity checks.

Both algorithms have linear time complexity: $O(|V| + |E|)$.

Figure 5.9 visually contrasts how BFS and DFS explore a graph starting from the same node u, with a search depth limit of $k = 3$.

BFS (shown in solid arrows) explores all immediate neighbors first—s_1,s_2,s_3,s_4—before moving to the next layer. This level-by-level strategy ensures that the shortest path is found in unweighted graphs and is useful when the goal is to reach nodes within a specific number of steps.

In contrast, DFS (shown in dotted arrows) dives deep into one branch before backtracking—visiting nodes like $s_4 \rightarrow s_5 \rightarrow s_6$ without first exploring all neighbors of u. This depth-first strategy is more memory-efficient and better suited for tasks like cycle detection or exhaustive path searches. While BFS gives broader but shallower coverage early on, DFS offers deep but narrow paths quickly, making each suitable for different types of graph problems.

5.1.3.1.3 *Shortest Path*

Shortest path algorithms compute the optimal path between two vertices, considering path cost or edge weights.

- BFS is used for unweighted graphs where each edge has equal cost. It finds the shortest path by expanding outward uniformly.
- Dijkstra's algorithm handles graphs with nonnegative edge weights, using a priority queue to always expand the nearest unvisited node [12].
- Bellman–Ford algorithm handles graphs with negative edge weights (but no negative cycles), iteratively relaxing all edges [13, 14].
- Floyd–Warshall algorithm computes all-pairs shortest paths using dynamic programming [15, 16].

Table 5.3 lists the comparisons of the shortest path algorithms.

The shortest path algorithms have applications in areas such as routing (e.g., GPS, network protocols), influence spread modeling, and query graph generation in Graph-RAG. We'll discuss more details and illustrations of the shortest path algorithms in the context of knowledge graphs in Section 6.3 of the book.

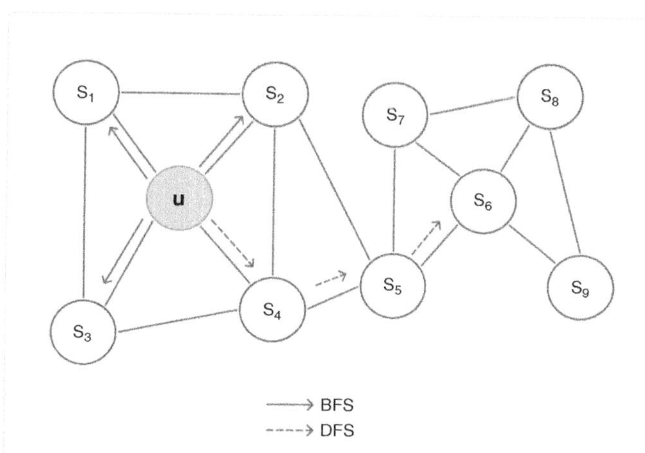

Figure 5.9 BFS and DFS search strategies from node u (k=3) [11] / Association for Computing Machinery (ACM).

Table 5.3 Shortest path algorithm comparisons.

Graph type	Algorithm	Time complexity	Space complexity
Unweighted	BFS	$O(V + E)$	$O(V)$
Positive weights	Dijkstra	$O((V + E) \log V)$	$O(V)$
Negative weights	Bellman–Ford	$O(VE)$	$O(V)$
All pairs	Floyd–Warshall	$O(V^3)$	$O(V^2)$

5.1.3.2 Structural Graph Analysis

Beyond pathfinding, graphs are often analyzed for their global structure. Cycle detection, for instance, helps identify anomalies or violations in DAGs, such as dependency loops in software or feedback loops in regulation networks. DFS-based cycle detection works well for both directed and undirected graphs by identifying back edges in the DFS tree [17].

Minimum spanning trees (MSTs) represent another important concept. They connect all nodes in an undirected graph with the smallest possible total edge weight. Kruskal's algorithm constructs an MST by greedily adding the smallest edge that doesn't form a cycle, using a union-find data structure [18]. Prim's algorithm grows the MST from a single starting node, always choosing the least expensive edge that extends the tree. These techniques are central in network design, clustering, and approximate matching [19].

5.1.3.3 Strongly Connected Components

Strongly connected components (SCCs) represent maximal subgraphs where every node can reach every other node through directed paths. In knowledge graphs, SCCs often indicate tightly coupled concept clusters, mutual dependencies, or definitional cycles that require collective understanding.

Kosaraju's algorithm employs an elegant two-pass strategy to identify SCCs [20]. The first pass performs DFS on the original graph, recording nodes in order of completion time. The second pass runs DFS on the reversed graph, processing nodes in reverse order of their completion times from the first pass. Each DFS tree in the second pass corresponds to one SCC. The algorithm's correctness stems from a key insight: the reversal and ordering ensure that we can't accidentally merge separate SCCs during the second pass.

Tarjan's algorithm achieves the same result in a single pass through clever bookkeeping [21]. As it performs DFS, it maintains a stack of nodes and tracks the lowest-indexed node reachable from each node's subtree. When a node's low-link value equals its index, we've found the root of an SCC and can pop all nodes above it on the stack. This single-pass approach makes Tarjan's algorithm more memory-efficient and often faster in practice.

5.1.3.4 Topological Sorting

Topological sorting brings order to DAGs by arranging vertices such that all edges point forward in the resulting sequence. This fundamental operation enables dependency analysis, causal reasoning, and workflow optimization in knowledge graphs where relationships encode precedence or logical ordering.

Kahn's algorithm approaches topological sorting through systematic node removal [22]. Beginning with nodes having no incoming edges (in-degree zero), the algorithm repeatedly removes these nodes and their outgoing edges, adding newly freed nodes to the processing queue. This process continues until all nodes are ordered or a cycle is detected. The intuition behind Kahn's algorithm mirrors real-world dependency resolution: we can only process tasks with satisfied prerequisites. In knowledge graphs, this translates to identifying foundational concepts that depend on nothing else, then building up to more complex ideas layer by layer.

The DFS-based approach to topological sorting leverages the natural ordering provided by depth-first traversal [23]. When we finish exploring all descendants of a node in DFS, we know that node can be placed after all its dependencies in the topological order. By recording nodes in reverse order of their DFS completion times, we obtain a valid topological sort. This method's elegance lies in its implicit handling of dependencies through the recursive structure of DFS itself.

5.1.4 Graph Algorithms

In this section, we'll briefly introduce the major graph algorithms. We are going to demonstrate the implementation examples of these graph algorithms in Section 5.3.3 of this book.

5.1.4.1 Centrality Measures

Centrality measures quantify the importance or influence of nodes within a graph structure. Popular centrality metrics include degree centrality, betweenness centrality, closeness centrality, eigenvector centrality, and PageRank.

5.1.4.1.1 Degree Centrality: Connection-based Importance

Degree centrality measures the number of edges a node has. Here is its mathematical definition: for a graph $G = (V, E)$ with n nodes:

- Unnormalized Degree Centrality:

$$C_D(v) = \deg(v)$$

- Normalized Degree Centrality:

$$C'_D(v) = \frac{\deg(v)}{n-1}$$

where $\deg(v)$ is the number of edges incident to node v.

5.1.4.1.2 Betweenness Centrality: Bridge Detection

In the information flow context, the betweenness centrality measures the influence a node has on information spread through the network. For a node v in graph G, betweenness centrality measures the fraction of shortest paths between all pairs of nodes that pass through v. Here is the mathematical definition of betweenness centrality:

$$C_B(v) = \sum_{s \neq v \neq t \in V} \frac{\sigma_{st}(v)}{\sigma_{st}}$$

where σ_{st} is the total shortest paths from s to t, and $\sigma_{st}(v)$ is the paths passing through v.

To compute the betweenness centrality, first compute the shortest paths for all node pairs (s,t), then count what fraction passes through v. Finally, Brandes' algorithm reduces this from $O(n^3)$ to $O(nm)$.

5.1.4.1.3 Closeness Centrality: Information Spread Efficiency

Closeness centrality measures how close a node is to all other nodes in the graph. Higher closeness centrality indicates that a node can reach all other nodes through shorter paths, making it efficient for information propagation.

Here is the mathematical definition of closeness centrality:

- Unnormalized Closeness Centrality:

$$C_C(v) = \frac{1}{\sum_{u \in V} d(u, v)}$$

where $d(u, v)$ is the shortest-path distance between u and v.

- Normalized Closeness Centrality:

$$C_C(v) = \frac{n - 1}{\sum_{u \neq v} d(u, v)}$$

where $d(u, v)$ is the shortest path distance.

5.1.4.1.4 *Eigenvector Centrality: Authority Through Connection Quality*

Eigenvector centrality gives each vertex a score that is proportional to the sum of the scores of its neighbors. Here is the mathematical definition of eigenvector centrality: for adjacency matrix A of G, the centrality of x_v of node v satisfies:

$$\lambda x_v = \sum_{u \in N(v)} x_u$$

or in matrix form (where λ is the largest eigenvalue of A) :

$$A\mathbf{x} = \lambda \mathbf{x}$$

interpretation: x_v is proportional to the sum of centralities of v's neighbors.

For directed graphs, eigenvector centrality x_i is defined as being proportional to the sum of the scores of the nodes that point to node v_i. In this case one problem might arise. Suppose a node v_A has only outgoing edges but no incoming ones. v_A thus has an eigenvector centrality value of zero, which is reasonable since it has no incoming edges. However, suppose another vertex v_B has only one incoming edge and this edge is from vertex v_A. v_B then also has a centrality value zero, which is not reasonable. This problem for directed graphs is addressed by using another centrality measure called PageRank.

5.1.4.1.5 *PageRank: Authority Measurement in Directed Networks*

PageRank enhances eigenvector centrality with a damping factor to ensure convergence and handle disconnected components: defined via a random walk with teleportation probability α (typically 0.85):

$$\mathrm{PR}(v) = \frac{1 - \alpha}{n} + \alpha \sum_{u \in N_{in}(v)} \frac{\mathrm{PR}(u)}{\text{Out-degree}(u)}$$

where $N_{in}(v)$ is nodes linking to v, and dangling nodes (out-degree=0) redistribute rank uniformly.

As we briefly mentioned the PageRank concept in Section 1.1.2.3, the PageRank score was first introduced by Google to rank websites in its search engine. The intuition is that if some important pages point to your page, then that is an indication that your page might be important as well. The out-degree in the denominator guarantees that pages that point to a huge number of other pages will not pass much centrality on to any of them. PageRank extends beyond simple connectivity to consider the importance of connected nodes, creating a recursive measure of authority and influence.

5.1.4.2 Community Detection Algorithms

Community detection is crucial for uncovering structure in complex graphs. It identifies clusters or subgroups where nodes are more densely connected to each other than to the rest of the network. These algorithms are widely used in social network analysis, recommendation systems, customer segmentation, and anomaly detection. The three most prominent algorithms are Louvain, Label Propagation, and Leiden.

- Louvain is a modularity-based algorithm that groups nodes into communities to maximize the modularity score [24]. It operates by assigning nodes to communities and merging them in a way that increases modularity, then repeating the process hierarchically. It's fast and works well on large graphs.
- Label propagation is a lightweight algorithm that starts by giving each node a unique label and iteratively updates each node's label to the most frequent label among its neighbors [25]. It's great for quick, unsupervised clustering when performance is a concern.
- Leiden is an enhancement over Louvain [24]. It ensures well-connected communities and avoids disconnected subgroups, making it more robust for large-scale analytics like fraud ring detection or influence group discovery.

5.1.4.3 Similarity Algorithms

Similarity algorithms evaluate how similar two nodes are based on structure or attributes. These are particularly important in recommendation systems, fraud detection (e.g., shared behavior), and link prediction. Key algorithms include node similarity, K-nearest neighbors (KNN), and Jaccard similarity.

- Node similarity calculates pairwise similarity scores between nodes based on common neighbors. It's typically used for collaborative filtering-style recommendations—"people who bought this also bought that."
- KNN finds the top-*k* most similar nodes to a given node. It's often used in customer segmentation or item-to-item recommendation.
- Jaccard similarity is a normalized similarity measure defined as the size of the intersection of neighbors divided by the union. It's simple and interpretable and often used for social network matching or document similarity.

5.1.4.4 Pathfinding Algorithms

As we briefly introduced the graph traversal methods in Section 5.1.3.1, the path algorithms are fundamental to graph traversal. They're used in logistics, routing, navigation, fraud analysis, and knowledge graphs. While widely used, their application tends to be more straightforward. Important examples are Dijkstra's shortest path, A* search, and BFS.

- Dijkstra's shortest pathfinds the path with the least cost between two nodes [12]. It's commonly used in route planning, supply chain optimization, and fraud path tracing.
- A* search enhances Dijkstra by adding heuristics, which makes it faster when you have some knowledge about the destination (e.g., geographic proximity) [26]. It's ideal in map-based routing or game AI.
- BFS finds all reachable nodes within a given number of hops. It's useful for proximity search or bounded explorations, like finding nearby users in a social graph.

5.1.4.5 Node Embeddings

Node embedding algorithms convert graph structure into vector representations for use in machine learning models. These embeddings capture structural patterns and relationships, enabling node classification, link prediction, and anomaly detection. Common algorithms include Node2Vec, GraphSAGE, and FastRP.

- Node2Vec performs random walks to generate sequences of nodes and then uses a skip-gram model (similar to Word2Vec) to generate embeddings [11]. It balances between structural and community similarity, useful in recommendation engines and fraud detection.
- GraphSAGE generates embeddings by sampling and aggregating features from a node's neighborhood [27]. It supports inductive learning, meaning it can embed new nodes. Great for real-time ML systems and graph-based feature engineering.
- FastRP is a high-performance alternative to Node2Vec, optimized for speed [28]. It's useful when you want fast embeddings for large graphs without sacrificing too much accuracy.

We are going to discuss more details of the node embedding methods in the context of knowledge graphs in Chapter 6 of the book (Section 6.3.2).

5.1.4.6 Topological Link Prediction

Thesc algorithms estimate the likelihood of future or missing connections based on graph topology [29]. They're used in social graph expansion, recommendation engines, and biological network completion. Common algorithms include common neighbors, Adamic–Adar, and preferential attachment.

- Common neighbors simply counts shared neighbors. It's effective for friend suggestions and item co-purchase prediction.
- Adamic–Adar gives more weight to less-connected common neighbors [30]. This is useful in sparse networks, like citation prediction or niche interest communities.
- Preferential attachment assumes nodes with higher degrees attract more connections [31]. It models real-world phenomena like "the rich get richer" and applies to viral content prediction or influencer dynamics.

5.1.4.7 DAG Algorithms

These are specialized algorithms for DAGs, such as workflows or dependency trees. They're critical in task scheduling, compiler design, and lineage tracking, but less broadly applicable across general graph workloads. Important ones include topological sort and SCC, which we briefly introduced in Section 5.1.3.

- Topological sort orders nodes so that for every edge from node A to B, A comes before B. It's used in build systems, job scheduling, and version control.
- SCC identifies subgraphs where every node is reachable from every other node. While commonly used in community detection, its value in DAGs lies in identifying cyclic dependencies or validating the acyclic property.

5.1.5 Applications of Graphs

The applications of graphs are diverse and impactful, spanning numerous domains due to their ability to represent complex relationships. In social networks, graphs model interpersonal connections, enabling insights into human behavior and societal structures. Platforms like Facebook and LinkedIn utilize graph-based algorithms to recommend friends or professional contacts, while researchers analyze community formation to study phenomena such as political polarization or viral marketing campaigns. Techniques like PageRank and centrality measures help identify influential individuals or groups, aiding in understanding how information or trends propagate through populations.

Transportation systems rely heavily on graph theory to optimize routes and infrastructure. Road maps, airline networks, and public transit systems are all represented as graphs, where intersections, cities, or stations act as nodes, and roads or routes form edges. Shortest-path algorithms, such as Dijkstra's or A*, power GPS navigation tools like Waze or Google Maps, calculating optimal travel paths in real time. Logistics companies employ graph-based solutions for delivery route optimization, minimizing costs and time—a classic example being the traveling salesman problem. Urban planners also use graphs to simulate traffic flow, detect congestion hotspots, and design efficient public transportation networks.

In computer networks, graphs model the intricate web of devices, servers, and routers that constitute the internet or local networks. Nodes represent devices, while edges denote communication channels or physical connections. Routing protocols, such as OSPF or BGP, leverage graph theory to dynamically determine the best paths for data transmission. Security analysts use graph analytics to map potential attack pathways, identifying vulnerabilities in network topologies. Distributed systems also benefit from graph models to balance computational loads across servers, ensuring efficient resource allocation.

Biological networks exemplify the power of graphs in unraveling life's complexity. Metabolic pathways, where enzymes and substrates form nodes connected by biochemical reactions, enable researchers to study cellular processes and design targeted drugs. Protein interaction networks help uncover mechanisms behind diseases like cancer, while neural networks—mapped in connectomics—shed light on brain function and disorders. Ecologists model ecosystems as food webs to study species interdependencies, predicting ecosystem responses to environmental changes. Even genetic regulatory networks, which illustrate how genes interact to control traits, rely on graph representations to advance personalized medicine and synthetic biology. These examples highlight how graph theory bridges abstract mathematics with tangible, real-world problem-solving across disciplines.

5.2 Graph Databases: Comparison and Analysis

In Section 5.1.5, we explored the fundamental concepts of graph theory, including graph definition, types of graphs, graph representation, basic operations on graphs, and applications of graphs. This section will demonstrate how graph databases implement these theoretical concepts, providing powerful tools for storing, managing, and querying graph-structured data.

Graph databases have gained significant popularity in recent years due to their ability to efficiently handle these complex data structures and relationships. They excel in scenarios where the connections between data points are as important as the data points themselves, such as in social networks, recommendation systems, fraud detection, and knowledge graphs.

This section provides an overview of some of the most popular graph databases in the current market, comparing their features, deployment options, and licensing models. We'll explore how each database implements graph concepts and what unique features they offer for managing graph data.

5.2.1 Notable Graph Databases

The graph database landscape has evolved significantly over the past decade, with several robust solutions now available to address diverse use cases across industries. This section examines five leading graph database systems:

1. Neo4j: A pioneer in the graph database space, known for its robust performance and comprehensive feature set [32].
2. NebulaGraph: A distributed graph database designed for handling large-scale data with high performance and scalability [33].
3. Amazon Neptune: A fully managed graph database service provided by amazon web services (AWS), offering seamless integration with other AWS services [34, 35].
4. Microsoft Azure Cosmos DB: A multimodel database service that includes graph database capabilities, known for its global distribution and low latency [36].
5. ArangoDB: A versatile multimodel database that supports graph, document, and key-value data models, providing flexibility for diverse use cases [37].

Each database implements graph data structures and algorithms differently, with distinct trade-offs that make them suitable for specific application scenarios. For each database, we will explore its key features, advantages, and potential drawbacks, providing a comprehensive analysis to help you determine which might be the best fit for your specific requirements.

5.2.1.1 Neo4j

Neo4j is one of the most well-known graph databases in the market. It is a property graph database that offers both open-source (community edition) and commercial (enterprise edition) licensing options. Neo4j can be deployed on-premises or in the cloud through their managed service, Neo4j Aura. It supports both single-server and clustered distributions, providing flexibility for various use cases.

At the core of Neo4j's feature set is its powerful Cypher query language, which allows for intuitive and efficient querying of graph data. The database ensures ACID compliance, making it suitable for applications that require strong consistency. Neo4j also offers full-text search capabilities and a visual query builder, enhancing its usability for both developers and data analysts.

One of Neo4j's major strengths lies in its maturity and wide adoption in the industry. This has fostered a strong community around the database, resulting in excellent documentation and a rich ecosystem of tools and integrations. Neo4j's powerful visualization tools are particularly noteworthy, allowing users to gain insights from complex graph structures easily. However, Neo4j is not without its challenges. Scaling can become complex when dealing with very large datasets, which might require careful planning and management. Additionally, while the community edition offers a robust set of features, many enterprise-grade features are only available in the paid enterprise edition, which may impact the total cost of ownership for large-scale deployments.

Index-free adjacency is a key feature of native graph databases that enables highly efficient data traversal, where each node directly references its neighboring nodes, allowing relationships to be accessed via memory pointer lookups rather than relying on index lookups. This results in constant-time performance ($\mathcal{O}(1)$) for traversing a graph, regardless of its size, avoiding the performance issues seen in non-native graph databases, which require index lookups for each connection, slowing query performance as the graph grows. In non-native graph databases (e.g., relational or document-based), nodes and relationships are stored separately, making traversals less efficient and relying on complex index-based joins, adding significant overhead, particularly when querying multiple levels of relationships. As a result, performance degrades as the complexity of the queries or the size of the graph increases. In contrast, native graph databases like Neo4j are designed from the ground up to handle interconnected data, optimizing both storage and processing for graph queries, leading to faster execution and better scalability even as data grows.

5.2.1.2 NebulaGraph

NebulaGraph is an open-source, distributed graph database designed for large-scale graphs with a focus on high-performance querying. It supports property graphs and is particularly well-suited for handling graphs with billions of vertices and edges. NebulaGraph can be deployed both on-premises and in the cloud, providing flexibility for different deployment scenarios.

One of NebulaGraph's standout features is its horizontal scalability. It can scale out by adding more nodes to the cluster, which is ideal for handling large and complex datasets. NebulaGraph's query language, nGQL, is similar to Cypher, making it accessible for developers familiar with other property graph databases like Neo4j. The database also integrates well with the broader big data ecosystem, supporting integration with platforms such as Spark and Flink.

NebulaGraph is highly optimized for performance, particularly in querying large-scale graphs with low-latency requirements. It also supports ACID transactions, which are crucial for applications that require strong consistency. The database offers high availability through its Raft-based consensus mechanism and provides fine-grained role-based access control, enhancing its security capabilities.

However, NebulaGraph is relatively new compared to more mature options like Neo4j, meaning that its community and ecosystem are still growing. This can result in fewer third-party tools and integrations being available. Additionally, while the database is designed for scalability and performance, it may require more expertise to set up and optimize compared to some out-of-the-box solutions.

5.2.1.3 Amazon Neptune

Amazon Neptune is a fully managed graph database service offered by AWS. It supports both property graph and RDF models, making it versatile for different graph use cases. As a commercial, cloud-native solution, Neptune is exclusively deployed on AWS infrastructure and is designed for distributed operations.

Neptune's tight integration with other AWS services is one of its standout features. This integration allows for seamless data pipelines and analytics workflows within the AWS ecosystem. The database also offers automatic backups and point-in-time recovery, enhancing data durability and disaster recovery capabilities.

Being a fully managed service, Neptune provides several advantages. It offers high availability and durability out of the box, with AWS handling the underlying infrastructure management. Scaling is also straightforward, with the ability to adjust resources as needed. The pay-as-you-go pricing model can be cost-effective for many use cases, especially those with variable workloads.

However, the convenience of a managed service comes with trade-offs. Users are locked into the AWS ecosystem, which may not be ideal for all organizations, especially those with multi-cloud strategies. Additionally, the managed nature of the service means users have limited control over the underlying infrastructure, which could be a concern for applications with specific optimization requirements.

5.2.1.4 Microsoft Azure Cosmos DB

Microsoft Azure Cosmos DB is a fully managed, multimodel database service offered by Microsoft Azure. While it supports multiple data models, including document and key-value, it also provides robust support for graph data through its Gremlin API. As a commercial, cloud-native solution, Cosmos DB is exclusively deployed on the Azure cloud infrastructure and is designed for distributed, global-scale operations.

One of Cosmos DB's standout features is its global distribution capability. It allows for seamless data replication across multiple Azure regions, enabling low-latency access to data from anywhere in the world. This makes Cosmos DB particularly well-suited for applications with a global user base. The database also offers automatic indexing, which can significantly simplify database management and optimize query performance across different data models.

Being a fully managed service, Cosmos DB provides several advantages in terms of operational simplicity. It offers seamless scalability, allowing users to easily adjust throughput and storage as their needs evolve. The integration with the broader Azure ecosystem is another significant benefit, enabling streamlined workflows with other Azure services like Azure functions, Azure kubernetes service, or Azure synapse analytics.

Cosmos DB also stands out for its flexibility in consistency models. It offers five well-defined consistency levels, allowing developers to make fine-grained trade-offs between consistency, availability, and latency. This can be particularly useful in distributed systems where traditional ACID guarantees may be too restrictive or unnecessary.

However, like other cloud-native managed services, Cosmos DB comes with certain limitations. The most significant is vendor lock-in to the Azure ecosystem, which may not align with all organizations' cloud strategies, especially those pursuing multi-cloud approaches. Additionally, while the pay-as-you-go pricing model can be cost-effective for many scenarios, it can become expensive for large-scale operations or poorly optimized queries. Careful capacity planning and query optimization are crucial to manage costs effectively with Cosmos DB.

5.2.1.5 ArangoDB

ArangoDB is another notable player in the multimodel database space, supporting graph, document, and key-value models. Like OrientDB, it offers both open-source (community edition) and commercial (enterprise edition) versions. ArangoDB can be deployed on-premises or in the cloud, and it supports both single-server and clustered distributions, providing flexibility for various deployment scenarios.

One of ArangoDB's distinguishing features is its ArangoDB query language (AQL), which provides a unified querying interface across all supported data models. This allows developers to seamlessly combine different data models within a single query, potentially simplifying complex data operations. ArangoDB also offers robust full-text search capabilities, enhancing its utility for applications that require advanced search functionalities alongside graph operations. The multimodel nature of ArangoDB brings significant advantages in terms of flexibility. It allows developers to work with different data models within a single database system, potentially reducing the complexity of application architectures that would otherwise require multiple specialized databases. ArangoDB has also demonstrated good performance and scalability, making it suitable for applications with demanding workloads.

However, as with other multimodel databases, ArangoDB's versatility comes with a trade-off. It may be less mature in some aspects compared to specialized graph databases like Neo4j. The learning curve can be steeper, as developers need to understand how to effectively utilize and combine different data models. Additionally, while ArangoDB has an active development community, its ecosystem of tools and integrations is smaller compared to some more established databases, which might limit some out-of-the-box functionality or third-party support.

5.2.2 Comparison Summary

When selecting a graph database, several factors come into play, including the specific requirements of your project, deployment preferences, scalability needs, and budget constraints. Neo4j remains a popular choice due to its maturity and strong community support, making it well-suited for a wide range of applications. For those deeply integrated with cloud ecosystems, Amazon Neptune and Azure Cosmos DB offer powerful, managed solutions with the benefits of seamless scaling and integration with other cloud services.

Multimodel databases like OrientDB and ArangoDB provide flexibility for projects that may require different data models, potentially simplifying overall system architecture. For open-source, highly scalable solutions, particularly those integrated with big data ecosystems, JanusGraph presents a compelling option.

Ultimately, the choice of graph database will depend on your specific use case, existing technology stack, performance requirements, and development team's expertise. It's often beneficial to prototype with a few options to determine which database best meets your needs in practice.

As we discussed various databases compared in Table 5.4, and for demonstration purposes, we will focus on **Neo4j**. Neo4j is known for its ease of use and accessibility, making it an ideal choice for deployment on personal laptops.

5.3 Introduction to Neo4j and Cypher

Cypher is the original query language for Neo4j, which has since evolved into the open standard openCypher. This development has influenced the creation of the graph query language (GQL) standard, which aspires to become a universal GQL. The evolution can be summarized as follows:

Neo4j Cypher → openCypher → GQL

Graph databases like Neo4j are powerful tools for storing and querying highly interconnected data. In this section, we will explore fundamental operations for interacting with graph databases using a query language that is based on the core elements of Neo4j Cypher, openCypher, and GQL.

Table 5.4 Comparison of graph databases.

Attribute	Neo4j	NebulaGraph	Amazon Neptune	Azure Cosmos DB	ArangoDB
Property graph	Yes	Yes	Yes	No	No
RDF support	No	No	Yes	Yes	No
Multimodel	No	No	Yes	Yes	Yes
Open source	Yes	Yes	No	No	Yes
Cloud deployment	Yes	Yes	Yes	Yes	Yes
On-prem deployment	Yes	Yes	No	No	Yes
Clustered support	Yes	Yes	Yes	Yes	Yes
Query language	Cypher	nGQL	SPARQL / Gremlin	Gremlin	AQL

5.3.1 Cypher Basic Operations

In this section, we are going to introduce the basic operations of Cypher. Each operation is illustrated with examples relevant to practical applications.

5.3.1.1 Creating Nodes

Nodes represent entities in your graph, such as users, products, or locations. To create simple nodes, use the following Cypher command:

```
CREATE (p:Person {name: "Alice", age: 30});
CREATE (p:Person {name: "Bob", age: 25});
```

These commands create nodes labeled `Person` with properties for `name` and `age`. Creating nodes is foundational for graph modeling as they represent the core entities of your data.

```
MERGE (p:Person {name: "Alice", age: 30})
```

MERGE works like a combination of MATCH and CREATE. It tries to match the pattern specified in the command. If it finds a match, it does nothing; if no match is found, it creates the node. In this example, MERGE will check for a Person node with name: "Alice" and age: 30. If such a node exists, MERGE does nothing. If it doesn't, MERGE creates the node. The match is based on both properties (name and age). This means if a Person node with name: "Alice" but a different age (e.g., age: 25) already exists, MERGE will create a new node instead of matching the existing one. MERGE is ideal for cases where you want to avoid duplicate entries but still want to create a node if it doesn't exist.

5.3.1.2 Creating Relationships

Relationships connect nodes and define how entities are related, which is crucial for building the graph structure. Here's how to create a relationship between two nodes:

```
MATCH (a:Person {name: "Alice"}), (b:Person {name: "Bob"})
CREATE (a)-[r:KNOWS {since: 2020}]->(b)
```

This command establishes a `KNOWS` relationship between Alice and Bob, with a `since` property indicating the year they met. Understanding relationships allows for richer queries and insights, especially in social networks or collaborative applications.

5.3.1.3 Querying Nodes

To retrieve nodes based on specific criteria, you can execute a query like this:

```
MATCH (p:Person)
WHERE p.age > 25
RETURN p.name, p.age
```

This query finds all Person nodes where the age is greater than 25 and returns their names and ages. By effectively querying nodes, you can extract meaningful insights from your graph data, such as identifying users above a certain age for targeted marketing campaigns.

5.3.1.4 Querying Relationships

To find connections between nodes, use the following query:

```
MATCH (a:Person)-[r:KNOWS]->(b:Person)
WHERE a.name = "Alice"
RETURN b.name, r.since
```

This command retrieves all individuals Alice knows, along with the year they met. This operation is essential in exploring relationships in social networks, allowing for the analysis of friend networks and their interactions.

5.3.1.5 Updating Properties

To modify the properties of an existing node or relationship, you can use:

```
MATCH (p:Person {name: "Alice"})
SET p.age = 31
```

This updates Alice's age to 31. You can also update multiple properties at once, like so:

```
MATCH (p:Person {name: "Alice"})
SET p.age = 31, p.location = "New York"
```

Updating properties helps maintain accurate and relevant data in your graph.

5.3.1.6 Adding Labels

You can assign additional labels to nodes to better categorize and manage them:

```
MATCH (p:Person {name: "Alice"})
SET p:Employee
```

This adds the Employee label to Alice's node. Labels can help organize data for more efficient querying and can often be added during node creation or property updates.

5.3.1.7 Removing Nodes and Relationships

To delete nodes or relationships, you can use:

```
MATCH (p:Person {name: "Charlie"})
DETACH DELETE p
```

This command removes Charlie and all associated relationships from the graph. It's crucial to use DETACH DELETE to avoid orphaned relationships and maintain graph integrity.

5.3.1.8 Aggregation and Counting

Graph queries can perform aggregations, such as counting relationships:

```
MATCH (p:Person)-[:KNOWS]->(friend)
RETURN p.name, COUNT(friend) AS friendCount
```

This query counts how many friends each person has. Other aggregation functions like AVG, SUM, and COLLECT can also provide deeper insights, such as the average age of friends or lists of friend names.

5.3.1.9 Pathfinding

To find paths between nodes, you can execute:

```
MATCH path = shortestPath((a:Person)-[:KNOWS*]-(b:Person))
WHERE a.name = "Alice" AND b.name = "Bob"
RETURN path
```

This command finds the shortest path of KNOWS relationships connecting Alice and Bob. Pathfinding is vital in applications like recommendation systems and network analysis, where understanding connections can drive insights.

These fundamental operations lay the groundwork for querying graphs. They enable you to create, read, update, and delete graph data, as well as perform complex operations such as aggregation and pathfinding. As you gain familiarity with these basics, you can delve into more advanced features specific to your chosen graph database system.

For further exploration of Cypher's capabilities, refer to the official Neo4j documentation on Cypher Introduction. Besides the Cypher basic operations, Neo4j also provides two often used features: APOC and GDS, which we are going to introduce in detail in the next sections.

5.3.2 Neo4j Utility Functions: APOC

APOC, short for "**awesome procedures on cypher,**" is a powerful library of utilities that supercharge Neo4j's Cypher query language. Think of it as a Swiss army knife for developers working with graph databases—offering a broad set of tools that fill in the gaps left by native Cypher.

From data import/export and dynamic Cypher scripting to graph refactoring and path expansion, APOC enables use cases that would otherwise be difficult, if not impossible, to implement cleanly.

APOC is split into two libraries:

- APOC Core—Officially supported by Neo4j, bundled with most distributions.
- APOC Extended—Community-maintained, offering experimental or specialized features.

All APOC procedures are written in Java and can be invoked directly from Cypher. You can find the complete documentation here: APOC Procedures & Functions.

APOC functions span across many practical areas. A breakdown of the key categories in the following:

5.3.2.1 Collection Utilities (apoc.coll)

This is the most widely used category in APOC. Cypher provides basic list handling, but real-world graph data often needs more powerful list operations. This category gives you functions for checking membership, combining lists, flattening nested structures, and set operations like union and intersection.

For example, `apoc.coll.flatten()` can turn [[1, 2], [3, 4]] into [1, 2, 3, 4]. `apoc.coll.union()` merges two lists without duplicates, and `apoc.coll.containsAll()` checks if a list includes all elements of another list. These are frequently used in data transformations, filters, and aggregations. They make Cypher much more flexible and expressive when working with collections.

5.3.2.2 General Utilities(apoc)

This category includes a grab bag of essential tools that add basic programming logic to Cypher. Cypher lacks native conditionals or dynamic query control, which is where these functions shine.

For example, `apoc.when()` allows if-else branching, and `apoc.do.when()` executes full Cypher blocks based on conditions. These are commonly used in procedures where logic flow depends on data properties. Other helpful functions like `apoc.version()` or `apoc.help()` support debugging and introspection. Without these, complex Cypher workflows would be clumsy or impossible to implement cleanly.

5.3.2.3 Import/Export Utilities (apoc.load, apoc.export)

These functions are critical for getting data in and out of Neo4j in practical applications.

`apoc.load.json()` is one of the most used APOC functions—it allows you to read JSON from files or APIs directly into Cypher. `apoc.load.csv()` is a more flexible and powerful alternative to Cypher's built-in LOAD CSV.

Export functions like `apoc.export.csv.all()` or `apoc.export.graphml.query()` allow exporting full graphs or query results in various formats for reporting, backups, or sharing. These are essential for integrations, data pipelines, and working with external systems.

5.3.2.4 Path Expansion (apoc.path)

This category gives you much more control over traversals than Cypher's MATCH patterns. With `apoc.path.expandConfig()`, you can define specific relationship types, directions, labels to filter, and depth in a single call. This is extremely useful in complex graphs with many relationship types.

Functions like `apoc.path.subgraphAll()` return the entire connected component from a node, and `apoc.path.spanningTree()` helps build structures like organization charts or category hierarchies. These functions are often used in data exploration, recommendations, or understanding the reachability in networks.

5.3.2.5 Graph Refactoring (apoc.refactor)

This category is essential for managing and evolving the structure of your graph. During data import, deduplication, or schema changes, these functions help transform nodes and relationships.

For example, `apoc.refactor.mergeNodes()` merges duplicate nodes while preserving relationships and properties. `apoc.refactor.rename.label()` lets you rename labels, which isn't possible in pure Cypher. These tools are commonly used during data cleaning, migrations, or after ingesting messy datasets.

5.3.2.6 Periodic/Batch Processing (apoc.periodic)

Cypher executes all at once, which can be a problem with very large data sets. APOC's periodic functions allow you to iterate over large datasets in smaller, more manageable chunks.

The most used function here is `apoc.periodic.iterate()`, which runs a large read query and processes the results in batches. It supports parallel execution, which significantly improves performance.

`apoc.periodic.submit()` lets you run jobs in the background, and `apoc.periodic.repeat()` enables scheduled repetition, making it useful for automating recurring tasks like cleanup or sync jobs.

5.3.2.7 Map and List Utilities (apoc.map, apoc.list)

These functions allow more advanced manipulation of maps (key-value data) and lists:

- `apoc.map.merge()` merges two or more maps, which is useful when combining properties from multiple sources.
- `apoc.map.removeKey()` removes keys from a map, useful for cleaning sensitive or unnecessary data.

On the list side, `apoc.list.sort()` helps when ordering items that aren't naturally sortable in Cypher. These utilities are often used for shaping and transforming data dynamically in complex queries.

5.3.2.8 Date and Time Utilities (apoc.date)

While Neo4j supports temporal types, this category adds more flexibility, especially for formatting and parsing dates.

`apoc.date.parse()` converts string dates into timestamps using custom formats, and `apoc.date.format()` does the reverse. This is particularly useful when dealing with data from external sources that don't follow Neo4j's native date format.

`apoc.date.currentTimestamp()` gives you the current time in milliseconds, which is useful for time-stamping or tracking durations.

5.3.2.9 Algorithm Utilities (apoc.algo)

Before the graph data science (GDS) library was released, APOC included basic graph algorithms like Dijkstra and shortest path. While still available, these are largely deprecated in favor of GDS, which is more powerful, accurate, and scalable.

Functions like `apoc.algo.shortestPath()` and `apoc.algo.dijkstra()` still work for small graphs or if you don't have GDS installed, but they're no longer the recommended approach for serious graph analytics.

5.3.2.10 Security Utilities (apoc.security)

These functions provide basic cryptographic operations, like hashing and generating secure random strings.

`apoc.security.sha256()` and `apoc.security.md5()` hash strings for basic security or data integrity. `apoc.security.rand()` is useful for generating secure random tokens or IDs. While not frequently used in every project, they are critical when you need them—such as in user authentication flows or anonymizing sensitive data.

5.3.2.11 Bitwise and Math Functions (apoc.bitwise, apoc.math)

These categories provide lower-level numeric functions that are rarely used in general Cypher queries but can be important in specific domains.

`apoc.bitwise.and()` or `apoc.bitwise.or()` are used when encoding flags or working with binary data. `apoc.math.round()`, `sqrt()`, and others fill in the gaps for numerical processing. These are mostly for edge cases in scientific or financial applications.

5.3.2.12 Aggregation Functions (apoc.agg)

These functions provide additional aggregation capabilities not built into Cypher. For example, `apoc.agg.percentiles()` calculates percentiles overvalues in a collection, and `apoc.agg.groupBy()` helps in summarizing grouped data. These can be helpful in analytics dashboards or reporting, though they are used less frequently than the core utilities above.

5.3.3 Neo4j Graph Analytics: GDS

The Neo4j **GDS** library is a powerful toolkit for analytics and machine learning on graph-structured data. It provides high-performance, parallelized implementations of common graph algorithms and exposes them as Cypher procedures. GDS also includes machine learning pipelines for tasks like link prediction and node classification.

5.3.3.1 Running Graph Algorithms

In Section 5.1.4, we've introduced the major graph algorithms. In this section, we are going to introduce the implementation details of graph algorithms using GDS.

The standard workflow for utilizing GDS algorithms involves the following structured approach:

5.3.3.1.1 *Step 1: Project a Graph*

Before executing any graph analysis, you must first project a graph from your Neo4j database into an in-memory graph. This in-memory representation enables efficient computation and is stored as a named graph in the graph catalog. By assigning a name to the graph during projection, you create a persistent reference that can be reused across multiple analytical tasks.

The graph catalog serves as a centralized repository for these named graphs, allowing you to:

- Manage graph instances via management operations (e.g., listing, dropping, or refreshing graphs).
- Avoid redundant projections by reusing the same named graph for subsequent algorithm runs, thereby optimizing performance.

This step is foundational because all GDS algorithms require an in-memory graph as input. Properly projecting and cataloging the graph ensures seamless execution of analyses, result inspection, and optional data persistence back to Neo4j.

5.3.3.1.2 *Step 2: Choose and Run an Algorithm*

All algorithms are exposed as Neo4j procedures. Syntax outline:

```
CALL gds[.<tier>].<algorithm>.<execution-mode>[.<estimate>](
  graphName: String,
  configuration: Map
)
```

GDS supports several categories of graph algorithms. Table 5.5 lists the key families and their most commonly used members of the graph algorithms.

5.3.3.1.3 *Step 3: Choose an Execution Mode*

When executing a GDS algorithm, select from three execution modes to control how results are returned or stored. Each mode serves distinct use cases:

- Stream Mode: Returns computed results directly as Cypher query results, enabling real-time analysis or integration into applications. This mode is ideal for scenarios where results are only needed temporarily, as it does not modify the in-memory graph or Neo4j database—functioning similarly to standard Cypher read operations.

Table 5.5 Graph algorithm categories of GDS.

Category	Key algorithms	Primary use cases
Centrality	PageRank, Betweenness, Eigenvector	Identify influential nodes
Community detection	Louvain, Label Propagation, Leiden	Find clusters/groups
Similarity	Node similarity, KNN, Jaccard	Measure node similarity
Pathfinding	Dijkstra, A*, BFS	Find shortest/optimal paths
Node embeddings	Node2Vec, GraphSAGE, FastRP	Convert nodes to ML-friendly vectors
Link prediction	Common neighbors, Adamic–Adar	Predict missing/future connections
DAG algorithms	Topological sort, SCC	Analyze dependencies in directed acyclic graphs

- Mutate Mode: Updates the in-memory graph in the graph catalog with algorithm outputs (e.g., node scores or relationship properties). By embedding results directly into the graph, this mode allows subsequent algorithms to leverage intermediate data without re-projecting the graph, optimizing performance for multistep workflows.
- Write Mode: Permanently stores algorithm results (e.g., centrality metrics or community labels) back into the Neo4j database. This mode behaves like standard Cypher write operations, ensuring results are retained for long-term use, integration with existing data, or further analysis outside the in-memory context.

5.3.3.2 Machine Learning Pipelines in GDS

GDS provides machine learning pipelines to support end-to-end workflows for predictive modeling on graphs, including tasks like link prediction and node classification. These pipelines combine feature extraction, model training, and prediction into a cohesive process, managed through dedicated catalogs:

- The pipeline catalog stores reusable pipeline configurations defining how features are extracted and models are structured.
- The model catalog retains trained models for easy retrieval and reuse.

The workflow begins by configuring a pipeline, specifying the features to extract from the graph and the type of machine learning model (e.g., logistic regression or random forest). This configuration is saved in the pipeline catalog for future reference. Next, the pipeline is used to train a model using labeled data from the graph, and the resulting model is stored in the model catalog.

Once trained, the model can generate predictions through three execution modes:

- Stream mode returns predictions immediately as Cypher query results, enabling real-time analysis or integration into applications.
- Mutate mode embeds predictions directly into the in-memory graph (stored in the graph catalog), allowing subsequent algorithms to leverage the results without re-projecting the graph.
- Write mode persists predictions to the Neo4j database, ensuring long-term storage and compatibility with standard Cypher write operations for downstream use.

This structured approach ensures seamless transitions from data preparation to model deployment, while catalogs and execution modes give users flexibility to optimize performance, retain results, or integrate predictions into broader workflows.

5.4 Practice: Flight Network Analysis and Optimization

The objective of this section is to demonstrate how to set up and analyze a flight graph using Neo4j. We'll walk through the entire process, from loading real-world aviation data and preparing the network for advanced graph analytics to applying key algorithms to solve practical logistics challenges.

5.4.1 Problem Setting

Imagine stepping into the role of Director of Global Logistics at a major international air cargo and passenger company. Your mission: optimize and analyze a complex network of airports and flight routes using graph algorithms.

5.4.2 Understanding the Network Structure

Our aviation network forms a directed graph, where each Airport node contains metadata such as geolocation and time zone, and each ROUTE relationship captures flight-specific attributes such as aircraft type, distance, number of stops, and loyalty rewards or penalties.

- **Nodes** represent airports, containing properties such as:
  - Geographic coordinates (latitude, longitude)
  - Airport identifiers (IATA, ICAO codes)
  - Location details (city, country, timezone)
  - Operational data (altitude, type)
- **Edges** represent flight routes with properties including:
  - Distance (in kilometers)
  - Equipment type
  - Number of stops
  - Airline information
  - Reward/deduction factors (for optimization scenarios)

While the network is primarily directed (flights have specific directions), we'll transform it as needed—creating DAGs for certain algorithms or treating it as undirected for analyses like MSTs.

5.4.3 Set Up Database and Load Data

Before performing analytics on the flight network, we need to establish our working environment, prepare the Neo4j database, and load the data. This section guides you through connecting to Neo4j and loading the aviation dataset.

5.4.3.1 Connecting to a Local Neo4j Server

Make sure the Neo4j server is set up, which is detailed in Appendix A. To connect to your local Neo4j server from the command line, you can use the Neo4j Cypher Shell.

1. Open Command Line Interface: Open a new terminal or command prompt.
2. Run Cypher Shell: Use the following command to connect to your local Neo4j instance:

```
docker exec -it neo4j-db bin/cypher-shell
```

   To exit Cypher shell:

```
:exit
```

3. Use Cypher Shell: Once connected, you can run Cypher queries directly from the command line. Here are some examples:
   - Inserting Data: To create nodes and relationships, use the following commands:

```
CREATE (a:Person {name: "Alice", age: 30}),
   (b:Person {name: "Bob", age: 25}),
   (a)-[:KNOWS {since: 2020}]->(b);
```

     This command creates two "Person" nodes (Alice and Bob) and a "KNOWS" relationship between them.
   - Querying Data: To retrieve data, use the following query:

```
MATCH (p:Person)
RETURN p.name, p.age;
```

     This command returns the names and ages of all "Person" nodes in the database.

- Deleting All Data: To remove all nodes and relationships in the database, use the following command:

```
MATCH (n)
DETACH DELETE n;
```

This command deletes all nodes and their relationships, effectively clearing the database.

5.4.3.2 Neo4j Tools for Visualization and Interaction

Neo4j provides several free and open-source tools to visualize and interact with your local server:

- Neo4j Browser: This is the web-based interface that comes with Neo4j. It allows you to run queries, visualize the graph, and explore your data interactively. You can access it at `http://localhost:7474`. Figure 5.10 is the snapshot of the previous section example.
- Neo4j Desktop: A desktop application that provides an integrated development environment for managing Neo4j projects, including multiple database instances, visualization tools, and project management features. You can download it from the Neo4j Desktop page.
- Graph App: Neo4j has various plugins and apps available in the Neo4j Desktop that enhance functionality. The Graph App allows you to visualize data in different formats and provides additional analysis tools.
- Browser-based Visualization: You can also use external visualization tools like GraphXR or Gephi, which can connect to Neo4j and provide advanced visualization capabilities.

5.4.3.3 Load the OpenFlights Dataset

To build a real-world flight network, we use the publicly available OpenFlights dataset, which includes airport and route data in CSV format. This dataset includes over 7,000 airports and 67,000 routes worldwide.

5.4.3.3.1 Step 1: Loading Airport Nodes

First, we'll create nodes for all airports that have associated routes:

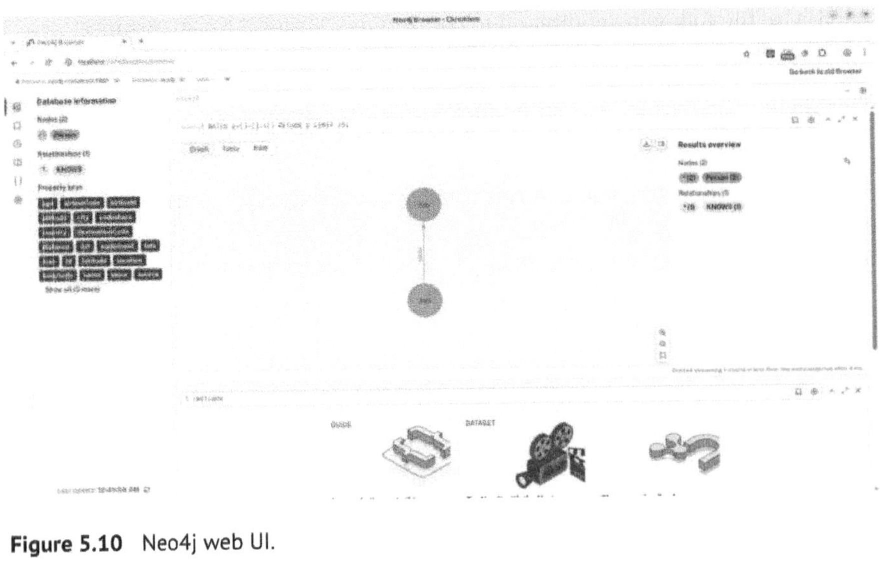

Figure 5.10 Neo4j web UI.

```
LOAD CSV FROM 'https://raw.githubusercontent.com/jpatokal/openflights/master/data/
    airports.dat' AS airportRow
WITH airportRow
WHERE airportRow[0] IS NOT NULL
CREATE (:Airport {
  id: toInteger(airportRow[0]),
  name: airportRow[1],
  city: airportRow[2],
  country: airportRow[3],
  iata: airportRow[4],
  icao: airportRow[5],
  latitude: toFloat(airportRow[6]),
  longitude: toFloat(airportRow[7]),
  altitude: toInteger(airportRow[8]),
  timezone: toFloat(airportRow[9]),
  dst: airportRow[10],
  tzDatabase: airportRow[11],
  type: airportRow[12],
  source: airportRow[13]
})
```

This query:

- Loads airport data from the CSV file.
- Filters out null entries.
- Creates Airport nodes with comprehensive properties.
- Ensures data type consistency (integers for IDs, floats for coordinates).

Expected Result: Added 7,698 labels, created 7,698 nodes, set 107,419 properties.

5.4.3.3.2 Step 2: Creating Flight Route Relationships

Next, we'll establish ROUTE relationships between airports, calculating distances using the Haversine formula:

```
LOAD CSV FROM 'https://raw.githubusercontent.com/jpatokal/openflights/master/data/
    routes.dat' AS routeRow
MATCH
(src:Airport {id: toInteger(routeRow[3])}),
(dst:Airport {id: toInteger(routeRow[5])})
WHERE src IS NOT NULL AND dst IS NOT NULL
AND src.latitude IS NOT NULL AND src.longitude IS NOT NULL
AND dst.latitude IS NOT NULL AND dst.longitude IS NOT NULL
WITH
src, dst, routeRow,
point({longitude: src.longitude, latitude: src.latitude}) AS srcPoint,
point({longitude: dst.longitude, latitude: dst.latitude}) AS dstPoint
CREATE (src)-[r:ROUTE]->(dst)
SET r += {
  airlineCode: routeRow[0],
  airlineId: toInteger(routeRow[1]),
  sourceCode: routeRow[2],
  destinationCode: routeRow[4],
  codeshare: routeRow[6],
  stops: toInteger(routeRow[7]),
  equipment: routeRow[8],
  distance: round(point.distance(srcPoint, dstPoint) / 1000.0)  // distance in km
}
```

This query performs several critical operations:

1. Matches existing airports by their IDs from the route data.
2. Validates coordinates to ensure distance calculation is possible.
3. Creates geographic points from latitude/longitude pairs.

4. Calculates distances using Neo4j's built-in Haversine distance function.
5. Establishes directed relationships with comprehensive route properties.

Expected Result: Set 533,993 properties, created 66,771 relationships.

5.4.3.3.3 *Understanding the Distance Calculation*

The `point.distance()` function implements the Haversine formula, which calculates the great-circle distance between two points on a sphere. This gives us the shortest distance between airports "as the crow flies," accounting for Earth's curvature. The result is divided by 1,000 to convert from meters to kilometers.

Figure 5.11 gives an illustration of the flight graph data in the Neo4j web UI. With your aviation network loaded and ready, you're now equipped to apply graph algorithms to real-world logistics challenges, which we'll discuss in the next section.

5.4.4 Graph Algorithms in Flight Networks

Business Context: As the Director of Global Logistics, you face complex challenges in optimizing flight networks spanning thousands of airports and routes. This section demonstrates how graph algorithms transform raw network data into actionable insights for route optimization, cost reduction, and operational efficiency.

We'll explore real-world applications using Neo4j's graph database, progressing from basic network exploration to sophisticated structural analysis. Each algorithm addresses specific business challenges faced by modern airlines.

5.4.4.1 Network Exploration and Path Discovery

The foundation of network optimization lies in understanding connectivity. We begin with fundamental pathfinding algorithms that answer critical questions about route availability and efficiency.

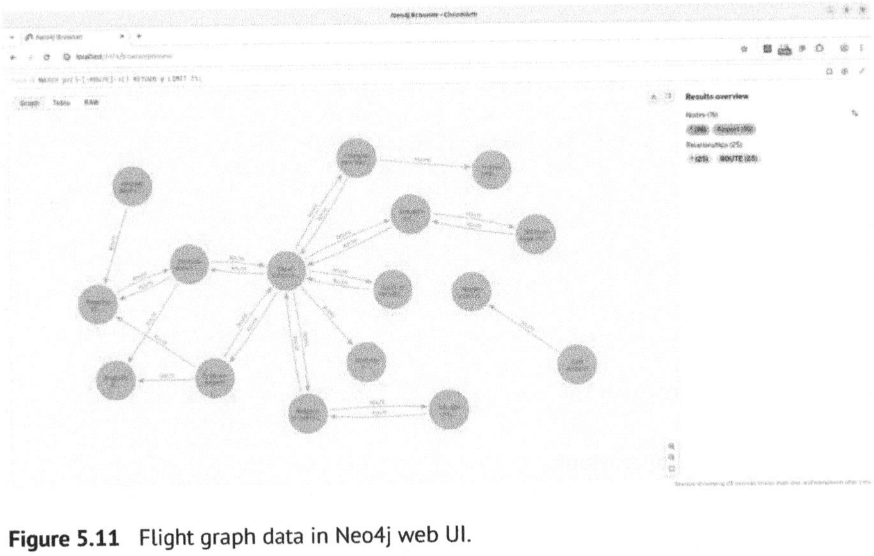

Figure 5.11 Flight graph data in Neo4j web UI.

5.4.4.1.1 Minimum Hop Routing with BFS

Your first task is to quickly measure how many flight hops are needed to move cargo from San Francisco International Airport (SFO) to Chhatrapati Shivaji Maharaj Airport (BOM).

Why BFS? BFS explores the graph level-by-level, ensuring the minimum number of hops. This is ideal for scenarios where cost per segment (rather than total distance) is the key factor—for example, when each flight incurs fixed handling fees regardless of distance.

Implementation:

```
MATCH path = shortestPath(
  (:Airport {iata: 'SFO'})-[:ROUTE*]->(:Airport {iata: 'BOM'})
)
RETURN
  [node IN nodes(path) | node.iata] AS itinerary,
  length(path) AS segments
```

Result:

itinerary	segments
["SFO", "CDG", "BOM"]	2

Business Insight: The shortest path requires 2 hops (edges), meaning two flights are needed to reach BOM from SFO. This information is crucial for:

- Estimating cargo handling time (each hop adds processing time).
- Calculating segment-based costs.
- Meeting customer delivery expectations.

5.4.4.1.2 Finding Alternative Routes

A follow-up question is: How many distinct paths exist with exactly 2 hops between SFO and BOM? This "counting all possible minimal-hop paths" problem helps identify:

- Backup routes for contingency planning
- Load balancing opportunities
- Competitive routing options

A solution using `allShortestPaths()`:

```
MATCH path = allShortestPaths(
  (:Airport {iata: 'SFO'})-[:ROUTE*]->(:Airport {iata: 'BOM'})
)
RETURN DISTINCT
  [node IN nodes(path) | node.iata] AS itinerary,
  length(path) AS segments
```

Results:

itinerary	segments
["SFO", "MUC", "BOM"]	2
["SFO", "CDG", "BOM"]	2
...	...

Business Value: While the minimal path length is 2 hops, multiple intermediate airports (e.g., Munich or Zurich) satisfy this condition. This provides flexibility in routing based on available capacity, options for different alliance partners, and risk mitigation through route diversity.

5.4.4.1.3 *Custom Pathfinding with DFS*

Sometimes you need more control over path exploration than built-in functions provide. Here's how to manually implement pathfinding with custom constraints.

Advanced Query with Distance Calculations:

```
MATCH path = (start:Airport {iata: 'SFO'})
             -[:ROUTE*..2]->
             (end:Airport {iata: 'BOM'})
WHERE ALL(n IN nodes(path)
          WHERE SINGLE(x IN nodes(path) WHERE x = n))
WITH path,
     reduce(total = 0, r IN relationships(path) |
            total + r.distance) AS total_distance
RETURN DISTINCT
       [node IN nodes(path) | node.iata] AS itinerary,
       length(path) AS segments,
       [rel IN relationships(path) |
        rel.distance] AS segment_distances,
       total_distance
ORDER BY total_distance ASC
```

[1] The `*..2` quantifier limits paths to 1 or 2 hops.
[2] `WHERE ALL(...)` ensures no repeated airports (avoids cycles).
[3] `reduce()` calculates total distance for each valid path.

Sample Results:

itinerary	segments	segment_distances	total_distance
["SFO", "ICN", "BOM"]	2	[9095.0, 5554.0]	14649.0
["SFO", "DXB", "BOM"]	2	[13035.0, 1928.0]	14963.0
...	...	...	...

5.4.4.1.4 *Constrained Search: Regional Network Analysis*

From SFO, find the shortest possible routes (up to 2 connecting flights) to all reachable airports where the total flight distance does not exceed 1,000 km. This represents regional network analysis for:

- Short-haul cargo operations
- Regional airline partnerships
- Emergency medical transport routes

Implementation:

```
MATCH path = (start:Airport {iata: 'SFO'})-[:ROUTE*..2]->(dest:Airport)
WITH path, dest, reduce(
  total = 0.0,
  r IN relationships(path) | total + r.distance
) AS total_distance
WHERE start <> dest AND total_distance <= 1000
ORDER BY total_distance ASC
WITH dest, COLLECT({
  itinerary: [n IN nodes(path) | n.iata],
  segments: length(path),
  segment_distances: [r IN relationships(path) | r.distance],
  total_distance: total_distance
})[0] AS best_path
RETURN
```

```
    best_path.itinerary AS itinerary,
    best_path.segments AS segments,
    best_path.segment_distances AS segment_distances,
    best_path.total_distance AS total_distance
```

[1] The query searches paths of 1 or 2 hops `ROUTE*..2`.
[2] The `WHERE` clause filters results to paths less than or equal to 1,000 km.
[3] For airports with multiple qualifying paths, the `COLLECT` and `[0]` constructs select the shortest-distance route.

Results:

itinerary	segments	segment_distances	total_distance
["SFO", "MRY"]	1	[124.0]	124.0
...	...	...	...
["SFO", "RNO", "SJC"]	2	[309.0, 303.0]	612.0
...	...	...	...

The key takeaways of the network exploration and path discovery are:

1. BFS excels at finding paths with minimum hops, crucial for segment-based cost optimization.
2. Multiple shortest paths provide operational flexibility and risk mitigation.
3. Custom path queries enable complex business rules and constraints.
4. Distance-constrained searches help define regional networks and service areas.

5.4.4.2 Analyzing Network Structure

Understanding the structural properties of your airline network reveals hidden patterns and optimization opportunities.

5.4.4.2.1 Cycle Detection with DFS/BFS

Inefficient loops such as LAX → SEA → JFK → LAX can cause delays and waste resources. Your goal is to detect these cycles within the network and propose improvements.

Why This Matters: Cycle detection helps identify:

- Redundancies in the network that can be streamlined.
- Circular routes that may indicate inefficient planning.
- Opportunities for direct flight additions.

Finding Largest Cycles by Distance: Detects the largest cycles by total distance within AA Airlines' route network inside the United States, starting and ending at SFO, limited to 3 flight segments.

```
MATCH (start:Airport {iata: 'SFO'})
MATCH path = (start)-[:ROUTE*..3]->(start)
WITH path, reduce(
  total = 0.0,
  r IN relationships(path) | total + r.distance
) AS total_distance
WHERE ALL(r IN relationships(path) WHERE r.airlineCode = 'AA')
  AND ALL(n IN nodes(path) WHERE n.country = 'United States')
ORDER BY total_distance DESC
RETURN
  [n IN nodes(path) | n.iata] AS route_iata_codes,
  [r IN relationships(path) | r.airlineCode] AS airline_codes,
```

```
    [r IN relationships(path) | r.distance] AS segment_distances_km,
    total_distance AS total_distance_km
LIMIT 3
```

Results:

route_iata_codes	airline_codes	segment_distances	total
["SFO", "JFK", "MIA", "SFO"]	["AA", "AA", "AA"]	[4156.0, 1759.0, 4158.0]	10073.0
["SFO", "MIA", "JFK", "SFO"]	["AA", "AA", "AA"]	[4158.0, 1759.0, 4156.0]	10073.0
["SFO", "PHL", "MIA", "SFO"]	["AA", "AA", "AA"]	[4052.0, 1636.0, 4158.0]	9846.0

5.4.4.2.2 SCC Analysis

Find groups of airports in your network that are mutually reachable via available routes. Understanding clusters of airports that frequently interchange routes helps identify SCCs—groups of airports that are mutually reachable.

Why SCC Matters: SCC breaks the network into regions that operate as cohesive units, which is crucial for:

- Regional optimization strategies.
- Identifying isolated network segments.
- Planning hub-and-spoke architectures.

Creating the Graph for SCC Analysis:

```
CALL gds.graph.drop('ssc', false);

MATCH (a:Airport)-[r:ROUTE]->(b:Airport)
WITH a AS source, b AS target, collect(r)[0] AS route
WITH gds.graph.project('ssc', source, target) AS g
RETURN g.graphName, g.nodeCount, g.relationshipCount;
```

Running the SCC Algorithm:

```
CALL gds.scc.write('ssc', {writeProperty: 'scc'})
YIELD componentCount;
```

componentCount
48

Analyzing SCC Groups:

```
MATCH (a:Airport)
WHERE a.scc IS NOT NULL
RETURN a.scc AS scc_group, count(*) AS airport_count
ORDER BY airport_count DESC, scc_group ASC;
```

Examining a Specific SCC:

```
MATCH (a:Airport)
WHERE a.scc = 3020
RETURN a.name, a.city, a.country, a.iata, a.icao
```

Business Insight: This SCC represents a group of remote Alaskan airports that form their own isolated network segment, requiring special consideration for service planning.

scc_group	airport_count
0	3147
1474	10
2952	4
...	...

a.name	a.city	a.country	a.iata	a.icao
"Unalaska Airport"	"Unalaska"	"United States"	"DUT"	"PADU"
"Akutan Seaplane Base"	"Akutan"	"United States"	"KQA"	"KQA"
"Atka Airport"	"Atka"	"United States"	"AKB"	"PAAK"
"Nikolski Air Station"	"Nikolski"	"United States"	"IKO"	"PAKO"

5.4.4.3 Working with DAGs

5.4.4.3.1 Creating and Analyzing DAG Subgraphs

DAGs are essential for modeling dependencies in airline operations, such as flight sequences, crew schedules, and maintenance workflows. We'll create a DAG from our network to demonstrate dependency-based algorithms.

Step-by-step DAG Creation:

Step 1: Reset Previous Data:

```
CALL gds.graph.drop('dag', false);
MATCH (n:AirportDAG) DETACH DELETE n;
```

Step 2: Generate DAG with Random Sampling:

To create a DAG, we randomly sample edges while ensuring no cycles exist. Different random seeds and edge counts affect the resulting structure.

```
MATCH (a1:Airport {country:"United States"})-[r:ROUTE]->(a2:Airport {country:"United
    States"})
with a1, a2, collect(r)[0] as route
ORDER BY apoc.hashing.fingerprint(a1.id + 'random_seed_12' + a2.id)
LIMIT 150 //  edge count
MERGE (newA1:AirportDAG {id: a1.id, iata: a1.iata})
MERGE (newA2:AirportDAG {id: a2.id, iata: a2.iata})
MERGE (newA1)-[:ROUTE_DAG {distance: route.distance}]->(newA2)
return newA1.iata, newA2.iata
ORDER by newA1.iata, newA2.iata
```

Step 3: Verify DAG Properties:

```
MATCH (a1)-[r:ROUTE_DAG]->(a2)
WITH gds.graph.project('dag', a1, a2) AS g
CALL gds.wcc.stats('dag')
YIELD componentCount as wcc
CALL gds.scc.stats('dag')
YIELD componentCount as scc
return g.nodeCount,scc, g.nodeCount = scc as isDag,  wcc
```

Step 4: Analyze DAG Structure:

```
CALL gds.wcc.stream('dag')
YIELD nodeId, componentId
with componentId, collect(gds.util.asNode(nodeId).iata) as iata
ORDER BY size(iata) desc, componentId, iata
return componentId, iata, size(iata)
```

Step 5: Find Complex Paths in the DAG:

This Cypher query finds all possible flight paths between every pair of distinct airports and then identifies the longest path and the total number of paths for each pair.

```
MATCH (origin:AirportDAG), (destination:AirportDAG)
WHERE id(origin) <> id(destination)
CALL {
  WITH origin, destination
  MATCH routePath = (origin)-[:ROUTE_DAG*1..]->(destination)
  ORDER BY length(routePath) DESC
  RETURN collect(routePath) AS allPaths
}
WITH origin, destination, allPaths, size(allPaths) AS totalPaths
WHERE totalPaths > 0
RETURN
  origin.iata AS originIATA,
  destination.iata AS destinationIATA,
  totalPaths,
  [airport IN nodes(allPaths[0]) | airport.iata] AS longestPathIATASequence
ORDER BY totalPaths DESC
```

Results:

originIATA	destinationIATA	totalPaths	longestPathIATASequence
"IND"	"GRK"	4	["IND", "IAD", "LAX", "DFW", "GRK"]
"SBA"	"SJC"	3	["SBA", "SEA", "DTW", "CMH", "MDW", "SAN", "SLC", "SJC"]
"SBA"	"RNO"	3	["SBA", "SEA", "DTW", "CMH", "MDW", "SAN", "SLC", "SJC", "RNO"]
...			

5.4.4.3.2 *Topological Sorting for Dependency Management*

Given the DAG as your company's promotion campaign network or flight dependency chain, you need to deliver cargo in the right order. Topological sorting helps plan shipments by ensuring all dependencies (e.g., meeting flight schedules) are met through a linear order of operations.

Implementation:

```
CALL gds.dag.topologicalSort.stream("dag", {
  computeMaxDistanceFromSource: true
})
YIELD nodeId, maxDistanceFromSource
RETURN gds.util.asNode(nodeId).iata AS iata,
  maxDistanceFromSource
ORDER BY maxDistanceFromSource, iata
```

Practical Applications:

- Scheduling maintenance windows ensuring aircraft availability.
- Planning crew rotations considering rest requirements.
- Coordinating ground services in the correct sequence.
- Managing cargo shipments with specific routing requirements.

5.4.4.4 Advanced Route Optimization

5.4.4.4.1 Bellman–Ford Algorithm: Handling Complex Reward Systems

Airlines often have complex loyalty programs where the "cost" of a route isn't simply distance or price. The Bellman–Ford algorithm excels at finding optimal paths even when some edges have negative weights.

Scenario: Premium Member Reward Optimization

As a premium member, you receive a fixed reward of 1000 points for every flight segment, regardless of distance. However, each segment also has a distance-based deduction, where 1 point equals 1 km flown.

Example Calculation: For a flight path from IND to GRK with stopovers, the points earned per leg:

- JFK → WAW: 2,500 – 6,855 = –4,355 points
- WAW → EVN: 2,500 – 2,231 = 269 points
- EVN → IKA: 2,500 – 794 = 1,706 points
- IKA → BOM: 2,500 – 2,802 = –302 points
- Total Points Accumulated = –4,355 + 269 + 1,706 – 302 = –2,682 points

The goal is to determine the path from IND to GRK that yields the maximum total reward points.

Technical Note: Cypher is declarative, not imperative—which makes implementing algorithms like Bellman–Ford directly (with mutable state like dist[v] = min(...)) challenging. However, GDS provides built-in functions to solve this.

Implementation Steps:

Step 1: Create Graph with Negative Weights

```
CALL gds.graph.drop('negative_weights', false);

MATCH (a)-[r:ROUTE_DAG]->(b)
with gds.graph.project('negative_weights', a, b, {
  sourceNodeProperties: a {.id},
  targetNodeProperties: b {.id},
  relationshipProperties: { points: r.distance-1000 }
}) as g
RETURN  g.graphName, g.nodeCount, g.relationshipCount;
```

g.graphName	g.nodeCount	g.relationshipCount
"negative_weights"	143	150

Step 2: Verify Edge Weights

```
CALL gds.graph.relationshipProperty.stream('negative_weights','points')
YIELD sourceNodeId, targetNodeId, propertyValue
WITH
  gds.util.asNode(sourceNodeId) AS source,
  gds.util.asNode(targetNodeId) AS target,
  propertyValue
RETURN
  source.iata AS start_iata,  target.iata AS end_iata,  propertyValue
ORDER BY propertyValue
```

Step 3: Find Optimal Path for Maximum Points

Check the negative weights for paths between IND and GRK:

```
MATCH (:AirportDAG {iata: 'IND'})-[r:ROUTE_DAG*]->(:AirportDAG {iata: 'GRK'})
UNWIND r AS rr
with distinct rr as rrr
```

start_iata	end_iata	propertyValue
"SHG"	"OBU"	−988.0
"LNY"	"MKK"	−956.0
...	...	...

```
WITH collect ( [id(startNode(rrr)), id(endNode(rrr))]) AS pairList
CALL gds.graph.relationshipProperty.stream('negative_weights', 'points')
YIELD sourceNodeId, targetNodeId, propertyValue
WHERE [sourceNodeId, targetNodeId] in pairList
RETURN
  gds.util.asNode(sourceNodeId).iata AS start_iata,
  gds.util.asNode(targetNodeId).iata AS end_iata,
  propertyValue
ORDER BY start_iata, end_iata
```

start_iata	end_iata	propertyValue
"ATL"	"GRK"	291.0
"BUF"	"ATL"	148.0
...	...	...

Step 4: Execute Bellman-Ford Algorithm

```
MATCH (source:AirportDAG {iata: 'IND'}), (target:AirportDAG {iata: 'GRK'})
CALL gds.bellmanFord.stream('negative_weights', {
  sourceNode: source,
  relationshipWeightProperty: 'points'
})
YIELD index, sourceNode, targetNode, totalCost, nodeIds, costs, route, isNegativeCycle
where targetNode=id(target)
RETURN
  index,
  gds.util.asNode(sourceNode).iata AS sourceNode,
  gds.util.asNode(targetNode).iata AS targetNode,
  - totalCost as maxPoints,
  [nodeId IN nodeIds | gds.util.asNode(nodeId).iata] AS nodeNames,
  costs,
  relationships(route),
  isNegativeCycle as isNegativeCycle
```

Result:

nodeNames	costs	relationships (route)	isNegativeCycle
["IND", "IAD", "TYS", "DFW", "GRK"]	[0.0, −234.0, −559.0, −318.0, −1101.0]	...	false

5.4.4.4.2 Dijkstra's Algorithm: Classic Shortest Path

Consider a scenario of "High-Priority Shipment Routing." A time-sensitive shipment must fly from San Francisco (SFO) to Mumbai (BOM). You need to choose the route that minimizes the total distance. Dijkstra's algorithm efficiently computes the shortest path when all edges are nonnegative.

Implementation:

```
MATCH (origin:Airport {iata: 'SFO'}), (destination:Airport {iata: 'BOM'})
CALL apoc.algo.dijkstra(origin, destination, 'ROUTE', 'distance')
YIELD path, weight
RETURN
  [airport IN nodes(path) | airport.iata] AS airportPath,
  [route IN relationships(path) | route.distance] AS segmentDistances,
  weight AS totalDistance
```

Result:

airportPath	segmentDistances	totalDistance
["SFO", "PEK", "LXA", "KTM", "BOM"]	[9505.0, 2625.0, 572.0, 1594.0]	14296.0

5.4.4.4.3 A Search Algorithm: Heuristic-based Optimization*

For premium services requiring faster computation than Dijkstra's algorithm, A* uses heuristics (e.g., "as-the-crow-flies" distance) to guide the search.

Implementation:

```
MATCH (origin:Airport {iata: 'SFO'}), (destination:Airport {iata: 'BOM'})
CALL apoc.algo.aStarConfig(origin, destination, 'ROUTE', {})
YIELD path, weight
RETURN
  [airport IN nodes(path) | airport.iata] AS airportPath,
  [route IN relationships(path) | route.distance] AS segmentDistances,
  weight AS totalDistance
```

Note: Default configuration uses: { weight = 'distance', y = 'latitude', x = 'longitude' }
Result:

airportPath	segmentDistances	totalDistance
`["SFO", "DXB", "BOM"]`	[13035.0, 1928.0]	14963.0

Trade-off: This algorithm is faster but may return a slightly suboptimal result (14296 vs. 14963 km).

5.4.4.5 Constructing Efficient Networks

5.4.4.5.1 MST–Building Emergency Networks

Design a robust emergency backup network for the United States that connects all airports with minimum infrastructure cost. The MST ensures every airport is reachable while minimizing total connection distance.

Graph Requirements:

- Graph Type: Typically applied to undirected graphs (we convert our directed graph by assuming symmetric distances).
- Purpose: Finds the subset of routes connecting every airport with minimum total distance (cost).

Implementation Steps:

Step 1: Reset and Prepare Data

```
CALL gds.graph.drop('mst', false);
CALL gds.graph.drop('verify_mst', false);

MATCH (n:AirportMST) DETACH DELETE n;
```

Step 2: Create Undirected Graph

```
MATCH (a:Airport {country: "United States"})-[r:ROUTE]->(b:Airport {country: "United
    States"})
with gds.graph.project('mst', a, b, {relationshipProperties: r {.distance}}, {
    undirectedRelationshipTypes: ['*']}) as g
RETURN  g.graphName, g.nodeCount, g.relationshipCount;
```

g.graphName	g.nodeCount	g.relationshipCount
"mst"	549	21036

Step 3: Run Prim's Algorithm from SFO

```
MATCH (n:Airport {iata: 'SFO'})
CALL gds.spanningTree.stream('mst', {
  sourceNode: n,
  relationshipWeightProperty: 'distance'
})
YIELD nodeId,parentId, weight
with gds.util.asNode(nodeId) as node, gds.util.asNode(parentId) AS parent, weight
merge (a:AirportMST {id: node.id, iata: node.iata})
merge (b:AirportMST {id: parent.id, iata: parent.iata})
merge (a)-[:ROUTE_MST {distance: weight}]-(b)
RETURN node.iata , parent.iata ,weight
ORDER BY node.iata , parent.iata
```

Sample Results:

node.iata	parent.iata	weight
"ABE"	"PHL"	88.0
"ABI"	"DFW"	254.0
...	...	...

Step 4: Verify MST Properties

```
MATCH (a1)-[r:ROUTE_MST]->(a2)
WITH gds.graph.project('verify_mst', a1, a2) AS g
CALL gds.wcc.stats('verify_mst')
YIELD componentCount as wcc
CALL gds.scc.stats('verify_mst')
YIELD componentCount as scc
return g.nodeCount,scc, g.nodeCount = scc as isDag,  wcc
```

Verification Results:

g.nodeCount	scc	isDag	wcc
541	541	true	1

5.4.4.5.2 Floyd–Warshall Algorithm: All-pairs Shortest Paths

Management requires a comprehensive dashboard showing the shortest possible distance between every pair of airports in the US network.

Algorithm Characteristics:

- Graph Compatibility: Works on both directed and undirected graphs.
- Weight Handling: Accommodates positive and moderate negative weights (without negative cycles).
- Purpose: Computes shortest paths between all pairs, providing a holistic network view.

Implementation:

Step 1: Create Graph for Floyd–Warshall

```
CALL gds.graph.drop('fw', false);

MATCH (a:Airport {country: "United States"})-[r:ROUTE]->(b:Airport {country: "United
    States"})
RETURN gds.graph.project('fw', a, b, {
  relationshipProperties: r {.distance}
});
```

Step 2: Run Floyd–Warshall Algorithm

```
CALL gds.allShortestPaths.stream('fw', {
  relationshipWeightProperty: 'distance'
})
YIELD sourceNodeId, targetNodeId, distance
WHERE gds.util.isFinite(distance) = true and  sourceNodeId <> targetNodeId
WITH gds.util.asNode(sourceNodeId).iata AS source, gds.util.asNode(targetNodeId).iata
    AS target, distance
RETURN source,  target, distance
ORDER BY distance DESC, source ASC, target ASC
LIMIT 10
```

Results (Longest Shortest-paths):

source	target	distance
"ADK"	"PSM"	9889.0
"PSM"	"ADK"	9889.0
...	...	...

Step 3: Verify Results with Dijkstra

```
MATCH (origin:Airport {iata: 'ADK'}), (destination:Airport {iata: 'PSM'})
CALL apoc.algo.dijkstra(origin, destination, 'ROUTE', 'distance')
YIELD path, weight
RETURN
  [airport IN nodes(path) | airport.iata] AS airportPath,
  [route IN relationships(path) | route.distance] AS segmentDistances,
  weight AS totalDistance
```

Verification Result:

airportPath	segmentDistances	totalDistance
["ADK", "ANC", "MSP", "MLI", "SFB", "PSM"]	[1915.0, 4047.0, 441.0, 1641.0, 1845.0]	9889.0

5.4.4.5.3 Summary

This comprehensive exploration of graph algorithms in airline network optimization demonstrates the power of graph databases for solving complex real-world problems. Through our journey from basic pathfinding to sophisticated network analysis, we've shown how:

1. **Pathfinding algorithms** (BFS, DFS) provide fundamental connectivity insights.
2. **Structural analysis** (cycle detection, SCC) reveals network patterns and inefficiencies.
3. **Dependency management** (topological sorting on DAGs) enables proper scheduling.
4. **Advanced routing** (Bellman–Ford, Dijkstra, A*) optimizes for different business objectives.
5. **Network construction** (MST, Floyd–Warshall) supports strategic infrastructure planning.

These algorithms, implemented through Neo4j's GDS library, transform raw network data into actionable business intelligence, enabling airlines to optimize operations, reduce costs, and improve service quality. The combination of graph theory and practical implementation provides a powerful toolkit for modern logistics management.

5.5 Summary

In this chapter, we've journeyed through the fundamental concepts and practical applications of graphs and graph databases, establishing a solid foundation for understanding how interconnected data can be modeled, stored, and analyzed. We began by exploring the diverse types of graphs—from simple undirected graphs to complex weighted, directed networks—and examined various representation methods, including adjacency matrices and lists, each with their own trade-offs between memory efficiency and computational performance.

Our exploration of graph operations revealed the power of traversal algorithms, from basic random walks to sophisticated pathfinding techniques like BFS and DFS. We dived into structural analysis methods, including SCCs and topological sorting, which enable us to understand the underlying patterns and hierarchies within complex networks. The comprehensive survey of graph algorithms—spanning centrality measures, community detection, similarity analysis, and node embeddings—demonstrated how graphs can reveal insights about influence, clustering, and relationships that would be hidden in traditional tabular data structures.

The practical aspects of graph data management came to life through our examination of popular graph databases, with Neo4j, NebulaGraph, Amazon Neptune, Azure Cosmos DB, and ArangoDB each offering unique strengths for different use cases. We mastered Cypher, the expressive query language that makes graph traversal intuitive, and explored advanced features through APOC's utility functions and GDS's powerful algorithm library. These tools transform complex graph operations into accessible, production-ready solutions.

Our hands-on practice with flight networks brought theory into practice, demonstrating how real-world networks can be analyzed using the full spectrum of graph algorithms. From exploring network connectivity with BFS and DFS to optimizing routes with Dijkstra's and Bellman–Ford algorithms, to constructing efficient networks with MSTs, we've seen how graph algorithms solve practical problems in transportation, logistics, and beyond.

As we move forward to Chapter 6 on Knowledge Graphs, remember that the graph fundamentals mastered here—the structures, algorithms, and database technologies—form the backbone of semantic data representation. The ability to model relationships explicitly, traverse connections efficiently, and apply sophisticated algorithms to networked data will prove invaluable as we explore how knowledge graphs combine these technical capabilities with semantic meaning to create intelligent information systems.

References

1 Douglas Brent West. *Introduction to Graph Theory*, volume 2. Upper Saddle River: Prentice Hall, 2001.

2 Kazimierz Kuratowski. Sur le problème des courbes gauches en topologie. *Fundamenta Mathematicae*, 15:271–283, 1930.

3 Harold W Kuhn. The Hungarian method for the assignment problem. *Naval Research Logistics Quarterly*, 2(1–2):83–97, 1955. https://doi.org/10.1002/nav.3800020109.

4 Claude Berge. *Hypergraphs: Combinatorics of Finite Sets*, volume 45. Elsevier, 1984.

5 Alain Bretto. *Hypergraph Theory*. An Introduction. Mathematical Engineering, volume 1, pages 209–216. Cham: Springer, 2013.

6 Herbert Edelsbrunner and John Harer. *Computational Topology: An Introduction*. American Mathematical Society, 2010.

7 Petter Holme and Jari Saramäki. Temporal networks. *Physics Reports*, 519(3):97–125, 2012.

8 Petter Holme and Jari Saramäki. *Temporal Network Theory*, volume 2. Springer, 2019.

9 Graham Klyne and Jeremy J Carroll. Resource Description Framework (RDF): Concepts and Abstract Syntax. W3C Recommendation, W3C, Feb 2004. https://www.w3.org/TR/2004/REC-rdf-concepts-20040210/.

10 Amit Singhal. Introducing the knowledge graph: Things, not strings. In *Official Google Blog*, May 2012. https://googleblog.blogspot.com/2012/05/introducing-knowledge-graph-things-not.html.

11 Aditya Grover and Jure Leskovec. node2vec: Scalable feature learning for networks. In *Proceedings of the 22nd ACM SIGKDD International Conference on Knowledge Discovery and Data Mining*, pages 855–864, 2016.

12 Edsger W Dijkstra. A note on two problems in connexion with graphs. *Numerische Mathematik*, 1(1):269–271, 1959. https://doi.org/10.1007/BF01386390.

13 Richard Bellman. On a routing problem. *Quarterly of Applied Mathematics*, 16(1):87–90, 1958. https://doi.org/10.1090/qam/102435.

14 LR Ford Jr. Network flow theory. Technical Report P-923, RAND Corporation, 1956. https://www.rand.org/pubs/papers/P923.html.

15 Robert W Floyd. Algorithm 97: Shortest path. *Communications of the ACM*, 5(6):345, 1962. https://doi.org/10.1145/367766.368168.

16 Stephen Warshall. A theorem on boolean matrices. *Journal of the ACM*, 9(1):11–12, 1962. https://doi.org/10.1145/321105.321107.

17 Thomas H Cormen, Charles E Leiserson, Ronald L Rivest, and Clifford Stein. *Introduction to Algorithms*. MIT Press, 3rd edition, 2009.

18 Joseph B Kruskal. On the shortest spanning subtree of a graph and the traveling salesman problem. *Proceedings of the American Mathematical Society*, 7(1):48–50, 1956. https://doi.org/10.2307/2033241.

19 Robert C Prim. Shortest connection networks and some generalizations. *Bell System Technical Journal*, 36(6):1389–1401, 1957. https://doi.org/10.1002/j.1538-7305.1957.tb01515.x.

20 Micha Sharir. A strong-connectivity algorithm and its applications in data flow analysis. In *Computers & Mathematics with Applications*, volume 7, pages 67–72. Elsevier, 1981.

21 Robert E Tarjan. Depth-first search and linear graph algorithms. *SIAM Journal on Computing*, 1(2):146–160, 1972.

22 Arthur B Kahn. Topological sorting of large networks. *Communications of the ACM*, 5(11):558–562, 1962.

23 Reinhard Diestel. *Graph Theory*. Springer, 5th edition, 2017. ISBN 9783662536216.

24 Vincent A Traag, Ludo Waltman, and Nees Jan Van Eck. From Louvain to Leiden: Guaranteeing well-connected communities. *Scientific Reports*, 9(1):5233, 2019.

25 Usha Nandini Raghavan, Réka Albert, and Soundar Kumara. Near linear time algorithm to detect community structures in large-scale networks. *Physical Review E*, 76(3):036106, 2007.

26 Peter E Hart, Nils J Nilsson, and Bertram Raphael. A formal basis for the heuristic determination of minimum cost paths. *IEEE Transactions on Systems Science and Cybernetics*, 4(2):100–107, 1968.

27 Will Hamilton, Zhitao Ying, and Jure Leskovec. Inductive representation learning on large graphs. *Advances in Neural Information Processing Systems*, 30:1024–1034, 2017.

28 Haochen Chen, Syed Fahad Sultan, Yingtao Tian, Muhao Chen, and Steven Skiena. Fast and accurate network embeddings via very sparse random projection. In *Proceedings of the 28th ACM International Conference on Information and Knowledge Management*, pages 399–408, 2019.

29 David Liben-Nowell and Jon Kleinberg. The link-prediction problem for social networks. *Journal of the American Society for Information Science and Technology*, 58(7):1019–1031, 2007.

30 Lada A Adamic and Eytan Adar. Friends and neighbors on the web. *Social Networks*, 25(3):211–230, 2003.

31 Albert-László Barabási and Réka Albert. Emergence of scaling in random networks. *Science*, 286(5439):509–512, 1999.

32 Ian Robinson, Jim Webber, and Emil Eifrem. *Graph Databases: New Opportunities for Connected Data*. O'Reilly Media, 2nd edition, 2015.

33 NebulaGraph Team. Nebulagraph documentation, 2023. https://docs.nebula-graph.io/.

34 Brad Bebee, Daniel Choi, Anurag Gupta, Andi Gutmans, Ankesh Khandelwal, Yashwanth Kiran, Sainath Mallidi, Bruce McGrath, Mike Personick, Karthik Rajan, et al. Amazon Neptune: Graph data management in the cloud. In *The Semantic Web–ISWC 2018 Posters & Demonstrations*. Springer, 2018.

35 Amazon Web Services. Amazon neptune user guide, 2023. https://docs.aws.amazon.com/neptune/latest/userguide/intro.html.

36 Microsoft Corporation. Azure cosmos DB documentation, 2023. https://docs.microsoft.com/en-us/azure/cosmos-db/.

37 ArangoDB GmbH. Arangodb white paper, 2022. https://www.arangodb.com/why-arangodb/white-paper/.

6

Knowledge Graphs

As the digital age continues to progress, the need to organize, interpret, and utilize vast amounts of data has never been more pressing. In this context, knowledge graphs (KGs) have emerged as a powerful tool, offering a way to represent information in a structured, interconnected manner that closely mirrors human understanding. A KG links concepts and entities in a way that allows machines to comprehend and reason about data, making it invaluable in domains ranging from search engines to artificial intelligence.

This chapter explores the fundamental principles, construction methods, and advanced techniques that make KGs invaluable for both human understanding and machine intelligence. We begin by examining the basic building blocks of a KG and explore these concepts using a hands-on example: the Marvel Universe KG. We then transition to practical aspects of graph construction, including schema design, data extraction techniques, and real-world implementation workflows. With the graph in place, we explore how to enrich its semantics and utility using graph algorithms (e.g., centrality, community detection, and link prediction), graph embeddings, and graph neural networks (GNNs). Finally, we examine how KGs and LLMs can complement one another—enabling automated knowledge extraction, ontology learning, and hybrid systems like Graph-RAG. Throughout the chapter, practical exercises and case studies (e.g., Wayfair Annotation Dataset [WANDS] product data in Neo4j) help reinforce the concepts and illustrate how KGs can be operationalized in real-world settings.

Whether you're interested in building semantic search systems, enhancing AI with structured knowledge, or simply understanding how platforms like Google organize the world's information, this chapter provides both theoretical foundations and hands-on practices—including real-world implementations using Neo4j and LlamaIndex. By the end, you'll understand not only how to construct and enrich KGs, but also how they serve as a bridge between human-readable knowledge and machine-processable intelligence, particularly in the era of generative AI.

6.1 Understanding KGs

A KG is a structured framework for organizing information as a network of interconnected entities—objects, events, or concepts—and their relationships. Entities in a KG are represented as nodes, while the relationships between them are denoted as edges, creating a dynamic and contextual representation of knowledge. This graph-based structure is ideal for encoding semantic meaning, enabling tasks such as reasoning, information retrieval, and decision-making.

Unlike traditional databases, which rely on rigid schemas, KGs excel at modeling complex and evolving relationships, offering versatility for real-world applications. They provide a foundation for solving problems in areas like semantic search, question answering, and AI-driven insights across diverse industries.

6.1.1 Core Structure and Components

In Chapter 5, we introduced the concept of general graphs. Building on that foundation, a KG represents a specialized type of graph that models real-world entities and their relationships with rich semantic context. Unlike simple network graphs, KGs encode meaning and domain knowledge into their structure.

Key features that distinguish KGs include:

- **Nodes (Entities):** Represent real-world objects, concepts, or ideas such as people, places, products, or events.
- **Edges (Relationships):** Define specific, typed connections between entities.
- **Properties:** Attach additional attributes to both nodes and edges.
- **Semantic Meaning:** Incorporate domain-specific knowledge, ontologies, and taxonomies.
- **Inference Capabilities:** Support deriving new knowledge through reasoning.

To illustrate the difference between a basic graph and a KG, consider these examples (shown in Figure 6.1):

- In a basic graph (social network):

  Person A – connected_to –> Person B

- In a KG:

  [Person A: name="John Smith", occupation="Teacher", years_experience=15]
  –works_with_since_2020–>
  [Person B: name="Mary Jones", occupation="Principal", institution="Lincoln High"]

Table 6.1 lists the comparisons of the characteristics of basic graph and KG. The KG provides rich context about both the entities and their relationships, enabling more sophisticated querying and inference.

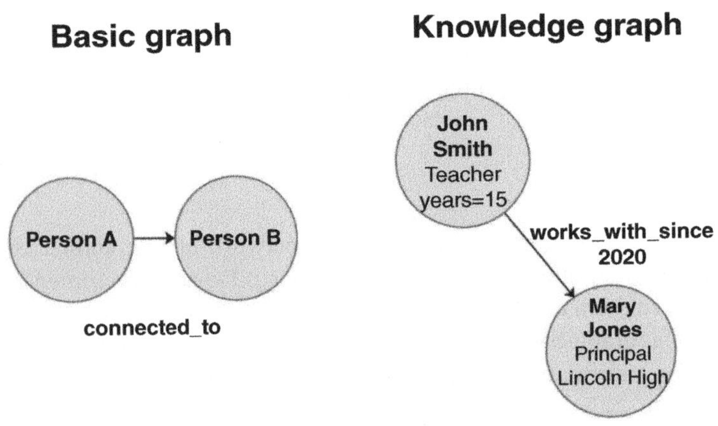

Figure 6.1 Comparison of basic graph and KG.

Table 6.1 Comparison of basic graph and KG.

Basic Graph	KG
Nodes: Unlabeled	Nodes: Labeled with properties (e.g., occupation)
Edges: Generic (e.g., "connected_to")	Edges: Typed and contextual (e.g., "works_with_since_2020")
No semantics	Supports reasoning (e.g., "manages" implies hierarchy)

KGs play a foundational role in many technologies that shape our digital experiences, quietly driving intelligence behind the scenes in everyday systems.

In search engines, for instance, Google's use of its KG revolutionized how users find information. Instead of simply matching keywords, the system understands the entities within a query and their relationships. This enables features like rich result panels, disambiguation of ambiguous terms, and contextually relevant suggestions—offering users more precise and meaningful answers.

Virtual assistants such as Alexa, Siri, and Google Assistant rely heavily on KGs to understand and interpret natural language queries. When a user asks, "What's the weather like in Paris this weekend?," the assistant parses the location, the time frame, and the weather concept—resolving each into structured knowledge before assembling an answer. This ability to contextualize information and infer intent is made possible by the underlying graph of entities and relationships.

In the realm of recommendation systems, companies like Netflix and Amazon use KGs to move beyond simplistic behavioral patterns. They incorporate rich relationships between users, items, genres, preferences, and even external metadata (such as actors, directors, or customer reviews), allowing them to suggest content that aligns more closely with a user's latent interests.

Healthcare systems are also being transformed by KGs. By connecting symptoms, diseases, drugs, treatments, and patient histories, medical KGs assist practitioners in diagnosing conditions, recommending treatments, or identifying potential side effects. These graphs enable decision-support tools that are both interpretable and context-aware.

In financial services, KGs help identify complex fraud patterns. By modeling transactions, account holders, locations, and behavioral signals as a network of interconnected entities, they allow for the detection of unusual or suspicious relationships—highlighting risks that might go unnoticed in traditional data tables.

Ultimately, KGs excel in environments where data is deeply interconnected and constantly evolving. They serve not only as a method of organizing information but also as a bridge between structured data and human-like reasoning—supporting systems that must respond to nuanced, context-dependent queries with speed and precision.

6.1.2 Example: Marvel Universe KG

To illustrate the structure and expressive power of a KG, let's examine a simplified example drawn from the Marvel Universe. This example (in Figure 6.2) shows how a wide variety of interconnected entities—such as characters, teams, locations, and artifacts—can be organized in a graph-based format to reflect the rich semantics of the domain.

The Marvel Universe presents unique modeling challenges that mirror real-world KG applications. Characters have multiple identities (Peter Parker/Spider-Man), belong to various teams simultaneously, possess evolving relationships, and interact with artifacts that have their own rich histories. These complexities showcase how KGs excel at representing multifaceted, interconnected information that would be difficult to capture in traditional database structures.

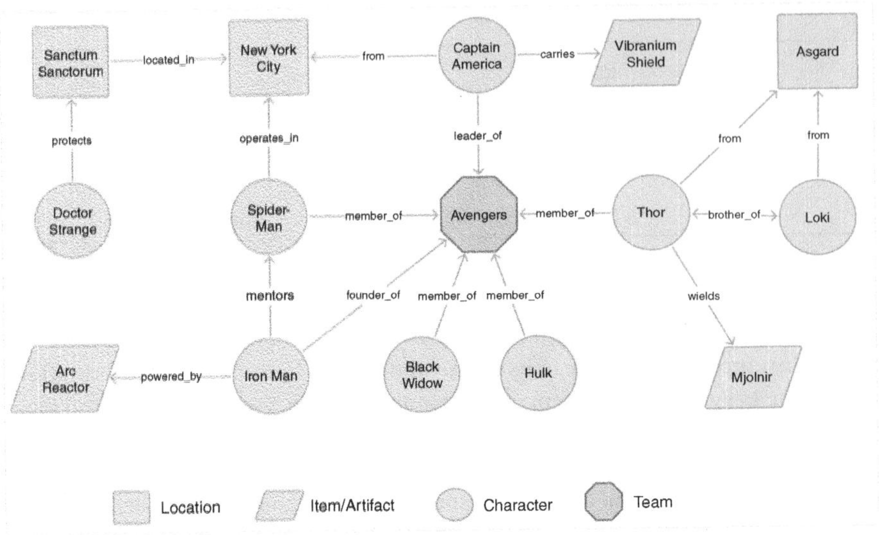

Figure 6.2 Marvel universe KG.

6.1.2.1 Building Marvel KG

6.1.2.1.1 Entity Types and Their Properties

In our Marvel KG, we define four primary entity types, each with distinct characteristics:

Characters (Circle Nodes): Characters form the heart of our graph, representing heroes, villains, and supporting figures. Each character node contains not just a name but can be enriched with properties such as:

- Real name and aliases (e.g., Tony Stark/Iron Man)
- Powers and abilities
- First appearance dates
- Current status (active, deceased, retired)

Our example includes: Spider-Man, Iron Man, Captain America, Thor, Hulk, Black Widow, Doctor Strange, and Loki.

Teams (Octagon Nodes): Team nodes represent organized groups that characters affiliate with. These nodes can include properties like:

- Founding date
- Base of operations
- Current roster size
- Team purpose or mission

Featured teams: Avengers.

Locations (Rectangle Nodes): Location nodes capture the geographical and dimensional spaces where stories unfold:

- New York City: The bustling metropolis home to many heroes.
- Asgard: The mystical realm of Norse gods.
- Sanctum Sanctorum: Doctor Strange's mystical headquarters.

Items/Artifacts (Parallelogram Nodes) These nodes represent significant objects with their own histories and powers:

- Mjolnir: Thor's enchanted hammer.
- Vibranium Shield: Captain America's iconic defensive weapon.
- Arc Reactor: Tony Stark's life-sustaining power source.

6.1.2.1.2 *Relationship Types: The Semantic Connections*

The true power of our KG emerges through typed relationships that carry semantic meaning:

Team Affiliations

- `MEMBER_OF`: Standard team membership (e.g., Spider-Man → Avengers)
- `FOUNDER_OF`: Indicates founding members (e.g., Iron Man → Avengers)
- `LEADER_OF`: Denotes leadership roles (e.g., Captain America → Avengers)

Item Relationships

- `WIELDS`: For weapons actively used in combat (e.g., Thor → Mjolnir)
- `CARRIES`: For signature equipment (e.g., Captain America → Vibranium Shield)
- `POWERED_BY`: For items integral to a character's abilities (e.g., Iron Man → Arc Reactor)
- `COLLECTED`: For items acquired or sought (e.g., Thanos → Infinity Stones)

Location Relationships

- `OPERATES_IN`: Primary area of activity (e.g., Spider-Man → New York City)
- `FROM`: Origin location (e.g., Thor → Asgard)
- `PROTECTS`: Guardianship roles (e.g., Doctor Strange → Sanctum Sanctorum)
- `LOCATED_IN`: Hierarchical location relationships (e.g., Sanctum Sanctorum → New York City)

Character Relationships

- `BROTHER_OF`: Family connections (e.g., Thor ↔ Loki)
- `MENTORS`: Teaching relationships (e.g., Iron Man → Spider-Man)
- Additional relationships could include `RIVALS_WITH`, `ALLIES_WITH`, or `CREATED_BY`

6.1.2.1.3 *Graph Data Structure*

The KG data can be stored in a graph data structure, illustrated in this code example:

```
const marvelKnowledgeGraph = {
    nodes: [
        // Characters with enriched properties
        {
            id: 'spider-man',
            label: 'Spider-Man',
            type: 'character',
            properties: {
                realName: 'Peter Parker',
                powers: ['web-slinging', 'spider-sense', 'enhanced strength'],
                firstAppearance: 'Amazing Fantasy #15 (1962)'
            }
        },
        {
            id: 'iron-man',
            label: 'Iron Man',
            type: 'character',
            properties: {
                realName: 'Tony Stark',
                occupation: 'Inventor, CEO',
                affiliation: 'Stark Industries'
            }
        },
        // ... additional nodes with properties
    ],

    edges: [
        {
            source: 'iron-man',
            target: 'avengers',
            type: 'founder_of',
            properties: {
                since: '1963',
                role: 'Founding Member'
```

```
                }
            },
            {
                source: 'thor',
                target: 'mjolnir',
                type: 'wields',
                properties: {
                    requirement: 'worthiness',
                    abilities_granted: ['flight', 'lightning control']
                }
            },
            // ... additional edges with properties
        ]
};
```

6.1.2.2 KG Representation

There are multiple ways, in which a KG can be represented.

6.1.2.2.1 Visual Representation

The visual graph representation (as shown in Figure 6.2) provides an intuitive understanding of the network structure. Colors distinguish entity types while edge styles and labels clarify relationship types. This representation excels at: identifying clusters and communities; spotting important nodes (high connectivity); understanding overall graph topology; and presenting information to nontechnical audiences.

6.1.2.2.2 Adjacency List Representation

An adjacency list efficiently stores the graph structure for computational purposes:

```
const adjacencyList = {
    'spider-man': [
        { node: 'avengers', relation: 'member_of' },
        { node: 'new-york', relation: 'operates_in' },
        { node: 'iron-man', relation: 'mentored_by' }
    ],
    'iron-man': [
        { node: 'avengers', relation: 'founder_of' },
        { node: 'arc-reactor', relation: 'powered_by' },
        { node: 'spider-man', relation: 'mentors' }
    ],
    // ... continues for all nodes
};
```

This representation is ideal for cases such as: graph traversal algorithms; finding neighbors of a node; memory-efficient storage; dynamic graph updates.

6.1.2.2.3 Adjacency Matrix Representation

For analytical operations, an adjacency matrix provides mathematical convenience:

```
             | spider-man | iron-man | avengers | new-york | ...
spider-man   |     0      |    0     |    1     |    1     | ...
iron-man     |     1      |    0     |    1     |    0     | ...
avengers     |     1      |    1     |    0     |    0     | ...
new-york     |     1      |    0     |    0     |    0     | ...
...          |    ...     |   ...    |   ...    |   ...    | ...
```

where:

- 1 indicates a relationship exists.
- 0 indicates no direct relationship.
- Can be extended to weighted matrices for relationship strength.
- Cannot have key-value properties on edges.

6.1.2.3 Query and Extend the KG

6.1.2.3.1 Querying Our Marvel KG

The power of this KG structure becomes apparent when we consider the types of queries it can answer:

Simple Queries:

- "Which heroes are members of the Avengers?"
- "What artifacts does Thor possess?"
- "Where does Spider-Man operate?"

Complex Queries:

- "Find all heroes who are based in New York City AND are members of the Avengers."
- "Which characters have connections to Asgard through relationships or origin?"
- "What is the shortest path of relationships between Spider-Man and Thanos?"

Inference Queries:

- "Which characters might be affected if the Arc Reactor technology is compromised?" (following `POWERED_BY` relationships)
- "Who could potentially wield Mjolnir?" (based on character attributes and relationships)

6.1.2.3.2 Extending the KG

This Marvel Universe example demonstrates key KG principles that scale to real-world applications:

- Entity Resolution: Just as Peter Parker and Spider-Man represent the same entity, real-world KGs must handle multiple representations of the same concept.
- Temporal Dynamics: Team memberships change over time—Iron Man wasn't always an Avenger. Production KGs often include temporal properties on relationships.
- Multi-hop Reasoning: Finding that Doctor Strange could help Thor involves traversing multiple relationships: Thor → Asgard (from), Asgard has mystical properties, Doctor Strange → Sanctum Sanctorum (protects), Sanctum Sanctorum deals with mystical threats.
- Schema Flexibility: New entity types (like "Events" for major storylines) or relationship types can be added without restructuring the entire graph.

By modeling the Marvel Universe as a KG, we've illustrated how this structure captures complex, interconnected information in a way that's both human-readable and computationally tractable. These same principles apply whether you're modeling fictional universes, corporate knowledge, scientific literature, or any other domain where understanding relationships is as important as understanding individual entities.

6.1.3 Types of KGs

Different types of KGs cater to varied use cases and data complexities. Some prominent examples are given in the following:

6.1.3.1 Encyclopedic KGs

These graphs aim to comprehensively capture facts about the real world by aggregating structured and semi-structured data from a wide range of public sources—encyclopedias, structured databases, and curated datasets. They have broad coverage of people, places, events, concepts, etc. They power general-purpose information retrieval and question-answering systems, such as digital assistants and semantic web tools. Some examples include Wikidata (used by Google's Knowledge Panel, Bing, and Siri) [1] and DBpedia (extracts structured content from Wikipedia) [2].

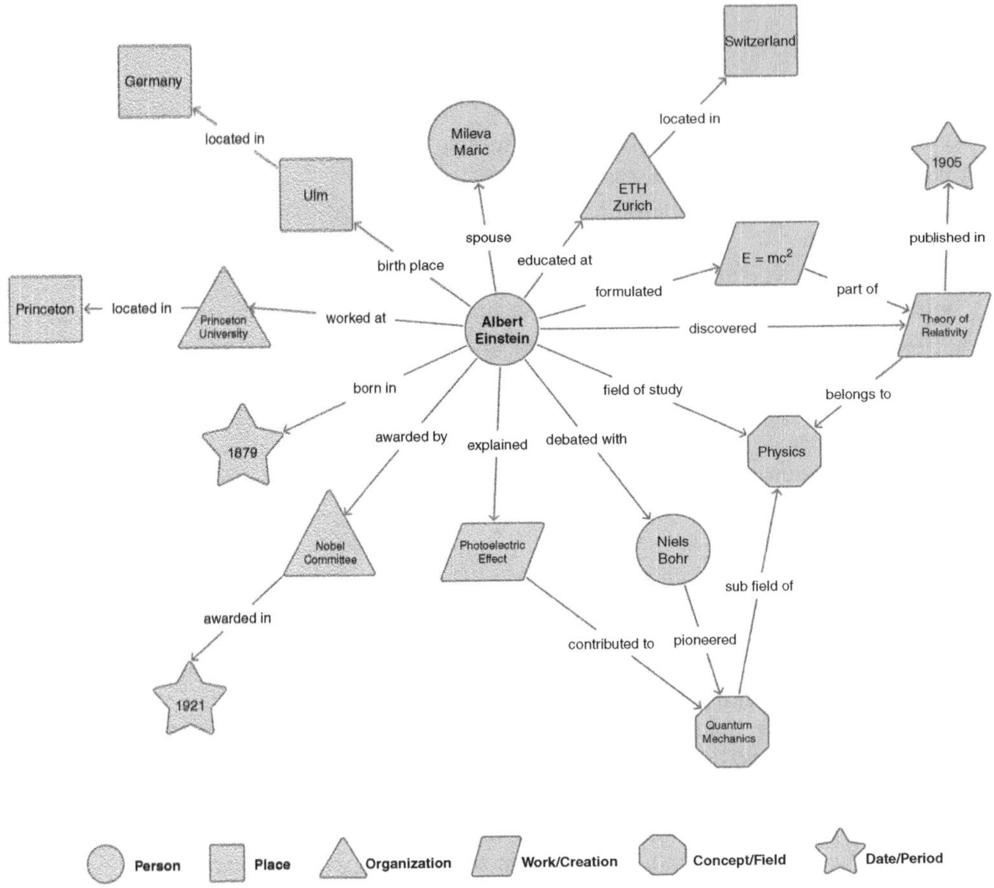

Figure 6.3 Example of encyclopedic knowledge: Einstein's life, work, and contributions.

Figure 6.3 illustrates an encyclopedic KG about Einstein's life, work and contributions, with nodes like "Albert Einstein," "Princeton University," "Physics," "1879," "Theory of Relativity" connected via "workedAt," "bornIn," "discovered," "fieldOfStudy," etc.

6.1.3.2 Commonsense KGs

These graphs represent implicit human knowledge—things we take for granted in everyday life but are rarely explicitly stated. They are usually lightweight graphs focused on everyday logic and often crowd-sourced or extracted from text corpora. Commonsense KGs encode facts like causality, affordance, intention, and typical event sequences. They are essential in reasoning, dialog systems, and grounded natural language understanding.

For example, Figure 6.4 illustrates a commonsense KG about "Daily Life and Weather," with nodes like "Summer," "Winter," "Sunny," "Rain," "Warm," "Cold" connected via "Causes," "CommonIn," "Makes," etc.

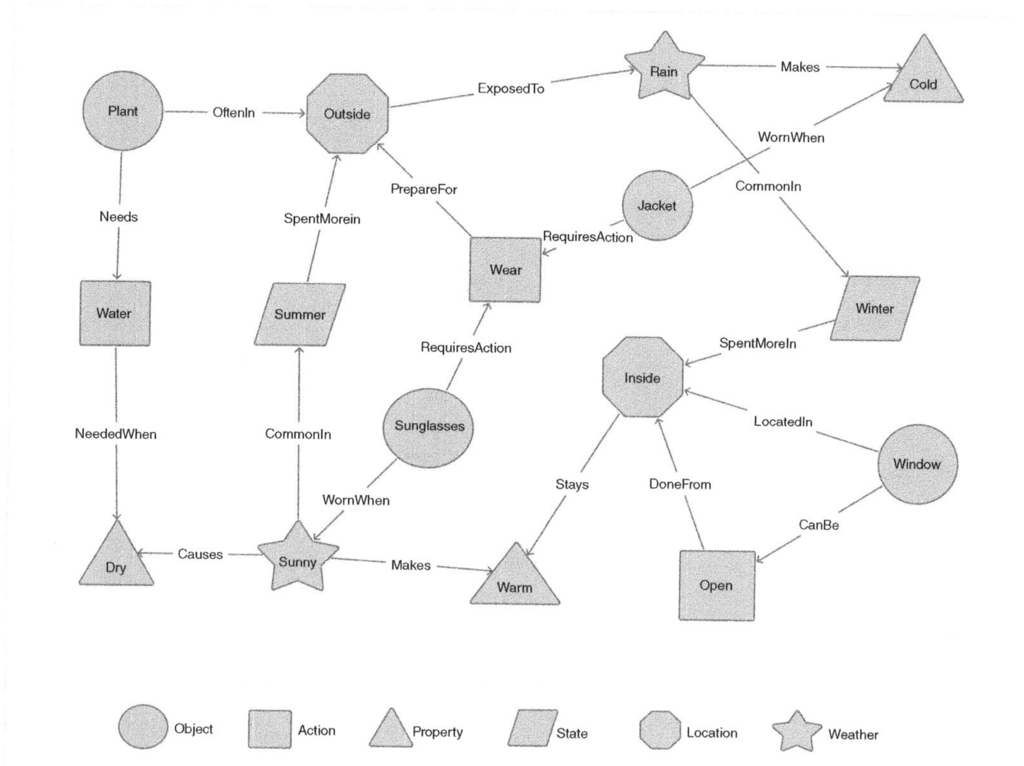

Figure 6.4 Commonsense KG: daily life and weather.

6.1.3.3 Domain-specific KGs

These graphs are tailored to specialized fields like medicine, law, finance, or engineering, and encode expert-curated facts and domain ontologies. They rely on structured schemas and contain curated, authoritative knowledge. While they cover narrower domains than encyclopedic KGs, they offer higher precision, consistency, and trustworthiness. They power expert systems, research tools, and regulatory compliance engines.

For example, the KG (Figure 3.12) we discussed in Section 3.3.2.3 is a medical KG which contains nodes like "COVID-19," "Remdesivir," "Lung Damage" with relations like "treats," "causes" and "symptom_of."

Figure 6.5 illustrates a legal KG which has nodes such as "Brown v. Board," "Equal protection," "Affirmative action policies" with relations like "established_precedent," "interpreted_by" and "influenced."

6.1.3.4 Multimodal KGs

Multimodal knowledge graphs (MMKGs) integrate structured information across diverse data types—including text, images, audio, and tabular data—into a unified semantic graph.

For example, in Figure 6.6, each node represents a different modality: medical reports (text), CT scans and X-rays (images), heart sounds (audio), lab results and vital signs (tables), and a central patient record that links them all.

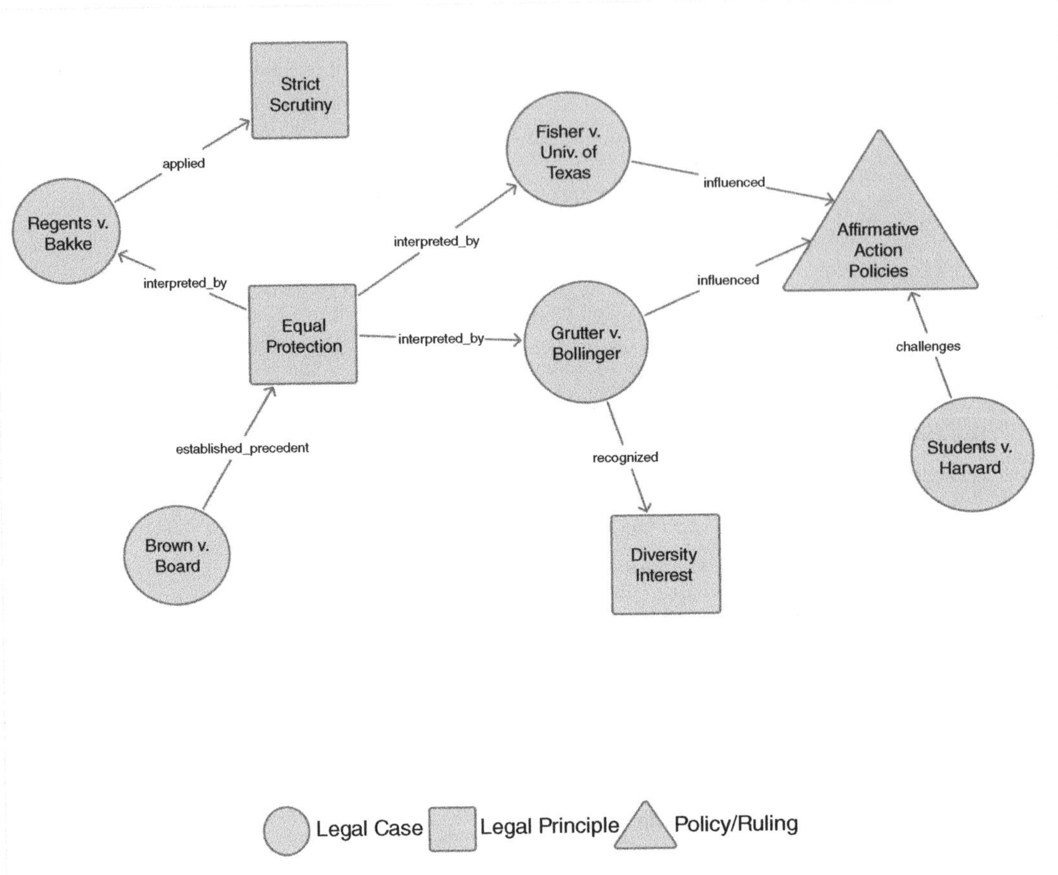

Figure 6.5 Legal domain—Case law knowledge graph.

MMKGs enable richer reasoning and retrieval by connecting heterogeneous evidence sources around a shared entity or event. In healthcare, this means an AI system can answer queries like "What evidence supports the diagnosis of pneumonia?" by jointly leveraging images, clinical notes, and structured vitals—bridging modalities to improve explainability and decision support.

6.1.4 Ontology in KGs

Behind every powerful KG lies an invisible but indispensable framework: the ontology. An ontology defines the formal structure of a KG—its conceptual vocabulary, categories, permissible relationships, and logical constraints. If the KG is the body of structured information, the ontology is its skeletal framework, guiding both construction and interpretation.

While data provides the what, and edges provide the how, the ontology defines the why and under what rules. It serves as a semantic contract, ensuring that knowledge represented in the graph remains consistent, interpretable, and extensible across use cases and systems.

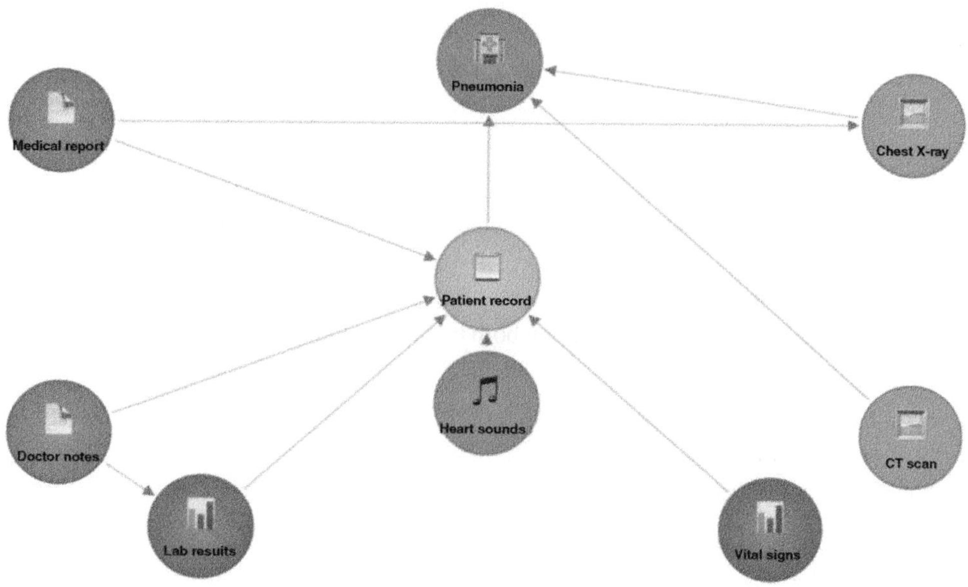

Figure 6.6 Multimodal knowledge graph.

6.1.4.1 Understanding Knowledge Organization

While often discussed together, taxonomy, schema, and ontology represent distinct but interconnected approaches to organizing knowledge. Each serves a unique purpose in information structuring and management.

6.1.4.1.1 Taxonomy: The Classification Backbone

A taxonomy functions as a hierarchical classification system—a sophisticated organizational framework that arranges concepts in parent-child relationships. Imagine it as a detailed family tree of concepts, where inheritance flows downward through the hierarchy. For example:

- Animal → Mammal → Dog → Golden Retriever
- Vehicle → Car → ElectricCar

Taxonomies excel at establishing "is-a" relationships but are limited to this single dimension of classification.

6.1.4.1.2 Schema: The Structural Framework

A schema provides the architectural blueprint for how data is organized and stored. Unlike taxonomy's hierarchical focus, schemas define:

- Entity types and their attributes
- Relationships between entities
- Constraints and validation rules

In a relational database schema, this might appear as:

```
CREATE TABLE Vehicle (
id INT PRIMARY KEY,
type VARCHAR(20),
max_speed INT,
requires_license BOOLEAN
);
```

Schemas are particularly valuable for data storage and retrieval systems, offering more flexibility than taxonomies but less semantic richness than ontologies.

6.1.4.1.3 *Ontology: The Semantic Universe*

An ontology encompasses and expands upon both taxonomy and schema to create a comprehensive semantic framework. It includes:

- The hierarchical structure of a taxonomy
- The structural rules of a schema
- Plus rich semantic relationships and constraints

For our vehicle example, an ontology would define:

```
:ElectricCar rdfs:subClassOf :Car .
:Car :requires :DriverLicense .
:Vehicle :has_max_speed xsd:integer .
:Car owl:disjointWith :Train .
```

Table 6.2 lists the comparisons of taxonomy, schema and ontology.

This layered approach allows for increasingly sophisticated knowledge representation, from basic classification to fully semantic, machine-interpretable models.

6.1.4.2 Core Components of an Ontology

An ontology comprises several core components that collectively define the structure and semantics of a domain. These components provide the schema—or blueprint—underlying a KG. In this section, we introduce each component and walk through a cohesive example to illustrate how they come together in practice.

Table 6.2 Comparisons of ontology vs. schema vs. taxonomy.

Aspect	Taxonomy	Schema	Ontology
Structure	Tree hierarchy	Tables/JSON schema	Graph (RDF/OWL)
Relations	is-a only	Foreign keys	Any (e.g., owns, partOf)
Expressiveness	Low (no properties)	Moderate (constraints)	High (logic, rules)
Use case	Classification	Data storage	Knowledge reasoning
Assumption	Closed-world	Closed-world	Open-world

A Running Example: A simple academic domain ontology —imagine we're designing an ontology for an academic institution. The domain includes people (e.g., students, professors), organizational units (e.g., departments, universities), and research outputs (e.g., papers, conferences). We'll use this example throughout the section to ground each component.

6.1.4.2.1 Classes (Concepts)

Classes represent the types or categories of things in a domain. These can be abstract (e.g., "Person," "Organization," "Event") or concrete (e.g., "Dog," "Book," "City"). They organize entities into a hierarchical structure where each class can have multiple subclasses and superclasses. Every individual (instance) in a KG is a member of one or more classes.

In our "academic domain" example:

```
Entity
|-- Person
|     |-- Student
|     +-- Professor
+-- Organization
      |-- Department
      +-- University
```

These classes define a hierarchy using the subClassOf relation. A Professor is a type of person, and a department is a kind of organization.

One tip when designing the ontology is to keep class definitions mutually exclusive unless necessary. If a TeachingAssistant can be both a Student and a Staff, consider creating a new class or using a property to capture role multiplicity.

6.1.4.2.2 Properties (Relations and Attributes)

Properties define how individuals are related to one another (object properties), or to literal values (data properties). They describe the structure of the graph and enrich it with meaning. There are usually three types of properties:

- Object Properties: Define relationships between two instances.
- Data Properties: Link an instance to a literal value.
- Annotation Properties: Add metadata, such as labels, comments, or versioning info.

In our "academic domain" example:

- Object Properties
  - teaches: links Professor → Course
  - enrolledIn: links Student → Course
  - partOf: links Department → University
- Data Properties
  - hasEmail: Person → string
  - startYear: Student → integer
  - gpa: Student → float
- Annotation Properties
  - rdfs:label: human-readable name
  - rdfs:comment: descriptive notes
  - dc:created: metadata (e.g., last modified)

6.1.4.2.3 Axioms and Constraints

Axioms define logical constraints or rules that govern the behavior of classes and properties. They allow automated reasoning engines to infer new facts, validate consistency, or detect contradictions. Axioms encode logical statements about classes and properties, supporting reasoning and validation. There are usually three types of constraints:

- Domain/Range Restrictions: Specify what types a property can apply to.
- Cardinality Constraints: Limit how many times a property can appear.
- Disjointness and Equivalence: Define logical relations.

In our "academic domain" example:

- Domain and Range
  - teaches:
    - Domain = Professor
    - Range = Course
      - This axiom restricts who can teach and what they can teach.
- Cardinality Constraints
  - hasEmail: exactly one per person
  - enrolledIn: zero or more courses
  - advises: a professor can advise up to 5 students
- Disjointness and Equivalence
  - Student $\perp$ Professor: no individual can be both
  - hasSupervisor $\equiv$ advisedBy: these are synonyms

6.1.4.3 Ontology Design Principles

Ontology is not just a metadata layer—it's the blueprint of intelligence within a KG. It defines what is allowed, what is inferred, and what can be trusted. Whether you're building a customer graph in e-commerce, a disease ontology for healthcare, or a product graph for retail—a well-crafted ontology ensures consistency, interpretability, and semantic richness.

Designing an ontology is a careful blend of scientific precision and creative structuring. A well-crafted ontology must begin with clarity, ensuring that all concepts and relationships are defined in a way that is unambiguous and easy to understand. This involves adopting meaningful and consistent naming conventions, along with thorough documentation. Ontologies should include human-readable labels and annotations, allowing both machines and humans to interpret the knowledge structure without confusion.

Beyond clear definitions, coherence is critical. An ontology must maintain logical consistency, avoiding contradictory definitions or circular dependencies. The relationships and constraints within the ontology should support valid inferencing. For example, if the ontology asserts that all mammals have lungs and that a dog is a mammal, it should logically entail that dogs have lungs. This capacity for reasoning is one of the core strengths of ontology-driven systems and must be preserved through rigorous validation.

Equally important is extensibility, the ability for an ontology to evolve alongside the needs of its users and applications. A modular design is often preferred, where general structures and concepts are separated from domain-specific extensions. For instance, a base ontology might define general concepts like person or location, while a specialized module could extend this for medical or legal domains. This modularity allows ontologies to grow organically without requiring complete reengineering.

Finally, reusability is a cornerstone of effective ontology development. Rather than constructing an ontology entirely from scratch, developers are encouraged to leverage established vocabularies. For instance, the friend of a friend (FOAF) vocabulary is commonly used to represent social relationships and personal information, while Schema.org provides a shared collection of structured data types that are used extensively across the web for enhancing search engine understanding. In more specialized domains, such as healthcare, the systematized nomenclature of medicine—clinical terms (SNOMED CT) offers a comprehensive, multilingual clinical healthcare terminology. By building on these established vocabularies, ontology designers can ensure greater interoperability, benefit from community best practices, and significantly reduce the time and effort required to model domain knowledge.

Altogether, these principles ensure that ontologies are not only technically sound but also sustainable, interoperable, and valuable in real-world applications.

6.2 Construction and Management of KGs

Building and managing a KG is a structured yet dynamic process, requiring careful attention to data sourcing, extraction, schema design, and evolution. A well-constructed KG integrates heterogeneous data into a semantically rich, queryable network, enabling advanced reasoning and AI applications. Below, we explore the key steps in KG construction, with illustrative examples and best practices.

6.2.1 Schema Design

In Section 6.1.4, we explored the conceptual foundations of ontology and established key design principles for creating robust semantic frameworks. Now we turn to the practical application of these concepts: how well-designed ontologies serve as the architectural foundation for constructing effective KGs. The transition from ontological theory to implementation requires careful consideration of schema design patterns, structural optimization, and evolutionary strategies.

6.2.1.1 From Ontology to Implementation

The relationship between ontology and schema design represents a critical bridge between conceptual modeling and practical KG construction. While an ontology defines the semantic universe of possibilities—what entities can exist, how they relate, and what constraints govern their interactions—the schema translates these abstract concepts into concrete implementation patterns that support efficient storage, querying, and reasoning.

A robust schema and ontology are fundamental for the usability, maintainability, and scalability of a KG. The schema serves as the operational blueprint that transforms ontological concepts into graph structures, ensuring that the resulting KG remains both semantically rich and computationally tractable.

For example:

- Ontology Defines:
  - "A Professor is a subclass of Person."
  - "Teaches links Professor to Course."

- Schema Implements:
  - Node labels: Professor, Course
  - Edge type: Teaches
  - Constraint: Professor must have at least one teaches relationship

This alignment ensures the KG is both semantically rich and computationally efficient.

6.2.1.2 Core Schema Components

6.2.1.2.1 Node Types: Entity Categories

Node types represent the concrete instantiation of ontological classes within the graph structure. These categories define the fundamental entities that populate the KG, each inheriting properties and constraints from their ontological definitions.

The primary entity categories are:

- People: Individuals, organizations, and collective entities with social or professional roles.
- Places: Geographic locations, administrative regions, and spatial contexts.
- Events: Temporal occurrences, processes, and state changes.
- Objects: Physical and abstract entities, artifacts, and conceptual constructs.
- Documents: Information resources, publications, and knowledge artifacts.

Each node type carries forward the semantic constraints defined in the ontology, ensuring that entities maintain their definitional integrity while supporting practical graph operations.

6.2.1.2.2 Edge Types: Relationship Patterns

Edge types operationalize the object properties defined in the ontology, creating the relational fabric that connects entities within the KG. These relationships must balance semantic expressiveness with computational efficiency. The fundamental relationship categories are listed in Table 6.3.

Table 6.3 Relationship categories and examples.

Category	Relationship type	Description	Example usage
Hierarchical	subClassOf	Taxonomic inheritance and specialization	Professor subClassOf Person
	partOf	Compositional and spatial containment	Department partOf University
	memberOf	Set membership and organizational belonging	Student memberOf Research-Group
Associative	relatedTo	General semantic association	Course relatedTo Department
	similarTo	Conceptual or functional similarity	Quiz similarTo Exam
	influencedBy	Causal or temporal dependencies	Research influencedBy Funding
Functional	hasRole	Role assignment and functional capacity	Person hasRole Professor
	participatesIn	Event participation and process involvement	Student participatesIn Conference
	produces	Creation and generation relationships	Professor produces Publication

Category	Relationship type	Description	Example usage
Temporal	precedes	Temporal ordering and sequence	Lecture precedes Exam
	during	Temporal containment and co-occurrence	Meeting during Semester
	triggers	Causal initiation and consequence	Deadline triggers Submission

Table 6.4 Attributes categories and examples.

Category	Attribute type	Description	Example values
Descriptive	Identity markers	Names, identifiers, labels	name: "John Smith," studentID: "CS2024001," label: "Professor"
	Physical properties	Dimensions, weight, color	height: 180 cm, weight: 75 kg, color: "blue"
	Qualitative characteristics	Type, category, status	type: "undergraduate," category: "research," status: "active"
Temporal	Creation/modification timestamps	When entities were created or last updated	createdAt: "2024-01-15T10:30:00Z," lastModified: "2024-06-08T14:22:00Z"
	Validity periods	Start and end dates for entity validity	validFrom: "2024-09-01," expire-sOn: "2025-05-31"
	Event durations/ frequencies	Time spans and recurring patterns	duration: "2 hours," frequency: "weekly," meetingTime: "MWF 10:00AM"
Quantitative	Measurements and metrics	Numerical values and units	enrollment: 150, budget: 50000.00, area: "1200 sq ft"
	Scores and ratings	Performance indicators and evaluations	gpa: 3.85, rating: 4.2, citation-Count: 47
	Statistical properties	Aggregated numerical data	averageGrade: 87.3, successRate: 0.92, variance: 12.4
Provenance	Source information	Origin and lineage tracking	source: "registrar_system," import-edFrom: "legacy_db," dataProvider: "IT_dept"
	Confidence/reliability	Trust and quality measures	confidence: 0.95, reliability: "high," verificationStatus: "confirmed"
	Version/change history	Tracking modifications over time	version: "2.1.3," revision: 15, changeLog: "updated contact info"

6.2.1.2.3 Attributes: Enriching Entity Descriptions

Attributes implement the data properties defined in the ontology, providing scalar values that enrich entity descriptions and enable precise querying and filtering. The major attribute categories and examples are listed in Table 6.4.

6.2.1.2.4 Constraints: Ensuring Semantic Integrity

Constraints translate ontological axioms into operational rules that maintain graph consistency and enable automated validation.

- Structural Constraints:
  - Cardinality restrictions: Limiting the number of relationships an entity can have.
  - Domain and range specifications: Constraining which entities can participate in specific relationships.
  - Disjointness rules: Preventing conflicting classifications.
- Semantic Constraints:
  - Type hierarchies: Enforcing inheritance relationships.
  - Symmetry and transitivity: Defining relationship properties.
  - Inverse relationships: Maintaining bidirectional consistency.

6.2.1.3 Schema Types: Knowledge Indexing vs. Knowledge Carriers

Modern KG architectures adopt two complementary schema patterns, each optimized for different use cases and operational requirements:

- Knowledge Indexing Schemas focus on search, retrieval, and navigation. These schemas prioritize keyword-centric properties and context-aware links, enabling efficient information lookup and discovery. Think of this as the schema for organizing a knowledge directory.
- Knowledge Carrier Schemas, by contrast, aim to represent rich semantics and logical entailments. These graphs serve as containers of domain expertise and support reasoning and inference. Here, ontological precision and axiomatic constraints play a larger role.

Both types are complementary: indexing schemas facilitate fast and broad access to data, while carrier schemas preserve its depth and structure.

6.2.1.4 Schema Design Considerations

A well-engineered schema should be:

- Scalable: Able to accommodate increasing volumes of nodes, edges, and data types without degradation in performance.
- Flexible: Designed to support evolving domains, allowing schema extension without breaking downstream applications.
- Maintainable: Easy to understand, document, and debug, with a clear mapping between schema elements and domain concepts.
- Efficient for Querying: Structured to optimize for frequent access patterns, such as neighborhood traversal, join operations, and subgraph matching.

KGs are living artifacts. As domains evolve, new entity types may emerge, relationships may be refined, and old structures may become obsolete. Supporting schema evolution is therefore critical. Best practices include:

- Versioning: Maintain clear version histories of schema changes to ensure traceability.
- Backward Compatibility: Where possible, design changes that do not break existing data or applications.
- Migration Strategies: Define systematic approaches for transforming older graph data to new schemas (e.g., mapping rules or transformation pipelines).
- Impact Analysis: Evaluate how schema changes propagate across retrieval systems, inference engines, analytics pipelines, and user-facing applications.

By grounding schema design in a well-defined ontology, and following principled design practices, KGs can remain robust, coherent, and extensible—supporting both human interpretation and machine reasoning across a wide variety of domains and applications.

6.2.2 Data Sources and Extraction Techniques

6.2.2.1 Data Sources

The foundation of any KG is its data, which can come from a variety of sources. Similar to the data source types categorized in Section 3.2.1, the data source for KG can be of the following types:

- Structured Data: Databases, spreadsheets, and APIs.
- Semi-structured Data: XML, JSON, and RDF formats.
- Unstructured Data: Textual documents, web pages, and multimedia content.

We'll list code examples of using different types of data sources to construct KG in Section 6.5.

6.2.2.2 Extraction Techniques

Transforming raw data into a structured KG requires converting information into triples—structured units in the form of (subject, predicate, object). These methods have evolved from simple rule-based systems to advanced AI-powered approaches. Let's explore the spectrum of techniques available for knowledge extraction.

6.2.2.2.1 Manual Extraction

Manual extraction refers to the process of building KGs through human effort—often involving domain experts, crowdworkers, or collaborative editing communities. Manual KG construction represents the most labor-intensive yet often most accurate approach to building structured knowledge repositories. This methodology relies on human expertise to identify, validate, and organize information into coherent graph structures, ensuring high-quality knowledge representation at the cost of scalability.

One prominent example is Wikidata [1], a collaboratively edited knowledge base that serves as the structured data backbone for Wikipedia. Entities in Wikidata are curated through a crowdsourcing model where contributors manually create and refine triples—structured facts in the form of (subject, predicate, object). Each entity in Wikidata often maps to a corresponding article in Wikipedia, allowing for the integration of structured and unstructured knowledge.

Another example is the Unified Medical Language System (UMLS) [3], developed by the U.S. National Library of Medicine. UMLS integrates and standardizes biomedical terminologies from multiple sources, providing a rich semantic network of concepts and relationships in the healthcare domain. Experts and organizations contribute to its curation to ensure consistency and comprehensiveness across medical vocabularies.

Manual methods, while not scalable for all use cases, are foundational in domains requiring a high degree of accuracy and semantic rigor. They also play a crucial role in bootstrapping KGs where automated tools may initially lack sufficient context or domain expertise.

6.2.2.2.2 Rule-based Methods

Rule-based systems were among the earliest techniques used for KG construction. These approaches rely on predefined linguistic patterns, regular expressions, or manually crafted templates to identify entities and extract relationships. While limited in flexibility, they offer precision and interpretability for well-structured domains.

Pattern matching uses predefined templates to identify and extract relationships from text.

- Common Patterns:

```
Pattern: "X founded Y"
Rule: Extract (X, founded, Y)
Example: "Larry Page founded Google" -> (Larry_Page, founded, Google)

Pattern: "X, CEO of Y"
Rule: Extract (X, is_CEO_of, Y)
Example: "Tim Cook, CEO of Apple" -> (Tim_Cook, is_CEO_of, Apple)

Pattern: "X acquired Y for Z"
Rule: Extract (X, acquired, Y) and (acquisition, value, Z)
Example: "Microsoft acquired GitHub for $7.5B" ->
         (Microsoft, acquired, GitHub)
         (acquisition_Microsoft_GitHub, value, "$7.5B")
```

- Advanced pattern Templates:

```
# Regular expression patterns
patterns = {
    'founding': r'(\w+)\s+(?:founded|established|started)\s+(\w+)(?:\s+in\s+(\d
    {4}))?',
    'location': r'(\w+)\s+(?:is located in|is based in|headquarters in)\s+(\w+)',
    'acquisition': r'(\w+)\s+(?:acquired|bought|purchased)\s+(\w+)\s+for\s+\$?([\d
    .]+[BMK]?)'
}

# Dependency parsing patterns
dependency_patterns = {
    'invention': {'subject': 'nsubj', 'verb': ['invented', 'discovered', 'created
    '], 'object': 'dobj'},
    'leadership': {'subject': 'nsubj', 'verb': ['leads', 'manages', 'runs'], '
    object': 'dobj'}
}
```

Some classic systems include:

- ConceptNet: Constructs a commonsense KG using hand-curated relationships [4].
- Freebase: Compiled general knowledge using human input and schema-defined rules [5].

While rule-based systems are simple and interpretable, they are often brittle—limited by vocabulary, grammar variation, and domain specificity.

6.2.2.2.3 Traditional ML Methods

As NLP matured, statistical and supervised learning approaches gained popularity for extraction tasks. These models rely on labeled data or distant supervision.

- Entity Recognition and Linking: Identifying key entities in unstructured text and linking them to existing nodes in the graph. For example, identifying "Apple Inc." in text and linking it to a KG node (dbr:Apple_Inc).
- Relation Extraction: Deriving relationships between entities using NLP methods. For example, detecting "Elon Musk founded SpaceX" → (Elon Musk, founded, SpaceX).
- Coreference Resolution: Resolving different mentions of the same entity across various data sources. For example, Resolving "he" in "Steve Jobs returned to Apple in 1997, and he revolutionized the iPhone" to Steve Jobs.

6.2.2.2.4 Deep Learning and LLM-based Methods

Neural models such as BERT have dramatically improved the accuracy of both entity and relation extraction tasks by leveraging contextual embeddings. They can capture contextual nuance, help disambiguate entities (e.g., distinguishing "Paris" the city from "Paris Hilton") and can be fine-tuned on NER, RE, or QA datasets.

LLMs represent a paradigm shift in knowledge extraction, offering flexibility and understanding previously impossible. They can extract structured triples from raw text via prompting. LLMs can perform zero-shot or few-shot knowledge extraction through prompting alone—no training or predefined rules required.

Zero-shot Extraction Example:

```
# Direct extraction with GPT-4/Claude
prompt = """
Extract all relationships from this text as (subject, predicate, object) triples:

"Marie Curie, born in Warsaw in 1867, was a pioneering physicist and chemist.
She discovered radium and polonium in 1898 while working at the University of Paris.
She was the first woman to win a Nobel Prize."

Output format:
(subject, predicate, object)
"""

# LLM Response:
(Marie_Curie, born_in, Warsaw)
(Marie_Curie, birth_year, 1867)
(Marie_Curie, profession, physicist)
(Marie_Curie, profession, chemist)
(Marie_Curie, discovered, radium)
(Marie_Curie, discovered, polonium)
(discovery, year, 1898)
(Marie_Curie, worked_at, University_of_Paris)
(Marie_Curie, achievement, first_woman_Nobel_Prize)
```

Few-shot Learning Example:

```
few_shot_prompt = """
Extract knowledge triples from text following these examples:

Example 1:
Text: "Amazon was founded by Jeff Bezos in Seattle in 1994."
Output:
(Amazon, founded_by, Jeff_Bezos)
(Amazon, founded_in_city, Seattle)
(Amazon, founded_in_year, 1994)

Example 2:
Text: "The Eiffel Tower, located in Paris, was completed in 1889."
Output:
(Eiffel_Tower, located_in, Paris)
(Eiffel_Tower, completion_year, 1889)

Now extract from:
Text: "Tesla, led by Elon Musk, manufactures electric vehicles in Fremont, California
    ."
"""
```

Table 6.5 lists the summary and comparisons of these different types of extraction methods. In Section 6.4.2, we will explore LLM-driven enrichment techniques in greater depth.

Table 6.5 Summary and comparisons of extraction methods for KG construction.

Method	Strengths	Weaknesses	Typical tools
Rule-based	Precise, interpretable	Brittle, low coverage	SpaCy, REANO, ConceptNet
ML-based	Scalable, more adaptive	Needs annotated data	CRFs, SVMs, neural RE models
DL-based	High accuracy, robust	Resource intensive	BERT, BiLSTM-CRF
LLM-based	Flexible, promptable, no training	May hallucinate, less consistent	GPT-4, Claude, AutoKG, Graph-RAG

6.3 KGs Analytics and Enrichment

KGs serve as powerful data structures for representing complex relationships, but their true value emerges through the quality, depth, and analytical insights they provide. While a basic KG captures explicit relationships between entities, enrichment transforms it into an intelligent system capable of discovering hidden patterns, predicting missing connections, and supporting sophisticated reasoning tasks.

The enrichment process leverages advanced computational techniques from data science and machine learning to extract new insights, identify latent relationships, and dynamically expand the graph's scope and utility. This transformation elevates KGs from static repositories to dynamic, intelligent systems that can adapt, learn, and provide increasingly sophisticated analytical capabilities.

KGs are often incomplete, noisy, or sparse, particularly in areas where data is constantly evolving or automatically extracted from unstructured sources like text. Enriching a KG is motivated by the need to improve its coverage by adding missing entities and relationships, since real-world information is dynamic and ever-changing. It also boosts the utility of the graph in tasks such as question answering, semantic search, and recommendation systems, while helping correct inaccuracies through validation processes. Moreover, enrichment allows the KG to adapt to specific domains like healthcare or finance, and enhances reasoning capabilities by incorporating inferred knowledge or linking to external resources. The core goal of enrichment is to make the graph more complete, accurate, and relevant by adding new facts from unstructured data, predicting missing links using machine learning, aligning entities with external knowledge bases, refining its schema, and ensuring it stays up to date with temporal and contextual changes.

This section explores three foundational approaches for KG enrichment: graph algorithms for structural analysis and pattern discovery, graph embeddings for vector-based representation learning, and GNNs for deep learning on graph-structured data. Each approach offers unique capabilities and can be combined synergistically to create comprehensive enrichment pipelines.

6.3.1 Graph Algorithms in KGs

In Section 5.1.4 of the book, we've introduced the essential algorithms in general graphs. In this section, we are going to discuss their specific use cases in KGs.

Graph algorithms form the analytical foundation for understanding and processing the complex structural relationships within KGs. These mathematical techniques reveal hidden patterns, quantify entity importance, and discover emergent properties that may not be immediately apparent from the raw graph structure.

Graph algorithms operate over a central graph representation that can be constructed from diverse data sources and exported to serve various downstream tasks. The architecture typically separates persistent storage from the computational layer, allowing algorithms to efficiently process large-scale graphs through operations like reading graph structures, performing mutations for graph enrichment, and streaming results for real-time applications.

By leveraging specialized graph algorithms—from simple traversals to complex neural message-passing schemes—KG systems can extract semantic patterns, identify influential entities, compute multi-hop inferences, and maintain consistency across millions of interconnected facts, ultimately bridging the gap between raw structured data and actionable intelligence.

6.3.1.1 Centrality Measures: Identifying Key Entities

We've learned the definitions of the centrality measures in Section 5.1.4.1 of the book. Here, we are going to discuss more details of the centrality measures in KGs.

6.3.1.1.1 Degree Centrality: Connection-based Importance

Here is the example Python code to compute the degree centrality of a KG:

```
import networkx as nx

G = nx.karate_club_graph()
degree_centrality = nx.degree_centrality(G)
```

Degree centrality represents the most intuitive measure of node importance, based purely on the number of connections.

- **In-degree centrality** counts incoming edges, identifying entities that are frequently referenced or depend upon by others. For example: in a research publication KG, a seminal paper like "Attention Is All You Need" would have high in-degree centrality because hundreds of subsequent papers cite it, indicating its foundational importance in the transformer architecture domain.
- **Out-degree centrality** counts outgoing edges, highlighting entities that connect to many others, often serving as central hubs or starting points for exploration. For example: in a product recommendation system, a category node like "Electronics" might have high out-degree centrality because it connects to thousands of specific products, making it a natural entry point for browsing.

In KG-based RAG systems, high in-degree nodes often contain authoritative information that should be prioritized during retrieval. High out-degree nodes serve as effective starting points for subgraph exploration, enabling comprehensive context gathering for complex queries. For example, a node with high in-degree such as "large language models" might be connected from many applications, papers, or related techniques, indicating it is a critical retrieval target for many queries. A node with high out-degree like "artificial intelligence" may point to a wide range of related topics, making it useful for exploratory queries or hierarchical topic expansion.

6.3.1.1.2 Betweenness Centrality: Bridge Detection

Here is the example Python code to compute betweenness centrality for a KG:

```
import networkx as nx

G = nx.karate_club_graph()
betweenness = nx.betweenness_centrality(G)
```

Betweenness centrality identifies nodes that serve as bridges between different parts of the graph, often revealing entities that connect disparate domains or facilitate information flow. A node's betweenness centrality is calculated as the fraction of shortest paths between all node pairs that pass through that node.

For example, in a biomedical KG, the concept "Inflammation" might have a high betweenness centrality value, because it connects various diseases (cardiovascular disease, arthritis, cancer), treatments (anti-inflammatory drugs, lifestyle interventions), and biological processes (immune response, tissue repair). This makes it a critical bridge entity for interdisciplinary medical reasoning.

In KG-based RAG systems, bridge entities identified through betweenness centrality are invaluable for cross-domain queries. When a user asks about connections between seemingly unrelated topics, these high-betweenness nodes often play a critical role in connecting different domains, topics, or clusters of knowledge. For example, a node like "neurosymbolic AI" might connect the symbolic reasoning and neural network branches of an AI KG, serving as a bridge between two traditionally separate research areas. In a medical KG, a condition like "Chronic Inflammation" might link various diseases, treatments, and symptoms across multiple specialties, making it a key retrieval point for interdisciplinary queries. Identifying such bridge entities can enhance the diversity and depth of information retrieved in a RAG pipeline.

6.3.1.1.3 *Closeness Centrality: Information Spread Efficiency*

Here is the example Python code to compute closeness centrality for a KG:

```
import networkx as nx

G = nx.karate_club_graph()
closeness = nx.closeness_centrality(G)
```

For example, in a social KG, a person with high closeness centrality can influence or receive information from the entire network more quickly than others, making them valuable for viral marketing or rapid information dissemination.

6.3.1.1.4 *Eigenvector Centrality: Authority Through Connection Quality*

Here is the example Python code to compute eigenvector centrality for a KG:

```
import networkx as nx

G = nx.karate_club_graph()
eigenvector_centrality = nx.eigenvector_centrality(G, max_iter=500)
```

The eigenvector centrality of a node can be large either because the node has many neighbors or because it has a few neighbors who have many neighbors or both. In the context of a social network graph, a person can be important because that person has many friends or knows a few people who have many friends.

6.3.1.1.5 *PageRank: Authority Measurement in Directed Networks*

Here is the example Python code to compute PageRank for a KG:

```
import networkx as nx

G = nx.karate_club_graph()
pagerank = nx.pagerank(G, alpha=0.85)
```

A node's PageRank score depends not only on the number of connections but also on the PageRank scores of its neighbors, creating a reinforcing measure of structural importance. For example, in a technology KG, "machine learning" might achieve a high PageRank score, not just because many topics connect to it, but because those connected topics are themselves important—including "artificial intelligence," "data science," and "neural networks." This creates a compound measure of significance.

In KG-based RAG systems, PageRank is used to rank entities based not just on the number of connections they have, but also on the importance of the nodes they are connected to. This allows the system to surface globally significant nodes that are influential within the broader structure of the graph, even if they don't have the highest number of direct links. It helps prioritize which nodes to retrieve when grounding answers with high-impact or authoritative information. For example, in a scientific KG, a foundational concept like "deep learning" might receive a high PageRank because it is referenced by other influential nodes such as key algorithms, landmark papers, and widely used frameworks. In a business KG, a company like "OpenAI" might rank highly due to strong connections to influential technologies, partnerships, and research domains. Using PageRank in RAG retrieval ensures that the most contextually and structurally important nodes are considered during the answer generation process.

Figure 6.7 illustrates a small network for "artificial intelligence research," where octagon nodes indicate high centrality, circle nodes indicate medium centrality, and rectangle nodes indicate low centrality. In research networks, highly cited papers have high degree centrality. High betweenness centrality papers are crucial for cross-domain knowledge transfer. Authoritative nodes connected to other authoritative nodes score the highest PageRank.

Table 6.6 lists the comparisons of the centrality measures in terms of their computational complexity and use cases.

6.3.1.2 Community Detection

Community formation is a common phenomenon across many types of networks—academic, social, biological, and enterprise. Identifying communities is vital for understanding group behavior, uncovering emergent phenomena, and guiding more informed decision-making.

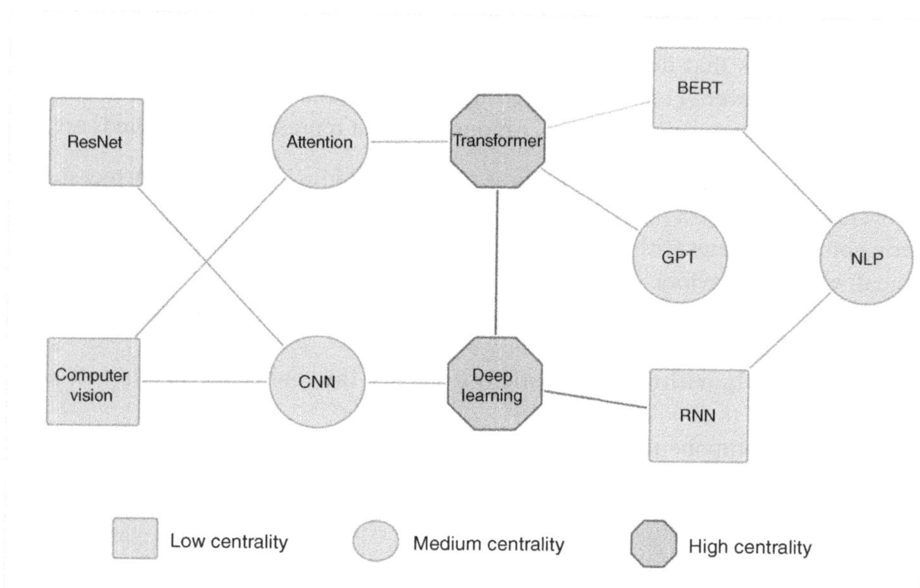

Figure 6.7 Centrality measure for "artificial intelligence research" graph.

Table 6.6 Comparative analysis of centrality measures, where n = |V| and m = |E|.

Measure	Computational complexity	Best for	Limitations	Selection guideline
Degree	$O(n + m)$	Local importance, hub identification	Ignores global structure	Identify highly connected entities
Betweenness	$O(nm + n^2 \log n)$	Bridge detection, information flow	Computationally expensive	Find critical connectors between communities
Closeness	$O(nm)$	Information spread, accessibility	Sensitive to disconnected components	Optimize information propagation
Eigenvector	$O(nm)$	Prestige networks, authority ranking	May not converge in directed graphs	Measure prestige in undirected networks
PageRank	$O(nm)$	Web-scale ranking, authority with stability	Parameter tuning required	Robust authority ranking in directed or mixed networks

At the core of community detection lies a simple principle: nodes within the same community tend to be more densely connected to each other than to nodes outside their group. These clusters can help reveal peer groups with shared preferences, suggest functional or topical segmentation, identify isolated subnetworks, and improve the interpretability of the overall graph structure. Furthermore, community detection is often used to prepare data for downstream tasks, such as recommendation systems, influence analysis, and fraud detection, or to produce network visualizations for domain experts. For example, in a pharmaceutical KG, community detection might identify tightly clustered groups of drugs that treat similar conditions or share biochemical mechanisms. These natural groupings could reveal potential drug repurposing opportunities or highlight gaps in treatment options.

6.3.1.2.1 Key Algorithms and Applications

There are various algorithms that detect communities, each based on different principles and assumptions. Some focus on overall relationship density, while others seek to optimize modularity or use label diffusion to form consensus clusters. Among the most representative methods are:

Triangle Count and Clustering Coefficient: These offer insight into local density by measuring the number of closed triplets (triangles) and the tendency of a node's neighbors to be connected. High clustering often correlates with tight-knit communities, useful for estimating stability and detecting "small-world" behaviors.

In a KG of academic collaborations, these measures could identify research teams with strong internal cooperation patterns.

Connected Components Algorithms: We've briefly discussed the connected components algorithm in Section 5.1.3.2 of the book:

- Strongly Connected Components (SCC): These identify groups of nodes where every node is reachable from every other node within the same group, following the direction of relationships. SCCs are especially useful in directed graphs for discovering internally consistent subgraphs.
- Weakly Connected Components (CC): Similar to SCCs but in undirected graphs, CCs help identify isolated clusters and provide a useful first step in understanding graph connectivity.

Label Propagation: This algorithm spreads node labels based on majority voting within a node's neighborhood, quickly forming clusters. It's fast, scalable, and ideal for large graphs or dynamic networks where quick segmentation is needed. For example, in an e-commerce KG with millions of products, label propagation might automatically discover product categories (shown as follows), based on shared attributes or co-purchasing patterns:

- Electronics Cluster: Smartphones, laptops, and accessories naturally group together through purchase correlations and complementary relationships.
- Home and Garden Cluster: Furniture, appliances, and gardening tools form communities based on co-purchase patterns and functional relationships.
- Sports and Fitness Cluster: Athletic equipment, supplements, and workout gear cluster through shared customer interests and product compatibility.

Label propagation requires minimal computation and memory, making it feasible for graphs with millions of nodes where more sophisticated algorithms would be computationally prohibitive.

Louvain Algorithm: One of the most widely used methods, Louvain detects communities by optimizing modularity—a score that compares intra-community edge density against a null model [6]. It not only reveals flat communities but can also uncover hierarchical relationships. For example, when analyzing a healthcare KG, it might first identify broad specialty clusters (e.g., cardiology, neurology), then reveal finer-grained groupings within each specialty. The process overview of the Louvain algorithm:

1. Local Optimization: Each node is initially assigned to its own community, then iteratively moved to neighboring communities that increase modularity.
2. Community Aggregation: Communities are collapsed into super-nodes, creating a new graph at a higher level of abstraction.
3. Iteration: The process repeats until no further modularity improvement is possible.

Each of these methods serves distinct analytical goals. For example, clustering coefficient metrics may help assess whether communities form naturally or are imposed; SCCs and CCs provide hard partitions useful in early stage exploration; and Louvain or label propagation support deeper segmentation with greater granularity. To better understand how these algorithms function, refer to Table 6.7, which summarizes their properties and use cases.

Table 6.7 Comparisons of community detection algorithms.

Algorithm type	Computational complexity	Community quality	Hierarchical structure	Best applications
Triangle count/clustering coefficient	O(m^1.5)	Measures local cohesion	No	Network analysis, small-world detection
Connected components	O(n + m)	Fundamental connectivity	No	Basic structure analysis, island detection
Strongly connected components	O(n + m)	Directed connectivity	No	Citation analysis, dependency tracking
Label propagation	O(m)	Fast approximation	No	Large-scale networks, rapid prototyping
Louvain modularity	O(m log n)	High-quality communities	Yes	Research networks, fraud detection

6.3.1.2.2 Validation and Benchmarking

Community Detection Validation: Beyond detection, validating the quality of communities is a crucial next step. Algorithms may produce varying clusters based on their assumptions, and results can differ significantly across methods. The validation of community detection results becomes particularly crucial when communities will inform downstream applications such as RAG systems, recommendation engines, or knowledge organization schemes. Unlike supervised learning tasks where ground truth labels provide clear evaluation criteria, community detection operates in an unsupervised context where the "correct" community structure may be subjective or domain-dependent.

Real-world KGs present additional validation challenges because they often exhibit less distinct community boundaries than idealized network models. In academic collaboration networks, researchers may contribute to multiple fields, creating overlapping communities. In biomedical KGs, diseases may share symptoms or treatments, leading to fuzzy community boundaries. These characteristics require validation approaches that can assess algorithm performance under realistic conditions.

Community Detection Benchmarks: Validation typically involves comparing detected communities against reference datasets with known group structures. The development of standardized benchmark networks has provided crucial tools for evaluating and comparing community detection algorithms. These synthetic networks offer the advantage of known ground truth community structure while allowing researchers to control various network properties that affect algorithm performance.

- The Girvan–Newman (GN) benchmark represents one of the earliest and most influential validation frameworks for community detection algorithms [7, 8]. This benchmark generates random networks with predefined community structure, where nodes are more likely to connect to others within their assigned community than to nodes in different communities. The GN benchmark creates relatively homogeneous networks where all nodes have similar degrees and all communities have similar sizes.

  In a typical GN benchmark network designed to evaluate KG community detection, each node might represent a research paper, with communities corresponding to different research fields such as "machine learning," "computer vision," and "natural language processing." The benchmark controls the mixing parameter μ, which determines the fraction of edges that connect nodes to different communities. When μ is low (around 0.1), communities are well-separated and easy to detect. As μ increases toward 0.5, community boundaries become increasingly blurred, testing the limits of algorithm performance.
- The Lancichinetti–Fortunato–Radicchi (LFR) benchmark addresses limitations of the GN model by generating more realistic networks that better reflect the heterogeneous structure observed in real-world KGs [9]. LFR networks incorporate power-law distributions for both node degrees and community sizes, creating networks with hubs (highly connected nodes) and communities of varying scales.

  Consider an LFR benchmark designed to simulate a biomedical KG where communities represent different disease categories. The power-law degree distribution ensures that some entities (such as common symptoms like "fever" or "pain") have many connections across communities, while others (such as rare genetic markers) have few connections within specific communities. The power-law community size distribution creates some large communities (such as "cardiovascular diseases") alongside many smaller, specialized communities (such as "rare metabolic disorders").

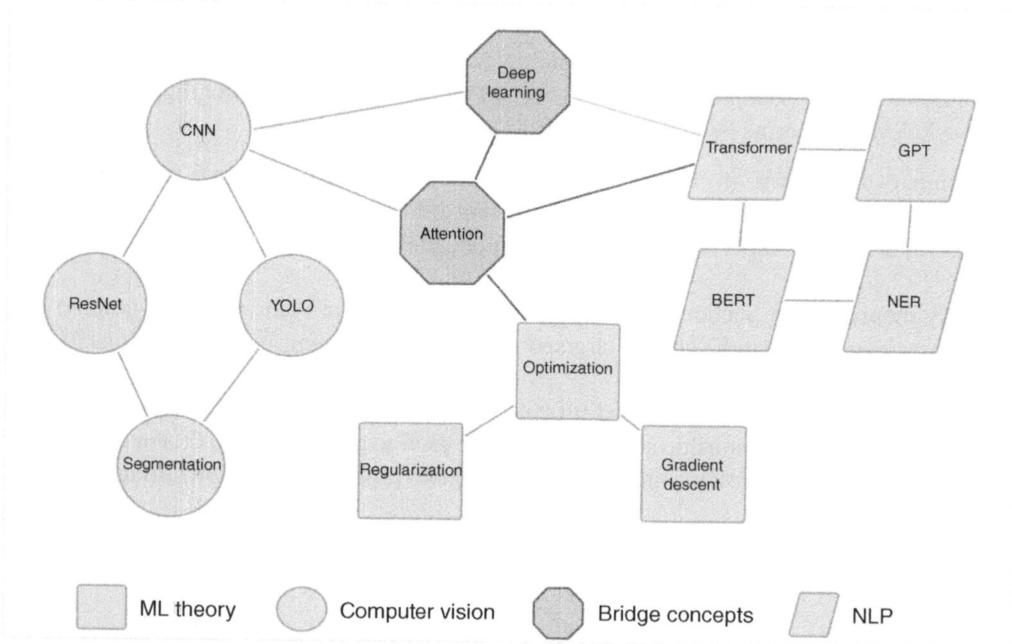

Figure 6.8 Research domain communities.

The choice between GN and LFR benchmarks significantly impacts validation results and should align with the characteristics of the target KG. For KGs with relatively uniform entity connectivity, such as taxonomic hierarchies or structured vocabularies, GN benchmarks provide appropriate validation. For KGs with scale-free characteristics, such as citation networks or social media KGs, LFR benchmarks offer more realistic evaluation conditions.

Figure 6.8 illustrates how community detection uncovers research clusters in an academic KG. Domains like computer vision (circle nodes), NLP (parallelogram nodes), and ML theory (rectangle nodes) form coherent communities, each visually distinguished by node shape. Bridge nodes, shown as octagon nodes, such as “deep learning” or “attention,” connect disparate topics, highlighting their interdisciplinary role. This view not only structures the knowledge space but also supports focused retrieval in downstream RAG applications.

Altogether, community detection and validation offer a structured lens through which to interpret complex graphs, enabling more meaningful organization, improved retrieval precision, and greater transparency in reasoning tasks.

6.3.1.3 Pathfinding

Pathfinding algorithms are essential for efficiently traversing KGs and uncovering meaningful connections between entities. These algorithms not only support direct search queries but also enable advanced reasoning, such as tracing causal or semantic pathways between seemingly unrelated concepts.

The most common pathfinding techniques focus on identifying the shortest path—the minimal sequence of nodes and edges connecting two entities. This is particularly important for semantic search, explanatory reasoning, and RAG.

To illustrate the impact of path-based search, consider the following comparison:

- Shortest Path (3 hops): User Query → Semantic Search → Graph Traversal → Search Results
- Traditional IR Path (7 hops): User Query → NLP Parsing → Keyword Matching → Text Indexing → Relevance Scoring → Ranking Algorithm → Search Results

This comparison, shown in Figure 6.9, demonstrates why shortest path algorithms are crucial. They reduce complexity, enable more efficient information access, and provide clearer rationales—making KGs especially powerful for applications in search, recommendation, and reasoning.

As briefly mentioned in Section 5.1.4.4, algorithms like Dijkstra's (for weighted graphs) and A* (for heuristic-guided search) enable targeted exploration, reducing retrieval complexity and improving the interpretability of results.

Dijkstra's Algorithm is a classic method for computing the shortest paths in weighted graphs. Each edge can represent relationship strength, semantic distance, or traversal cost. For example, in a social KG, finding the shortest weighted path between two professionals might surface:

- A direct connection: Person A → Works_With → Person B (weight: 0.9, strong confidence)
- An indirect path: Person A → Colleagues_With → Person C → Collaborated_With → Person B (total weight: 0.7, mediated via Person C)

In large graphs where exhaustive search is computationally expensive, A* Search enhances pathfinding by incorporating heuristics to guide the search toward promising regions. For example, in a policy KG, A* might use semantic similarity to prioritize exploring connections between "climate change" and "economic policy" via intermediate nodes like "carbon taxation" or "sustainable development goals."

Finding relationships between entities, reasoning about connections in KG-based RAG systems, shortest path and traversal techniques are used to efficiently navigate the graph and retrieve relevant information. These methods help identify the most direct or meaningful connections between entities, which is especially valuable in semantic search and reasoning tasks. In a biomedical graph, a shortest path from a symptom to a treatment might pass through an intermediate condition, providing a mechanistic explanation for therapeutic relevance. In a legal graph, tracing a path between two court cases may surface shared legal principles or precedents, enhancing explainability in legal retrieval systems.

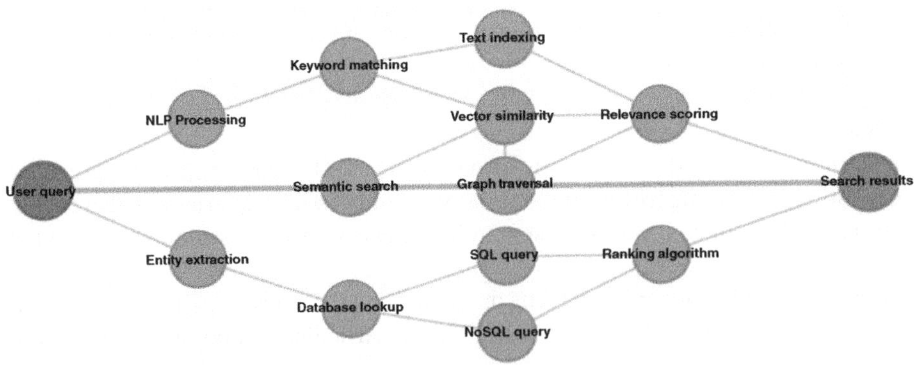

Figure 6.9 Path comparison: from "user query" to "search results."

6.3.1.4 Link Prediction

While traditional KGs model existing relationships between entities, they often suffer from incompleteness due to data limitations or evolving knowledge. Link prediction addresses this challenge by identifying potential edges—relationships that are not currently in the graph but are likely to exist. These inferred links enhance the coverage and utility of the KG, supporting proactive discovery, semantic enrichment, and better downstream performance.

Figure 6.10 visualizes link prediction in a sample academic research graph. Initially, disconnected nodes are connected by predicted links shown in dashed lines, indicating high-probability associations. This academic collaboration network contains 20 entities including researchers, papers, research topics, institutions, and conferences. Link prediction identifies potential future collaborations based on shared research interests, common co-authors, institutional connections, and publication patterns. High-confidence predictions suggest researchers who are likely to collaborate due to complementary expertise and established networks.

6.3.1.4.1 Structural Prediction Methods

Link prediction relies on structural similarity metrics to estimate the likelihood of missing edges in a graph. These methods analyze node neighborhoods, degrees, and connection patterns to infer potential relationships. Below, we explore five key approaches—common neighbors, Adamic–Adar, Jaccard similarity, preferential attachment, and resource allocation index—with mathematical definitions, intuitive explanations, and comparative insights.

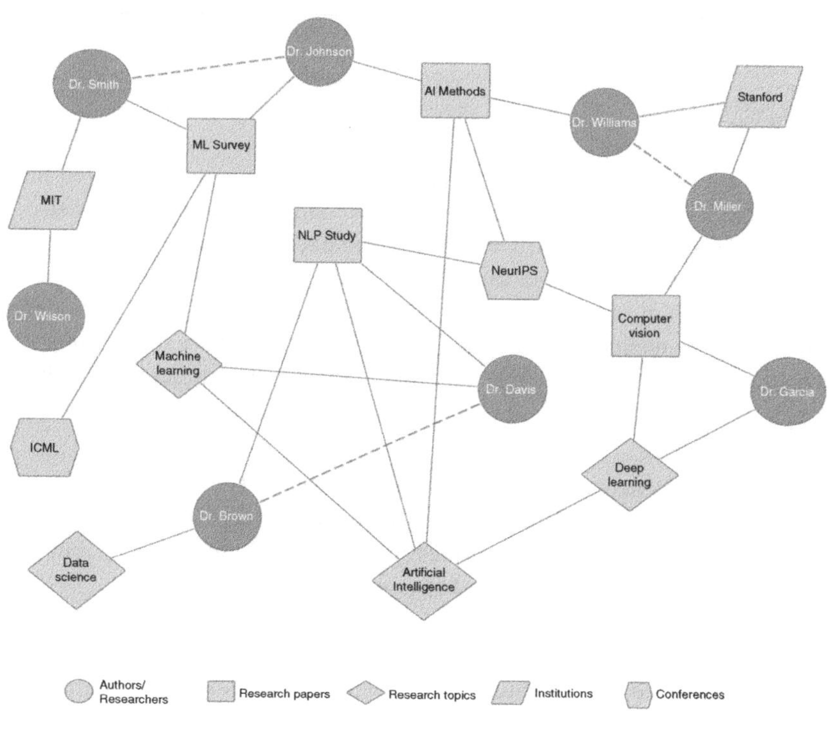

Figure 6.10 Link prediction example in academic research network analysis.

One of the simplest yet effective methods is the common neighbors metric, which counts the number of neighbors shared by two nodes:

$$\text{CN}(u, v) = |\Gamma(u) \cap \Gamma(v)|$$

where $\Gamma(u)$ denotes the set of neighbors of node u.

The intuition of this metric is that nodes with many shared connections are more likely to form a link. This assumes that relationships cluster in communities. For example, in a co-authorship network, two researchers who share multiple collaborators are likely to co-author a paper in the future. This approach is computationally simple and effective in densely connected graphs. Though it can potentially suffer from biasing toward high-degree nodes (since they naturally share more neighbors) and ignoring the "quality" of shared connections.

The Adamic–Adar index refines this idea by weighting common neighbors inversely to their degree:

$$\text{AA}(u, v) = \sum_{z \in \Gamma(u) \cap \Gamma(v)} \frac{1}{\log |\Gamma(z)|}$$

In other words, rarer shared neighbors contribute more to the prediction score than highly connected ones. This distinction helps emphasize meaningful, domain-specific niche relationships. In a biomedical network, for instance, if two drugs share a common protein interaction that is otherwise rarely connected in the graph, this could indicate a nonobvious but clinically significant relationship.

Jaccard similarity is another metric that measures the overlap of neighbors relative to the total unique neighbors:

$$\text{Jaccard}(u, v) = \frac{|\Gamma(u) \cap \Gamma(v)|}{|\Gamma(u) \cup \Gamma(v)|}$$

This metric normalizes for node degree, favoring pairs with a high fraction of shared neighbors rather than absolute counts. For example, two users in a recommendation system with 5 shared friends out of 10 total unique friends (Jaccard = 0.5) are more likely to connect than two users with 5 shared friends out of 100 (Jaccard = 0.05). Jaccard similarity works well for nodes with varying degrees and is interpretable as it ranges from 0 to 1. Though it still ignores the node degrees, in cases such as a hub and a low-degree node may share few neighbors but are still related.

The preferential attachment method assumes that high-degree nodes attract more links:

$$\text{PA}(u, v) = |\Gamma(u)| \times |\Gamma(v)|$$

It models the "rich-get-richer" effect observed in growing networks. For example, in citation networks, a highly cited paper is more likely to be referenced by new papers than lesser-known ones. This metric captures the scale-free network dynamics and is computationally trivial, but overlooks local clustering and tends to ignore predicting links between low-degree nodes.

6.3.1.4.2 Machine Learning Approaches

Recent advances have introduced sophisticated machine learning models that combine structural features with attribute information. These models typically employ:

- Feature engineering from graph topology (common neighbors, path features).
- Embedding-based approaches that learn latent relationship patterns.
- Hybrid models incorporating both structural and semantic information.

GNNs have emerged as particularly powerful tools, as they can simultaneously learn from node attributes, edge properties, and global graph structure. These models excel at capturing complex relationship patterns that simple heuristics might miss.

We are going to discuss more details of graph embedding and GNN methods in Sections 6.3.2 and 6.3.3. In combination with embedding-based or learning-based models, structural link prediction can be enhanced by integrating both topological patterns and semantic information, resulting in a more robust and insightful enrichment strategy.

6.3.1.4.3 Applications in KG Completion

In biomedical discovery, link prediction techniques have been used to identify potential drug-disease associations, predict adverse drug reactions, and uncover molecular interactions. For example, by analyzing shared pathways, gene expressions, or phenotypic similarities, researchers can forecast novel treatment opportunities or repurposing candidates. A high-scoring edge between a drug and a disease node might suggest unexplored therapeutic potential worth clinical investigation.

In recommendation systems, link prediction plays a key role in matching users with relevant items, services, or content. A bipartite graph of users and items can be enriched by predicting links based on user behavior patterns or item co-occurrence. If User A and User B have interacted with similar items and User B has shown strong interest in Item X, the system may predict that User A will also engage with Item X, even if that connection doesn't yet exist in the data.

In scientific literature, predicting future citations or collaborations can help forecast the evolution of research fields. By modeling topical similarity, co-authorship networks, or citation contexts, KGs can identify potential references or research alliances. For instance, two papers with highly overlapping keywords and citation patterns may be likely to cite each other in future work.

These applications illustrate the value of link prediction in making KGs more dynamic, intelligent, and capable of supporting inference. Rather than passively reflecting known information, an enriched graph can anticipate and suggest connections, enabling discovery across scientific, industrial, and digital domains.

6.3.2 Graph Embeddings

We discussed word embeddings such as Word2Vec, GloVe, and FastText in Section 2.1.2.2, which are widely used to represent natural language tokens in a continuous vector space. These embeddings allow models to capture semantic similarity and syntactic structure in text data.

In the case of graph-structured data, a similar approach can be applied. Just as word embeddings convert discrete tokens into dense vectors suitable for neural models, graph embeddings translate nodes, edges, or entire subgraphs into continuous vector representations. These embeddings are designed to capture both the structural topology and semantic properties of the graph, enabling effective integration with downstream machine learning tasks such as classification, clustering, and link prediction.

Raw graphs are discrete, sparse, and often high-dimensional, making them difficult to use directly in most machine learning pipelines. Graph embeddings provide a compact and information-rich representation that supports:

- Structural Preservation: Encodes local (e.g., node degree, neighbors) and global (e.g., communities, paths) structure.
- Semantic Encoding: Captures attribute information and relational patterns, particularly in KGs.

- Compatibility with ML: Embeddings allow graph elements to be fed into standard ML models such as logistic regression, neural networks, or clustering algorithms.

Graph embeddings are essential tools that convert symbolic graph structures into continuous representations, enabling learning and reasoning at scale. As illustrated in Table 6.8, they can be broadly categorized into:

6.3.2.1 Node-level Embedding

Node-level embeddings focus on learning vector representations for individual nodes while preserving their structural roles and relationships within the graph. Node2Vec and DeepWalk are commonly used random-walk-based methods that capture node similarity based on graph traversal paths.

6.3.2.1.1 DeepWalk

DeepWalk, introduced in "DeepWalk: Online Learning of Social Representations"[10], pioneered the application of language modeling techniques to graphs. The key insight is treating random walks on graphs as sentences and nodes as words.

As cited from the paper, Figure 6.11 illustrates DeepWalk's output on the Karate Club network. Figure 6.11a shows the raw graph; Figure 6.11b shows a 2D projection of the learned embeddings, where nodes naturally cluster according to community structure.

This is the DeepWalk algorithm overview:

1. Generate random walks starting from each node.
2. Treat walk sequences as sentences in a corpus.
3. Apply skip-gram model to learn embeddings that predict context nodes.

The resulting embeddings capture structural equivalence—nodes with similar roles in the graph (e.g., hub nodes, bridge nodes) obtain similar representations.

Table 6.8 Comparisons of node-level embeddings and graph-level embeddings.

Type	Techniques	Example Models	Applications
Node-level embeddings	Random walk, GNNs	DeepWalk, Node2Vec, GCN	Node classification, link prediction
Graph-level embeddings	Aggregation, KG modeling	TransE, TransH, RotatE	Graph classification, KG completion

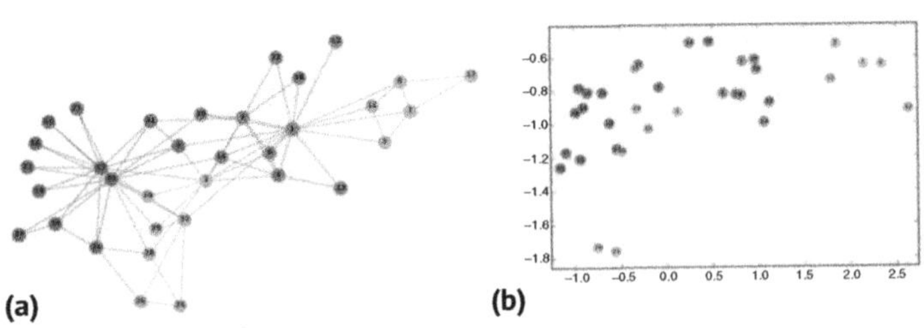

Figure 6.11 Node embedding representations of a graph. *Source*: [10]/Association for Computing Machinery (ACM).

6.3.2.1.2 Node2Vec

Node2Vec [11] extends DeepWalk by introducing controlled random walks that can interpolate between BFS and DFS behaviors. The Node2Vec algorithm introduces two tunable parameters:

- Return parameter (p): Controls the likelihood of immediately revisiting a node
- In-out parameter (q): Balances exploration between local neighborhoods (BFS-like) and distant nodes (DFS-like)

Transition probability from node v to x:

- If x is the previous node: weight = 1/p
- If x is connected to previous node: weight = 1
- If x is not connected to previous node: weight = 1/q

This flexibility allows Node2Vec to capture both homophily (similar nodes cluster together) and structural equivalence (nodes with similar roles have similar embeddings).

There are also other notable node embedding methods such as:

- LINE (large-scale information network embedding) [12]: Preserves both first-order (direct connections) and second-order (shared neighbors) proximity.
- SDNE (structural deep network embedding) [13]: Uses deep autoencoders to capture nonlinear graph structure.

6.3.2.2 Graph-Level Embeddings

While node embeddings focus on individual nodes, graph-level embeddings are designed for entire graphs or subgraphs. These are essential for tasks like graph classification, graph similarity, and graph-based retrieval.

KGs present unique challenges as they contain multi-relational data. Multi-relational data refers to directed graphs whose nodes correspond to entities and edges of the form (head, label, tail). KG embedding models aim to encode both entities and relations into a continuous space such that the semantics of relational triples (head, relation, tail) are preserved. Let a triple be denoted as (h,r,t), where: h is head entity, r is relation, and t is tail entity. The goal is to learn embeddings $h, r, t \in R$ such that a scoring function assigns high scores to true triples and low scores to false ones.

Several specialized embedding methods have been developed:

TransE (Translating Embeddings): TransE [14] models relationships as translations in the embedding space. The assumption of TransE is: relations are modeled as translations in the embedding space:

$$\mathbf{h} + \mathbf{r} \approx \mathbf{t}$$

where h, r, and t are embeddings for head entity, relation, and tail entity, respectively.

TransE uses a simple scoring function:

$$f(h, r, t) = -||\mathbf{h} + \mathbf{r} - \mathbf{t}||_2$$

It is efficient and effective for 1-to-1 relations. For example, for the triple (Paris, isCapitalOf, France), the embedding of "Paris" plus the embedding of "isCapitalOf" should closely match the embedding of "France." However, TransE struggles with 1-to-many, many-to-1, and many-to-many relationships.

TransH (Translation on Hyperplanes): TransH [15] addresses TransE's limitation with 1-to-N or N-to-N relations. It projects entities onto a relation-specific hyperplane before translation:

$$\mathbf{h}_\perp + \mathbf{r} \approx \mathbf{t}_\perp$$

where each relation r defines a normal vector w_r to the hyperplane:

$$h_\perp = h - w_r^T h w_r$$

$$t_\perp = t - w_r^T t w_r$$

$$h_\perp + d_r \approx t_\perp$$

where d_r is the translation vector on the hyperplane.

This allows the same entity to have different representations for different relations, better handling complex relationship patterns. TransH improves relationship modeling when entities participate in multiple roles across different contexts.

RotatE (Rotation Embeddings): RotatE[16] represents relations as rotations in complex vector space:

$$\mathbf{t} = \mathbf{h} \circ \mathbf{r}$$

where $\circ$ is element-wise complex multiplication, where $\circ$ denotes element-wise (Hadamard) product in complex space, and $|r_i| = 1$ for all elements.

It captures various relation patterns such as symmetric relations ($r = 1$), inverse relations ($r^{-1} = \bar{r}$) and composition patterns ($r_1 \circ r_2 = r_3$). RotatE is more expressive than TransE and TransH, achieving state-of-the-art performance on KG completion tasks.

There are also other graph-level embedding approaches:

- ComplEx [17]: Uses complex-valued embeddings to handle asymmetric relations while maintaining computational efficiency
- DistMult [18]: Simplifies the model by using diagonal matrices for relations, trading expressiveness for efficiency
- ConvE [19]: Applies 2D convolutions over embeddings, introducing nonlinearity for better expressiveness

Table 6.9 lists the comparisons of the KG embedding models. Besides the above KG embedding models, modern approaches like graph convolutional networks (GCNs), graph attention networks (GATs) and GraphSAGE learn embeddings through message-passing mechanisms, which we'll explore in detail in Section 6.3.3.

6.3.3 Graph Neural Networks

GNNs represent a major advancement in the field of graph-based machine learning. They extend classical neural networks to non-euclidean domains, enabling models to reason directly over graph-structured data. This has opened up new possibilities in KG reasoning, entity resolution, recommendation systems, drug discovery, and more.

In this section, we explore the fundamental concepts, representative architectures, and real-world applications of GNNs, particularly in the context of KGs.

Table 6.9 Comparison of knowledge graph embedding models.

Model	Representation style	Handles asymmetry	Nonlinearity	Strengths
TransE	Vector translation	✗	✗	Simple, scalable
TransH	Projection + translation	✔	✗	Better multi-relational modeling
RotatE	Complex rotation	✔	✗	Captures inverse/symmetric patterns
DistMult	Bilinear (diagonal matrix)	✗	✗	Lightweight, efficient
ComplEx	Complex-valued embedding	✔	✗	Expressive and efficient
ConvE	2D convolution	✔	✔	Highly expressive, nonlinear

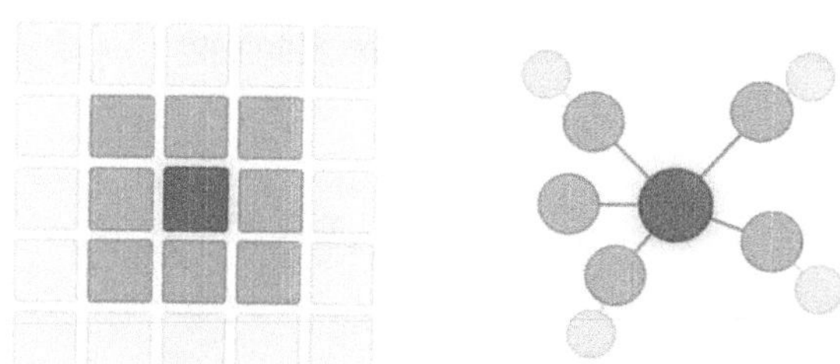

Figure 6.12 Left: regular grid structure (for CNN); Right: irregular graph structure (for GNN).

6.3.3.1 GNN Concepts

To understand GNNs, consider the analogy with convolutional neural networks (CNNs), as illustrated in Figure 6.12. CNNs majorly process regular grid structures (images) while GNNs process irregular graph structures.

Both CNNs and GNNs follow a layered aggregation paradigm, but differ in how they define neighborhood structures. As illustrated in Figure 6.13, while CNNs rely on regular spatial grids and fixed patches, GNNs aggregate from graph-defined neighbors that can vary in number and connectivity. This flexibility allows GNNs to generalize neural architectures to domains like KGs, social networks, and molecular graphs, where irregularity is the norm.

At the core of all GNNs lies the message passing paradigm. At each layer of a GNN, nodes update their embeddings by aggregating information from their local neighborhood—this includes both structural information about the graph and feature information of the nodes. After k iterations (layers), each node's embedding contains information from its k-hop neighborhood.

Message passing is the process where nodes iteratively update their representations by aggregating information from their neighbors, with the following steps:

1. Message Creation: Each neighbor creates a "message" based on its current state using the source node's features, edge attributes, and a learned transformation.
2. Aggregation: Nodes combine messages from all neighbors using permutation-invariant functions (mean, sum, max)
3. Update: The node updates its representation, integrating aggregated messages

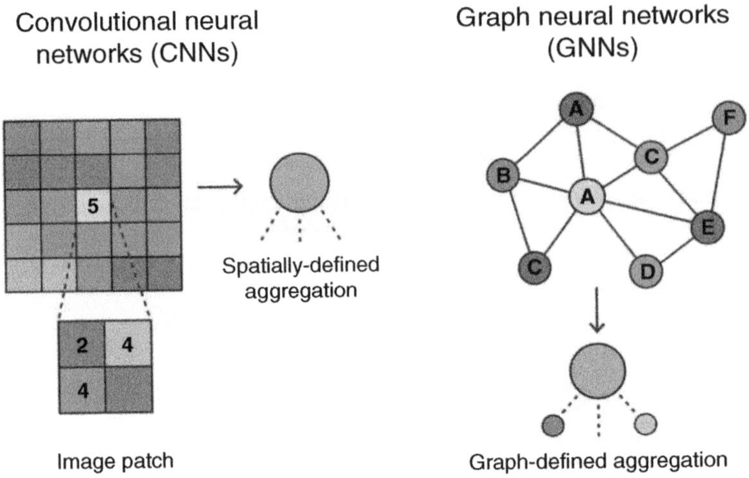

Figure 6.13 Left: CNNs aggregate information from regular spatial grids; Right: GNNs aggregate information from neighboring nodes.

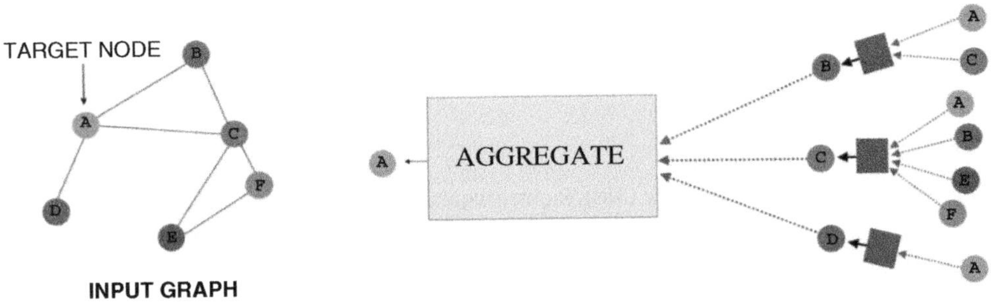

Figure 6.14 Aggregate messages from a node's local neighbors. *Source*: [20]/William L. Hamilton.

The message passing process can be expressed as follows:

$$\begin{aligned}\mathbf{h}_u^{(k+1)} &= \text{UPDATE}^{(k)}\left(\mathbf{h}_u^{(k)}, \text{AGGREGATE}^{(k)}(\{\mathbf{h}_v^{(k)}, \forall v \in \mathcal{N}(u)\})\right)\\ &= \text{UPDATE}^{(k)}\left(\mathbf{h}_u^{(k)}, \mathbf{m}_{\mathcal{N}(u)}^{(k)}\right)\end{aligned}$$

where $h_u^{(k)}$ denotes the embedding representation of node u at layer k; $N(u)$ denotes node u's neighbors set; $m_{N(u)}^{(k)}$ denotes the message that is aggregated from node u's graph neighborhood at layer k.

This process repeats across multiple layers, allowing information to propagate across the graph in a controlled manner.

Figure 6.14 illustrates an overview of how a single node aggregates messages from its local neighborhood. The model aggregates messages from A's local graph neighbors (i.e., B, C, and D), and in turn, the messages coming from these neighbors are based on information aggregated from their respective neighborhoods, and so on. This visualization shows a two-layer version of a message-passing model.

6.3.3.2 GNN Architectures

Below are some of the most widely adopted GNN architectures, each optimized for different modeling scenarios.

6.3.3.2.1 Graph Convolutional Networks

GCNs generalize the idea of convolution from grids to graphs. It was first brought up by Kipf and Welling [21] and has become one of the most popular baseline GNN models.

GCNs perform normalized feature aggregation from neighboring nodes, followed by a learned linear transformation. At each layer k, a node v's representation is updated based on its neighbors $N(v)$ and itself. The layer-wise propagation rule is:

$$\mathbf{h}_v^{(k+1)} = \sigma\left(\sum_{u\in\mathcal{N}(v)\cup\{v\}} \frac{1}{\sqrt{d_v d_u}}\mathbf{W}^{(k)}\mathbf{h}_u^{(k)}\right)$$

where:

- $\mathbf{h}_v^{(k)}$ is the feature vector of node v at layer k,
- $\mathcal{N}(v)$ is the set of neighbors of v,
- $d_u = |\mathcal{N}(u)|$ is the degree of node u,
- $\mathbf{W}^{(k)}$ is a learnable weight matrix at layer k,
- σ is a nonlinear activation function (e.g., ReLU).

This propagation rule ensures that each node receives a normalized, weighted sum of its neighbors' features, including itself. This "Self-Loop" which explicitly incorporates the nodes' own features is critical for preserving identity information. The normalization term $\frac{1}{\sqrt{d_v d_u}}$ stabilizes training and prevents nodes with higher degrees from dominating the aggregation.

GCNs have a minimalistic design, making them computationally efficient and easy to implement. GCNs lay the foundation for many advanced GNN architectures. By leveraging local graph structure and performing feature aggregation through normalized neighborhood averaging, GCNs offer a powerful yet efficient method for learning from relational data.

6.3.3.2.2 Graph Attention Networks

A popular strategy for enhancing the expressive power of GNNs is to integrate attention mechanisms into the aggregation step. First introduced in the context of machine translation [22], attention mechanisms enable models to dynamically assign importance weights to input elements—allowing selective focus on relevant features.

In the context of graphs, this concept translates to learning the relative importance of a node's neighbors during message aggregation. Rather than treating all neighbors equally, as in traditional GCNs, GATs allow the model to weigh each neighbor's contribution differently.

Attention-based Aggregation: The seminal work [23] introduced the GAT architecture, which formalized this idea. In GAT, each node computes attention coefficients for its neighbors, determining how much influence each should have in the update:

$$\alpha_{vu} = \text{softmax}\left(a^T[Wh_v||Wh_u]\right)$$

where α_{vu} reflects the importance of node u to node v, and a^T is a learned attention vector, and W is a learned matrix.

The updated node embedding is a weighted sum of the transformed neighbor features:

$$h'_v = \sigma\left(\sum_{u \in \mathcal{N}(v)} \alpha_{vu} W h_u\right)$$

This architecture enables the model to learn which neighbors are most relevant for a particular task, increasing its representational capacity.

Multi-head Attention: GATs often employ multi-head attention, inspired by Transformer architectures [24]. Multiple attention heads compute separate attention-weighted combinations of neighbor features, and the outputs are either concatenated or averaged to form the final node representation. This provides additional flexibility and robustness, similar to how multi-head attention helps in capturing diverse semantic aspects in Transformer-based NLP models.

GATs and Transformers: A Connection: While we've introduced the Transformer architecture in Section 2.1.2.3 of the book, an emerging trend in GNN research is to borrow architectural ideas from Transformers, particularly by applying self-attention mechanisms across graph nodes. One practical strategy is to start with a Transformer model and restrict its attention matrix using the graph's adjacency matrix—ensuring that nodes only attend to their direct neighbors.

This approach allows GNNs to benefit from the highly optimized transformer infrastructure available in modern deep learning libraries. However, it comes with computational trade-offs:

- Standard GATs scale linearly with the number of edges $\mathcal{O}(|V| + |E|)$,
- Whereas full self-attention mechanisms in transformers scale quadratically with the number of nodes $\mathcal{O}(|V|^2)$.

Thus, GATs strike a balance between the structure-awareness of GNNs and the flexibility of attention-based models, while avoiding the heavy computational cost of full graph-wide attention.

In summary, GATs represent a powerful extension of message-passing GNNs by introducing learnable attention mechanisms that allow fine-grained control over neighbor aggregation. They are particularly effective in scenarios where (1) the graph is noisy or heterogeneous, (2) certain neighbors are more relevant than others, or (3) where the task involves capturing nuanced local patterns.

6.3.3.2.3 GraphSAGE

While models like GCNs work well for transductive settings—where all nodes are known during training—many real-world applications demand inductive capabilities, where the model must generalize to previously unseen nodes or subgraphs. To address this, GraphSAGE (graph sample and aggregate) was proposed by Hamilton et al. [25] as an efficient and scalable GNN architecture for inductive representation learning on large graphs.

Message Passing and Sampling: At each layer, GraphSAGE updates a node's embedding by aggregating feature vectors from a randomly sampled subset of neighbors, rather than the full neighborhood. This approach reduces memory and computation cost, especially in graphs with high-degree nodes. The update for a node v at layer k+1 is given by:

$$\mathbf{h}_v^{(k+1)} = \sigma\left(\mathbf{W}^{(k)} \cdot \text{AGGREGATE}^{(k)}\left(\{\mathbf{h}_v^{(k)}\} \cup \{\mathbf{h}_u^{(k)}, \forall u \in \mathcal{S}_v\}\right)\right)$$

where:

- $\mathcal{S}_v \subseteq \mathcal{N}(v)$ is a sampled subset of neighbors,
- $\text{AGGREGATE}^{(k)}$ is a permutation-invariant aggregator function,
- $\mathbf{W}^{(k)}$ is a trainable weight matrix,
- σ is a nonlinear activation (e.g., ReLU),
- $\mathbf{h}_v^{(0)}$ is initialized with input node features.

Aggregator Functions: GraphSAGE supports a variety of aggregation functions, such as:

- Mean aggregator: Averages neighbor embeddings.

$$\text{AGGREGATE}^{(k)} = \text{mean}\left(\mathbf{h}_u^{(k)}\right)$$

- LSTM aggregator: Applies an LSTM to the sequence of neighbor embeddings.
- Pooling aggregator: Applies a nonlinear transformation and element-wise max or mean.

$$\text{AGGREGATE}^{(k)} = \max\left(\text{ReLU}(\mathbf{W}_{\text{pool}} \cdot \mathbf{h}_u^{(k)})\right)$$

Each function trades-off complexity, expressiveness, and scalability.

In summary, GraphSAGE introduced a paradigm shift in graph learning by enabling scalable, inductive inference. GraphSAGE is particularly suited for large and evolving graphs, where full-graph computation is infeasible. GraphSAGE's inductive framework has inspired several innovations to address scalability and heterogeneity challenges:

- Cluster-GCN [26] partitions large graphs into dense subgraphs for mini-batch training, reducing memory overhead while preserving connectivity patterns.
- GraphSAINT [27] introduces probability-based sampling strategies (e.g., random walk, degree-based) to minimize variance in stochastic training.
- Heterogeneous GraphSAGE [28] extends the paradigm to multi-relational graphs by designing type-specific aggregators for nodes and edges.

These variations demonstrate the adaptability of the GraphSAGE framework across a wide range of data regimes, from web-scale homogeneous graphs to richly structured heterogeneous networks.

By combining graph algorithms, embeddings, and neural networks, KG enrichment creates intelligent systems capable of sophisticated reasoning, pattern discovery, and predictive analytics. These techniques transform static knowledge repositories into dynamic, learning systems that continuously improve their understanding and predictive capabilities.

The synergy between these approaches—using graph algorithms for structural analysis, embeddings for efficient representation, and neural networks for complex pattern recognition—enables the construction of knowledge-enhanced AI systems that can reason, learn, and adapt in ways that closely approximate human intelligence.

6.4 LLMs and KGs

In Chapter 2 of the book, we've introduced details of LLMs. LLMs and KGs represent two distinct yet complementary approaches to organizing and utilizing knowledge. LLMs process and generate language using implicit patterns derived from vast text corpora, while KGs provide highly curated, structured, and interconnected information. Together, they can create a powerful symbiosis, where the strengths of one compensate for the limitations of the other.

6.4.1 Comparisons and Unifications

To understand how KGs and LLMs can complement each other, it is important to highlight their respective strengths and limitations. Table 6.10 compares KGs and LLMs across key dimensions such as knowledge, accuracy, interpretability, and language understanding.

This comparison reveals complementary strengths: KGs excel in structured, verifiable knowledge representation, while LLMs demonstrate superior language understanding and knowledge generalization capabilities.

The unification of KGs and LLMs aims to harness the strengths of both systems, creating a framework that overcomes their individual limitations. KGs can provide the structured, accurate data that LLMs lack, improving the LLMs' ability to generate precise and reliable outputs. Conversely, LLMs can enrich KGs with the ability to process and understand natural language, making them more accessible and versatile in interpreting queries and textual data.

In this chapter, we'll focus on how LLMs can be used to enrich and expand KGs, particularly in areas such as knowledge completion, entity discovery, and relation extraction. The complementary direction—using KGs to enhance LLMs—will be the focus of Chapter 7.

6.4.2 Leveraging LLMs to Enrich KGs

The integration of LLMs into KG construction and maintenance pipelines represents a paradigm shift in how we build and evolve structured knowledge bases. This section explores comprehensive approaches for using LLMs to enhance KGs across multiple dimensions.

6.4.2.1 Automated Knowledge Extraction

LLMs have revolutionized the extraction of structured knowledge from unstructured text, dramatically reducing the manual effort traditionally required for KG construction.

6.4.2.1.1 Entity Recognition and Typing

LLMs can identify novel entities and concepts from unstructured text—entities that do not yet exist in the KG. For example, an LLM reading news articles can detect a newly launched drug and link it to relevant diseases and manufacturers, enriching a biomedical KG.

Table 6.10 Comparisons of KGs and LLMs.

	Knowledge graphs	Large language models
Knowledge	Structural knowledge	Implicit knowledge
	Domain-specific knowledge	Lacking domain-specific knowledge
	Evolving knowledge	Lacking new knowledge
	Incomplete/unseen facts	General knowledge
Accuracy	Accuracy and decisiveness	Hallucination and indecisiveness
Interpretability	Interpretability	Black-box
Language understanding	Lacking language understanding	Language processing

Here is the Python pseudo code for an entity extraction example:

```
prompt = """
Extract all entities from the following text and classify them:
"Marie Curie discovered polonium in 1898 at the University of Paris."

Format: Entity | Type | Confidence
"""

# LLM Output:
# Marie Curie | Person/Scientist | 0.98
# Polonium | Chemical Element | 0.95
# 1898 | Date | 0.99
# University of Paris | Institution/University | 0.97
```

6.4.2.1.2 *Relation Extraction*

LLMs can help identify semantic relationships between entities that are expressed in natural language but not yet reflected in the KG. LLMs can label triples with relation types like partOf, influencedBy, or subsidiaryOf.

For example, from the sentence "OpenAI was backed by Microsoft," an LLM can generate the triple (OpenAI, backedBy, Microsoft).

Here is the Python pseudo code for a relation extraction example:

```
prompt = """
Identify relationships in this text:
"Tim Cook succeeded Steve Jobs as Apple's CEO in 2011, transforming the company's
    focus toward services."

Extract: (Subject, Predicate, Object, Temporal Context, Confidence)
"""

# LLM extracts:
# (Tim Cook, succeeded, Steve Jobs, 2011, 0.95)
# (Tim Cook, position, CEO of Apple, since 2011, 0.93)
# (Tim Cook, transformed, Apple's focus, after 2011, 0.88)
# (Apple, focus_shifted_to, services, under Tim Cook, 0.85)
```

6.4.2.1.3 *Multi-hop Reasoning for Knowledge Extraction*

LLMs can perform multistep reasoning to extract implicit facts:

Here is the Python pseudo code for a multi-hop extraction example:

```
python
context = """
The Eiffel Tower was completed in 1889 for the Paris Exposition.
Gustave Eiffel's company designed and built the tower.
The exposition celebrated the French Revolution's centennial.
"""

# LLM can infer:
# (Eiffel Tower, commemorates, French Revolution centennial)
# (Gustave Eiffel, associated_with, 1889 Paris Exposition)
# (French Revolution, occurred_in, 1789) [inferred from centennial]
```

6.4.2.2 KG Completion

LLMs can infer missing entities or relationships in a KG by leveraging their understanding of context and co-occurrence patterns.

6.4.2.2.1 Link Prediction Using LLMs

LLMs can predict missing links by leveraging their vast training data and reasoning capabilities. For example, if the KG has the facts (Steve Jobs, founded, Apple) and (Apple, headquarteredIn, Cupertino), an LLM may infer (Steve Jobs, livedIn, Cupertino) even if this triple is not explicitly stated.

This can be framed as a link prediction task using prompting or fine-tuning techniques that generate or rank candidate triples based on a given context.

Here is the Python pseudo code for a link prediction example:

```
def llm_link_prediction(kg, entity1, entity2):
    """
    Use LLM to predict likely relationships between entities
    """
    existing_facts = kg.get_entity_context(entity1, entity2)

    prompt = f"""
    Given these facts about {entity1} and {entity2}:
    {existing_facts}

    What relationships likely exist between them?
    Consider: professional, temporal, causal, and hierarchical relations.
    Provide confidence scores.
    """

# LLM predicts missing relations with reasoning
    return llm.generate(prompt)
```

6.4.2.2.2 Triple Classification and Validation

LLMs can assess the validity of proposed triples, acting as intelligent filters.

Here is the Python pseudo code for a triple validation example:

```
candidate_triple = ("Einstein", "born_in", "Germany")

validation_prompt = f"""
Evaluate this statement: {candidate_triple}
Consider:
1. Factual accuracy
2. Temporal validity
3. Geographical/political context
4. Common knowledge vs. specialized knowledge

Provide: Validity (True/False), Confidence (0-1), Explanation
"""

# LLM provides nuanced validation considering historical context
```

6.4.2.3 Ontology Learning and Schema Enrichment

LLMs can automatically extend and refine KG schemas by learning from data patterns and domain knowledge.

6.4.2.3.1 Hierarchical Relationship Discovery

Here is the Python pseudo code for a hierarchical relationship discovery example:

```
def discover_concept_hierarchy(concepts, domain):
    """
    Use LLM to organize concepts into hierarchical structures
    """
    prompt = f"""
    Organize these {domain} concepts into a hierarchical taxonomy:
    {concepts}

```

```
    Provide:
    - Parent-child relationships
    - Sibling relationships
    - Multiple inheritance where applicable
    - Justification for each relationship
    """

    return llm.build_taxonomy(prompt)
```

6.4.2.3.2 *Property and Constraint Learning*

LLMs can infer logical constraints and property characteristics.

Here is the Python pseudo code for a learning property constraints example:

```
property_analysis = """
Analyze the property 'hasCapital':
- Domain: Country
- Range: City
- Functional: Yes (one capital per country)
- Inverse: isCapitalOf
- Temporal: Can change over time
- Cardinality: 1-to-1 at any given time
"""
```

6.4.2.4 Contextual Disambiguation and Coreference Resolution

LLMs are capable of resolving coreferences and entity ambiguities, which is essential for maintaining a coherent and de-duplicated graph structure. This ability improves the quality and consistency of KGs constructed from noisy or ambiguous text.

6.4.2.4.1 *Entity Disambiguation*

Here is the Python pseudo code for an entity disambiguation example:

```
def disambiguate_entity(mention, context, candidate_entities):
    """
    Use LLM to select correct entity from candidates
    """
    prompt = f"""
    Mention: "{mention}"
    Context: "{context}"

    Candidates:
    {[f"- {e.name}: {e.description}" for e in candidate_entities]}

    Select the correct entity based on:
    1. Contextual clues
    2. Temporal consistency
    3. Domain relevance
    4. Co-occurring entities

    Return: Selected entity, confidence, reasoning
    """


    return llm.disambiguate(prompt)
```

6.4.2.4.2 *Advanced Coreference Resolution*

Here is the Python pseudo code for a coreference resolution example:

```
text = """
Apple announced its new iPhone. The company's CEO, Tim Cook, said the device would
     revolutionize communication. He emphasized that their commitment to privacy
     remains paramount.
```

```
"""

# LLM resolves:
# "its" -> Apple
# "The company" -> Apple
# "the device" -> new iPhone
# "He" -> Tim Cook
# "their" -> Apple
```

In summary, by empowering KGs with language-centric reasoning from LLMs, we can build dynamic, rich, and adaptive knowledge repositories that far surpass the capabilities of static, manually curated graphs.

6.4.3 Using KGs to Enhance LLMs: A Preview of Graph-RAG

While we've explored how LLMs can enrich KGs by extracting entities, relations, and missing knowledge from unstructured data, the synergy between these two systems is bidirectional.

In the reverse direction, KGs can play a crucial role in enhancing the accuracy, reliability, and interpretability of LLMs. KGs offer structured, explicit, and curated information that complements the implicit knowledge encoded in LLMs' parameters. When integrated into LLM workflows, KGs help ground model outputs in factual, traceable sources—reducing hallucination and improving contextual reasoning.

This hybrid approach is the foundation of Graph-RAG (Graph-based Retrieval-Augmented Generation). In a Graph-RAG system:

- The user's query is first transformed into a semantic vector via an embedding model.
- This vector is then used to retrieve relevant nodes or subgraphs from a KG.
- The retrieved structured information is passed to the LLM to guide factually accurate and contextually rich generation.
- In some setups, the LLM also contributes back to the KG by identifying new entities or relations, closing the feedback loop.

Figure 6.15 illustrates this workflow, where a user prompt is semantically embedded, matched to graph content, and then used to augment the LLM's generative reasoning.

Such KG-augmented LLMs enable powerful capabilities in:

- Explainable AI Systems: By grounding responses in explicit knowledge paths, the system can provide clear explanations for its outputs—crucial for high-stakes domains like healthcare or finance.
- Multi-hop Reasoning: KGs enable LLMs to perform complex reasoning across multiple relationship chains, answering questions that require connecting disparate pieces of information.
- Personalized Knowledge Access: User-specific or organization-specific KGs can provide tailored context, enabling LLMs to generate highly relevant, personalized responses.

By combining the symbolic reasoning strengths of KGs with the natural language understanding of LLMs, Graph-RAG opens new possibilities for trustworthy and intelligent AI.

We will explore this architecture in greater detail in Part III of this book, where Chapter 7 is devoted to methods, design patterns, and applications of KG-augmented LLMs.

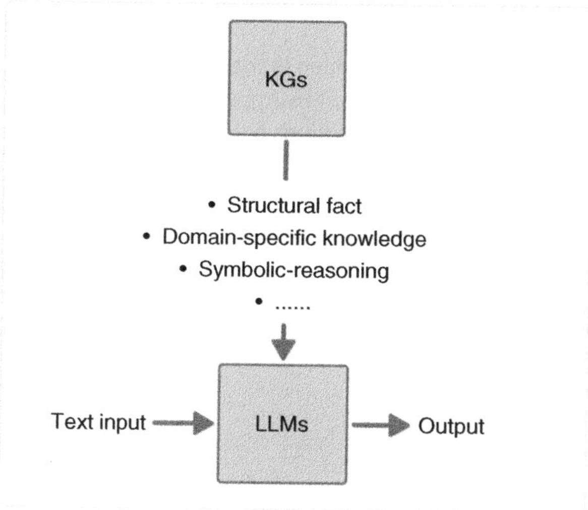

Figure 6.15 KG-enhanced LLMs. *Source*: [29]/IEEE.

6.5 Practice: Constructing KG from the WANDS Dataset

The objective of this section is to demonstrate how to construct KGs programmatically using the LlamaIndex and Neo4j integration. This method, which complements manual graph creation via Cypher queries (as described in Section 5.3), provides a key benefit: the resulting graph is compatible with the LlamaIndex framework, enabling seamless integration with RAG pipelines. However, it is important to note that the default property graph in LlamaIndex only accepts text-based input, so a custom extractor must be defined to incorporate other data types.

For this demonstration, we will use the WANDS dataset introduced in Section 1.3.3. The dataset contains information about products, search queries, and their mappings. Our objective is to build a KG that connects products to queries, product classes, and category hierarchies. We aim to model three main types of relationships within a unified graph structure:

- **Product (entity: attributes)**—*exact match/partial match/irrelevant* (relation)—**Query (entity)**
- **Product (entity: attributes)**—*product_class* (relation)—**Product (entity: attributes)**
- **Product (entity: attributes)**—*category_hierarchy (root/root-1/root-2)* (relation)—**Product (entity: attributes)**

To achieve this, we will divide the process into two steps:

1. Implementing a custom extractor to parse each row of the dataset into vertex-edge-vertex triples.
2. Using the LlamaIndex Neo4j integration to ingest these triples into the Neo4j database.

This structured approach ensures that our KG is both semantically rich and fully compatible with downstream LLM applications via LlamaIndex.

6.5.1 LlamaIndex Neo4j Integration

LlamaIndex offers several community-driven integrations, one of which is with Neo4j. This section outlines how to connect to a local Neo4j instance (set up as described in Appendix A) and execute Cypher queries using the LlamaIndex Neo4j integration. Follow the steps below to get started.

6.5.1.1 Connect to the Neo4j Server

To establish a connection to the Neo4j server, use the `Neo4jPropertyGraphStore` class from the LlamaIndex library:

```
from llama_index.graph_stores.neo4j import Neo4jPropertyGraphStore

graph_store = Neo4jPropertyGraphStore(
    url="bolt://localhost:7687",
    username=None,
    password=None,
)
```

6.5.1.2 Test the Connection

To verify that the connection works correctly, run a Cypher query to insert sample data into the Neo4j database:

```
query = r"""
    CREATE (a:Person {name: "Alice", age: 30}),
           (b:Person {name: "Bob", age: 25}),
           (a)-[:KNOWS {since: 2020}]->(b);
"""
graph_store.structured_query(query)
```

Next, retrieve the inserted data using another Cypher query:

```
query = """
    MATCH (p:Person)
    RETURN p.name, p.age;
"""
result = graph_store.structured_query(query)
print(result)
```

The result will be returned as a list of dictionaries, where each dictionary corresponds to a node and includes its properties. For example:

```
[{'p.name': 'Alice', 'p.age': 30}, {'p.name': 'Bob', 'p.age': 25}]
```

This format allows for easy access to query results when integrating with other Python-based workflows or applications.

6.5.2 Extract Query-product Triples

The query dataframe is split into two parts (75% training, 25% testing), as described in Section 1.3.3.2:

```
from sklearn.model_selection import train_test_split

df_query_train, df_query_test = train_test_split(
    df_query, test_size=0.25, random_state=42
)
```

Below are two helper functions (`*_node_name()`). Since both products and queries may share the same ID space (i.e., duplicate IDs exist across different node types), we add type-specific prefixes to distinguish them.

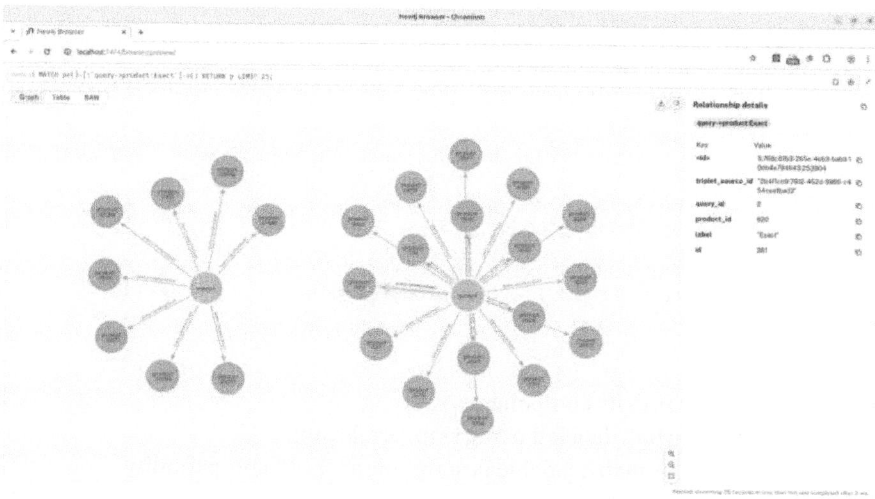

Figure 6.16 Exact label mapping query to product.

```
def product_node_name(product_id):
    return "product:" + product_id

def query_node_name(query_id):
    return "query:" + query_id
```

Using the training portion of the query dataset (`df_query_train`), the following Python class builds a graph connecting products to queries based on exact match labels from `label.csv`. It creates triples only for rows labeled "Exact" and "Partial." An example graph is shown in Figure 6.16.

```
class LabelExtractor(TransformComponent):
    def __call__(self, nodes: Sequence[BaseNode], **kwargs: Any) -> Sequence[BaseNode]:
        for node in nodes:
            label_id, query_id, product_id, label = node.get_content().split("\t")   # [1]

            if label in ["Irrelevant"]:                                               # [2]
                continue

            query, product = (
                df_query.loc[df_query["query_id"] == int(query_id)].squeeze(),
                df_product.loc[df_product["product_id"] == int(product_id)].squeeze(),
            )
            node.set_content(                                                         # [3]
                f"below product {label} matches query {query.to_json()} => {product.to_json()}"
            )

            existing_nodes = node.metadata.pop(KG_NODES_KEY, [])                      # [4]
            existing_relations = node.metadata.pop(KG_RELATIONS_KEY, []) # [4]
            metadata = node.metadata.copy()                                           # [4]

            subj_node = EntityNode(                                                   # [5]
                label="query",
                name=query_node_name(query_id),
                properties=metadata | query.to_dict(),
            )
            obj_node = EntityNode(                                                    # [6]
                label="product",
                name=product_node_name(product_id),
                properties=metadata | product.to_dict(),
```

```
            )
            rel_node = Relation(                                        # [7]
                label=f"query->product:{label}",
                source_id=subj_node.id,
                target_id=obj_node.id,
                properties=metadata
                | df_label.loc[df_label["id"] == int(label_id)].squeeze().to_dict(),
            )
            existing_nodes.extend([subj_node, obj_node])                # [8]
            existing_relations.append(rel_node)                         # [8]

            node.metadata[KG_NODES_KEY] = existing_nodes                # [9]
            node.metadata[KG_RELATIONS_KEY] = existing_relations        # [9]

        return nodes
```

[1] Split each row of the CSV file into its components.
[2] Only include "Exact" and "Partial" labeled pairs; skip "Irrelevant."
[3] Replace raw CSV data with a human-readable description for LLM compatibility.
[4] Preserve existing node metadata before appending new graph elements.
[5] Create a query node using the prefixed ID and relevant metadata.
[6] Create a product node similarly, with prefixed ID and metadata.
[7] Define a directed relation from query to product with associated properties.
[8] Append the new nodes and relations to the existing lists.
[9] Update the node's metadata with the extended list of nodes and relations.

6.5.3 Extract Product-class Triples

This Python code extracts KG triples from the `product.csv` file. Each triple represents a relationship between a product and its corresponding product class, labeled as `"product->product_class''`. The process works as follows:

1. For each row (treated as a node) in the CSV file, the code parses fields such as product ID, name, and class.
2. If the `product_class` field is missing or invalid (as determined by the `missing_data()` function), the row is skipped.
3. Two entity nodes are created:
   - A subject node representing the product.
   - An object node representing the product class.
4. A directed relation is established between them using the fixed label `"product->product_class''`.
5. These components—subject, predicate, object—are stored in the node's metadata for later use in constructing the full KG.

An example is shown in Figure 6.17.

```
class ProductClassExtractor(TransformComponent):
    def __call__(self, nodes: Sequence[BaseNode], **kwargs: Any) -> Sequence
    [BaseNode]:
        for node in nodes:
            (
                product_id,
                product_name,
                product_class,
                category_hierarchy,
                product_description,
                product_features,
```

```
                rating_count,
                average_rating,
                review_count,
            ) = node.get_content().split("\t")
            if missing_data(product_class):
                continue

            existing_nodes = node.metadata.pop(KG_NODES_KEY, [])
            existing_relations = node.metadata.pop(KG_RELATIONS_KEY, [])
            metadata = node.metadata.copy()

            subj_node = EntityNode(
                label="product",
                name=product_node_name(product_id),
                properties=metadata
                | df_product.loc[df_product["product_id"] == int(product_id)]
                .squeeze()
                .to_dict(),
            )
            obj_node = EntityNode(
                label="product_class",
                name=product_class_node_name(product_class),
                properties=metadata,
            )
            rel_node = Relation(
                label="product->product_class",
                source_id=subj_node.id,
                target_id=obj_node.id,
                properties=metadata,
            )
            existing_nodes.extend([subj_node, obj_node])
            existing_relations.append(rel_node)

            node.metadata[KG_NODES_KEY] = existing_nodes
            node.metadata[KG_RELATIONS_KEY] = existing_relations

        return nodes
```

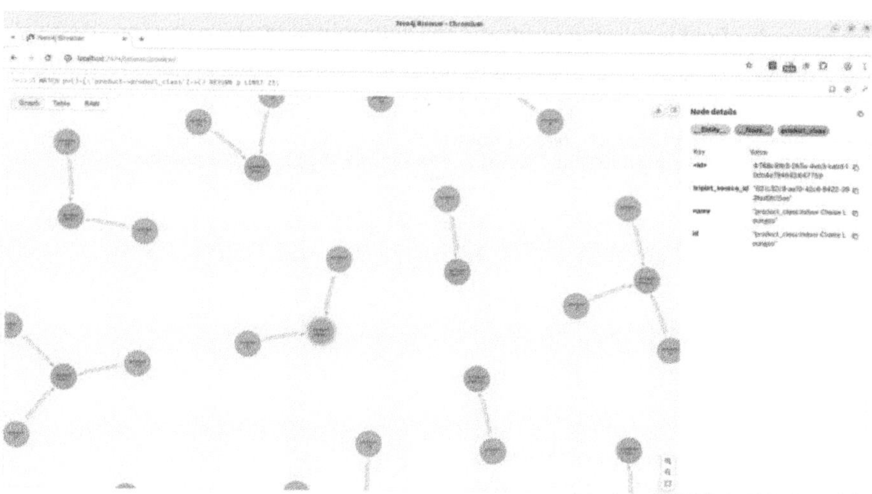

Figure 6.17 Product-class triples.

6.5.4 Extract Product-category Triples

This code extracts hierarchical relationships between products and their categories from the `product.csv` file. It constructs two types of KG triples:

- Links are from products to their full category path.
- Recursive links show the parent-child relationships within the category hierarchy itself.

For each row in the CSV (treated as a node), the code parses the `category_hierarchy` field (e.g., `"Furniture/Bedroom Furniture/Beds & Headboards/Beds/Twin Beds"`). This string is split by the delimiter `"/"` into individual levels. Then:

- A triple is created linking the product to the full category path.
- The code recursively removes the last category level and creates parent links all the way up to the root. For example:

  `"Furniture/Bedroom Furniture/Beds & Headboards/Beds/Twin Beds"` →
  `"Furniture/Bedroom Furniture/Beds & Headboards/Beds"` →
  `"Furniture/Bedroom Furniture/Beds & Headboards"` →
  `"Furniture/Bedroom Furniture"` →
  `"Furniture"`

  We use the full path for each intermediate node in the hierarchy because we found that the same category string may appear in different branches. Using full paths ensures uniqueness and allows us to distinguish between such duplicates.

Each triple consists of:

- A subject (either a product or category),
- A relation (e.g., `"product->category_hierarchy"` or `"parent"`),
- An object (the connected category).

If the `category_hierarchy` field is missing or invalid (as determined by the `missing_data()` function), the row is skipped to ensure only valid hierarchies are added to the graph. An example is shown in Figure 6.18.

```
class CategoryHierarchyExtractor(TransformComponent):
    def __call__(self, nodes: Sequence[BaseNode], **kwargs: Any) -> Sequence
     [BaseNode]:
        for node in nodes:
            (
                product_id,
                product_name,
                product_class,
                category_hierarchy,
                product_description,
                product_features,
                rating_count,
                average_rating,
                review_count,
            ) = node.get_content().split("\t")
            if missing_data(category_hierarchy):
                continue

            existing_nodes = node.metadata.pop(KG_NODES_KEY, [])
            existing_relations = node.metadata.pop(KG_RELATIONS_KEY, [])
            metadata = node.metadata.copy()

            sep = "/"
            parts = category_hierarchy.split(sep)
```

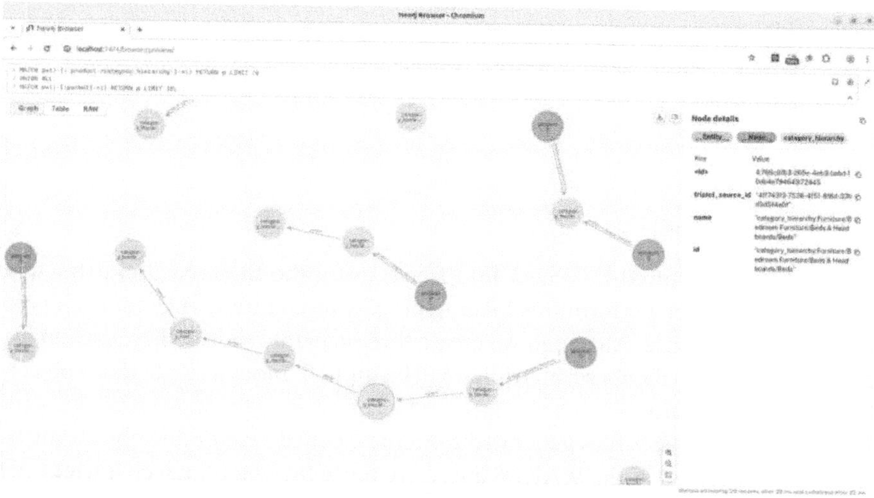

Figure 6.18 Product-category triples.

```
            # link `product` to `category_hierarchy`
            subj_node = EntityNode(
                label="product",
                name=product_node_name(product_id),
                properties=metadata
                | df_product.loc[df_product["product_id"] == int(product_id)]
                .squeeze()
                .to_dict(),
            )
            obj_node = EntityNode(
                label="category_hierarchy",
                name=category_hierarchy_node_name(sep.join(parts)),
                properties=metadata,
            )
            rel_node = Relation(
                label="product->category_hierarchy",
                source_id=subj_node.id,
                target_id=obj_node.id,
                properties=metadata,
            )
            existing_nodes.extend([subj_node, obj_node])
            existing_relations.append(rel_node)

            # link `category_hierarchy` to tree root
            parts.pop()
            while parts:
                parent_node = EntityNode(
                    label="category_hierarchy",
                    name=category_hierarchy_node_name(sep.join(parts)),
                    properties=metadata,
                )
                rel_node = Relation(
                    label="parent",
                    source_id=obj_node.id,
                    target_id=parent_node.id,
                    properties=metadata,
                )
                existing_nodes.append(parent_node)
                existing_relations.append(rel_node)

                obj_node = parent_node
```

```
                parts.pop()

        node.metadata[KG_NODES_KEY] = existing_nodes
        node.metadata[KG_RELATIONS_KEY] = existing_relations

    return nodes
```

6.5.5 Load Documents into Neo4j

We now proceed to load the structured CSV data into Neo4j using the three custom extractors defined in the previous sections. We perform this loading process three times, once for each type of KG relationship we want to capture. Although we create three separate index objects, when we later reload the data from the Neo4j database, all triples will be unified under a single index object, preserving the complete KG structure.

The documents are parsed from CSV files generated from the cleaned DataFrame objects introduced in Section 1.3.3.2. As specified in the WANDS guide, we reuse the default tab character (`\t`) as the field separator.

Since we plan to generate per-column vector embeddings in the indexing phase (Section 4.3.1), we disable node-level embeddings in Neo4j by setting `embed_kg_nodes=False`. Additionally, because each CSV row contains text within the model's context window length, we do not need to split the text. Therefore, we use `SimpleFileNodeParser()` to directly convert rows into nodes.

```
index_label = PropertyGraphIndex.from_documents(
    PandasCSVReader(col_joiner="\t", concat_rows=False).load_data(
        StringIO(df_label_train.to_csv(index=False))
    ),
    kg_extractors=[LabelExtractor(), ImplicitPathExtractor()],
    property_graph_store=graph_store,
    embed_kg_nodes=False,
    show_progress=True,
    transformations=[SimpleFileNodeParser()],
)

index_product_class = PropertyGraphIndex.from_documents(
    PandasCSVReader(col_joiner="\t", concat_rows=False).load_data(
        StringIO(df_product.to_csv(index=False))
    ),
    kg_extractors=[ProductClassExtractor(), ImplicitPathExtractor()],
    property_graph_store=graph_store,
    embed_kg_nodes=False,
    show_progress=True,
    transformations=[SimpleFileNodeParser()],
)

index_category_hierarchy = PropertyGraphIndex.from_documents(
    PandasCSVReader(col_joiner="\t", concat_rows=False).load_data(
        StringIO(df_product.to_csv(index=False))
    ),
    kg_extractors=[CategoryHierarchyExtractor(), ImplicitPathExtractor()],
    property_graph_store=graph_store,
    embed_kg_nodes=False,
    show_progress=True,
    transformations=[SimpleFileNodeParser()],
)
```

6.5.6 Key Takeaways

- This section details a method for programmatically constructing a KG from the structured WANDS dataset using the `LlamaIndex` and `Neo4j` integration.

- The core of the process involves creating custom extractors that parse each row of the dataset and convert it into a set of nodes and relationships, bypassing the need for an LLM to extract this information from unstructured text.
- Three distinct types of relationships are modeled:
  - Product to Query: Links products to queries based on exact and partial matches.
  - Product to Product Class: Connects products to their categorical class.
  - Product to Category Hierarchy: Links products to their full category path and establishes a recursive parent relationship between successive categories in the hierarchy.
- The generated triples (subject, predicate, object) are ingested into a `Neo4jPropertyGraphStore` using the `PropertyGraphIndex.from_documents()` method.
- This approach leverages LlamaIndex's framework to create a semantically rich KG that is fully compatible with downstream RAG pipelines, improving the quality of LLM-based applications.

6.6 Practice: Construct KG from Unstructured Data

The objective of this section is to demonstrate how to build a KG from **unstructured data** using an LLM. Unlike the previous section where we imported **structured data** (from a CSV Dataframe) into Neo4j, this method requires extracting knowledge triples directly from plain text. LlamaIndex provides a streamlined way to perform this extraction and store the resulting triples in a property graph. For this process, we'll reuse the LlamaIndex settings described in Section 4.2.

Below is the implementation of constructing a KG from a source of unstructured data – contents extracted from the White House website:

```
from llama_index.core import PropertyGraphIndex
from llama_index.readers.web import SimpleWebPageReader
from llama_index.core.node_parser import HTMLNodeParser
from llama_index.core.node_parser import SentenceSplitter

index = PropertyGraphIndex.from_documents(
    documents=SimpleWebPageReader().load_data(          # [1]
        ["https://www.whitehouse.gov/administration/melania-trump/"]
    ),
    transformations=[                                   # [2]
        HTMLNodeParser(tags=["p", "h1"]),
        SentenceSplitter(paragraph_separator="\n", chunk_size=512, chunk_overlap=0),
    ],
    property_graph_store=graph_store,                   # [3]
    show_progress=True,
)
```

`[1]` Document created by loading a webpage.
`[2]` First extract content from <p> and <h1> tags, then split into smaller text chunks.
`[3]` Reuse the Neo4j connection defined in Section 6.5.1.1.

For each node, LlamaIndex invokes the LLM to extract up to 10 knowledge triplets in the form `(subject, predicate, object)` from the text. Here's an example prompt (one-shot) used for one node:

user Some text is provided below. Given the text, extract up to 10 knowledge triplets in the form of (subject, predicate, object). Avoid stopwords.

Example:Text: Alice is Bob's mother.Triplets:

```
(Alice, is mother of, Bob)
Text: Philz is a coffee shop founded in Berkeley in 1982.
Triplets:
(Philz, is, coffee shop)
(Philz, founded in, Berkeley)
(Philz, founded in, 1982)
---------------------
Text: tag: p

In her time away from the White House, Mrs. Trump established Fostering the Future,
a BE BEST initiative, which provides college–level scholarships to those aging out of
the foster care system. The scholarship funds are provided by a portion of the revenue
generated from her Web 2 and Web 3 platforms. Her platforms utilize blockchain
technology, marking her as a pioneer within the industry while empowering the next
generation of young Americans with the resources needed to excel. Mrs. Trump
consistently implements innovative approaches to advance her advocacy for children
and the causes she supports, demonstrating a forward–thinking commitment to
making a positive impact.
In 2024, Mrs. Trump released her memoir, MELANIA, which quickly became #1
onThe New York Times best sellers list.

Triplets:

assistant
 1. (Mrs. Trump, established, Fostering the Future)
 2. (Fostering the Future, is part of, BE BEST initiative)
 3. (Fostering the Future, provides, college–level scholarships)
 4. (scholarship funds, provided by, revenue from Web 2 and Web 3 platforms)
 5. (Web 2 and Web 3 platforms, utilize, blockchain technology)
 6. (Mrs. Trump, marked as, pioneer in industry)
 7. (Mrs. Trump, empowers, next generation of young Americans)
 8. (Mrs. Trump, implements, innovative approaches to advance advocacy for children)
 9. (Mrs. Trump, released, memoir MELANIA in 2024)
10. (MELANIA, became #1 on, New York Times best sellers list)
```

The resulting KG is visualized in Figure 6.19.

6.6.1 Key Takeaways

- Building a KG from unstructured data (like plain text) is a different process than using structured data. It requires a Large Language Model (LLM) to perform the extraction of entities and relationships.
- The LlamaIndex framework automates this process. The `PropertyGraphIndex.from_documents()` method handles the entire workflow:
  1. **Document Loading:** Loads data from a source, like a webpage using `SimpleWebPageReader`.
  2. **Text Preprocessing:** Uses `HTMLNodeParser` to extract relevant content (e.g., from <p> and <h1> tags) and `SentenceSplitter` to divide the text into smaller, manageable chunks. This is crucial for fitting the text within the LLM's context window.

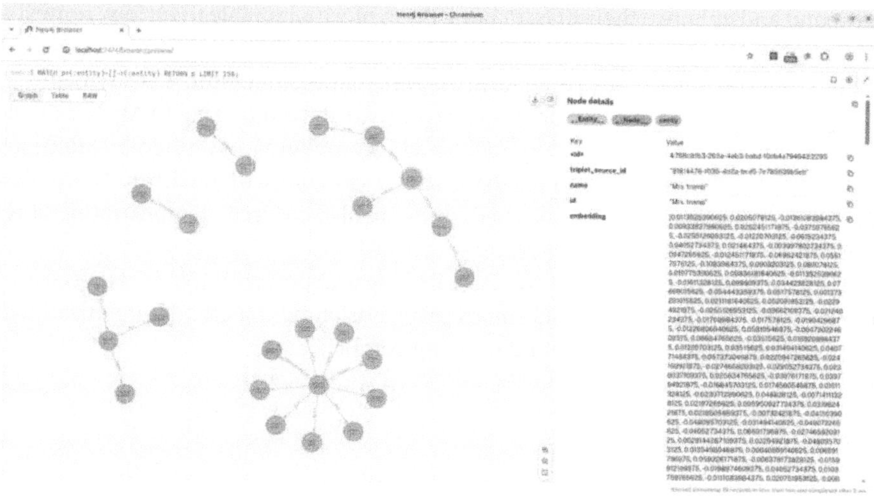

Figure 6.19 Extracted knowledge triples from unstructured text.

3. LLM-based Extraction: For each text chunk, LlamaIndex sends a prompt to an LLM, asking it to extract knowledge triplets in the format of `(subject, predicate, object)`.
4. Graph Storage: Stores the extracted triplets as nodes and relationships in a graph database, such as Neo4j, using a preconfigured `graph_store`.

- The final result is a visualized KG that represents the key information and relationships from the original unstructured text in a structured format, enabling complex querying and reasoning.

6.7 Summary

This chapter provides a foundational introduction to KGs, focusing on their construction, enrichment, and integration with LLMs to support advanced data-driven applications. KGs are data structures that connect information through relationships, enabling the discovery of meaningful connections and insights. The chapter begins by explaining the basic concepts behind KGs and their significance in organizing complex data in a structured and accessible way.

The chapter then explores methods for constructing and managing KGs, detailing essential techniques for organizing and linking data efficiently. Proper construction and management of KGs ensure that data is structured for easy retrieval and effective analysis, which is crucial for organizations seeking to leverage their datasets fully.

A significant portion of the chapter focuses on enriching KGs using graph algorithms. Algorithms like node embedding, centrality, and clustering can deepen the quality of a KG, allowing it to represent complex relationships and patterns within data. These techniques enhance the utility of KGs, making them more valuable for applications requiring in-depth data relationships, such as recommendation systems and predictive analytics.

The integration of large language models with KGs is another important topic covered in this chapter. This unification allows for a more powerful approach to knowledge representation by combining the reasoning capabilities of LLMs with the structured nature of KGs. By leveraging LLMs, organizations can expand and enhance their KGs, making them more adaptable to various contexts and enabling better insights.

A practical case study demonstrates how to transform different data sources into a KG. It includes a detailed walkthrough of converting unstructured text into structured data, using LLMs to extract subject, predicate, object triples and apply them to search and recommendation systems. Additionally, methods for converting tabular data into database structures compatible with KGs are discussed. Techniques for keyword and synonym-based queries are also covered, demonstrating how to optimize KGs for improved search and retrieval.

Overall, this chapter emphasizes the value of KGs as a flexible and powerful tool in data ecosystems. By integrating KGs with LLMs and using advanced graph enrichment methods, organizations can create sophisticated systems for managing and utilizing complex data.

References

1 Denny Vrandečić and Markus Krötzsch. Wikidata: A free collaborative knowledgebase. *Communications of the ACM*, 57(10):78–85, 2014.

2 Sören Auer, Christian Bizer, Georgi Kobilarov, Jens Lehmann, Richard Cyganiak, and Zachary Ives. DBpedia: A nucleus for a web of open data. In *Proceedings of the 6th International Semantic Web Conference (ISWC)*, pages 722–735. Springer, 2007. https://doi.org/10.1007/978-3-540-76298-0_52.

3 Olivier Bodenreider. The unified medical language system (UMLS): Integrating biomedical terminology. *Nucleic Acids Research*, 32(suppl_1):D267–D270, 2004.

4 Robyn Speer, Joshua Chin, and Catherine Havasi. Conceptnet 5.5: An open multilingual graph of general knowledge. In *Proceedings of the AAAI Conference on Artificial Intelligence*, volume 31, 2017.

5 Kurt D Bollacker, Colin Evans, Praveen Paritosh, Tim Sturge, and Jamie Taylor. Freebase: A collaboratively created graph database for structuring human knowledge. In *Proceedings of the SIGMOD Conference*, pages 1247–1250. ACM, 2008. https://doi.org/10.1145/1376616.1376746.

6 Vincent A Traag, Ludo Waltman, and Nees Jan Van Eck. From Louvain to Leiden: Guaranteeing well-connected communities. *Scientific Reports*, 9(1):5233, 2019.

7 Michelle Girvan and Mark EJ Newman. Community structure in social and biological networks. *Proceedings of the National Academy of Sciences*, 99(12):7821–7826, 2002. https://doi.org/10.1073/pnas.122653799.

8 Mark EJ Newman and Michelle Girvan. Finding and evaluating community structure in networks. *Physical Review E*, 69(2):026113, 2004. https://doi.org/10.1103/PhysRevE.69.026113.

9 Andrea Lancichinetti, Santo Fortunato, and Filippo Radicchi. Benchmark graphs for testing community detection algorithms. *Physical Review E*, 78(4):046110, 2008. https://doi.org/10.1103/PhysRevE.78.046110.

10 Bryan Perozzi, Rami Al-Rfou, and Steven Skiena. Deepwalk: Online learning of social representations. In *Proceedings of the 20th ACM SIGKDD International Conference on Knowledge Discovery and Data Mining*, pages 701–710, 2014.

11 Aditya Grover and Jure Leskovec. node2vec: Scalable feature learning for networks. In *Proceedings of the 22nd ACM SIGKDD International Conference on Knowledge Discovery and Data Mining*, pages 855–864, 2016.

12 Jian Tang, Meng Qu, Mingzhe Wang, Ming Zhang, Jun Yan, and Qiaozhu Mei. Line: Large-scale information network embedding. In *Proceedings of the 24th International Conference on World Wide Web*, pages 1067–1077, 2015.

13 Daixin Wang, Peng Cui, and Wenwu Zhu. Structural deep network embedding. In *Proceedings of the 22nd ACM SIGKDD International Conference on Knowledge Discovery and Data Mining*, pages 1225–1234, 2016.

14 Antoine Bordes, Nicolas Usunier, Alberto Garcia-Duran, Jason Weston, and Oksana Yakhnenko. Translating embeddings for modeling multi-relational data. *Advances in Neural Information Processing Systems*, 26:2787–2795, 2013.

15 Zhen Wang, Jianwen Zhang, Jianlin Feng, and Zheng Chen. Knowledge graph embedding by translating on hyperplanes. In *Proceedings of the AAAI Conference on Artificial Intelligence*, volume 28, 2014.

16 Zhiqing Sun, Zhi-Hong Deng, Jian-Yun Nie, and Jian Tang. Rotate: Knowledge graph embedding by relational rotation in complex space. *arXiv preprint arXiv:1902.10197*, 2019.

17 Théo Trouillon, Johannes Welbl, Sebastian Riedel, Éric Gaussier, and Guillaume Bouchard. Complex embeddings for simple link prediction. In *International Conference on Machine Learning*, pages 2071–2080. PMLR, 2016.

18 Bishan Yang, Wen-tau Yih, Xiaodong He, Jianfeng Gao, and Li Deng. Embedding entities and relations for learning and inference in knowledge bases. *arXiv preprint arXiv:1412.6575*, 2014.

19 Tim Dettmers, Pasquale Minervini, Pontus Stenetorp, and Sebastian Riedel. Convolutional 2d knowledge graph embeddings. In *Proceedings of the AAAI Conference on Artificial Intelligence*, volume 32, 2018.

20 William L Hamilton. *Graph Representation Learning*. Morgan & Claypool Publishers, 2020.

21 Thomas N Kipf and Max Welling. Semi-supervised classification with graph convolutional networks. *arXiv preprint arXiv:1609.02907*, 2016.

22 Dzmitry Bahdanau, Kyunghyun Cho, and Yoshua Bengio. Neural machine translation by jointly learning to align and translate. *arXiv preprint arXiv:1409.0473*, 2014.

23 Petar Velickovic, Guillem Cucurull, Arantxa Casanova, Adriana Romero, Pietro Lio, and Yoshua Bengio. Graph attention networks. *Statistics*, 1050(20):10–48550, 2017.

24 Ashish Vaswani, Noam Shazeer, Niki Parmar, Jakob Uszkoreit, Llion Jones, Aidan N Gomez, Łukasz Kaiser, and Illia Polosukhin. Attention is all you need. *Advances in Neural Information Processing Systems*, 30:5998–6008, 2017.

25 Will Hamilton, Zhitao Ying, and Jure Leskovec. Inductive representation learning on large graphs. *Advances in Neural Information Processing Systems*, 30:1024–1034, 2017.

26 Wei-Lin Chiang, Xuanqing Liu, Si Si, Yang Li, Samy Bengio, and Cho-Jui Hsieh. Cluster-GCN: An efficient algorithm for training deep and large graph convolutional networks. In *Proceedings of the 25th ACM SIGKDD International Conference on Knowledge Discovery & Data Mining*, pages 257–266, 2019.

27 Hanqing Zeng, Hongkuan Zhou, Ajitesh Srivastava, Rajgopal Kannan, and Viktor Prasanna. Graphsaint: Graph sampling based inductive learning method. *arXiv preprint arXiv:1907.04931*, 2019.

28 Ziniu Hu, Yuxiao Dong, Kuansan Wang, and Yizhou Sun. Heterogeneous graph transformer. In *Proceedings of the Web Conference 2020*, pages 2704–2710, 2020.

29 Shirui Pan, Linhao Luo, Yufei Wang, Chen Chen, Jiapu Wang, and Xindong Wu. Unifying large language models and knowledge graphs: A roadmap. *IEEE Transactions on Knowledge and Data Engineering*, 36(7):3580–3599, 2024.

Part III

Integrate RAG with Graph

The convergence of knowledge graphs (KGs) and retrieval-augmented generation (RAG) marks a pivotal evolution in how intelligent systems retrieve, reason, and generate information. By combining the structured, semantically rich representations of knowledge graphs with the flexibility and power of large language models, we unlock new capabilities in contextual understanding, precise retrieval, and explainable generation (illustrated in Figure PIII.1). Part III explores this integration in depth, providing both architectural blueprints and practical strategies for building and evaluating Graph-RAG systems.

We begin with Chapter 7, which introduces the Graph-RAG paradigm—a framework that augments traditional RAG pipelines with graph-structured data to enable more semantically accurate and context-aware retrieval. This chapter lays the foundation by examining core components such as graph-based indexing, multilevel retrieval granularity, traversal techniques, and generation strategies that leverage graph structure. Through these mechanisms, we explore how Graph-RAG overcomes the limitations of traditional retrieval and paves the way for graph-aware reasoning and generation.

Chapter 8 shifts from conceptual architecture to implementation, providing hands-on case studies that demonstrate how to build KG-RAG systems using frameworks like LlamaIndex and LangChain. Real-world data from e-commerce, Wikipedia, and Kaggle are used to illustrate graph construction, embedding strategies, and query orchestration in an operational RAG pipeline. These practical guides equip readers with the tools to transform tabular and unstructured data into graph-augmented workflows.

Finally, Chapter 9 addresses the evaluation of Graph-RAG systems—an essential aspect often overlooked in complex artificial intelligence pipelines. We introduce a multidimensional framework for assessing retrieval precision, generation quality, graph-specific metrics, and system performance. In addition to covering automated and human-centric evaluation methodologies, this chapter presents a curated suite of benchmarks tailored to the unique demands of graph-enhanced RAG systems.

Together, these chapters form a cohesive guide to understanding, building, and validating KG-RAG systems, setting the stage for cutting-edge applications that blend symbolic structure with neural intelligence.

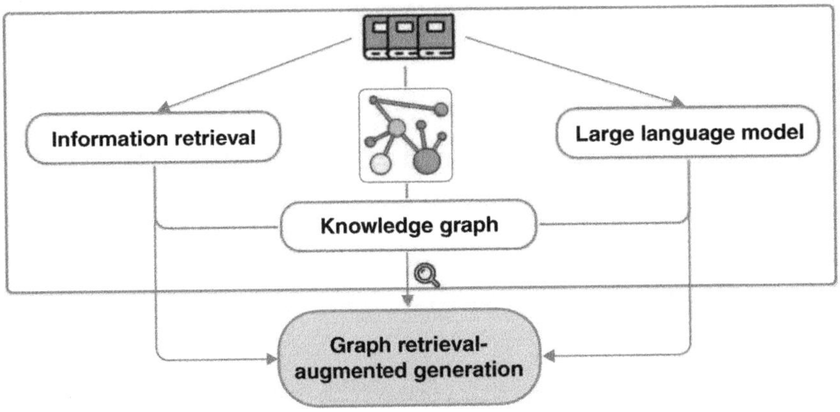

Figure PIII.1 Integrating knowledge graphs into retrieval-augmented generation systems.

7

Graph-based Retrieval-augmented Generation

In the evolving landscape of artificial intelligence, the integration of graph structures into retrieval-augmented generation (RAG) models is ushering in a new era of contextual understanding and information retrieval. This chapter dives into the paradigm of graph-based RAG (Graph-RAG) systems, where the power of knowledge graphs (KGs) is harnessed to enhance both the accuracy and relevance of retrieved information. By leveraging the intricate relationships and entities represented in KGs, we can significantly improve the quality of the data that feeds into the generation process, resulting in more informed and contextually aware responses.

We begin by exploring the pivotal role that KGs play in RAG systems. As we examine their contribution to enhancing retrieval accuracy and contextual relevance, we will outline the specific advantages they bring to the table—such as improved semantic understanding and the ability to capture nuanced relationships between concepts. Following this foundational discussion, we will dive into the design of Graph-RAG models, focusing on architecture, design principles, and integration strategies that allow for effective synergy between traditional RAG frameworks and graph-based structures.

The chapter will also cover essential implementation techniques, including data preprocessing, graph construction, and the critical phases of training and fine-tuning RAG models. By providing practical use cases, we will illustrate how these concepts can be applied in real-world scenarios, such as search and recommendation systems. As we work through a straightforward demo leveraging the graph developed in Part II of the book, you will gain insights into how querying and generating information can be enhanced through graph-based methodologies, setting the stage for a more intuitive and efficient interaction with knowledge-driven systems.

7.1 Introduction to Graph-RAG

Building upon our discussions of RAG fundamentals (Chapter 3) and KG construction (Chapter 6), we now examine how these technologies converge in Graph-RAG systems. This framework represents an evolution beyond traditional RAG approaches, addressing their fundamental limitations while opening new possibilities for knowledge-intensive applications.

7.1.1 The Graph-RAG Paradigm

Graph-RAG is an advanced form of RAG that integrates KGs directly into the retrieval process. Instead of searching through large collections of unstructured text, Graph-RAG leverages structured, interlinked graph data to surface semantically relevant and contextually precise information.

This retrieved content is then provided as context to a language model, enabling more accurate and explainable generation.

As illustrated in Figure 7.1 [1], the key difference lies in the retrieval backend: while traditional RAG systems query flat text chunks based on vector similarity, Graph-RAG performs retrieval over entities, relations, and graph neighborhoods, unlocking deeper semantic reasoning.

7.1.2 Why Graph-RAG?

While the foundational concepts of Graph-RAG lay the groundwork for integrating graph structures with language models, understanding the motivation behind this integration is crucial.

In the following sections, we examine the specific limitations of traditional RAG architectures and introduce how KGs—through their structured, semantically rich representations—address these challenges. This comparison will highlight the compelling need for graph-enhanced retrieval systems in complex, information-dense applications.

7.1.2.1 Limitations of Traditional RAG

Traditional RAG systems offer a pragmatic approach to customizing large language models (LLMs) with domain knowledge by retrieving external content relevant to a given query. However, in complex, dynamic, or specialized domains, traditional RAG faces several critical limitations.

There are four major limitations of traditional RAG systems [2]:

7.1.2.1.1 Shallow Semantic Understanding

Traditional RAG systems often rely on keyword matching or vector similarity, which may capture surface-level semantics but fail to grasp complex queries that require multistep reasoning or domain-specific logic. It retrieves only directly relevant documents, missing intermediate connections, and cannot trace logical paths between concepts. For instance, answering "What are the environmental impacts of the supply chain for electric vehicle batteries?" requires connecting Battery Composition→ Raw Material Sources→ Extraction Methods→ Environmental Effects—a reasoning chain traditional RAG cannot navigate.

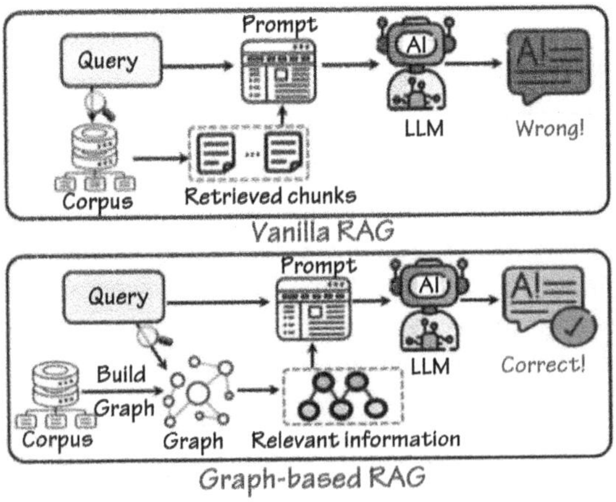

Figure 7.1 Overview of traditional (vanilla) RAG and Graph-RAG. *Source:* [1]/Association for Computing Machinery (ACM).

7.1.2.1.2 Fragmented Context from Distributed Sources

Domain knowledge is often spread across textbooks, manuals, reports, and papers—each with different formatting, granularity, and focus. When such sources are chunked and embedded for RAG, critical contextual cues (e.g., hierarchy, causality) are lost, leading to flat and fragmented retrieval results.

7.1.2.1.3 Limited Long-context Integration

Even when RAG retrieves useful content, LLMs have finite context windows (e.g., 4K–32K tokens), and long-range dependencies across documents are often truncated. This compromises the coherence and completeness of generated responses, especially in knowledge-dense domains.

7.1.2.1.4 Retrieval Inefficiency and Scalability Challenges

As the size of the external corpus increases, retrieval quality declines due to noise and sparse domain signals. The retrieval time grows proportionally with corpus size. Duplicated information across documents wastes precious context space. Traditional RAG becomes computationally expensive and difficult to scale, as the model searches through large volumes of semantically loose text chunks.

7.1.2.2 How Graph-RAG Addresses These Challenges

Graph-RAG addresses these limitations by structuring knowledge prior to retrieval, enabling precise, high-context reasoning.

7.1.2.2.1 Semantic Structure Preservation

Graphs represent explicit relationships between concepts and their semantic meaning. By organizing knowledge into nodes (entities) and edges (relations), KGs preserve concept hierarchies, taxonomies, and causal relationships—allowing for deeper and more structured contextualization during generation.

For example, Figure 7.2 illustrates how Graph-RAG preserves the rich semantic structure of real-world knowledge through typed entities and multifaceted relationships. The example demonstrates

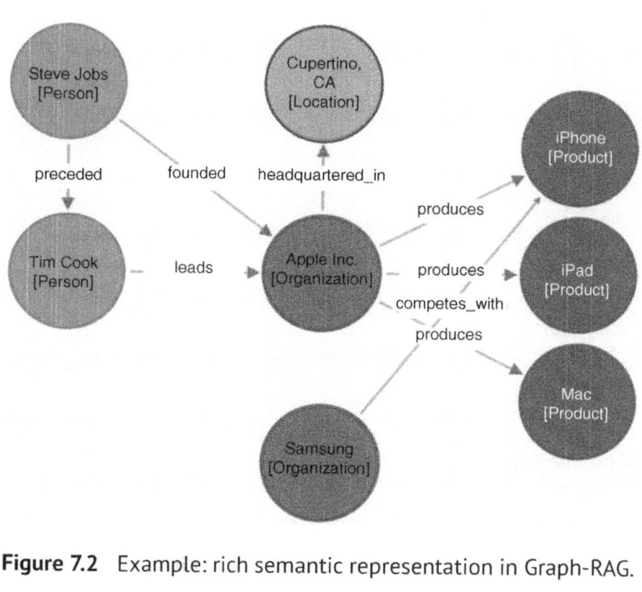

Figure 7.2 Example: rich semantic representation in Graph-RAG.

Apple Inc. as a central organization node with multiple relationship types radiating outward—from founding relationships (Steve Jobs) to geographical connections (Cupertino, CA) and product associations (iPhone). Particularly noteworthy is the preservation of competitive relationships, shown through Apple Inc.'s competes_with connection to Samsung. This captures market dynamics that would be lost in traditional text-based retrieval.

7.1.2.2.2 *Unified Knowledge Representation*

Where conventional RAG flattens documents into isolated chunks, Graph-RAG reconstructs the conceptual hierarchy. This proves particularly valuable when knowledge spans multiple sources with varying reliability—research papers, technical manuals, and clinical reports can be integrated while preserving their interrelationships.

Figure 7.3 illustrates how information from different sources (research paper and clinical trial) is unified in the Graph-RAG framework. Traditional RAG systems would retrieve these as separate, unconnected pieces of information, leaving the burden of connection to the user or LLM. Graph-RAG's entity resolution process recognizes that "Drug A" and "compound A-127" refer to the same pharmaceutical entity, merging them into a single node with both names preserved (the alias relationship). This unified representation not only connects the drug to the disease through a clear "treats" relationship but also enriches it with quantitative efficacy data (40% reduction) from the clinical trial. The system transforms fragmented knowledge into actionable intelligence.

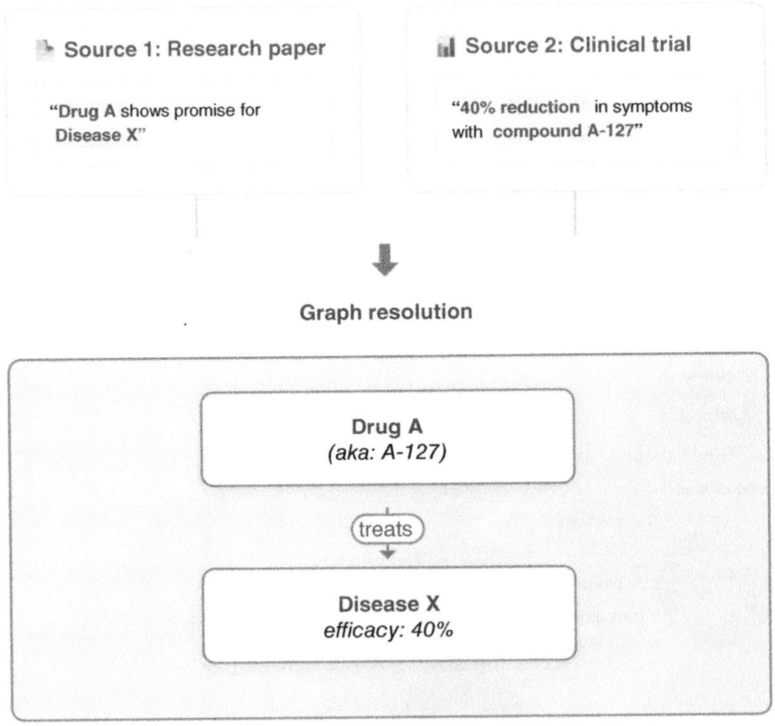

Figure 7.3 Cross-source knowledge integration example.

7.1.2.2.3 Advanced Reasoning Capabilities

The graph structure supports multi-hop reasoning, enabling models to retrieve relevant chains of information, not just isolated facts.

For example, consider the input query: "What are the environmental impacts of lithium mining for EV batteries?" Traditional RAG may use direct retrieval only, fetching text chunks such as "Lithium is used in EV batteries," while a Graph-RAG multi-hop reasoning can provide a more complete reasoning chain.

As illustrated in Figure 7.4, Graph-RAG navigates through multiple relationships to answer complex queries. The system traverses from "EV Battery" → "Lithium" → "Lithium Mine" → "Environmental Impacts," discovering specific impacts like water usage and habitat loss. Each hop reveals new connections that traditional RAG would miss.

Another example is for the input query: "What caused the 2008 financial crisis?". Through simple vector similarity search, traditional RAG may return text chunks where there is no connection between facts, thus missing causal relationships and temporal sequences. Graph-RAG, instead, is able to reveal the complete causal chain.

As illustrated in Figure 7.5, the upper part of the image shows the traditional RAG which retrieves isolated text chunks based on keyword similarity, missing the causal connections between events; the lower part of the image shows Graph-RAG which reveals the complete causal chain from Housing Bubble → Subprime Crisis → CDO Crash → Lehman Collapse, enabling comprehensive understanding of complex events.

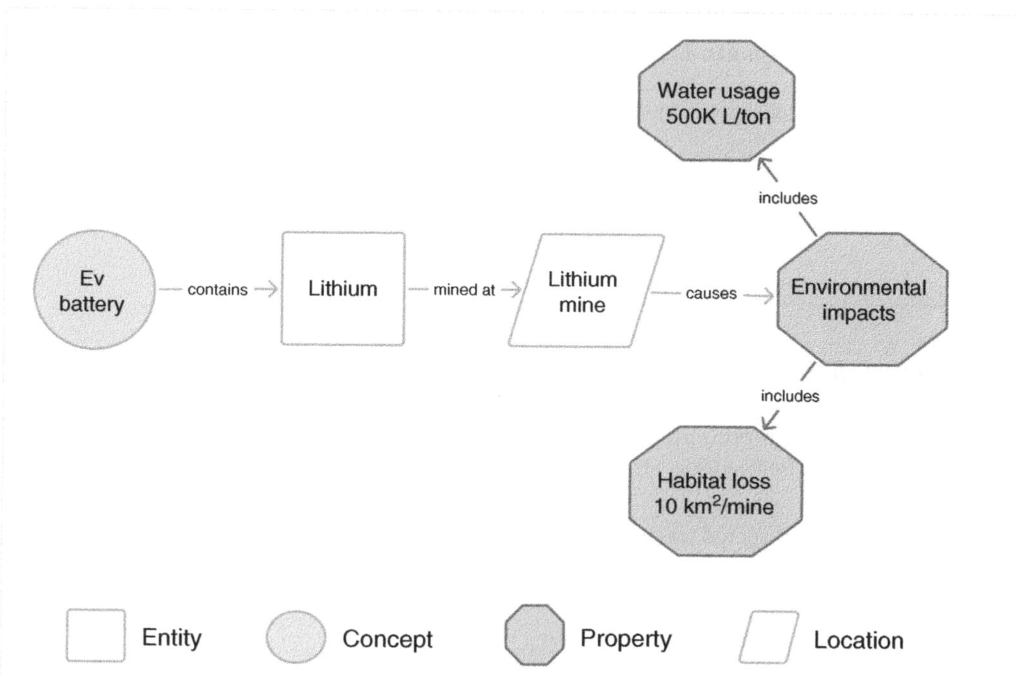

Figure 7.4 Graph-RAG traversal to answer the query: "What are the environmental impacts of lithium mining for EV batteries?"

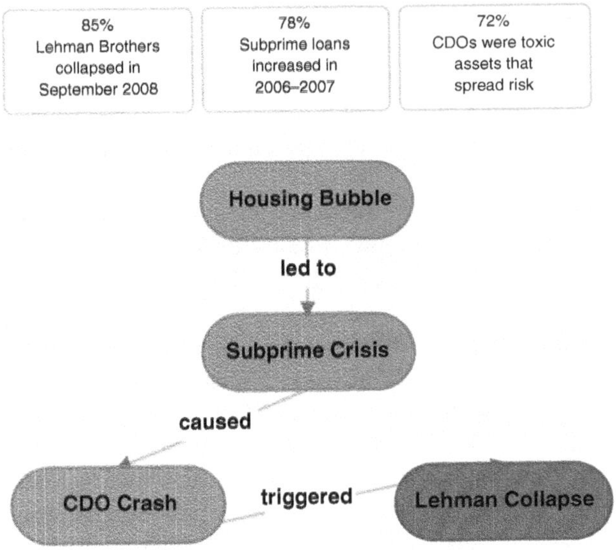

Figure 7.5 Graph-RAG traversal to answer the query: "What caused the 2008 financial crisis?"

Table 7.1 Evolution from RAG to Graph-RAG.

Traditional RAG mindset	Graph-RAG paradigm
Documents	Concepts and relations
Text search	Graph traversal
Similarity	Semantic reasoning
Static DB	Dynamic KG
Black box	Interpretable paths

7.1.2.2.4 Greater Efficiency and Scalability

While LLMs face inherent context length limitations, Graph-RAG's structured approach allows for more efficient knowledge representation. Rather than truncating documents, it extracts and connects key entities and relationships, effectively "compressing" domain knowledge without losing critical connections.

Because graphs enable targeted traversal, Graph-RAG can restrict retrieval to relevant subgraphs, reducing unnecessary computation and improving inference latency—even in large knowledge domains.

Table 7.1 lists the Graph-RAG paradigm shift:

- From Documents to Concepts: Instead of retrieving text chunks, Graph-RAG retrieves interconnected concepts and their relationships.
- From Search to Reasoning: Rather than finding similar content, Graph-RAG navigates knowledge to answer questions through logical inference.

- From Static to Dynamic: Unlike fixed document stores, KGs evolve continuously as new relationships are discovered and validated.
- From Opaque to Transparent: Graph structures make the retrieval and reasoning process interpretable and verifiable.

In summary, Graph-RAG represents a powerful evolution of the RAG framework—infusing it with structured, interconnected knowledge to overcome limitations of shallow text retrieval. By modeling semantic relationships explicitly, Graph-RAG enables more accurate, explainable, and domain-aware generation from LLMs.

7.2 Architecture and Components of Graph-RAG

The Graph-RAG workflow consists of three fundamental stages (illustrated in Figure 7.6), each designed to leverage the unique properties of graph-structured knowledge:

- **Graph-based Indexing:** Transforms unstructured documents into a structured KG.
- **Graph-guided Retrieval:** Intelligently traverses the graph to extract relevant subgraphs.
- **Graph-enhanced Generation:** Synthesizes retrieved graph structures into coherent, grounded responses.

Each of these modules transforms how queries interact with data, ensuring that retrieved information reflects not just textual proximity but semantic relationships embedded in graph structures.

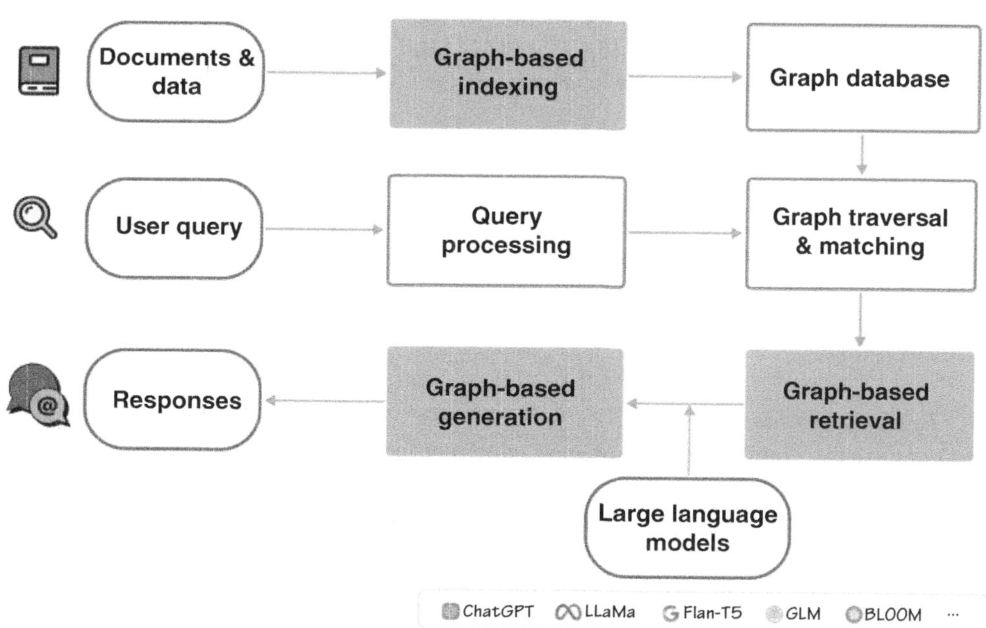

Figure 7.6 Workflow of Graph-RAG components.

7.2.1 Graph-based Indexing

Graph-based indexing represents the foundational stage where raw documents undergo transformation into queryable KGs. This process combines sophisticated natural language processing (NLP) techniques with advanced graph construction algorithms to create rich, interconnected knowledge representations that far exceed the capabilities of traditional vector-based indexing.

7.2.1.1 Construct KG

The initial phase of Graph-RAG involves converting unstructured or semi-structured documents into structured graph representations. This transformation process, detailed in Section 6.2, encompasses several critical stages that ensure the resulting KG captures both explicit and implicit relationships within the source material.

The construction of the KG usually follows a systematic pipeline:

1. Schema Definition: Establish node types, edge types, and property constraints
2. Triple Generation: Convert extracted information into (subject, predicate, object) triples
3. Graph Assembly: Integrate triples into a coherent graph structure
4. Quality Assurance: Validate consistency, remove contradictions, and ensure completeness
5. Graph Enrichment: Add metadata, confidence scores, and provenance information

Modern Graph-RAG systems employ various approaches for graph construction.

7.2.1.1.1 Graph Construction from Corpus

These approaches utilize advanced NLP pipelines for entity recognition and relationship extraction, apply coreference resolution to maintain entity consistency across documents, and use temporal and spatial reasoning to capture context-dependent relationships.

For example, systems like StructRAG [3], LightRAG [4], and GraphReader [5] automatically extract entities and relationships from raw text corpora.

7.2.1.1.2 Integration with Existing KGs

These approaches leverage preexisting knowledge bases like DBpedia, Wikidata, or domain-specific ontologies for the KG construction. They have the benefit of reducing the computational overhead of graph construction, ensuring consistency with established knowledge representations, and enabling immediate access to rich, curated relationships.

For example, systems such as KnowGPT [6], ToG [7, 8], and RoG [9] utilize existing KGs for the Graph-RAG workflow.

7.2.1.1.3 Hybrid Approaches

The hybrid approaches combine both strategies: construct from corpus and integrate with existing graphs. They can augment existing graphs with corpus-specific knowledge, which maintains domain expertise while incorporating new information and enable dynamic graph evolution as new documents are processed.

For example, advanced systems like GoR [10], MedGraph-RAG [11], and CodexGraph [12] utilize the hybrid strategy for KG construction.

7.2.1.2 Integrate with Graph Database

Once a KG has been constructed from raw documents or external sources, the next critical step in the Graph-RAG pipeline is its integration into a graph database. This stage ensures that the structured graph representation is persistently stored, efficiently indexed, and queryable via graph-native interfaces, enabling downstream retrieval and reasoning processes.

Graph-native databases provide the ideal infrastructure for this purpose. As detailed in Section 5.2, these databases are purpose-built to handle highly interconnected data and support expressive query languages such as Cypher, SPARQL, Gremlin, or AQL. Unlike traditional relational or document databases that treat relationships as foreign keys or nested structures, graph databases store connections as first-class citizens, optimizing for traversal-based workloads, multi-hop reasoning, and relationship-centric queries.

7.2.1.2.1 Roles of Graph Databases in Graph-RAG

In a typical Graph-RAG system, the graph database plays several key roles:

- Storage and Indexing: Once triples or paths are assembled, the database stores them as native graph structures, using optimized adjacency representations to ensure fast lookups and traversals. Features like index-free adjacency (e.g., in Neo4j) significantly accelerate graph queries compared to index-based joins used in non-native systems.
- Query Execution: Graph databases provide declarative query languages (e.g., Cypher, SPARQL) that enable expressive retrieval patterns such as:
  - 1-hop or multi-hop neighborhood retrieval;
  - path queries with type constraints;
  - subgraph pattern matching and reasoning paths;
  - graph projection and filtering based on metadata or edge types.
- Semantic Search and Reasoning: Through rule-based inferencing (common in RDF stores like Neptune) or property graph traversals (as in Neo4j or NebulaGraph), databases support entity resolution, fact verification, and logical reasoning—critical components of a Graph-RAG retrieval engine.
- Hybrid Knowledge Grounding: For systems integrating constructed graphs with external KGs (e.g., DBpedia, Wikidata), the database acts as a unified graph layer, enabling federated queries and hybrid reasoning.

7.2.1.2.2 Choosing the Right Graph Database

As detailed in Chapter 5, the choice of database significantly impacts the performance and scalability of a Graph-RAG system.

For example: Neo4j is ideal for rapid development and visual query prototyping, especially on local or single-node deployments; NebulaGraph provides horizontally scalable infrastructure for billion-scale graphs; Amazon Neptune and Azure Cosmos DB offer fully managed cloud-native options with strong integration into cloud ecosystems; ArangoDB enables flexible modeling for multimodal KGs.

The integration must also consider compatibility with Cypher, SPARQL, or Gremlin depending on the system's schema, query needs, and external KG alignment.

By integrating KGs into native graph databases, Graph-RAG systems gain not only performance and scalability, but also semantic expressiveness, logical explainability, and retrieval agility—all critical to enabling sophisticated question answering (QA), reasoning, and generative tasks in knowledge-intensive domains.

7.2.2 Graph-based Retrieval

Instead of simple vector similarity matching in traditional RAG, graph-based retrieval leverages the rich structure of KGs to provide contextually relevant information. This component focuses on extracting the most relevant graph components to serve as context for language generation.

7.2.2.1 Retrieval Granularity

Unlike traditional RAG systems, Graph-RAG retrieves relational knowledge from a prebuilt KG. Graph-RAG supports multiple levels of retrieval granularity: *nodes*, *triples*, *paths*, and *subgraphs*.

7.2.2.1.1 Node-level Retrieval

Node-level retrieval extracts a single entity from the KG along with all its associated properties and attributes. This is the most granular level of retrieval, focusing on comprehensive information about one specific concept. Retrieve nodes $v \in V$ directly linked to the query entities.

For example, as shown in Figure 7.7a, the retrieval result is a single-node "Remdesivir" and the associated attributes. This can be cases of responding to the following example query questions:

- Fact Checking: "What is Remdesivir?"
- Entity Lookup: "Show me details about this drug"
- Property Verification: "Is Remdesivir FDA approved?"
- Quick Reference: "What type of drug is Remdesivir?"

The corresponding Cypher Code is:

```
MATCH (d:Drug {name: 'Remdesivir'}) RETURN d
```

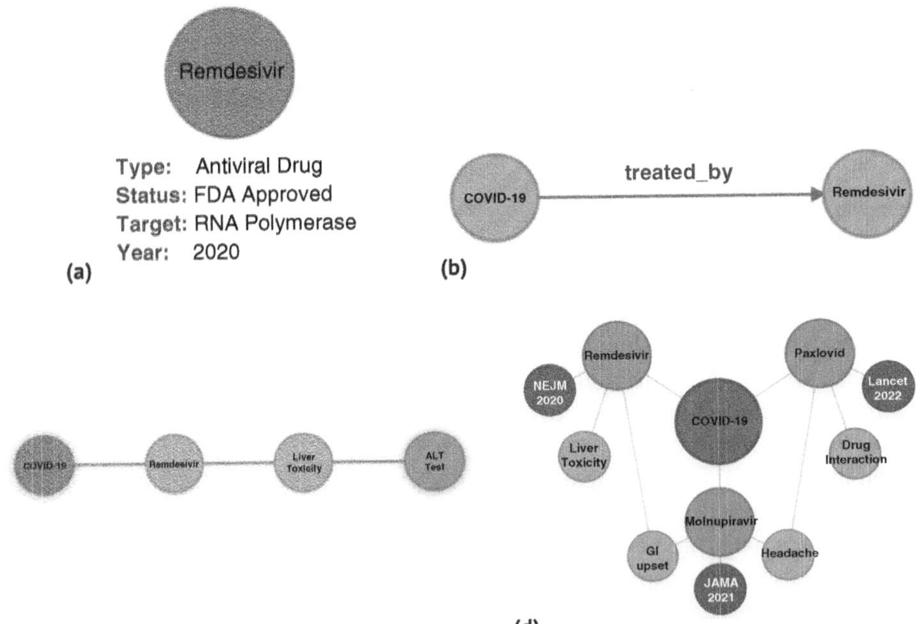

Figure 7.7 Graph retrieval granularities for the "COVID-19" input query. (a) node-level, (b) triple-level, (c) path-level, and (d) graph-level.

7.2.2.1.2 Triple-level Retrieval

Triplet retrieval extracts atomic facts from the KG in the form of subject-predicate-object relationships. This represents the fundamental building block of knowledge representation, capturing a single relationship between two entities. Extract subject-predicate-object triples $(h, r, t) \in E$ involving query-relevant entities.

For example, as shown in Figure 7.7b, the retrieval result is a triplet: (COVID-19, treated_by, Remdesivir). This can be cases of responding to the following example query questions:

- Relationship Queries: "What treats COVID-19?"
- Fact Verification: "Does Remdesivir treat COVID-19?"
- Simple Associations: "Which drug is associated with this disease?"
- Direct Connections: "Show me the treatment relationship"

The corresponding Cypher Code is:

```
MATCH (d:Disease)-[r:TREATED_BY]->(drug:Drug) WHERE d.name = 'COVID-19' RETURN d, r,
    drug
```

7.2.2.1.3 Path-level Retrieval

Path-based retrieval extracts connected sequences of entities and relationships, enabling multi-hop reasoning across the KG. This allows the system to follow logical chains of connections to answer complex queries that require understanding relationships between multiple concepts. Retrieve simple or complex paths $p = (v_1, v_2, \ldots, v_k)$ that link multiple entities across relations.

For example, as shown in Figure 7.7c, the retrieval result is a path: COVID-19 → Remdesivir → Liver Toxicity → ALT Test. This can be cases of responding to the following example query questions:

- Causal Chains: "How does COVID-19 treatment lead to liver monitoring?"
- Multi-hop Queries: "What tests monitor side effects of COVID-19 drugs?"
- Process Understanding: "Show the complete treatment pathway"
- Impact Analysis: "What are the downstream effects of this treatment?"

The corresponding Cypher Code is:

```
MATCH path = (d:Disease {name: 'COVID-19'})-[:TREATED_BY]->(drug)-[:HAS_SIDE_EFFECT
    ]->(effect)-[:MONITORED_BY]->(test) RETURN path
```

7.2.2.1.4 Subgraph Retrieval

Subgraph retrieval extracts entire neighborhoods of interconnected entities, providing comprehensive context around a query. This includes all relevant nodes within a certain distance, their relationships, and associated metadata, enabling holistic understanding of complex topics. Extract dense subgraphs $S_q = (V_q, E_q) \subseteq G$ that match a semantic constraint or pattern.

For example, as shown in Figure 7.7d, the retrieval result is a subgraph related to COVID-19, its treatments and treatments' side effects. This can be cases of responding to the following example query questions:

- Comprehensive Analysis: "Show me everything about COVID-19 treatments"
- Research Overview: "All drugs, studies, and side effects for this disease"

- Decision Support: "Complete treatment landscape with evidence"
- Knowledge Exploration: "What's the full context around this topic?"

The corresponding Cypher Code is:

```
MATCH (d:Disease {name: 'COVID-19'})-[:TREATED_BY]->(drug:Drug) OPTIONAL MATCH (drug)
    -[:EVALUATED_IN]->(study:Study) OPTIONAL MATCH (drug)-[:HAS_SIDE_EFFECT]->(effect:
    SideEffect) RETURN d, drug, study, effect
```

7.2.2.1.5 Hybrid Granularities

While node, triple, path, and subgraph retrieval each provide distinct levels of semantic granularity, real-world queries often require combining multiple retrieval granularities to construct a richer and more precise context for generation. Hybrid retrieval strategies integrate different granularity levels to address user intents that cannot be fulfilled by a single scope of graph information.

Figure 7.8 illustrates an adaptive retrieval pipeline for the hybrid granularity. The hybrid system also utilized a "query-aware" strategy, as different query types benefit from specific hybrid

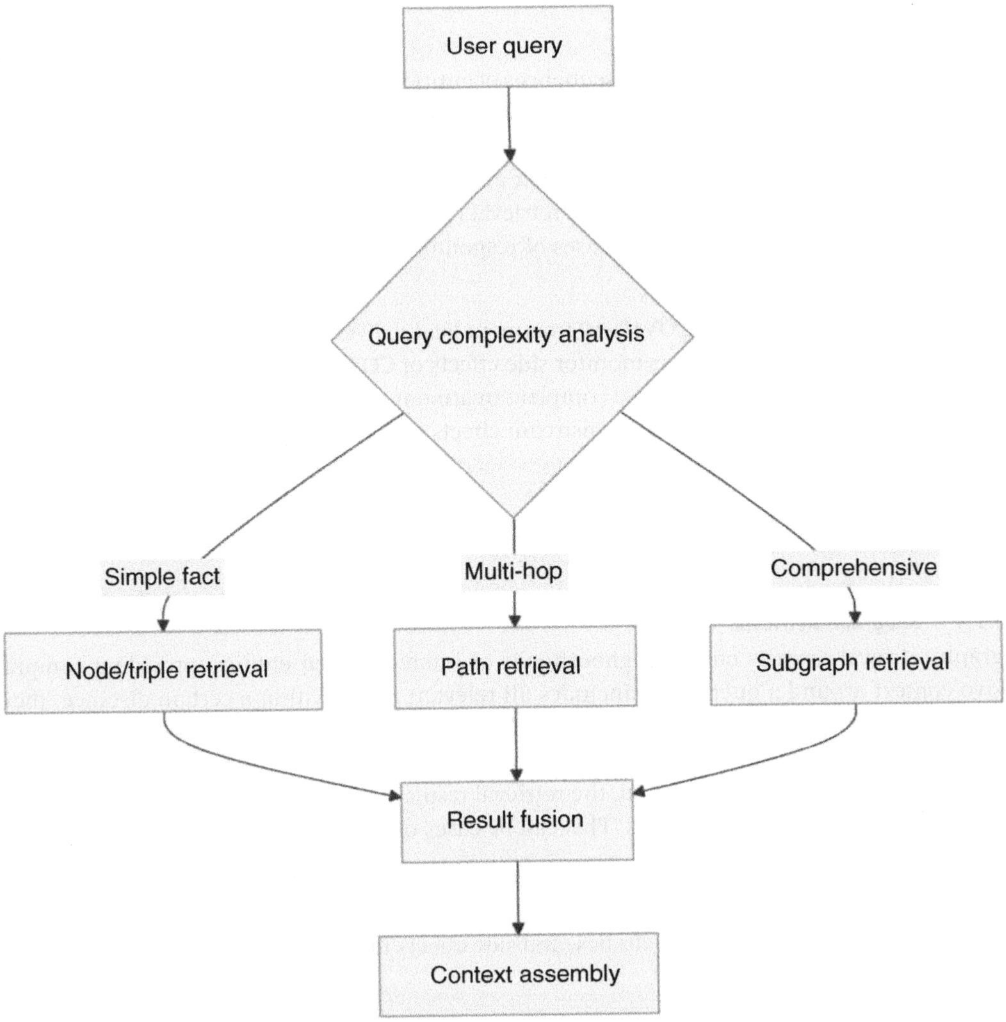

Figure 7.8 Hybrid granularity retrieval.

approaches. The system can automatically determine the optimal combination of retrieval granularities by analyzing query characteristics. Retrieval strategy orchestration can be rule-based (e.g., triggering different levels based on intent classification) or adaptive using LLM-driven routing or reinforcement learning (RL).

Some common hybrid patterns are:

- Node + Triple Combination: retrieves an entity with its immediate relationships, providing both detailed attributes and direct connections.
- Path + Subgraph Hybrid: combines specific reasoning chains with local context, useful for queries requiring both logical flow and comprehensive understanding:
- Progressive Retrieval: starts with lightweight retrieval (e.g., node or triple) and progressively expands (e.g., path or subgraph) based on initial retrieved results.

For example, a medical question like: "What are the drugs used to treat COVID-19, and what are the studies validating their side effects?" may require:

- triples to identify treatments: (COVID-19, TREATED_BY, Remdesivir)
- paths to trace downstream effects: Remdesivir → Liver Toxicity → ALT Test
- subgraphs to include surrounding evidence and metadata: (Remdesivir)–[:EVALUATED_IN]→(NEJM Study)

Hybrid retrieval also facilitates response explanation, enabling a model to cite both atomic facts and broader context in its answer. By combining these units, Graph-RAG can produce answers that are both accurate and evidence-rich, tailored to the depth and complexity of each query.

7.2.2.2 Graph Traversal Methods for Retrieval

Graph traversal is essential for retrieving relevant contextual knowledge in Graph-RAG systems. As discussed in Section 5.1.3.1, traversal algorithms like BFS, DFS, and Random Walks form the basis for exploring graph structures. In Graph-RAG, however, these methods are adapted and extended to support semantic-aware, query-driven exploration of KGs, enabling the construction of informative subgraphs and the extraction of evidence relevant to the input query.

In particular, sampling-based traversal strategies are useful when the graph is too large to explore exhaustively or when the system needs to dynamically construct a representative but manageable context subgraph. In the following section, we describe the key traversal and sampling techniques used in graph-based retrieval.

7.2.2.2.1 Classical Traversal Techniques for Graph-RAG

BFS and DFS are two classical traversal techniques in Graph-RAG retrieval:

- BFS is effective for collecting shallow, semantically connected neighborhoods around a query node. It is often used to gather fact triples, descriptive properties, or immediate relations for entity-centric summarization.
- DFS is suitable for tasks like path reasoning, causal inference, or subgraph extraction where longer relational chains are of interest, e.g., tracing influence in academic or legal KGs.

7.2.2.2.2 Random Walk and Sampling-based Traversal

Random walk algorithms are particularly effective for stochastic sampling of large graphs. They help balance exploration (diversity) and exploitation (relevance), and support probabilistic

Table 7.2 Random walk variants for graph sampling.

Variant	Description	Graph-RAG application	Reference
Simple random walk	Equal probability for all neighbors	Basic graph sampling, unbiased exploration	[13]
Random walk with restart	Returns to start node with probability α	query-focused retrieval	[14]
Personalized PageRank (PPR)	Computes node importance relative to seed nodes	Query-specific node ranking, relevance propagation	[15]
Biased random walk	Weighted transitions based on node/edge properties	Targeted exploration of high-relevance regions	[16, 17]
Forest fire sampling	Spreads like fire with "burning" probability	Discovering tightly connected communities	[18]
Snowball sampling	Expands by including all neighbors at each step	Comprehensive local structure retrieval	[19]
Rejection sampling walk	Accepts/rejects nodes based on criteria	Constraint-satisfying path discovery	[20]
Node2Vec walk	Balances BFS/DFS with parameters p, q	Flexible exploration for embedding generation	[21]
Metropolis-hastings walk	Ensures uniform stationary distribution	Unbiased sampling for graph statistics	[22, 23]

evidence aggregation in RAG pipelines. The following sampling methods (in Table 7.2) extend the basic random walk idea for more targeted or representative subgraph extraction.

These graph sampling techniques can be combined or tuned depending on the retrieval objective. For instance, snowball sampling is effective for constructing connected, progressively expanding subgraphs, while forest fire sampling introduces natural graph sparsity and clustering, helpful in surfacing dense evidence clusters without overloading the context [24].

7.2.2.2.3 Semantic-aware Traversal

Unlike traditional graph traversal that treats all edges equally, Graph-RAG incorporates semantic awareness into its traversal decisions. This approach considers:

- Edge Type Prioritization: Different relationship types carry varying semantic importance. For instance, in a medical KG, "causes" relationships might be prioritized over "mentioned-with" relationships when answering causal queries.
- Node Relevance Scoring: Nodes are dynamically scored based on their semantic similarity to the query, influencing traversal decisions.
- Context-dependent Path Selection: The traversal strategy adapts based on the query type—exploratory queries might favor breadth, while specific fact-finding queries optimize for depth along relevant paths.

Graph-RAG dynamically adjusts traversal strategies based on query characteristics. Table 7.3 lists the traversal strategies for different query types.

In summary, by leveraging a mix of deterministic, probabilistic, and semantic-aware traversal methods, Graph-RAG can efficiently extract contextual subgraphs tailored to both the user query and the structure of the KG. These strategies enable flexible, scalable, and interpretable retrieval pipelines, making them a core component of advanced RAG architectures.

Table 7.3 Traversal strategies for different query types.

Query type	Traversal strategy	Optimization focus
Factual	Depth-first along high-confidence paths	Precision and directness
Exploratory	Breadth-first with diversity sampling	Coverage and variety
Comparative	Parallel traversal from multiple seeds	Structural similarity
Aggregative	Exhaustive traversal within boundaries	Completeness
Causal	Directed path following causal edges	Temporal consistency

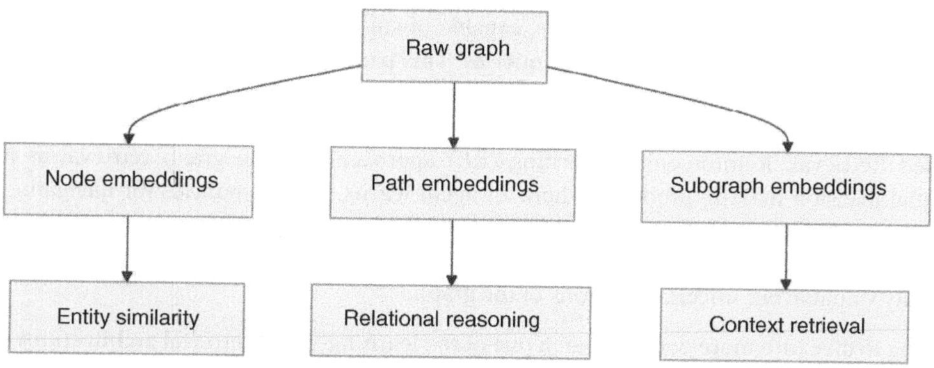

Figure 7.9 Different granularities of graph embeddings.

7.2.2.3 Advanced Retrieval Techniques

While the traversal-based methods discussed in the last section have provided fundamental mechanisms for navigating KGs, modern Graph-RAG systems increasingly leverage advanced neural and learning-based techniques to enhance retrieval quality, efficiency, and semantic understanding. These methods combine the structural advantages of graphs with the representational power of deep learning, enabling more sophisticated query understanding and context extraction.

Graph embeddings serve as the foundation for learning-based retrieval, transforming discrete graph structures into continuous vector spaces where semantic similarity can be efficiently computed. As detailed in Section 6.3.2, these techniques encode both structural patterns and semantic relationships into dense representations, enabling Graph-RAG systems to leverage efficient vector operations while preserving the rich relational information inherent in KGs. In the context of embedding-based retrieval, graph embeddings operate at multiple granularities, each suited to different retrieval scenarios (illustrated in Figure 7.9).

Node Embeddings: Node embeddings learn vector representations for individual entities in the KG. Each node is represented as a low-dimensional vector that captures both its local topology and semantic context. These embeddings can be precomputed (e.g., using Node2Vec or GCNs) or learned end-to-end during training. In Graph-RAG, node embeddings are useful for computing query-to-node relevance and selecting seed candidates.

Path Embeddings: Instead of encoding isolated nodes, path embeddings capture the semantics of relational chains (e.g., Drug → Treats → Disease → Causes → Symptom). These embeddings are crucial for multi-hop reasoning and fact composition tasks.

Subgraph Embeddings: For complex queries, relevant information may be distributed across a small subgraph. Techniques such as GraphSAGE or pooling-based GNNs can encode these subgraphs into fixed-length vectors to support subgraph-level matching, retrieval, and ranking.

Beyond embeddings, a growing body of research explores end-to-end learning-based retrieval architectures [2], which can be broadly categorized into three main classes:

- GNN-based Retrieval: Unlike traditional traversal methods constrained by predefined exploration patterns, GNNs dynamically propagate information across the entire graph structure, synthesizing both local neighborhood patterns and global graph context to create rich, query-aware representations that adapt to the specific information needs of each query.
- LLM-based Retrieval: LLMs can be directly used in the retrieval loop. These models are fine-tuned or prompted to act as soft retrievers, capable of suggesting relevant entities or paths within a graph based on natural language queries. This paradigm shifts the retrieval burden from symbolic traversal or learned embeddings to generative language reasoning, which is especially powerful when operating over noisy, incomplete, or multimodal graphs.
- RL-based Retrieval: Reinforcement learning (RL) approaches frame graph retrieval as a sequential decision-making problem, where an agent learns optimal policies for navigating the KG based on query requirements and retrieval objectives. This paradigm enables adaptive exploration strategies that balance exploitation of known relevant paths with exploration of potentially valuable but uncertain regions of the graph.

We are going to dive into more details of each one of the learning-based retrieval architectures.

7.2.2.3.1 GNN-based Retrieval

As introduced in Section 6.3.3, GNNs enable learned, message-passing-based reasoning over KGs. Unlike traditional traversal methods that follow predefined paths, GNNs allow information to propagate dynamically across the entire graph structure, incorporating both local neighborhood patterns and global graph context to create rich, query-aware representations.

GNN Architectures for Graph Encoding:

In GNNs, each node iteratively aggregates features from its neighbors. This process encodes the structure and attributes of the graph into node representations, allowing Graph-RAG to model relational dependencies and context-aware relevance. Some common GNN architectures include GCN, GAT, and GraphSAGE. Different architectures offer unique advantages:

- GCNs: Excel at capturing local patterns through spectral convolution, proving effective for entity-centric retrieval where immediate neighborhoods contain key information.
- GATs: Introduce dynamic neighbor weighting based on query context, transforming retrieval from static to query-aware exploration. Essential when the same entity has different relevant connections depending on the query focus.
- GraphSAGE: Provides inductive learning for evolving graphs, learning functions to generate embeddings for unseen nodes. Critical for dynamic domains where new entities emerge constantly.

While the previous architectures provide the algorithmic foundation, complete Graph-RAG systems combine these building blocks with additional components to create end-to-end retrieval solutions. Some examples include: GNN-Ret [25] and GNN-RAG [26], which integrate GNN layers within broader retrieval frameworks that support QA, subgraph ranking, and reasoning over multi-hop paths. These systems not only encode node and edge semantics but also incorporate modules for query interpretation, graph construction, relevance scoring, and integration with language models for generation.

GNN Training Strategies:

Effective training of GNNs is critical to enabling intelligent, structure-aware retrieval in Graph-RAG systems. Depending on the availability of labels, domain specificity, and task complexity, GNNs can be trained using supervised, contrastive, or self-supervised paradigms. Each approach encodes graph signals differently, offering trade-offs between precision, generalizability, and annotation cost.

1. Supervised Learning

   Supervised training is the most direct and accurate method when labeled data is available. Loss functions commonly used include cross-entropy, margin ranking loss, or pairwise hinge loss, depending on the formulation.

   The GNN is trained on graph-level or node-level tasks with explicit supervision signals, such as:

   - Node Classification: Predicting a node's category (e.g., whether a paper node discusses "graph neural networks").
   - Relevance Ranking: Assigning relevance scores to candidate subgraphs or paths given a query (e.g., ranking drug-disease paths for a biomedical question).

   For example: Given a query "What are the side effects of Remdesivir?," the GNN is trained to score paths like:

   Remdesivir → has_side_effect → Liver Toxicity

   as more relevant than unrelated or distractor paths like:

   Remdesivir → manufactured_by → Gilead Sciences

2. Contrastive Learning

   Contrastive learning teaches GNNs to differentiate positive graph structures (relevant to the query) from negative ones (plausible but irrelevant), even when no explicit labels are provided. The GNN is trained using a contrastive loss (e.g., InfoNCE or triplet loss) to embed positive subgraphs closer to the query than negatives in representation space. This method is robust to noise and ambiguity and often improves generalization in retrieval-intensive settings.

   This is especially useful in domains with subtle semantic variation or incomplete labeling.

   - Positive Samples: Paths or subgraphs known to contain the answer.
   - Negative Samples: Hard negatives that share entities or relations but do not satisfy the query intent.

   For example: For the question "What tests monitor Remdesivir side effects?," the model may contrast:

   - Positive: Remdesivir → has_side_effect → Liver Toxicity → monitored_by → ALT Test
   - Negative: Remdesivir → clinical_trialed_in → Study 2021

3. Self-supervised Pretraining

   When labeled data is scarce, self-supervised learning offers a powerful alternative by leveraging the inherent structure of the KG. These methods define proxy tasks where the supervision signal is generated from the data itself. Such pretraining enables the GNN to learn high-quality node and path embeddings without manual annotation. These embeddings can then be fine-tuned on downstream retrieval tasks using minimal supervision.

Table 7.4 Summary of GNN training strategy trade-offs.

Strategy	Requires labels	Strengths	Typical use cases
Supervised	✔ Yes	High accuracy with ground truth; interpretable	Domain-specific retrieval, high-precision QA
Contrastive	! Weak/no labels	Robust to ambiguity; good generalization	Open-domain QA, entity disambiguation
Self-supervised	✗ None	Scalable, low-resource pretraining	Cold-start graphs, large unlabeled corpora

Common self-supervised tasks include:

- Link Prediction: Predicting whether an edge exists between two nodes (e.g., recovering missing treats links).
- Masked Node Feature Recovery: Inferring missing node attributes from their neighborhood (similar to masked language modeling in bidirectional encoder representations from transformers [BERT]).
- Contextual Path Completion: Predicting the next node or relation in a multi-hop path.

For example, a task of "Mask the edge Remdesivir → ? → Liver Toxicity" will need to GNN to be trained to recover the edge "has_side_effect."

Table 7.4 lists the summary of these three training strategies and their trade-offs.

7.2.2.3.2 LLM-based Retrieval

LLMs can serve as powerful neural retrievers when integrated into Graph-RAG architectures. Rather than relying solely on symbolic traversal or static embeddings, LLMs are capable of dynamically interpreting user queries, understanding context, and generating retrieval instructions that are semantically aligned with the underlying graph structure. Some examples of recent Graph-RAG systems which use LLMs directly as retrievers are: ToG [7], Microsoft-Graph-RAG [27], KGP [28], and TQA-KG [29].

Retrieval Mechanism:

The retrieval mechanism typically begins with comprehensive query interpretation, where the LLM analyzes natural language input to extract key entities, relationships, and underlying intent. When faced with ambiguous queries, the LLM generates multiple interpretations, each potentially leading to different retrieval paths. Modern systems employ few-shot prompting with carefully curated examples of query-to-graph mappings to improve accuracy and handle domain-specific terminology.

- In embedding-based retrieval, the LLM generates dense query representations that are matched against precomputed graph embeddings. This approach often leverages contrastive learning to align textual and graph representations in a shared embedding space.
- In symbolic retrieval, the LLM generates explicit retrieval paths by generating sequences of entities and relations. Example methods include beam-expanded paths, multi-hop reasoning templates, or keyword-based Cypher filters to identify relevant subgraphs explicitly.

Graph-side Augmentation:

LLM-based retrievers benefit significantly from graph preprocessing techniques that make structural information more accessible to language models. For example, community detection algorithms identify densely connected regions within the graph, which are then summarized into natural language descriptions by the LLM. These summaries serve as high-level indices, enabling rapid identification of relevant graph regions before detailed exploration begins.

Frequently occurring relation paths undergo transformation into natural language templates. For instance, the path structure:

$$\texttt{person} \xrightarrow{\text{authored}} \texttt{paper} \xrightarrow{\text{cites}} \texttt{paper} \xrightarrow{\text{authored}} \texttt{person}$$

becomes the template "co-authorship through citation networks," helping the LLM recognize and leverage common reasoning patterns.

Specialized models provide rich semantic annotations for graph nodes. For example, SapBERT [30] enhances biomedical entity representations, while CodeBERT [31] enriches programming constructs with contextual embeddings that capture domain-specific semantics. These annotations include synonym sets for entity disambiguation, type hierarchies for ontological classification, and cross-lingual representations for multilingual KGs.

In summary, these approaches are especially effective in low-resource or high-ambiguity settings, where schema constraints are minimal, input phrasing varies, or domain knowledge is sparsely populated. LLM-based retrieval thus excels in tasks such as open-domain QA, conversational search, and exploratory querying over heterogeneous graphs.

7.2.2.3.3 RL-based Retrieval

RL offers a fundamentally different approach to retrieval: rather than scoring static representations, it trains an agent to navigate the KG in a goal-directed manner. This agent makes a sequence of decisions, traversing from one node to another, guided by feedback from the environment.

RL Framework for Graph Retrieval:

The RL framework models graph retrieval as a Markov Decision Process (MDP) defined by the tuple (S, A, P, R, γ), where S represents the state space, A the action space, P the transition probabilities, R the reward function, and γ the discount factor. The state representation $s_t \in S$ at time step t encodes multiple components:

$$s_t = [\phi(v_t), \psi(q), h_t, b_t, \mathcal{E}_t]$$

where $\phi(v_t)$ represents the encoding of current nodes, $\psi(q)$ captures the query embedding, h_t maintains the exploration history, b_t tracks the remaining exploration budget, and $\mathcal{E}_t$ contains partial evidence collected.

The action space of such an agent corresponds to the set of possible outgoing edges from the current node. At each step, the agent selects the next hop based on both prior context and an evolving reward signal—such as the semantic alignment of a retrieved fact with the query, or downstream generation fidelity. Edge selection actions choose which outgoing edges to traverse from current nodes, while node expansion decisions determine whether to explore a node's neighborhood or proceed to other regions. The agent can also execute backtracking operations to revisit previously explored areas when alternative paths might yield better results. Termination actions allow the agent to stop exploration once sufficient evidence has been gathered, and meta-actions enable switching between different exploration strategies such as breadth-first, depth-first, or best-first search patterns.

Training Methodologies:
The training for RL-based retrieval is typically framed as a multi-episode optimization problem, where the agent receives rewards based on retrieval outcomes. Objectives may include maximizing precision and recall, improving answer F1 scores, or minimizing path cost and traversal depth.

The training process of the retriever follows an episodic protocol where each episode begins with a query-answer pair. The agent explores the graph within a limited step budget, receiving rewards based on retrieval quality and efficiency. The reward function typically combines multiple objectives:

$$R(s_t, a_t) = r_{\text{answer}} + \lambda_1 \cdot r_{\text{efficiency}} + \lambda_2 \cdot r_{\text{diversity}} + \lambda_3 \cdot r_{\text{intermediate}}$$

where r_{answer} rewards correct information retrieval, $r_{\text{efficiency}}$ penalizes excessive exploration, $r_{\text{diversity}}$ encourages discovering multiple evidence paths, and $r_{\text{intermediate}}$ provides partial credit for reaching relevant intermediate nodes.

Curriculum learning strategies structure the training progression from simple to complex tasks. Initial stages focus on direct entity matching and single-hop relations, gradually advancing to two-hop reasoning with clear relation paths. Advanced stages introduce complex queries requiring inference across multiple evidence sources, culminating in adversarial queries designed to test system robustness and generalization capabilities.

State-of-the-art systems such as the Spider framework [32] implement a hierarchical RL approach to detect essential subgraphs adaptive to each query. Its architecture comprises:

- A High-level policy that identifies promising regions of the graph (i.e., sets of nodes likely relevant to the query).
- A Low-level policy that performs fine-grained navigation within those regions to select decisive nodes and connections.

By decomposing the search into hierarchical steps, Spider reduces the action space dramatically and focuses inference on the most impactful graph segments, and thus reduces computational overhead while maintaining accuracy, especially in large-scale KGs.

In summary, RL-based retrieval proves particularly effective in challenging scenarios such as temporal reasoning, where agents learn to prioritize recent information and track entity state changes over time. In sparsely connected KGs, RL agents develop efficient exploration strategies that exploit long-range connections and leverage hub nodes as navigation waypoints. By learning optimal retrieval behaviors, RL-based agents complement embedding and LLM-based methods with explicit decision-making and interpretability. Their use is growing in applications such as biomedical reasoning, regulatory policy analysis, and multi-turn dialog systems.

7.2.2.4 Retrieval Strategies

While traversal techniques define how information is accessed within a graph, retrieval strategies govern the *higher-level orchestration* of multiple retrieval methods, combining them adaptively to meet complex query requirements. This section explores the fundamental approaches to retrieval strategy design, from query routing and hybrid architectures to multi-round iterative processes and post-retrieval refinement.

7.2.2.4.1 Query Routing and Decomposition in Graphs

Query routing in Graph-RAG systems refers to the process where incoming queries are dynamically analyzed and directed to the most appropriate retrieval mechanism or region of the KG. This approach improves efficiency by avoiding exhaustive graph searches and ensures that retrieval

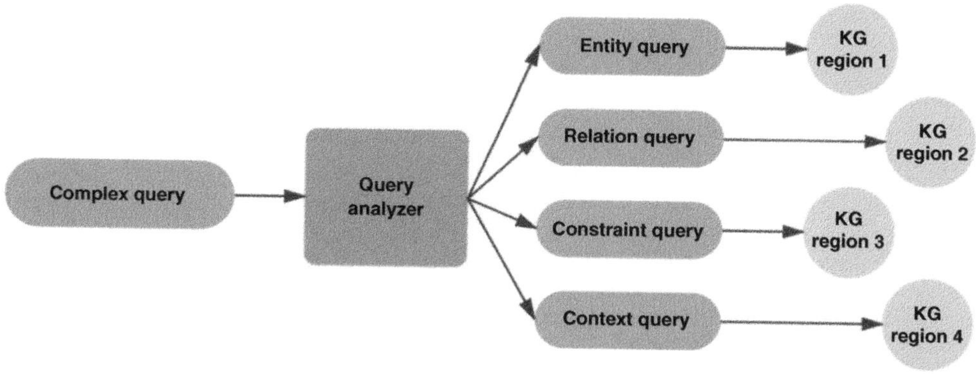

Figure 7.10 Query routing and decomposition.

efforts are focused on the most relevant portions of the knowledge base. Figure 7.10 illustrates an example workflow of the query routing and decomposition process.

Complex queries often contain multiple facets that require different retrieval approaches. Query decomposition breaks down these multifaceted queries into manageable sub-queries, each targeting specific aspects of the KG. Once queries are decomposed, the system must identify relevant subgraphs containing the necessary information.

Consider a medical diagnosis query: "What are the treatment options for elderly patients with diabetes who also have kidney complications?" This query naturally decomposes into several components: entity identification (elderly patients, diabetes, kidney complications), relation extraction (has_condition, has_complication, treatment_for), and constraint specification (age_group = elderly, comorbidity = true).

Modern Graph-RAG systems increasingly leverage LLMs as intelligent routers that understand query intent and select appropriate retrieval strategies. The LLM router first analyzes query intent to determine whether the query requires factual retrieval, multi-hop reasoning, aggregation, or exploratory search. Each intent type maps to different retrieval strategies optimized for that specific need. Based on this analysis, the router selects the appropriate retrieval granularity: node-level retrieval for specific facts, path-level retrieval for relationships, or subgraph retrieval for comprehensive context.

7.2.2.4.2 Hybrid Retrieval Strategies

Hybrid retrieval strategies reflect a key design philosophy in Graph-RAG systems: combining complementary methods to optimize both recall and precision. No single retrieval approach—whether symbolic, semantic, or structural—can universally address all query types or reasoning needs. Hybrid strategies address this limitation by integrating multiple retrieval and reasoning techniques across different stages of the pipeline: from query routing, to graph traversal, to encoding and reasoning.

Embedding-level Hybridization:

At the retrieval layer, systems can combine "symbolic methods" for precise structural matching (pathfinding, subgraph isomorphism) and "semantic methods" (GNN encodings, LLM embeddings) for conceptual understanding.

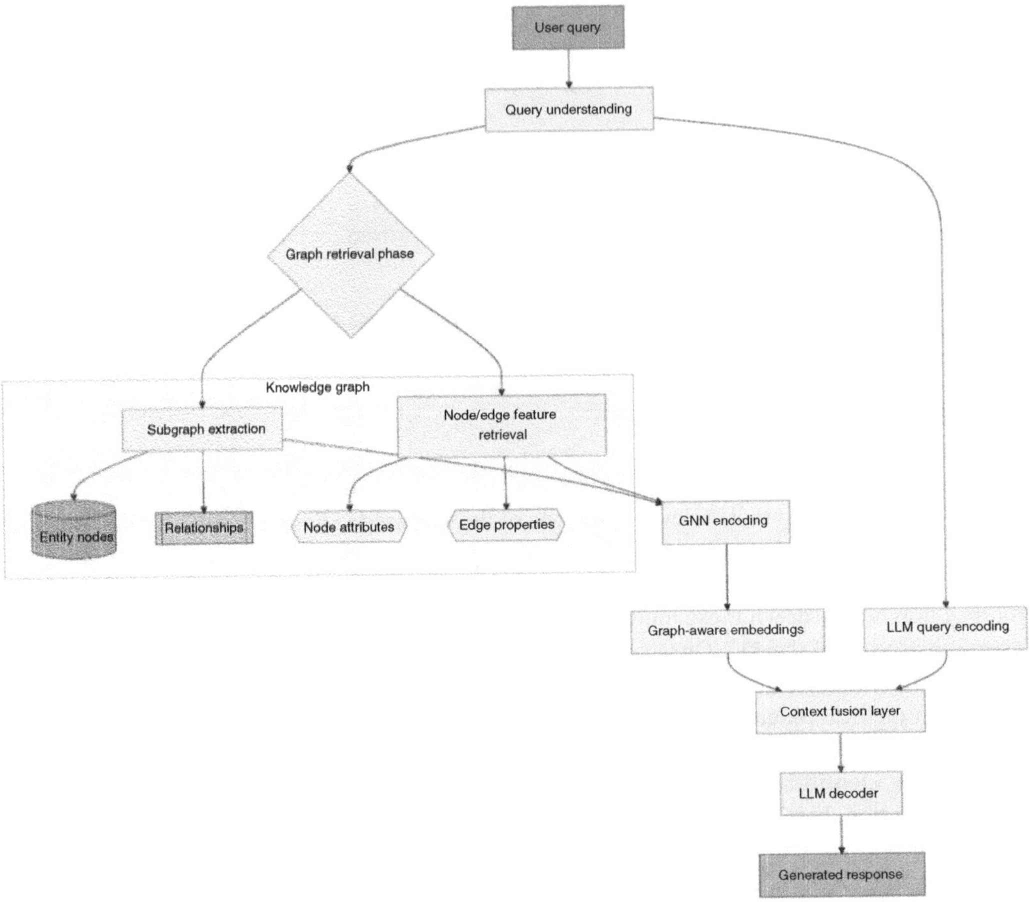

Figure 7.11 Hybrid architecture of Graph-RAG workflow.

As shown in Figure 7.11, a typical hybrid workflow integrates both GNN-based graph traversal, and LLM-based embedding queries. Structural methods like symbolic matching or subgraph pattern recognition are effective for precise relational reasoning and exact entity matching. In contrast, semantic methods use neural encoders to capture soft similarities and bridge terminology gaps, enabling the retrieval of conceptually relevant knowledge even when explicit structure is absent or noisy. By fusing graph embeddings (e.g., GNN-based) with LLM-derived query representations—via concatenation or cross-attention—the system gains the capacity to reason over both topological context (e.g., multi-hop paths, relation weights) and semantic intent (e.g., latent topic or concept alignment).

Hybrid Structure Routing:
Importantly, hybrid strategies can be implemented at multiple architectural levels. For instance, StructRAG [3] demonstrates a hybrid architecture that operates not only at the retrieval level but also at the knowledge structuring and reasoning level. The StructRAG framework introduces a hybrid structure router that selects the optimal representation format (e.g., table, tree, graph) based

on task requirements. It then uses a scattered knowledge structurizer to convert unstructured documents into structured representations. Finally, a structured knowledge utilizer decomposes complex questions and aligns them with the structured knowledge for downstream reasoning.

7.2.2.4.3 Multi-round and Iterative Retrieval

Multi-round retrieval strategies enable dynamic exploration of KGs through iterative refinement, proving particularly valuable in conversational settings and complex reasoning tasks. These strategies recognize that comprehensive understanding often requires multiple passes through the KG, with each round building upon insights from previous iterations.

For example, Graph-RAG systems like DialogGSR [33], Graph-CoT [34] employ intermediate LLM generations to guide subsequent retrievals (e.g., refining search paths based on user feedback or inferred intent), which are particularly effective for exploratory tasks.

Systems implementing stepwise guidance use intermediate generation results to inform subsequent retrieval rounds. Based on generated content, the system formulates new retrieval queries to gather supporting evidence or explore related concepts. Figure 7.12 illustrates the iterative process of multi-round retrieval.

For example, if the initial generation mentions a specific protein in a biological pathway, the next retrieval round might focus on that protein's interactions and regulatory mechanisms. This process continues iteratively, with each round refining the understanding until the generation reaches a satisfactory conclusion or exhausts a predetermined number of rounds.

Convergence criteria determine when to stop iterating. Common convergence indicators include information gain falling below a threshold, confidence scores reaching acceptable levels, computational budget exhaustion, or explicit user satisfaction signals. The system must also handle cases where convergence is not achieved, providing partial results with appropriate caveats about completeness.

7.2.2.4.4 Post-retrieval Processing and Verification

Post-retrieval processing represents a critical stage where retrieved information undergoes verification, disambiguation, and organization before generation. This stage ensures faithfulness to the KG and enhances the quality of generated outputs, particularly important in domains where accuracy is paramount.

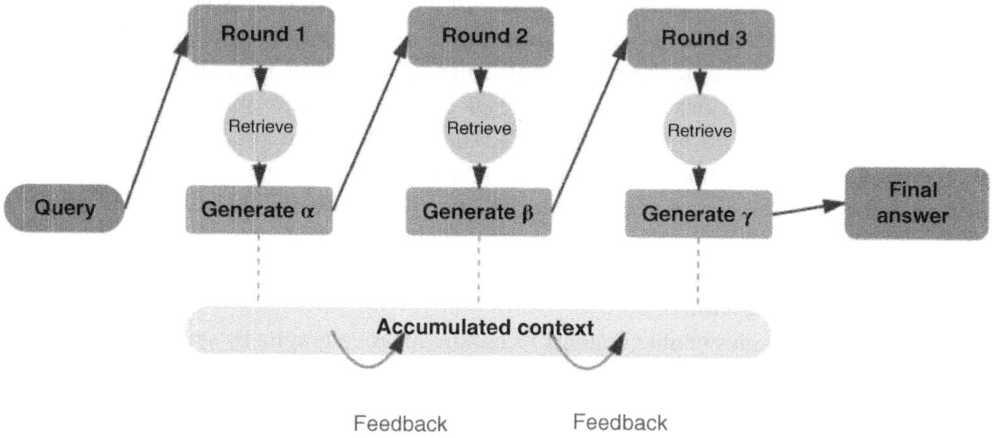

Figure 7.12 Multi-round iterative retrieval process.

Claim Verification:
In domains such as biomedicine or legal applications, claim verification becomes essential for maintaining trust and accuracy. The system actively searches for contradictory information, implementing a form of adversarial verification. When retrieving supporting evidence for a claim, it simultaneously searches for information that might contradict or qualify that claim. Source attribution maintains provenance metadata for each piece of retrieved information, enabling generated content to include appropriate citations and confidence indicators.

For example, the Chain-of-Knowledge (CoK) [35] system is a prompting strategy designed to improve the reasoning quality of LLMs. To ensure the reliability of these reasoning steps, CoK introduces a F2-verification process that checks each knowledge triple for factual accuracy and logical consistency. If the reasoning chain is found to contain errors, the system identifies which part is wrong and prompts the model to revise its answer. This improves the overall faithfulness and trustworthiness of the model's output. Experiments show that CoK outperforms standard CoT approaches in tasks requiring commonsense, factual, symbolic, and arithmetic reasoning.

Relation Disambiguation:
Context resolves ambiguous relations—"treats" means "provides treatment" for drug-disease pairs but "processes" for facility-condition pairs. Type constraints and domain ontologies provide disambiguation rules. Temporal and contextual variations require sophisticated handling.

7.2.2.4.5 Performance Optimizations
Effective retrieval requires careful performance optimization for scale and quality. These optimizations span from low-level caching strategies to high-level architectural decisions about resource allocation and distributed processing.

Caching and Indexing Strategies:
Query result caching stores frequently accessed subgraphs and query results to eliminate redundant computation. Effective caching strategies must balance cache size with hit rates, implementing sophisticated eviction policies that consider both recency and frequency of access. Cache invalidation strategies must ensure that cached results remain fresh, particularly important in dynamic KGs where information changes regularly.

Embedding precomputation and indexing enable rapid similarity search during retrieval. Node and subgraph embeddings are computed offline and stored in specialized index structures optimized for high-dimensional similarity search. These indices must support efficient updates as the KG evolves, balancing index maintenance costs with query performance benefits.

Hot path optimization identifies common retrieval patterns through query log analysis and creates specialized optimizations for these patterns. This might include precomputed results for frequent queries, specialized indices for common graph patterns, or even restructuring parts of the graph to optimize common access patterns. The challenge lies in identifying truly stable patterns that justify optimization investment vs. transient trends that might change.

Scalability and Resource Management:
Progressive sampling strategies enable systems to handle massive graphs by starting with small, high-confidence samples and expanding as needed. This approach allows systems to return quick approximate results for simple queries while still supporting comprehensive analysis when required. The sampling strategy must ensure that initial samples are representative and that expansion follows principled approaches to maintain statistical properties.

Resource-aware scheduling adapts retrieval complexity based on available computational resources and response time requirements. Under high load, the system might reduce exploration depth or rely more heavily on cached results, while during low-load periods, it can perform more comprehensive retrieval. This adaptive behavior requires sophisticated monitoring and prediction of resource availability and query complexity.

For truly massive KGs, distributed retrieval becomes necessary. Retrieval strategies must coordinate across graph partitions while maintaining consistency and completeness. This introduces challenges in maintaining global graph properties, handling cross-partition queries efficiently, and ensuring that distributed retrieval does not introduce biases or gaps in coverage. Advanced systems implement sophisticated partition strategies that minimize cross-partition queries while maintaining balanced load distribution.

In summary, the choice and implementation of retrieval strategies significantly impact Graph-RAG system performance, making this an active area of research with substantial practical implications. As KGs continue to grow in size and complexity, and as user expectations for intelligent retrieval increase, the development of sophisticated retrieval strategies remains crucial for the advancement of Graph-RAG systems.

7.2.3 Graph-based Generation

The final stage of the Graph-RAG pipeline transforms retrieved graph structures into coherent natural language responses. Unlike traditional RAG systems that concatenate text passages, graph-based generation must preserve the semantic richness of nodes, edges, and paths while producing fluent, factually grounded text. This section examines how Graph-RAG systems bridge structured graph representations and natural language through context construction, prompting strategies, reasoning mechanisms, and model adaptation.

7.2.3.1 Context Generation with Graph Structure

Transforming retrieved graph components into language model inputs requires careful consideration of graph semantics, token constraints, and relational preservation.

7.2.3.1.1 Graph-aware Context Window Construction

Building context windows from graph data differs fundamentally from text-based approaches. The system must select relevant graph portions while respecting token limits.

Subgraph Selection and Prioritization:

Selection begins by identifying query-relevant graph portions. Edge types receive weights based on semantic importance—in medical queries, `treats` and `causes` relations outweigh generic `related_to` edges. Confidence scores guide inclusion, ensuring only high-quality information enters the generation context.

For the query "What are the treatments for COVID-19 and their side effects?" the system constructs:

```
COVID-19 --[treated_by]--> Remdesivir--[has_side_effect]-->
Liver Toxicity--[requires_monitoring]--> ALT Test
```

Context Optimization:
Token constraints require intelligent compression. Path compression retains key entities while removing redundant intermediates. Semantic clustering groups related facts—multiple drug side effects consolidate into single blocks. Dynamic truncation uses priority queues to include the most relevant elements until reaching token limits.

7.2.3.1.2 Graph Linearization Techniques

Converting nonlinear graphs into sequential text requires strategies that preserve relational semantics.

Path-based Linearization:
Graph paths transform into logical statements. Simple paths like `Aspirin -[prevents]-> Heart Attack` become "Aspirin prevents heart attacks."

Complex paths require careful handling:

```
Diabetes--[increases_risk]--> Kidney Disease--[diagnosed_by]
--> Creatinine Test
```

linearizes to "Diabetes increases the risk of kidney disease, which is diagnosed through creatinine tests."

Hierarchical Linearization:
Complex subgraphs organize hierarchically from general to specific. Disease information starts with the condition, branches to treatments with mechanisms and side effects, then includes prevention methods. This natural hierarchy translates into coherent paragraphs maintaining relationship logic.

Template-based Linearization:
Domain templates ensure consistent verbalization. For example:

- Medical: "{Drug} treats {Disease} but may cause {SideEffect}"
- Legal: "{Case1} established that {Principle}, applied in {Case2}"
- Scientific: "{Reactant1} and {Reactant2} yield {Product}"

7.2.3.2 Graph-enhanced Prompting

Effective prompting strategies are crucial for helping language models interpret and verbalize graph-structured information accurately. The prompt must bridge the gap between the graph's symbolic representation and the natural language capabilities of the model.

7.2.3.2.1 Structured Prompt Engineering

Multi-level Context:
Prompts provide context at multiple abstraction levels:

```
System: You are a medical expert explaining knowledge graph
information.
```

```
Context: Entities include [Disease], [Drug], [SideEffect]
Relations: treated_by, has_side_effect, monitored_by
Task: Explain the treatment and precautions clearly.
```

Relation-aware Prompting:
Different relations require different verbalization. Causal relations need mechanism explanations, temporal relations require sequence attention, hierarchical relations benefit from relationship clarification.

7.2.3.2.2 Dynamic Prompt Construction

Query-adaptive Prompting:
Prompts adapt to query types. Factual queries emphasize precision: "Provide direct answers using only verified graph information." Explanatory queries encourage elaboration: "Explain why these connections matter." Comparative queries structure analysis: "Compare entities based on their relationships."

Confidence Integration:
Edge confidence scores guide language:

- High confidence (> 0.9): "According to strong evidence ..."
- Medium confidence (0.6–0.9): "Studies suggest ..."
- Low confidence (< 0.6): "Preliminary findings indicate ..."

7.2.3.3 Reasoning and Inference in Generation

Graph-RAG systems enhance language model reasoning by grounding inference in structured knowledge paths and relationships. This grounding provides a logical scaffold that guides the generation process toward factually accurate and logically coherent outputs.

7.2.3.3.1 Multi-hop Reasoning

Chain-of-Thought from Paths:
Graph paths form natural reasoning chains. The path

```
Gene_X--[upregulates]-->Protein_Y--[activates]-->
Pathway_Z--[involved_in]-->Disease_A
```

generates: "Gene X upregulates Protein Y, which activates Pathway Z. Since this pathway is involved in Disease A, Gene X mutations may contribute to disease development."

Logical Rule Application:
Domain rules guide generation:

- Medical: "If Drug A interacts with Drug B, and patient takes Drug B, then avoid Drug A."
- Legal: "If Case X overturns Case Y, then Y's precedents no longer apply."

7.2.3.3.2 Causal and Temporal Reasoning

Causal Chain Verbalization:
The system explains causal mechanisms. From

```
Smoking --[causes]--> Lung_Damage --[leads_to]--> COPD
```

it generates: "Smoking damages lung tissue by destroying alveoli. This damage progressively leads to COPD, making breathing increasingly difficult."

Temporal Sequences:
Time-based relationships maintain chronology: "Symptoms began on day 0. Diagnosis confirmed day 7. Treatment started day 10. Recovery achieved by day 28."

7.2.3.3.3 Uncertainty Handling

When graphs contain contradictions, generation acknowledges multiple perspectives. If studies show conflicting drug effectiveness, the system generates: "Evidence remains mixed. Study A showed effectiveness while Study B found limited benefit. Further research is needed."

7.2.3.4 Fine-tuning LLMs with KG-augmented Datasets

Adapting language models to work effectively with graph-structured inputs requires specialized training approaches that teach models to interpret and reason over graph representations.

7.2.3.4.1 Training Data Construction

Align graph triples or multi-hop paths with expert-written explanations, question-answer pairs, or instruction-style prompts.

Synthetic Data Generation:
Synthetic data generation creates large-scale training sets by pairing graph structures with corresponding natural language. For example:

- Triple Verbalization: Each triple gets multiple expressions.
- Path Explanation: Multi-hop paths receive step-by-step annotation.
- Subgraph Summarization: Complex structures pair with summaries.

Instruction Datasets:
Instruction tuning datasets teach models to follow graph-based instructions through diverse examples. These might include explaining relationship paths between entities, inferring conclusions from subgraph patterns, or answering questions grounded in graph context. Each example pairs a specific instruction with the relevant graph structure and the expected response, teaching models to adapt their generation strategy based on the task.

Examples teach graph-based instruction following:

```
Instruction: Explain the path from A to B
Graph: A --[r1]--> C --[r2]--> B
Response: A connects to B through C via...
```

7.2.3.4.2 *Training Methods*

Multitask Learning:

Multitask learning frameworks train models on diverse graph-related tasks simultaneously. Graph-to-text generation teaches fluent verbalization of structured data. Text-to-graph extraction develops the inverse capability of identifying graph structures in text. Graph completion tasks teach models to reason about missing information. Graph-grounded QA combines all these skills in an applied setting. This diverse training helps models develop robust graph understanding capabilities.

Specialized Architectures:

Specialized model architectures incorporate graph-specific components to better handle structured inputs. InstructGLM [36] enhances instruction following with graph-aware attention mechanisms that prioritize relevant paths during generation. GraphGPT [37] introduces graph-conditioned generation where hidden states are modulated by graph structure embeddings. LLAGA [38] jointly trains on textual and graph modalities, learning unified representations that can seamlessly work with both.

7.2.3.5 Summary

Graph-based generation advances structured knowledge accessibility through natural language. By preserving graph semantics throughout generation, Graph-RAG systems achieve superior factual grounding compared to text-based approaches. Current techniques—from context construction to model adaptation—establish foundations while pointing toward future developments in neural-symbolic generation, interactive exploration, and cross-modal integration. Chapter 11 will explore frontier Graph-RAG methods pushing these boundaries further.

7.3 Applications

Graph-RAG's ability to combine structured knowledge representation with natural language understanding has led to transformative applications across diverse domains and tasks. The technology's unique strengths—multi-hop reasoning, relationship-aware retrieval, and explainable generation—make it particularly valuable for complex, knowledge-intensive applications where traditional RAG systems fall short.

7.3.1 Downstream Tasks

7.3.1.1 Question Answering

Question answering represents the most prominent application of Graph-RAG, where systems must navigate complex knowledge structures to provide accurate, grounded responses. Graph-RAG excels in several QA subcategories:

Multi-hop QA requires connecting information across multiple entities and relationships. For instance, answering "What is the population of the capital of the country where the inventor of the telephone was born?" requires traversing: Alexander Graham Bell → Scotland → Edinburgh → Population. Graph-RAG systems like GNN-Ret achieve significant improvements by modeling these reasoning chains explicitly.

Knowledge-intensive QA involves questions requiring deep domain expertise. In biomedical QA, Graph-RAG can trace complex pathways from symptoms to diseases to treatments, incorporating confidence scores and supporting evidence from clinical studies. Systems retrieve not just direct answers but entire reasoning subgraphs that explain the connection between query and answer.

Conversational QA benefits from Graph-RAG's ability to maintain context across dialog turns. The graph structure helps track entity mentions, resolve coreferences, and build upon previous interactions, enabling more coherent multi-turn conversations.

7.3.1.2 Information Retrieval

Graph-RAG revolutionizes information retrieval by moving beyond keyword matching to semantic and structural search:

Semantic search leverages graph embeddings to find conceptually related information even when lexical overlap is minimal. A search for "heart disease prevention" might retrieve information about exercise, diet, and stress management by following relationships in a health KG.

Faceted search uses graph structure to enable multidimensional filtering. Users can simultaneously constrain searches by entity type, relationship type, temporal bounds, and confidence levels—capabilities impossible with traditional document retrieval.

Exploratory search supports open-ended information discovery by suggesting related concepts and alternative paths through the KG. This proves invaluable for researchers exploring new domains or investigating complex phenomena.

7.3.1.3 Other Tasks

Text Summarization: Graph-RAG enhances summarization by identifying key entities and their relationships, ensuring summaries capture not just main points but also critical connections between concepts.

Fact Verification: By tracing claims through KGs, Graph-RAG systems can verify facts against structured knowledge bases, providing evidence paths that support or refute statements.

Recommendation Systems: Graph-based recommendations leverage both user-item interactions and rich item relationships, enabling explanations like "recommended because users who bought X also purchased Y, and Y is functionally similar to Z, which you previously viewed."

Knowledge Base Completion: Graph-RAG can identify missing relationships in KGs by analyzing patterns and using LLMs to propose and validate new connections.

7.3.2 Application Domains

7.3.2.1 E-commerce

In e-commerce, Graph-RAG transforms product discovery and customer service:

Product KGs connect items through multiple relationship types: compatibility, similarity, brand hierarchies, and usage scenarios. When customers ask complex questions like "What accessories do I need for underwater photography with this camera?," Graph-RAG navigates relationships between cameras, lenses, housings, and lighting equipment.

Conversational commerce uses Graph-RAG to handle sophisticated product inquiries, comparison shopping, and personalized recommendations. The system can explain why products are suggested based on user history, product relationships, and contextual factors.

Supply chain intelligence applies Graph-RAG to trace component dependencies, identify alternative suppliers, and assess risk propagation through supply networks.

7.3.2.2 Legal

Legal applications demand high accuracy and explainability, making Graph-RAG particularly valuable:

Case law analysis uses Graph-RAG to trace precedent chains, identify relevant statutes, and understand how legal principles evolved through court decisions. Lawyers can query complex relationships like "cases that applied the doctrine established in X but distinguished it based on Y."

Contract intelligence leverages Graph-RAG to analyze contract relationships, identify conflicts between clauses, and ensure consistency across document sets. The system can answer questions about obligations, rights, and dependencies across multiple contracts.

Regulatory compliance employs Graph-RAG to navigate complex regulatory frameworks, tracking how requirements flow from legislation to regulations to specific compliance actions.

7.3.2.3 Finance

Financial applications benefit from Graph-RAG's ability to model complex relationships and dependencies:

Risk assessment uses graph-based analysis to trace risk propagation through financial networks. Graph-RAG can identify hidden exposures by following ownership chains, derivative relationships, and counterparty connections.

Investment research applies Graph-RAG to connect market events, company relationships, and economic indicators. Analysts can explore questions like "How might supply chain disruptions in semiconductor manufacturing affect our technology portfolio?"

Fraud detection leverages graph patterns to identify suspicious transaction networks, with Graph-RAG explaining the reasoning behind fraud alerts by showing relationship patterns that match known fraud schemes.

7.3.2.4 Biomedical

Healthcare and life sciences represent one of the most impactful domains for Graph-RAG:

Clinical decision support uses Graph-RAG to navigate relationships between symptoms, diseases, treatments, and drug interactions. Physicians can query complex scenarios involving multiple conditions and medications, receiving evidence-based recommendations with clear reasoning paths.

Drug discovery applies Graph-RAG to explore relationships between proteins, pathways, diseases, and compounds. Researchers can identify drug repurposing opportunities by finding unexpected connections through the biomedical KG.

Precision medicine leverages Graph-RAG to integrate patient data with genomic information, clinical trials, and treatment outcomes, enabling personalized treatment recommendations based on similar patient cohorts.

7.3.2.5 Academic and Scientific Research

Graph-RAG accelerates scientific discovery across disciplines:

Literature review uses Graph-RAG to map research landscapes, identify seminal papers through citation networks, and discover interdisciplinary connections. Researchers can query relationships between concepts, methods, and findings across thousands of papers.

Hypothesis generation employs Graph-RAG to identify unexplored connections between scientific concepts, potentially revealing new research directions by analyzing patterns in existing knowledge.

Collaboration discovery applies Graph-RAG to academic networks, helping researchers find potential collaborators based on complementary expertise and research interests.

7.3.2.6 Other Domains

Education: Adaptive learning systems use Graph-RAG to model knowledge dependencies, create personalized learning paths, and answer student questions with appropriate prerequisite context.

Journalism: Investigative reporters use Graph-RAG to trace connections between entities in complex stories, identifying patterns in public records, financial disclosures, and document leaks.

Urban Planning: Smart city applications employ Graph-RAG to model relationships between infrastructure, services, and citizen needs, answering complex queries about urban system interactions.

Cultural Heritage: Museums and archives use Graph-RAG to connect artifacts, historical events, and cultural contexts, enabling rich exploratory experiences for visitors and researchers.

We will explore these applications in greater detail in Chapter 11 of Part IV, where we present comprehensive case studies, implementation patterns, and performance benchmarks for Graph-RAG systems across these domains. The chapter will include practical examples of graph schema design, query patterns, and integration strategies for domain-specific applications.

7.4 Summary

This chapter provided a comprehensive overview of Graph-RAG, highlighting its core motivations, architecture, and real-world applications. We began by introducing the fundamental paradigm of Graph-RAG and why it represents a meaningful evolution beyond traditional RAG systems. Conventional RAG pipelines, which rely solely on unstructured text, struggle with factual consistency and multi-hop reasoning. In contrast, Graph-RAG addresses these challenges by incorporating structured, relational knowledge from KGs, enabling more faithful, interpretable, and domain-adapted generation.

The chapter then dived into the three core components of Graph-RAG architecture:

- Graph-based Indexing: transforming unstructured documents into queryable KGs with schema-aware construction and quality assurance.
- Graph-guided Retrieval: employing traversal algorithms (BFS, DFS, random walks), neural techniques (GNNs, LLMs), and hybrid strategies to extract relevant subgraphs at varying granularities—from single nodes to interconnected subgraphs.
- Graph-enhanced Generation: linearizing retrieved structures into LLM-consumable prompts, incorporating metadata for factuality, enabling complex reasoning, and fine-tuning LLMs on graph-augmented datasets to improve fluency and grounding.

Finally, we concluded with a look at application scenarios across various tasks and domains. Graph-RAG proves especially powerful in knowledge-intensive settings like biomedical QA, legal document analysis, financial question answering, and academic research—where structure, traceability, and factual accuracy are paramount.

These systematic introductions of Graph-RAG components help lay foundations for real-world application demonstrations and more advanced topics in later chapters.

References

1 Yingli Zhou, Yaodong Su, Youran Sun, Shu Wang, Taotao Wang, Runyuan He, Yongwei Zhang, Sicong Liang, Xilin Liu, Yuchi Ma, et al. In-depth analysis of graph-based rag in a unified framework. *arXiv preprint arXiv:2503.04338*, 2025.

2 Qinggang Zhang, Shengyuan Chen, Yuanchen Bei, Zheng Yuan, Huachi Zhou, Zijin Hong, Junnan Dong, Hao Chen, Yi Chang, and Xiao Huang. A survey of graph retrieval-augmented generation for customized large language models. *arXiv preprint arXiv:2501.13958*, 2025.

3 Zhuoqun Li, Xuanang Chen, Haiyang Yu, Hongyu Lin, Yaojie Lu, Qiaoyu Tang, Fei Huang, Xianpei Han, Le Sun, and Yongbin Li. Structrag: Boosting knowledge intensive reasoning of LLMs via inference-time hybrid information structurization. *arXiv preprint arXiv:2410.08815*, 2024.

4 Zhaolei Guo, Lanjing Xia, Yixing Yu, Tianshu Ao, and Chao Huang. LightRAG: Simple and fast retrieval-augmented generation. *arXiv preprint arXiv:2410.05779 2*, 2024.

5 Shilong Li, Yancheng He, Hangyu Guo, Xingyuan Bu, Ge Bai, Jie Liu, Jiaheng Liu, Xingwei Qu, Yangguang Li, Wanli Ouyang, et al. Graphreader: Building graph-based agent to enhance long-context abilities of large language models. *arXiv preprint arXiv:2406.14550*, 2024.

6 Qinggang Zhang, Junnan Dong, Hao Chen, Daochen Zha, Zailiang Yu, and Xiao Huang. Knowgpt: Knowledge graph based prompting for large language models. *Advances in Neural Information Processing Systems*, 37:6052–6080, 2024.

7 Jiashuo Sun, Chengjin Xu, Lumingyuan Tang, Saizhuo Wang, Chen Lin, Yeyun Gong, Lionel M Ni, Heung-Yeung Shum, and Jian Guo. Think-on-graph: Deep and responsible reasoning of large language model on knowledge graph. *arXiv preprint arXiv:2307.07697*, 2023.

8 Shengjie Ma, Chengjin Xu, Xuhui Jiang, Muzhi Li, Huaren Qu, and Jian Guo. Think-on-graph 2.0: Deep and interpretable large language model reasoning with knowledge graph-guided retrieval. *arXiv e-prints*, pages arXiv–2407, 2024.

9 Linhao Luo, Yuan-Fang Li, Gholamreza Haffari, and Shirui Pan. Reasoning on graphs: Faithful and interpretable large language model reasoning. *arXiv preprint arXiv:2310.01061*, 2023.

10 Haozhen Zhang, Tao Feng, and Jiaxuan You. Graph of records: Boosting retrieval augmented generation for long-context summarization with graphs. *arXiv preprint arXiv:2410.11001*, 2024.

11 J Wu, J Zhu, and Y Qi. Medical graph rag: Towards safe medical large language model via graph retrieval-augmented generation. *arXiv preprint arXiv:2408.04187*, 2024.

12 Xiangyan Liu, Bo Lan, Zhiyuan Hu, Yang Liu, Zhicheng Zhang, Fei Wang, Michael Shieh, and Wenmeng Zhou. Codexgraph: Bridging large language models and code repositories via code graph databases. *arXiv preprint arXiv:2408.03910*, 2024.

13 László Lovász and Paul Erdos. Random walks on graphs: A survey. *Combinatorica*, 16(1):1–46, 1996.

14 Lawrence Page, Sergey Brin, Rajeev Motwani, and Terry Winograd. The pagerank citation ranking: Bringing order to the web. Technical report, Stanford Infolab, 1999.

15 Taher H Haveliwala. Topic-sensitive pagerank. In *Proceedings of the 11th International Conference on World Wide Web*, pages 517–526, 2002.

16 László Lovász. Random walks on graphs. *Combinatorics, Paul Erdos is Eighty*, 2(1–46):4, 1993.

17 G Ben Arous and Alexander Fribergh. Biased random walks on random graphs. *Probability and Statistical Physics in St. Petersburg*, 91:99–153, 2016.

18 Jure Leskovec, Jon Kleinberg, and Christos Faloutsos. Graphs over time: Densification laws, shrinking diameters and possible explanations. In *Proceedings of the Eleventh ACM SIGKDD International Conference on Knowledge Discovery in Data Mining*, pages 177–187, 2005.

19 Jure Leskovec and Christos Faloutsos. Sampling from large graphs. In *Proceedings of the 12th ACM SIGKDD International Conference on Knowledge Discovery and Data Mining*, pages 631–636, 2006.

20 James Cook, Krishnaram Kenthapadi, and Nina Mishra. Group chats on twitter. In *Proceedings of the 22nd International Conference on World Wide Web*, pages 225–236, 2013.

21 Aditya Grover and Jure Leskovec. node2vec: Scalable feature learning for networks. In *Proceedings of the 22nd ACM SIGKDD International Conference on Knowledge Discovery and Data Mining*, pages 855–864, 2016.

22 Nicholas Metropolis, Arianna W Rosenbluth, Marshall N Rosenbluth, Augusta H Teller, and Edward Teller. Equation of state calculations by fast computing machines. *The Journal of Chemical Physics*, 21(6):1087–1092, 1953.

23 W Keith Hastings. Monte carlo sampling methods using markov chains and their applications. *Biometrika*, 57(1):97–109, 1970.

24 Wendy Ran Wei. *On Estimation Problems in Network Sampling*. PhD thesis, The Ohio State University, 2016.

25 Zijian Li, Qingyan Guo, Jiawei Shao, Lei Song, Jiang Bian, Jun Zhang, and Rui Wang. Graph neural network enhanced retrieval for question answering of llms. *arXiv preprint arXiv:2406.06572*, 2024.

26 Yuxin Dong, Shuo Wang, Hongye Zheng, Jiajing Chen, Zhenhong Zhang, and Chihang Wang. Advanced rag models with graph structures: Optimizing complex knowledge reasoning and text generation. In *2024 5th International Symposium on Computer Engineering and Intelligent Communications (ISCEIC)*, pages 626–630. IEEE, 2024.

27 Darren Edge, Ha Trinh, Newman Cheng, Joshua Bradley, Alex Chao, Apurva Mody, Steven Truitt, Dasha Metropolitansky, Robert Osazuwa Ness, and Jonathan Larson. From local to global: A graph rag approach to query-focused summarization. *arXiv preprint arXiv:2404.16130*, 2024.

28 Yu Wang, Nedim Lipka, Ryan A Rossi, Alexa Siu, Ruiyi Zhang, and Tyler Derr. Knowledge graph prompting for multi-document question answering. In *Proceedings of the AAAI Conference on Artificial Intelligence*, volume 38, pages 19206–19214, 2024.

29 Mengliang He, Aimin Zhou, and Xiaoming Shi. Enhancing textbook question answering with knowledge graph-augmented large language models. In *The 16th Asian Conference on Machine Learning (Conference Track)*, 2024.

30 Fangyu Liu, Ehsan Shareghi, Zaiqiao Meng, Marco Basaldella, and Nigel Collier. Self-alignment pretraining for biomedical entity representations. *arXiv preprint arXiv:2010.11784*, 2020.

31 Zhangyin Feng, Daya Guo, Duyu Tang, Nan Duan, Xiaocheng Feng, Ming Gong, Linjun Shou, Bing Qin, Ting Liu, Daxin Jiang, et al. Codebert: A pre-trained model for programming and natural languages. In *Findings of the Association for Computational Linguistics: EMNLP 2020*, pages 1536–1547. Association for Computational Linguistics. November 2020. https://aclanthology.org/2020.findings-emnlp.139/.

32 Tiesunlong Shen, Erik Cambria, Jin Wang, Yi Cai, and Xuejie Zhang. Insight at the right spot: Provide decisive subgraph information to graph llm with reinforcement learning. *Information Fusion*, 117:102860, 2025.

33 Jinyoung Park, Minseok Joo, Joo-Kyung Kim, and Hyunwoo J Kim. Generative subgraph retrieval for knowledge graph-grounded dialog generation. *arXiv preprint arXiv:2410.09350*, 2024.

34 Bowen Jin, Chulin Xie, Jiawei Zhang, Kashob Kumar Roy, Yu Zhang, Zheng Li, Ruirui Li, Xianfeng Tang, Suhang Wang, Yu Meng, et al. Graph chain-of-thought: Augmenting large language models by reasoning on graphs. *arXiv preprint arXiv:2404.07103*, 2024.

35 Jianing Wang, Qiushi Sun, Xiang Li, and Ming Gao. Boosting language models reasoning with chain-of-knowledge prompting. *arXiv preprint arXiv:2306.06427*, 2023.

36 Ruosong Ye, Caiqi Zhang, Runhui Wang, Shuyuan Xu, and Yongfeng Zhang. Language is all a graph needs. *arXiv preprint arXiv:2308.07134*, 2023.

37 Jiabin Tang, Yuhao Yang, Wei Wei, Lei Shi, Lixin Su, Suqi Cheng, Dawei Yin, and Chao Huang. Graphgpt: Graph instruction tuning for large language models. In *Proceedings of the 47th International ACM SIGIR Conference on Research and Development in Information Retrieval*, pages 491–500, 2024.

38 Runjin Chen, Tong Zhao, Ajay Jaiswal, Neil Shah, and Zhangyang Wang. Llaga: Large language and graph assistant. *arXiv preprint arXiv:2402.08170*, 2024.

8

Practice: Graph-RAG Implementations

Having explored the theoretical foundations of graph retrieval in retrieval-augmented generation (RAG) in the previous chapter, we now shift our focus to practical implementations that bring these concepts into action.

In this chapter, we build upon the Wayfair ANnotation Dataset (WANDS) example introduced earlier—specifically, the WANDS RAG system described in Section 4.3 and its corresponding knowledge graph (KG) representation detailed in Section 6.5. Using this example, we will demonstrate how to retrieve information from a KG for use in RAG pipelines.

Additionally, we will explore how LangChain can be used to implement KG-based RAG systems using both plain text and CSV dataframe inputs, highlighting the flexibility of such approaches.

8.1 Graph-RAG on WANDS Dataset with LlamaIndex

Building on the graph construction from Section 6.5, our objective is to complete the KG-RAG pipeline by integrating graph-based retrieval and large language models (LLM)-driven generation components. To ensure consistency and facilitate reproducibility, we will reuse configurations and components established in prior sections throughout this section:

- **Data Preparation:** As outlined in Section 1.3.3 and Appendix B
- **LLM, Chroma, and Neo4j Setup:** Configuration details from Appendix A
- **Embedding Vector Index:** Based on the indexing strategy discussed in Section 4.3.1
- **Graph Index:** As constructed in Section 6.5

This modular approach enables seamless integration of retrieval and generation stages with previously built components, ensuring a robust and maintainable KG-RAG system.

8.1.1 Querying with Neo4j and Chroma

Given a user query, we can identify relevant nodes in the KG using embedding-based search. To map the query to concepts such as product classes or categories, we utilize an embedding vector index. Once we locate relevant nodes in the graph, we can traverse relationships using graph triples. Based on previously defined triplets—such as "product-class-product" and "product-category-product"—we have implemented two primary retrieval strategies. For more complex use cases, we can also explore paths like "product-query-product," leveraging historical query logs encoded as graph triples. Let's begin with the simpler case: retrieval by product class.

8.1.1.1 Retrieval: By Product Class

Before proceeding with retrieval steps, we first need to load the prebuilt graph index created in Section 6.5.5:

```
from llama_index.core import PropertyGraphIndex

index_kg = PropertyGraphIndex.from_existing(property_graph_store=graph_store)
```

Here, `graph_store` is reused from Section 6.5.1.1.

The core idea of retrieval by product class is straightforward:

1. Given a user query string, compute its embedding. Find the top 1 similar product class using embedding similarity.

   The foundation of effective KG-RAG begins with the ability to locate relevant vertices within a KG. Once a graph is constructed, simple keyword-based queries can help locate specific vertices. However, this approach has significant limitations—queries must match exact terms, making it impractical for real-world scenarios involving synonyms, typos, or conceptual overlaps. To overcome these constraints, we employ embeddings to find similar vertices based on semantic similarity rather than exact keyword matches.
2. Retrieve all products belonging to that class as candidate results.
3. Compute the embedding similarity between the query and each candidate product.
4. Return the top-*k* most similar products.

To implement this logic, we utilize LlamaIndex's `CustomPGRetriever`, which allows us to subclass and define custom retrieval behavior. We define our own `ProductClassRetriever` and implement the logic within the `custom_retrieve()` method.

```
from llama_index.core.indices.property_graph import (
    CustomPGRetriever,
    CUSTOM_RETRIEVE_TYPE,
)
from llama_index.core.base.embeddings.base import SimilarityMode, similarity
from llama_index.core.schema import TextNode, NodeWithScore
from llama_index.core import Settings, VectorStoreIndex
from llama_index.vector_stores.chroma import ChromaVectorStore
import chromadb

class ProductClassRetriever(CustomPGRetriever):
    def init(
        self,
        collection_name="product_class",                    # [1]
        cypher_query="""
MATCH (product:product)-[rel:`product->product_class`]->(productClass:product_class)
WHERE productClass.name = $x
RETURN product
""",                                                         # [2]
        columns_to_return=[
            "product_id",
            "product_name",
            "product_class",
            "category_hierarchy",
            "product_description",
            "rating_count",
            "average_rating",
            "review_count",
        ],
        top_k=3,
        **kwargs,
    ) -> None:
        self.collection_name = collection_name
        self.index_embed = VectorStoreIndex.from_vector_store(
            vector_store=ChromaVectorStore(
```

```
                chroma_collection=chromadb.HttpClient().get_or_create_collection(
                    collection_name                                     # [1]
                )
            )
        )
        self.cypher_query = cypher_query
        self.columns_to_return = columns_to_return
        self.top_k = top_k

    def custom_retrieve(self, query_str: str) -> CUSTOM_RETRIEVE_TYPE:
        p_c = self.index_embed.as_retriever().retrieve(query_str)[0].text   # [3]

        candidate_product_list = self.graph_store.structured_query(
            self.cypher_query, {"x": f"{self.collection_name}:{p_c}"}
        )                                                               # [2]
        candidate_product_text_node = [
            TextNode(                                                   # [4]
                text=json.dumps(
                    {
                        key: value
                        for key, value in p["product"].items()
                        if key in self.columns_to_return
                    }
                )
            )
            for p in candidate_product_list
        ]
        query_embedding = Settings.embed_model.get_text_embedding(query_str)
        node_list = [
            NodeWithScore(                                              # [5]
                node=text_node,
                score=similarity(                                       # [5]
                    query_embedding,
                    Settings.embed_model.get_text_embedding(text_node.get_text()),
                    SimilarityMode.DOT_PRODUCT,
                ),
            )
            for text_node in candidate_product_text_node
        ]

        return sorted(node_list, key=lambda node: -node.score)[: self.top_k]   # [6]
```

[1] Load the vector store index previously built in Section 4.3.1.
[2] Define a Cypher query to retrieve all products belonging to a given product class.
[3] Use the vector index to find the closest matching product class based on the input query.
[4] Convert the returned dictionary list into a list of `TextNode` objects, filtering only relevant fields.
[5] Calculate the similarity score between the query and each candidate product using dot product, converting `TextNode` objects into `NodeWithScore` objects.
[6] Sort the results by score and return the top-k results.

Once the retriever class is defined, we register it with the graph index and perform a test retrieval:

```
retriever_product_class = index_kg.as_retriever(
    sub_retrievers=[ProductClassRetriever(index_kg.property_graph_store)]
)

query_str = df_query[df_query["query_id"] == 0]["query"].item()
nodes = retriever_product_class.retrieve(query_str)
```

These retrieved nodes are now ready to be passed to the generation stage for final response creation.

8.1.1.2 Generation

Since the query strings in WANDS are typically composed of simple keywords, we can enhance them into a more natural, descriptive language to improve LLM comprehension. This helps the model better interpret user intent and generate meaningful responses. Below is the prompt template used to guide the response generation process:

```
text_qa_template = """
Role: You are a professional customer service assistant specializing in product search
    assistance.

Task: Generate a helpful response to the user's query using the provided product
    candidates from our inventory database.

Input Information:
1. User Query: "{query_str}"
2. Candidate Products Found: {context_str}

Response Requirements:
1. Analyze the user's query to understand their needs and intent.
2. Match the query to the most relevant products from the candidate list provided.
3. Compose a clear, friendly, and natural-sounding response in conversational English.
4. Include specific product details (name, key features, etc.) from the candidate list
    when applicable.
5. If multiple products are relevant, present them in a prioritized manner.
6. If no suitable products are listed, politely inform the user and suggest possible
    alternatives or next steps.
7. Avoid technical jargon and ensure the response is easy to understand.
8. Keep the reply concise but thorough, maintaining a professional tone.

Additional Guidelines:
- Use bullet points or numbering to organize product recommendations if appropriate.
- Highlight any standout features that directly address the user's query.
- Include a call-to-action (e.g., "Would you like more details about any of these
    options?").
- Check for consistency between the query terms and product descriptions.
"""
```

The generation step closely follows the RAG synthesis approach outlined in Section 4.2.2.3. We use LlamaIndex's `get_response_synthesizer` to generate a final answer by combining the user query and retrieved nodes (candidate products).

```
from llama_index.core import get_response_synthesizer
from llama_index.core.response_synthesizers import ResponseMode
from llama_index.core import PromptTemplate

response_synthesizer = get_response_synthesizer(
    response_mode=ResponseMode.SIMPLE_SUMMARIZE,
    text_qa_template=PromptTemplate(text_qa_template),
    structured_answer_filtering=True,
)
response = response_synthesizer.synthesize(query=query_str, nodes=nodes)
print(response)
```

This synthesized response is now ready to be returned to the user or further processed for presentation purposes.

8.1.1.3 Retrieval: By Product Category

The retrieval logic for product category closely follows the same pattern used for product class. The only required changes are updating the vector store collection name and adjusting the Cypher query to reflect the `category_hierarchy` relationship. Here's how we instantiate the retriever with these adjustments:

```
retriever_category_hierarchy = index_kg.as_retriever(
```

```
    sub_retrievers=[
        ProductClassRetriever(
            index_kg.property_graph_store,
            collection_name="category_hierarchy",
            cypher_query="""
MATCH (product:product)-[rel:`product->category_hierarchy`]->(categoryHierarchy:
     category_hierarchy)
WHERE categoryHierarchy.name = $x
RETURN product
""",
        )
    ]
)
```

This allows us to retrieve products based on their association with a specific category hierarchy, using the same embedding-based retrieval mechanism as before.

8.1.1.4 Retrieval: By Seed Product and Seed Query

The idea behind this retrieval strategy is as follows:

1. Given a user query string, find the top 1 most similar product based on product name embeddings.
2. Identify the queries that led to that seed product using an `Exact` label relationship in the graph.
3. Retrieve all products associated with those seed queries.
4. Rerank the resulting candidate products using product description embeddings.
5. Return the top-*k* results.

Below is the implementation of the `SeedRetriever`, which extends `CustomPGRetriever` from LlamaIndex to support custom logic:

```
class SeedRetriever(CustomPGRetriever):
    def init(
        self,
        collection_name_seed="product_name",
        collection_name_rerank="product_description",
        cypher_query="""
MATCH (p:product)
WHERE p.id IN $x
MATCH (q:query)-[:`query->product:Exact`]->(p)
MATCH (q)-[:`query->product:Exact`]->(product:product)
RETURN DISTINCT product
""",                                                              # [1]
        columns_to_return=[
            "product_id",
            "product_name",
            "product_class",
            "category_hierarchy",
            "product_description",
            "rating_count",
            "average_rating",
            "review_count",
        ],
        top_k=3,
        **kwargs,
    ) -> None:
        self.index_embed_seed = VectorStoreIndex.from_vector_store(
            vector_store=ChromaVectorStore(
                chroma_collection=chromadb.HttpClient().get_or_create_collection(
                    collection_name_seed
                )
            )
        )                                                         # [2]
```

```
        self.collection_rerank = chromadb.HttpClient().get_or_create_collection(
            collection_name_rerank
        )                                                              # [3]

        self.cypher_query = cypher_query
        self.columns_to_return = columns_to_return
        self.top_k = top_k

    def custom_retrieve(self, query_str: str) -> CUSTOM_RETRIEVE_TYPE:
        seed_product_ids = [
            x.metadata["product_id"]
            for x in self.index_embed_seed.as_retriever(
                similarity_top_k=1,
            ).retrieve(query_str)                                      # [1]
        ]

        candidate_products = self.graph_store.structured_query(
            self.cypher_query, {"x": [f"product:{s}" for s in seed_product_ids]}
        )
        prefix_length = len("product:")
        candidate_product_ids = [
            int(x["product"]["id"][prefix_length:]) for x in candidate_products
        ]                                                              # [4]

        query_embedding = Settings.embed_model.get_text_embedding(query_str)
        node_list = self.collection_rerank.query(
            query_embeddings=query_embedding,
            n_results=self.top_k,
            where={"product_id": {"$in": candidate_product_ids}},
        )                                                              # [3]

        return [str(x["product_id"]) for x in node_list["metadatas"][0]] # [5]
```

[1] Cypher query to find products under the same queries as the seed product.

[2] Use the seed embedding index to find the most similar product by name.

[3] Use the reranking embedding index to filter and rank products by description similarity.

[4] Convert graph nodes to product IDs using metadata injected during graph construction.

[5] Return product IDs as strings to conform to the `CUSTOM_RETRIEVE_TYPE` interface.

To use the retriever and obtain the retrieved product DataFrame:

```
retriever_seed = index_kg.as_retriever(
    sub_retrievers=[SeedRetriever(index_kg.property_graph_store)]
)
product_id_node_list = retriever_seed.retrieve(query_str)
product_id_list = [int(n.text) for n in product_id_node_list]
retrieved_products = df_product[df_product["product_id"].isin(product_id_list)]
```

Comparing this retriever implementation with those described in Sections 8.1.1.1 and 8.1.1.3, we observe that the retrieval logic is the most dynamic and variable component in the overall RAG pipeline. While the rest of the workflow—including generation and synthesis—remains relatively stable across different strategies, the retrieval mechanism can vary significantly depending on the use case or data relationships.

This observation also applies to recommendation systems, where retrieval methods can be tailored to different signals such as collaborative filtering, semantic similarity, or behavioral patterns. For brevity and to avoid redundancy, we do not repeat the full implementation details for recommendation scenarios here.

8.1.2 Removing Chroma Dependency

Modern graph databases have evolved to better support semantic search workflows. Neo4j, starting from version 5.23.0, now includes native vector storage capabilities, eliminating the need for external vector stores like Chroma. This integration simplifies the system architecture by co-locating embeddings with graph data, enabling embedding-based queries directly within the database. With this feature, we can identify semantically similar nodes and then traverse the graph to explore related entities—all within a single system.

Building on the embedding configuration introduced in Sections 6.5.5 and 6.6, enabling embedding functionality in our code is straightforward. For example, during property graph index creation:

```
embed_kg_nodes=True,
```

To verify that embeddings have been successfully added to the graph nodes, you can inspect the Neo4j web interface (see Figure 6.19), where the embedding vectors should appear as properties of the corresponding nodes.

Once embeddings are enabled, we can retrieve semantically relevant nodes using a vector-aware retriever. Here's an example implementation:

```
from llama_index.core.indices.property_graph import VectorContextRetriever
vector_retriever = VectorContextRetriever(
    index.property_graph_store,
    similarity_top_k=1,
)
retriever = index.as_retriever(sub_retrievers=[vector_retriever])
nodes = retriever.retrieve("BE BEST")
print(nodes[0].score)
print(nodes[0].text)
```

Let's walk through what this code does:

1. A `VectorContextRetriever` is instantiated with the property graph store and configured to return the top matching node (`similarity_top_k=1`).
2. We bind the retriever to the graph index using `index.as_retriever()`.
3. Finally, we execute a semantic query for "BE BEST" and retrieve the most relevant paragraph along with its similarity score.

When executed, this query yields strong results. A similarity score of 0.8183 indicates a high degree of semantic alignment. The retrieved content is also contextually rich and relevant:

0.8182978630065918
Here are some facts extracted from the provided text:

Fostering the future —> Provides —> College—level scholarships
Mrs. trump —> Established —> Fostering the future

In her time away from the White House, Mrs. Trump established Fostering the Future, a BE BEST initiative, which provides college—level scholarships to those aging out of the foster care system. The scholarship funds are provided by a portion of the revenue generated from her Web 2 and Web 3 platforms. Her platforms utilize blockchain technology, marking her as a pioneer within the industry while empowering the next generation of young Americans with the resources needed to excel. Mrs. Trump consistently implements innovative approaches to advance her advocacy for children and the causes she supports, demonstrating a forward—thinking commitment to making a positive impact.
In 2024, Mrs. Trump released her memoir, MELANIA, which quickly became #1 on The New York Times best sellers list.

This demonstrates how embedding-based retrieval overcomes the limitations of exact keyword matching, enabling more flexible and semantically meaningful exploration of KGs.

8.1.3 Key Takeaways

- KG-RAG pipelines leverage KGs for enhanced retrieval, combining the structured knowledge of a graph with the generative power of an LLM.
- LlamaIndex offers a modular approach to building these pipelines, allowing for the reuse of existing components like data, LLMs, and vector/graph indices.
- Retrieval strategies can be customized to suit different use cases. In LlamaIndex, this is achieved by subclassing `CustomPGRetriever` and implementing `custom_retrieve` methods for different retrieval logic (e.g., by product class or category, or by using a seed product and query).
- Modern graph databases like Neo4j (v5.23.0+) eliminate the need for external vector stores like Chroma by integrating native vector search, simplifying the architecture and enabling semantic queries directly within the graph database.
- The retrieval stage is the most dynamic part of the RAG pipeline; while the generation and synthesis steps remain relatively stable, the retrieval logic must be tailored to the specific data relationships and use case.

8.2 Graph-RAG on Wiki and Kaggle Data with LangChain

The objective of this section is to demonstrate how to implement RAG with KG capabilities using the LangChain framework. We walk through examples that combine structured and unstructured data sources—specifically Wikipedia text and Kaggle CSV datasets—to showcase how LangChain enables effective question answering by integrating retrieval from diverse data formats.

The examples illustrate the end-to-end pipeline, including data loading, transformation into a KG, retrieval augmentation, and response generation. These hands-on demonstrations aim to equip developers with practical insights into building real-world, graph-enhanced RAG applications.

8.2.1 Wikipedia Text Search and Question Answering

The system functions through two primary stages: the **indexing phase** and the **query phase**.

8.2.1.1 Indexing Phase

The indexing phase involves connecting to a graph database, preparing documents, and populating it with structured data extracted from text.

8.2.1.1.1 Set Up LLM and Graph Database

We reuse the LLM servers defined in Appendix A. First, we prepare the chat model connection:

```
from langchain_openai import ChatOpenAI

llm = ChatOpenAI(
    base_url="http://localhost:4000/v1",
    model_name="FAKE",
    api_key="EMPTY",
)
```

Next, we establish a connection to the Neo4j graph database:

```
from langchain_neo4j import Neo4jGraph

graph = Neo4jGraph(
    url="bolt://localhost:7687",
    username="EMPTY",
    password="EMPTY",
    enhanced_schema=True,
)
```

8.2.1.1.2 Prepare Documents

We download a Wikipedia summary and split it into individual paragraphs:

```
import wikipedia

wp_summary = wikipedia.page(
    title="Tariffs in the second Trump administration",
    auto_suggest=False,
).summary.split("\n")
```

Then, we convert each paragraph into a LangChain Document object, one per paragraph:

```
from langchain_core.documents import Document

documents = [
    Document(page_content=paragraph)
    for paragraph in wp_summary
    if paragraph.strip() != ""
]
```

8.2.1.1.3 Documents to Triplets in Database

We use an LLM-based transformer to extract entities and relationships from the documents and insert them into the graph database:

```
from langchain_experimental.graph_transformers import LLMGraphTransformer

llm_transformer = LLMGraphTransformer(llm)
graph_documents = llm_transformer.convert_to_graph_documents(documents)
graph.add_graph_documents(
    graph_documents,
    baseEntityLabel=True,
    include_source=True,
)
```

Verify Graph Schema: After inserting the KG data, we can verify the schema:

```
graph.refresh_schema()
print(graph.schema)
```

An example output might look like this:

```
Node properties:
...
- **Person**
  - `id`: STRING Available options: ['Donald Trump', 'Trump']
...
Relationship properties:

The relationships:
...
(:Person)-[:RAISED_TARIFF]->(:Concept)
...
```

LLM Conversation Analysis: For each document, the system conducts one round of LLM interaction to extract structured triplets. Below is an example of the prompt used:

system

Knowledge Graph Instructions for GPT−4
1. Overview
You are a top−tier algorithm designed for extracting information in structured formats to build a knowledge graph.
Try to capture as much information from the text as possible without sacrificing accuracy. Do not add any information that is not explicitly mentioned in the text.
– **Nodes** represent entities and concepts.
– The aim is to achieve simplicity and clarity in the knowledge graph, making it accessible for a vast audience.
2. Labeling Nodes
– **Consistency**: Ensure you use available types for node labels.
Ensure you use basic or elementary types for node labels.
– For example, when you identify an entity representing a person, always label it as **'person'**. Avoid using more specific terms like 'mathematician' or 'scientist'. – **Node IDs**: Never utilize integers as node IDs. Node IDs should be names or human−readable identifiers found in the text.
– **Relationships** represent connections between entities or concepts.
Ensure consistency and generality in relationship types when constructing knowledge graphs. Instead of using specific and momentary types such as 'BECAME_PROFESSOR', use more general and timeless relationship types like 'PROFESSOR'. Make sure to use general and timeless relationship types!
3. Coreference Resolution
– **Maintain Entity Consistency**: When extracting entities, it's vital to ensure consistency. If an entity, such as "John Doe", is mentioned multiple times in the text but is referred to by different names or pronouns (e.g., "Joe", "he"), always use the most complete identifier for that entity throughout the knowledge graph. In this example, use "John Doe" as the entity ID.
Remember, the knowledge graph should be coherent and easily understandable, so maintaining consistency in entity references is crucial.
4. Strict Compliance
Adhere to the rules strictly. Non−compliance will result in termination.

user Tip: Make sure to answer in the correct format and do not include any explanations. Use the given format to extract information from the following input: During his second presidency, Donald Trump, president of the United States, triggered a global trade war after he enacted a series of steep tariffs affecting nearly all goods imported into the country. From January to April 2025, the average applied US tariff rate rose from 2.5% to an estimated 27%——the highest level in over a century. After changes and negotiations the rate was estimated at 18.6% as of August 2025. By July 2025, tariffs represented 5% of federal revenue compared to 2% historically.

assistant


```
{
  "nodes": [
    {
      "id": "Donald Trump",
      "type": "person"
    },
```

```
    ...
  ],
  "relationships": [
    {
      "source_node_id": "Donald Trump",
      "source_node_type": "person",
      "target_node_id": "United States",
      "target_node_type": "country",
      "type": "president_of"
    },
    ...
  ]
}
```

The `system` prompt also includes a tool of function type called `DynamicGraph` for generating structured output. The LLM responds with a function call containing parameters structured accordingly.

8.2.1.2 Query Phase

The query phase involves retrieving information from the graph database using a predefined chain that connects the LLM with the structured knowledge stored in Neo4j.

8.2.1.2.1 Define and Invoke the Query Chain

We define a Cypher-based QA chain using LangChain's `GraphCypherQAChain` and invoke it with a natural language question:

```
from langchain_neo4j import GraphCypherQAChain

chain = GraphCypherQAChain.from_llm(
    llm,
    graph=graph,
    verbose=True,
    allow_dangerous_requests=True,
    return_intermediate_steps=True,
)
chain.invoke(
    {
        "query": "What items did Trump raise tariffs on?",
    }
)
```

8.2.1.2.2 LLM Conversation Analysis

During execution, the LLM generates a valid Cypher query based on the provided schema and user input. Below is an example interaction:

user Task:Generate Cypher statement to query a graph database.
Instructions:
Use only the provided relationship types and properties in the schema.
Do not use any other relationship types or properties that are not provided.
Schema:
Node properties:
– **Person**
  – `id`: STRING Available options: ['Donald Trump', 'Trump']

```
...
– **Concept**
  – `id`: STRING Available options: ['Steel', 'Aluminum', 'Copper', 'Imported Cars', '
Domestic Manufacturing', 'National Security', 'Income Taxes', 'Trade Deficits', 'Us
Tariffs']
...
Relationship properties:

The relationships:
...
(:Person)–[:RAISED_TARIFF]–>(:Concept)
...
Note: Do not include any explanations or apologies in your responses.
Do not respond to any questions that might ask anything else than for you to construct
a Cypher statement.
Do not include any text except the generated Cypher statement.

The question is:
What items did Trump raise tariffs on?
```

assistant

```
MATCH (p:Person {id: "Trump"})–[:RAISED_TARIFF]–>(c:Concept) RETURN c.id
```

After executing the Cypher query against the Neo4j database, the system receives a structured result. The LLM then uses this result to generate a human-readable response:

user

```
You are an assistant that helps to form nice and human understandable answers.
The information part contains the provided information that you must use to construct
an answer.
The provided information is authoritative, you must never doubt it or try to use your
internal knowledge to correct it.
Make the answer sound as a response to the question. Do not mention that you based
the result on the given information.
Here is an example:

Question: Which managers own Neo4j stocks?
Context:[manager:CTL LLC, manager:JANE STREET GROUP LLC]
Helpful Answer: CTL LLC, JANE STREET GROUP LLC owns Neo4j stocks.

Follow this example when generating answers.
If the provided information is empty, say that you don't know the answer.
Information:
[{'c.id': 'Steel'}, {'c.id': 'Aluminum'}, {'c.id': 'Copper'}]

Question: What items did Trump raise tariffs on?
Helpful Answer:
```

assistant

```
Steel, Aluminum, Copper.
```

8.2.1.3 Key Takeaways

- The system for Wikipedia text search and question answering operates in two phases: indexing and querying.
- The indexing phase involves connecting an LLM and a graph database (like Neo4j), then using an LLM to convert raw text documents into a structured KG of nodes and relationships. This is done using a tool like `LLMGraphTransformer` to extract entities and their connections from each paragraph.
- The query phase uses a `GraphCypherQAChain` to translate natural language questions into Cypher queries, which are executed against the graph database.
- The LLM plays a dual role: first, it extracts structured data to build the KG, and second, it interprets the results of the Cypher query to generate a concise, human-readable answer for the user.
- This process demonstrates a full RAG pipeline that leverages a graph database for structured, precise retrieval before a final generation step.

8.2.2 Answering Kaggle CSV Recommendation Questions

This section presents a comprehensive methodology for enhancing document objects, constructing both graph-based and vector-based indexes, and utilizing a ReAct [1] agent to answer movie recommendation questions based on a Kaggle CSV dataset.

8.2.2.1 Indexing Phase

During the indexing phase, two complementary indexes are constructed: a KG and a vector index. These structures are derived from the dataset's "title" column and provide different but synergistic capabilities for retrieval and reasoning.

We selected the most popular movie-related dataset available on Kaggle, titled "Netflix Shows." After downloading the CSV file, we loaded it into a Pandas DataFrame, performed basic data cleaning, and sampled the data to create a reduced version used in our experiments, stored in the variable `df_movie_us_sample`.

8.2.2.1.1 Graph Index

The graph index utilizes the inherent subject-predicate-object relationships present in each row of the DataFrame. This approach eliminates the need for an LLM to extract relationships from unstructured text, allowing us to directly transform structured rows into `GraphDocument` objects. Below is the implementation:

Transforming Rows into Graph Documents: We define a function that converts each row of the DataFrame into a `GraphDocument`, consisting of nodes representing entities such as movies, directors, actors, genres, and countries, along with their associated relationships.

```
from langchain_community.graphs.graph_document import GraphDocument, Node,
    Relationship
from langchain_core.documents.base import Document

def row_graph_document(row):
    def explode(l):
        if isinstance(l, str) and len(l) > 0:
            return [x.strip() for x in l.split(",") if x.strip()]
        else:
            return []
```

```
    m = Node(
        id=row["title"],
        type="movie",
        properties={
            "show_id": row.name,
            "date_added": row["date_added"],
            "duration": row["duration"],
            "description": row["description"],
        },
    )
    d = [Node(id=director, type="director") for director in explode(row["director"])]
    a = [Node(id=actor, type="actor") for actor in explode(row["actors"])]
    c = [Node(id=country, type="country") for country in explode(row["country"])]
    y = Node(id=row["release_year"], type="release_year")
    r = Node(id=row["rating"], type="rating")
    g = [Node(id=genre, type="genre") for genre in explode(row["genres"])]

    nodes = [m, *d, *a, *c, y, r, *g]
    relationships = [
        *[Relationship(source=m, target=x, type="director") for x in d],
        *[Relationship(source=m, target=x, type="actor") for x in a],
        *[Relationship(source=m, target=x, type="country") for x in c],
        Relationship(source=m, target=y, type="release_year"),
        Relationship(source=m, target=r, type="rating"),
        *[Relationship(source=m, target=x, type="genre") for x in g],
    ]
    source = Document(page_content=row["title"] + "\n\n" + row["description"])
    return GraphDocument(nodes=nodes, relationships=relationships, source=source)

graph_documents = df_movie_us_sample.apply(row_graph_document, axis=1)
```

Storing the Graph in a Database: We reuse the previously established Neo4j database connection described in Section 8.2.1.1 to store the generated graph documents:

```
graph.add_graph_documents(graph_documents)
```

Verifying the Graph Schema: Using the schema verification method introduced in Section 8.2.1.1, we can inspect the resulting structure. An example of the extracted schema is shown below:

```
Node properties:
- **movie**
  - `show_id`: STRING Example: "s1397"
  - `id`: STRING Example: "Homefront"
  - `date_added`: STRING Example: "January 18, 2021"
  - `duration`: STRING Example: "100 min"
  - `description`: STRING Example: "When an ex-DEA agent is widowed, he moves with his
    "
...
- **actor**
  - `id`: STRING Example: "Jason Statham"
...
Relationship properties:

The relationships:
...
(:movie)-[:ACTOR]->(:actor)
...
```

8.2.2.1.2 Vector Index

To construct the vector index, we embed the "title" column using a preconfigured embedding model detailed in Appendix A, and store the resulting embeddings in an in-memory vector database.

Embedding Titles: We initialize the embedding model using the local LLM server configured for compatibility with the OpenAI API:

```
from langchain_openai import OpenAIEmbeddings

embedding = OpenAIEmbeddings(
    base_url="http://localhost:4000/v1",
    model="FAKE",
    api_key="EMPTY",
)
```

Building the Vector Database: Using the titles from the sampled movie dataset, we create an in-memory vector store:

```
from langchain_core.vectorstores.in_memory import InMemoryVectorStore

vector = InMemoryVectorStore.from_texts(
    texts=df_movie_us_sample["title"],
    embedding=embedding,
)
```

8.2.2.2 Query Phase

To handle user questions about movie recommendations, we employ a ReAct-based agent that utilizes custom tools and interacts with an LLM. The LLM instance is reused from Section 8.2.1.1.

8.2.2.2.1 Defining Tools

We define three key tools to support the recommendation process:

- Extract Movie Titles from Queries
- Validate Movie Titles
- Recommend Movies Based on Shared Actors

Below is the implementation of these tools using LangChain:

```
from langchain_core.tools import tool
from pydantic import Field, BaseModel

@tool
def extract_movie_titles_from_query(
    user_query: str = Field(description="A user's query about movies"),
) -> list[str]:
    """
    Extract mentioned movie titles from a user's query.
    """

    class ExtractedMovieTitles(BaseModel):
        """
        Represents a list of movie titles extracted from a query.
        """

        titles: list[str] = Field(
            description="A list of movie titles mentioned in the query"
        )

    response = llm.with_structured_output(ExtractedMovieTitles).invoke(
        "Extract all movie titles mentioned in the following text:\n" + user_query
    )
    return response.titles

@tool
```

```
def find_valid_movies_in_database(
    movie_titles: list[str] = Field(description="A list of movie titles to validate"),
) -> list[str]:
    """
    Validate movie titles against the database using embeddings.
    """
    return [
        document[0].page_content
        for document in movie_title_retriever.batch(movie_titles)
    ]

@tool
def recommend_movies_with_shared_actors(
    movie_titles: list[str] = Field(description="A list of movie titles"),
) -> list[dict]:
    """
    Recommend movies that share common actors with the provided movie titles.
    """
    cypher_query = """
    MATCH (m1:movie)-[:ACTOR]->(a:actor)
    WHERE m1.id IN $titles
    WITH a, COLLECT(m1) AS inputMovies
    WHERE SIZE(inputMovies) = SIZE($titles)  // Ensure actor appears in all input
     movies
    MATCH (m2:movie)-[:ACTOR]->(a)
    WHERE NOT m2 IN inputMovies  // Exclude input movies
    RETURN a.id AS actorName, COLLECT(m2.id) AS recommendedMovies
    """
    recommendations = graph.query(cypher_query, {"titles": movie_titles})
    return recommendations
```

8.2.2.2.2 Creating the ReAct Agent

Once the tools are defined, we create the ReAct agent using LangGraph's prebuilt utility:

```
tools = [
    extract_movie_titles_from_query,
    find_valid_movies_in_database,
    recommend_movies_with_shared_actors,
]

from langgraph.prebuilt import create_react_agent

movie_recommendation_agent = create_react_agent(llm, tools=tools)
```

8.2.2.2.3 Answering the User's Question

We can now run the agent to answer a natural language question. Here's how we invoke it:

```
inputs = {"messages": [("user", question)]}
print_stream(movie_recommendation_agent.stream(inputs, stream_mode="values"))
```

An example execution trace is shown below:

```
================================ Human Message =================================

Sandie mentioned watching 'Garfield Force' and 'Garfield Gets Fun Fest,' though she
     isn't sure about the exact titles. She wants recommendations for another movie
     featuring the same actors. Could you verify her watch history in the database and
     suggest a movie for her to watch next?
================================== Ai Message ==================================
Tool Calls:
  extract_movie_titles_from_query (chatcmpl-tool-d94d3c266c164b368184c2e88be260e2)
 Call ID: chatcmpl-tool-d94d3c266c164b368184c2e88be260e2
  Args:
```

```
    user_query: Sandie mentioned watching 'Garfield Force' and 'Garfield Gets Fun Fest
    ,' though she isn't sure about the exact titles. She wants recommendations for
    another movie featuring the same actors. Could you verify her watch history in the
     database and suggest a movie for her to watch next?
================================= Tool Message =================================
Name: extract_movie_titles_from_query

["Garfield Force", "Garfield Gets Fun Fest"]
================================== Ai Message ==================================
Tool Calls:
  find_valid_movies_in_database (chatcmpl-tool-ff5b8f81d20541e79216d722d41debd1)
 Call ID: chatcmpl-tool-ff5b8f81d20541e79216d722d41debd1
  Args:
    movie_titles: ['Garfield Force', 'Garfield Gets Fun Fest']
================================= Tool Message =================================
Name: find_valid_movies_in_database

["Garfield's Pet Force", "Garfield's Fun Fest"]
================================== Ai Message ==================================
Tool Calls:
  recommend_movies_with_shared_actors (chatcmpl-tool-d8ad093f78c84688972d19bf5f15435a)
 Call ID: chatcmpl-tool-d8ad093f78c84688972d19bf5f15435a
  Args:
    movie_titles: ["Garfield's Pet Force", "Garfield's Fun Fest"]
================================= Tool Message =================================
Name: recommend_movies_with_shared_actors

[{"actorName": "Greg Eagles", "recommendedMovies": ["Garfield Gets Real"]}, {"
    actorName": "Audrey Wasilewski", "recommendedMovies": ["Garfield Gets Real"]}, {"
    actorName": "Stephen Stanton", "recommendedMovies": ["Garfield Gets Real"]}, {"
    actorName": "Frank Welker", "recommendedMovies": ["Garfield Gets Real"]}, {"
    actorName": "Gregg Berger", "recommendedMovies": ["Garfield Gets Real"]}, {"
    actorName": "Neil Ross", "recommendedMovies": ["Garfield Gets Real"]}, {"actorName
    ": "Wally Wingert", "recommendedMovies": ["Garfield Gets Real"]}, {"actorName": "
    Jennifer Darling", "recommendedMovies": ["Garfield Gets Real"]}]
================================== Ai Message ==================================



Based on the movies "Garfield's Pet Force" and "Garfield's Fun Fest" in Sandie's watch
     history, all the actors from these films are also featured in **"Garfield Gets
    Real"**. This movie is a strong recommendation for Sandie, as it shares the same
    cast and aligns with her viewing preferences.

Would you like to confirm this recommendation or explore additional options?
```

8.2.2.2.4 LLM Conversation Analysis

This section provides a detailed breakdown of the multi-round interaction between the ReAct agent and the LLM during the movie recommendation process. Each round involves either calling a tool or generating a final answer based on the results returned from previous steps.

Round 1: Extract Movie Titles from Query The LLM receives the user's question along with available tools and initiates the first function call to extract movie titles from the query:

user Sandie mentioned watching 'Garfield Force' and 'Garfield Gets Fun Fest,' though she isn't sure about the exact titles. She wants recommendations for another movie featuring the same actors. Could you verify her watch history in the database and suggest a movie for her to watch next?

assistant

```
{
  "json": {
```

```
        "tool_calls": [
          {
            "type": "function",
            "id": "chatcmpl-tool-d94d3c266c164b368184c2e88be260e2",
            "function": {
              "name": "extract_movie_titles_from_query",
              "arguments": {
                "user_query": "Sandie mentioned watching 'Garfield Force' and 'Garfield
Gets Fun Fest,' though she isn't sure about the exact titles. She wants
recommendations for another movie featuring the same actors. Could you verify her
watch history in the database and suggest a movie for her to watch next?"
              }
            }
          }
        ]
      }
    }
```


Inside this function, a secondary prompt is sent to extract structured output using the `ExtractedMovieTitles` schema:

user Extract all movie titles mentioned in the following text:
Sandie mentioned watching 'Garfield Force' and 'Garfield Gets Fun Fest,' though she isn't sure about the exact titles. She wants recommendations for another movie featuring the same actors. Could you verify her watch history in the database and suggest a movie for her to watch next?

assistant


```
{
  "titles": [
    "Garfield Force",
    "Garfield Gets Fun Fest"
  ]
}
```


Round 2: Validate Movie Titles After extracting the movie titles, the result is appended to the chat history and passed back to the LLM. This time, the LLM selects the next tool: `find_valid_movies_in_database`.

tool: extract_movie_titles_from_query


```
[
  "Garfield Force",
  "Garfield Gets Fun Fest"
]
```


assistant


```
{
  "json": {
```

```
      "tool_calls": [
        {
          "type": "function",
          "id": "chatcmpl–tool–ff5b8f81d20541e79216d722d41debd1",
          "function": {
            "name": "find_valid_movies_in_database",
            "arguments": {
              "movie_titles": [
                "Garfield Force",
                "Garfield Gets Fun Fest"
              ]
            }
          }
        }
      ]
    }
  }
```


Round 3: Recommend Movies with Shared Actors Once the valid movies are retrieved, the agent appends the result to the history and sends it to the LLM again. This time, the LLM invokes the final tool: `recommend_movies_with_shared_actors`.

tool: find_valid_movies_in_database


```
  [
    "Garfield's Pet Force",
    "Garfield's Fun Fest"
  ]
```


assistant


```
  {
    "json": {
      "tool_calls": [
        {
          "type": "function",
          "id": "chatcmpl–tool–d8ad093f78c84688972d19bf5f15435a",
          "function": {
            "name": "recommend_movies_with_shared_actors",
            "arguments": {
              "movie_titles": [
                "Garfield's Pet Force",
                "Garfield's Fun Fest"
              ]
            }
          }
        }
      ]
    }
  }
```

Round 4: Final Answer Generation Finally, after receiving the list of recommended movies based on shared actors, the agent passes the result to the LLM one last time. Instead of calling another tool, the LLM generates a natural language response:

> **tool: recommend_movies_with_shared_actors**
>
> [{"actorName":"Greg Eagles","recommendedMovies":["Garfield Gets Real"]},{" actorName":"Audrey Wasilewski","recommendedMovies":["Garfield Gets Real"]},{" actorName":"Stephen Stanton","recommendedMovies":["Garfield Gets Real"]},{" actorName":"Frank Welker","recommendedMovies":["Garfield Gets Real"]},{" actorName":"Gregg Berger","recommendedMovies":["Garfield Gets Real"]},{" actorName":"Neil Ross","recommendedMovies":["Garfield Gets Real"]},{"actorName":" Wally Wingert","recommendedMovies":["Garfield Gets Real"]},{"actorName":"Jennifer Darling","recommendedMovies":["Garfield Gets Real"]}]
>
> **assistant**
>
> Based on the movies "Garfield's Pet Force" and "Garfield's Fun Fest" in Sandie's watch history, all the actors from these films are also featured in **"Garfield Gets Real"**. This movie is a strong recommendation for Sandie, as it shares the same cast and aligns with her viewing preferences.
>
> Would you like to confirm this recommendation or explore additional options?

8.2.2.3 Key Takeaways

- This approach combines a KG and a vector index to answer movie recommendation questions from a structured CSV dataset.
- The indexing phase builds two indexes: a graph index by transforming structured data rows into nodes and relationships (eliminating the need for an LLM to extract them from text), and a vector index by embedding movie titles.
- The query phase uses a ReAct agent to handle complex, multistep questions by defining and orchestrating specialized tools.
- Three tools are implemented: one to extract movie titles from a user's query, one to validate those titles against a vector store, and a third to recommend movies from the graph database based on shared actors.
- The agent's multi-round interaction with the LLM demonstrates how it can chain tool calls sequentially to resolve a request, starting with title extraction, followed by validation, and finally, fetching recommendations before synthesizing a final answer.

8.3 Summary

This chapter has explored the practical implementation of KG-RAG systems using a combination of tools and datasets. The focus was on two main case studies: LlamaIndex Query WANDS KG and LangChain KG-RAG on Wiki and Kaggle data. These examples demonstrated how KGs, vector stores, and LLMs can be integrated to build effective retrieval-augmented generation pipelines for knowledge-intensive applications.

In the LlamaIndex Query WANDS KG section, we implemented several retrieval strategies based on product class, product category, and seed products or queries. These approaches used embedding-based search combined with graph traversal to locate relevant nodes within the KG. We also transitioned away from external vector stores like Chroma by leveraging Neo4j's native vector storage capabilities introduced in version 5.23.0. This allowed embeddings to be stored directly alongside graph data, simplifying the architecture and enabling semantic similarity searches without additional dependencies. By activating embeddings in the property graph index, we showed how contextually relevant nodes could be retrieved, moving beyond traditional keyword-based matching.

The LangChain KG-RAG section covered two sub-use cases: one involving Wikipedia text for question answering and another using Kaggle CSV data for recommendation tasks. In both scenarios, raw data was transformed into structured graph representations through triplet extraction and schema verification. For Wikipedia, documents were converted into graph entities and relationships, while for Kaggle, CSV rows were mapped into graph documents. Vector indices were built from titles or key phrases to support fast semantic retrieval. During querying, LLMs were used to interpret user questions, extract relevant information from the graph, and generate natural language responses. A ReAct agent was employed in the Kaggle example to iteratively reason over graph and vector indices, demonstrating how multistep logic can enhance answer accuracy.

Key techniques across both frameworks included KG construction, embedding integration, and LLM augmentation. Structuring unstructured data into graph triplets enabled richer relational reasoning. Embeddings powered semantic search, allowing for more flexible and meaningful retrieval than exact matches. LLMs played a central role throughout the pipeline—whether for extracting triplets, analyzing conversations, or synthesizing final answers.

The overall workflow emphasized modularity, scalability, and improved user interaction. Separating indexing from querying made it easier to adapt the system to new domains. Using graph databases and vector stores together allowed efficient handling of large datasets. And by incorporating LLMs, the system delivered more natural and context-aware outputs.

Looking ahead, potential improvements include combining graph and vector retrieval into hybrid search methods, supporting real-time updates to the knowledge base, integrating multimodal data types, and optimizing performance through model fine-tuning and query execution enhancements.

In conclusion, this chapter illustrates how modern tools and techniques can be combined to build robust, scalable, and intelligent KG-based RAG systems. These systems are well-suited for complex tasks such as recommendation engines, conversational agents, and enterprise knowledge management solutions.

Reference

1 Shunyu Yao, Jeffrey Zhao, Dian Yu, Nan Du, Izhak Shafran, Karthik Narasimhan, and Yuan Cao. React: Synergizing reasoning and acting in language models. In *International Conference on Learning Representations (ICLR)*, 2023.

9

Graph-RAG Evaluations

The evaluation of graph retrieval-augmented generation (Graph-RAG) systems presents unique challenges that extend beyond traditional RAG assessment. While conventional RAG systems focus primarily on retrieval accuracy and generation quality, Graph-RAG introduces additional dimensions through its integration of structured knowledge representations, multi-hop reasoning capabilities, and graph traversal mechanisms. This chapter provides a comprehensive framework for evaluating these hybrid systems, ensuring that both their retrieval-generation pipeline and graph-specific enhancements are thoroughly assessed.

9.1 Performance Metrics Framework

Evaluating Graph-RAG systems requires a multifaceted approach that captures the interplay between graph structure, retrieval quality, and generation accuracy. As illustrated in Figure 9.1, we organize our metrics into four primary categories: retrieval-specific, generation quality, graph-specific, and system performance, each addressing a critical aspect of system performance.

9.1.1 Retrieval-specific Metrics

The retrieval component of Graph-RAG systems must be evaluated not only for traditional document retrieval but also for its ability to navigate graph structures effectively. These metrics assess how well the system identifies and extracts relevant graph elements in response to queries.

9.1.1.1 Precision and Recall

In the context of Graph-RAG, precision and recall take on nuanced meanings when applied to graph structures. Unlike traditional document retrieval, we must consider multiple levels of granularity: nodes, edges, and complete subgraphs. These metrics become particularly nuanced when dealing with multi-hop queries:

- Node-level Precision: Ratio of relevant nodes in the retrieved subgraph
- Edge-level Recall: Coverage of important relationships in the knowledge graph (KG)
- Subgraph Precision and Recall: Accuracy and recall of the entire retrieved graph structure

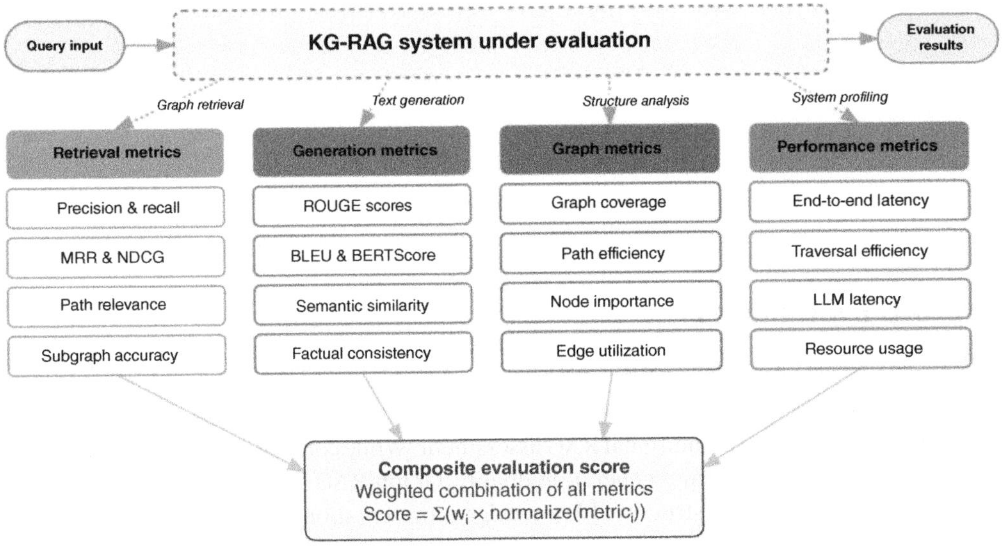

Figure 9.1 Graph-RAG evaluation framework overview.

Here are the mathematical formula definitions:

Node-level Precision:

$$P_{\text{node}} = \frac{|R_{\text{nodes}} \cap G_{\text{nodes}}|}{|R_{\text{nodes}}|}$$

where R_{nodes} is the set of retrieved nodes and G_{nodes} is the set of ground truth relevant nodes.

Edge-level Recall:

$$R_{\text{edge}} = \frac{|R_{\text{edges}} \cap G_{\text{edges}}|}{|G_{\text{edges}}|}$$

where R_{edges} is the set of retrieved edges and G_{edges} is the set of ground truth relevant edges.

Subgraph *F*1-score:

$$F1_{\text{subgraph}} = 2 \times \frac{P_{\text{subgraph}} \times R_{\text{subgraph}}}{P_{\text{subgraph}} + R_{\text{subgraph}}}$$

Example: Consider a query: "What medications interact with aspirin?" Ground Truth Subgraph:

```
Nodes: {Aspirin, Warfarin, Ibuprofen, Drug_Interaction_1,
Drug_Interaction_2}
Edges: {(Aspirin, interacts_with, Warfarin),
(Aspirin, interacts_with, Ibuprofen)}
```

Retrieved subgraph:

```
Nodes: {Aspirin, Warfarin, Ibuprofen, Acetaminophen,
Drug_Interaction_1}
Edges: {(Aspirin, interacts_with, Warfarin),
(Aspirin, interacts_with, Acetaminophen)}
```

- Node precision = 4/5 = 0.80
- Edge recall = 1/2 = 0.50

9.1.1.2 Mean Reciprocal Rank

Mean Reciprocal Rank (MRR) evaluates the ranking quality of retrieved graph elements, particularly important when the system returns multiple candidate paths or subgraphs.

Here are the mathematical formula definitions:

$$\text{MRR} = \frac{1}{|Q|}\sum_{i=1}^{|Q|}\frac{1}{\text{rank}_i}$$

where

- $|Q|$ is the total number of queries and rank_i is the rank position of the first correct answer for query i

Extended for Multi-path Queries:

$$\text{MRR}_{\text{path}} = \frac{1}{|Q|}\sum_{i=1}^{|Q|}\max_{j\in\text{paths}_i}\frac{\text{relevance}_j}{\text{rank}_j}$$

For example, for a query about "protein synthesis pathway":

```
Path 1 (Rank 1): DNA --> RNA --> Protein (Correct)
Path 2 (Rank 2): DNA --> Mutation --> Disease (Incorrect)
Path 3 (Rank 3): DNA --> Transcription --> RNA (Partially Correct)
```

$\text{MRR} = 1/1 = 1.0$ (since the correct path is ranked first)

Another example for multiple queries:

```
Query 1: Correct path ranked 1st. Reciprocal rank = 1/1 = 1.
Query 2: Correct path ranked 3rd. Reciprocal rank = 1/3.
Query 3: Correct path not found. Reciprocal rank = 0.
```

For these three queries, $\text{MRR} = (1 + 1/3 + 0)/3 = 4/9 \approx 0.44$.

This metric is especially valuable for evaluating systems that return ranked lists of reasoning paths through the KG, favoring systems that place the most relevant information higher up.

9.1.1.3 Normalized Discounted Cumulative Gain

Normalized Discounted Cumulative Gain (NDCG) captures graded relevance scores, crucial for Graph-RAG systems where different paths may have varying degrees of usefulness. Discounted Cumulative Gain (DCG@k) considers the position-discounted relevance of retrieved graph elements up to rank k and NDCG@k normalizes DCG@k by IDCG@k.

Here are the mathematical formula definitions:

$$\text{DCG@k} = \sum_{i=1}^{k}\frac{2^{rel_i} - 1}{\log_2(i+1)}$$

$$\text{NDCG@k} = \frac{\text{DCG@k}}{\text{IDCG@k}}$$

where

- rel_i is the relevance score of the item at position i
- IDCG@k is the DCG@k for the ideal ranking

Graph-adapted NDCG:

$$\text{NDCG}_{\text{graph}}@\text{k} = \frac{\sum_{i=1}^{k} \frac{\text{PathRelevance}_i \times \text{NodeImportance}_i}{\log_2(i+1)}}{\sum_{i=1}^{k} \frac{\text{MaxRelevance}_i}{\log_2(i+1)}}$$

9.1.1.4 Path Relevance Score

A graph-specific metric that evaluates the semantic coherence and relevance of retrieved paths:

Path Length Optimization: Shorter paths often indicate more direct and concise relationships, generally preferred. This can be quantified by a penalty for longer paths, e.g., 1/Path Length or an exponential decay.

For example: If a query asks "Where was Einstein born?," an optimal path might be

```
(Einstein)-[BORN_IN]->(Ulm)
```

A suboptimal path might be

```
(Einstein)-[HAD_PARENTS]->(Pauline Koch)-[BORN\_IN]->
(Cannstatt)-[NEAR]->(Ulm)
```

The shorter, direct path is more relevant.

Semantic Drift Measurement: How well the path maintains topical relevance throughout its traversal. This can be measured by calculating the semantic similarity between node/edge embeddings along the path and the original query embedding.

For example: A path that starts with "Apple Inc." and correctly moves to "Steve Jobs" (founder) has low semantic drift. A path that moves from "Apple Inc." to "Apple fruit" then to "fruit nutrition" has high semantic drift, even if connected.

Confidence Score Aggregation Along the Path: If edges or nodes have associated confidence scores (e.g., from an extraction process), this aggregates them (e.g., product, average, minimum) to reflect the overall reliability of the path.

For example: A path

```
(Entity A)-[rel1:0.9]->(Entity B)-[rel2:0.8]->(Entity C)
```

might have an aggregate confidence of $0.9 \times 0.8 = 0.72$ (product).

Here are the mathematical formula definitions:

$$\text{PathRelevance} = \alpha \cdot \frac{1}{\text{length}} + \beta \cdot \text{SemanticCoherence} + \gamma \cdot \text{ConfidenceAgg}$$

where

- α, β, γ is the weighting parameters (typically $\alpha + \beta + \gamma = 1$),
- length is the number of hops in the path
- $\text{SemanticCoherence} = \frac{1}{\text{length} - 1} \sum_{i=1}^{\text{length}-1} \text{sim}(\text{node}_i, \text{node}_{i+1})$
- $\text{ConfidenceAgg} = \prod_{i=1}^{\text{length}-1} \text{edgeConfidence}_i$

Example Calculation: Figure 9.2 gives an illustration of how to compute the path relevance score:
Path: "COVID-19" → "causes" → "Pneumonia" → "treated_by" → "Antibiotics"

- Length = 3
- Semantic Similarities: 0.8, 0.7
- Edge Confidences: 0.95, 0.60
- $\text{PathRelevance} = 0.3(1/3) + 0.4(0.75) + 0.3(0.57) = 0.571$

9.1.1.5 Subgraph Retrieval Accuracy

Measures how well the retrieved subgraph captures the complete and minimal context needed for answering a query. This is often assessed qualitatively or through set-based metrics comparing retrieved vs. gold subgraphs.

- Completeness: Does the subgraph contain all necessary information (nodes and edges) required to answer the query accurately and comprehensively?
  Assessment: Compare the set of essential nodes/edges for the answer to those in the retrieved subgraph.
- Minimality: Is the subgraph free from irrelevant nodes and edges that do not contribute to the answer? Over-retrieval can introduce noise and distract the LLM.
  Assessment: Identify nodes/edges in the retrieved subgraph that are not part of the ground truth needed for the answer.
- Connectivity: Are the retrieved elements properly connected within the subgraph, maintaining the semantic relationships inherent in the KG?
  Assessment: Verify that the retrieved subgraph forms a connected component (or a set of relevant connected components) necessary for logical reasoning.

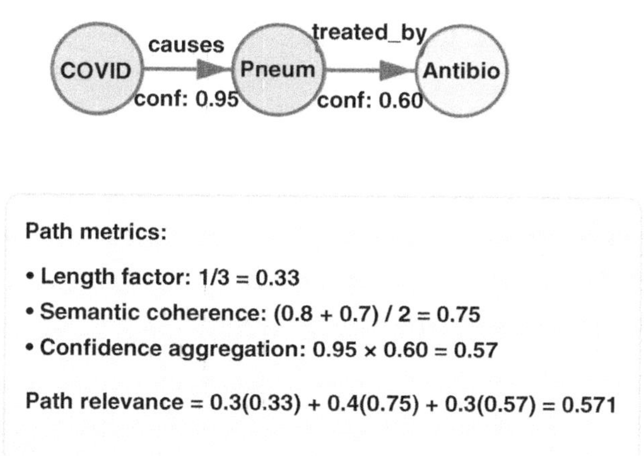

Figure 9.2 Example of path relevance score.

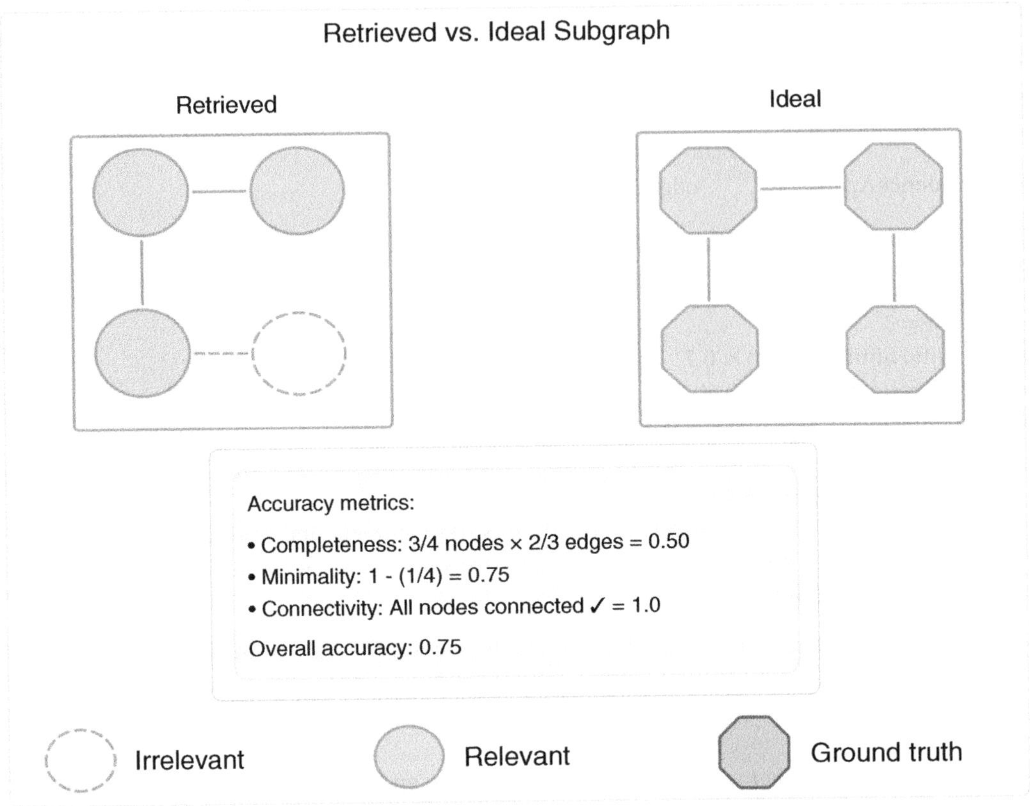

Figure 9.3 Example of subgraph retrieval accuracy.

Here are the mathematical formula definitions:

- Completeness score:

$$\text{Completeness} = \frac{|\text{EssentialNodes}_{\text{retrieved}}|}{|\text{EssentialNodes}_{\text{total}}|} \times \frac{|\text{CriticalEdges}_{\text{retrieved}}|}{|\text{CriticalEdges}_{\text{total}}|}$$

- Minimality score:

$$\text{Minimality} = 1 - \frac{|\text{IrrelevantElements}|}{|\text{TotalElements}|}$$

- Connectivity score:

$$\text{Connectivity} = \frac{|\text{ConnectedComponents}|}{|\text{IdealComponents}|}$$

Figure 9.3 gives an example of how to compute the subgraph retrieval accuracy.

9.1.2 Generation Quality Metrics

The generation phase must be evaluated for both linguistic quality and factual grounding in the retrieved graph structure.

9.1.2.1 ROUGE Scores

ROUGE (Recall-oriented Understudy for Gisting Evaluation) metrics assess the overlap between generated and reference texts:

- ROUGE-N: Measures the overlap of n-grams (sequences of n words).

$$\text{ROUGE-}N = \frac{\sum_{n\text{-gram}\in\text{Generated}} \text{CountMatch}(n\text{-gram})}{\sum_{n\text{-gram}\in\text{Reference}} \text{Count}(n\text{-gram})}$$

  - ROUGE-1: Unigram (single word) overlap.
  - ROUGE-2: Bigram (two-word sequence) overlap.
- ROUGE-L: Measures the longest common subsequence (LCS) between the generated and reference texts, capturing sentence-level structural similarity.

$$\text{ROUGE-}L = \frac{2 \times \text{LCS(Generated, Reference)}}{\text{Length(Generated)} + \text{Length(Reference)}}$$

For Graph-RAG, we extend ROUGE to consider graph-grounded content, measuring how well key entities and relationships from the graph (e.g., "Christopher Nolan," "directed," "Inception") appear in the generated text, not just arbitrary words.

For example:

- Query: "Who directed Inception?"
- Reference Answer: "Christopher Nolan directed Inception."
- Generated Answer: "Inception was directed by Christopher Nolan."
- ROUGE-1 would be high due to word overlap ("directed," "Inception," "Christopher," "Nolan"). ROUGE-L would also be high due to the common subsequence.

These are the mathematical formulation of ROUGE scores:

Standard ROUGE Formulas:

ROUGE-*N*

$$\text{ROUGE-}N = \frac{\sum_{S\in\text{References}} \sum_{\text{gram}_n\in S} \text{Count}_{\text{match}}(\text{gram}_n)}{\sum_{S\in\text{References}} \sum_{\text{gram}_n\in S} \text{Count}(\text{gram}_n)}$$

ROUGE-*L*: Longest Common Subsequence:

$$\text{ROUGE-}L = \frac{(1+\beta^2)\cdot R_{\text{lcs}} \cdot P_{\text{lcs}}}{R_{\text{lcs}} + \beta^2 \cdot P_{\text{lcs}}}$$

where

- $R_{\text{lcs}} = \frac{\text{LCS}(X, Y)}{m}$ (Recall)
- $P_{\text{lcs}} = \frac{\text{LCS}(X, Y)}{n}$ (Precision)
- m, n = lengths of reference and candidate

Graph-extended ROUGE

$$\text{ROUGE-}G = \lambda \cdot \text{ROUGE-}N + (1-\lambda) \cdot \text{EntityCoverage}$$

where,

$$\text{EntityCoverage} = \frac{|\text{Entities}_{\text{generated}} \cap \text{Entities}_{\text{graph}}|}{|\text{Entities}_{\text{graph}}|}$$

For example, for the generated text results: "Metformin is the first-line medication for type 2 diabetes treatment and helps control blood glucose levels effectively."

And the reference text results: "Metformin is considered the first-line treatment for type 2 diabetes and helps manage blood sugar levels in patients."

The overlapping n-grams are: "Metformin," "is first-line," "type 2 diabetes"

So the ROUGE scores are:

```
ROUGE-1: 0.73
ROUGE-2: 0.58
ROUGE-L: 0.69
```

9.1.2.2 BLEU Score

BLEU (Bilingual Evaluation Understudy) measures precision of n-gram overlap between a generated text and a set of reference texts, incorporating a brevity penalty. It's particularly useful for evaluating the fluency and structural similarity of generated responses that incorporate technical terms or specific phrasing derived from the KG.

Here are the mathematical formula definitions:

$$\text{BLEU} = \text{BP} \cdot \exp\left(\sum_{n=1}^{N} w_n \log p_n\right)$$

Where:

- $\text{BP} = \begin{cases} 1 & \text{if } c > r \\ e^{1-r/c} & \text{if } c \leq r \end{cases}$ (Brevity Penalty)
- p_n = Precision for n-grams
- w_n = Weights (typically $w_n = 1/N$)

Graph-aware BLEU:

$$\text{BLEU}_{\text{graph}} = \text{BP} \cdot \exp\left(\sum_{n=1}^{N} w_n \log p_n^{\text{graph}}\right)$$

where p_n^{graph} includes bonus weights for correctly used graph terminology.

For example:

- Reference: "Christopher Nolan directed the movie Inception."
- Generated: "The film Inception was directed by Christopher Nolan."
- BLEU would assess how many unigrams, bigrams, etc., match, and penalize if the generated text is too short compared to the reference.

9.1.2.3 BERTScore

BERTScore leverages contextual embeddings from pretrained language models (like BERT) to capture semantic similarity beyond surface-level matching. This is crucial for Graph-RAG systems that may paraphrase graph information in various ways, where exact word overlap (like ROUGE/BLEU) might not reflect true semantic equivalence.

$$\text{BERTScore} = F1(\text{precision\_BERT}, \text{recall\_BERT})$$

BERTScore leverages contextual embeddings to capture semantic similarity beyond surface-level matching.

Here are the mathematical formula definitions:

$$\begin{aligned} P_{\text{BERT}} &= \frac{1}{|x|} \sum_{x_i \in x} \max_{y_j \in y} x_i^T y_j \\ R_{\text{BERT}} &= \frac{1}{|y|} \sum_{y_j \in y} \max_{x_i \in x} x_i^T y_j \\ F_{\text{BERT}} &= 2 \cdot \frac{P_{\text{BERT}} \cdot R_{\text{BERT}}}{P_{\text{BERT}} + R_{\text{BERT}}} \end{aligned}$$

where x_i and y_j are BERT embeddings of tokens.

For example:

- Reference: "The capital of France is Paris."
- Generated: "Paris is the French capital city."
- ROUGE/BLEU might be lower due to different phrasing, but BERTScore would likely be high because the sentences are semantically very similar.

9.1.2.4 Semantic Similarity Scores

Using advanced embedding models (e.g., Sentence-BERT, OpenAI embeddings) to measure the overall semantic alignment between the generated content and both the query intent and the retrieved graph information.

This typically involves computing the cosine similarity between the embeddings of the generated text and a reference answer or the relevant retrieved graph context.

$$\text{CosineSimilarity}(\mathbf{A}, \mathbf{B}) = \frac{\mathbf{A} \cdot \mathbf{B}}{\| \mathbf{A} \| \cdot \| \mathbf{B} \|}$$

where A and B are sentence embeddings.

For example: An embedding model could confirm that "Leonardo DiCaprio acted in the movie" is semantically similar to KG triples like (Leonardo DiCaprio)-[ACTED_IN]->(Movie).

9.1.2.5 Factual Consistency

A critical metric for Graph-RAG that measures whether generated statements are truly supported by the retrieved graph, directly addressing the hallucination problem. This often involves fine-grained fact-checking.

Entity Consistency: Are all mentioned entities in the generated text present and correctly identified in the retrieved graph? For example: If the answer states "Nolan directed Interstellar," check if "Interstellar" and "Nolan" exist as entities in the retrieved subgraph.

Relationship Accuracy: Are stated relationships (e.g., "X works for Y," "A is a type of B") explicitly supported by edges in the retrieved graph? For example: If the answer claims "Inception is a horror film," verify if a (Inception)-[HAS_GENRE]->(Horror) edge exists.

Attribute Correctness: Are entity properties (attributes) accurately represented based on the graph data? For example: If the answer states "Nolan's birth year is 1970," check the (Christopher Nolan)-[BIRTH_YEAR]->(1970) attribute in the graph.

Here are the mathematical formula definitions:

- Entity consistency:

$$\text{EC} = \frac{\sum_{e \in \text{Entities}_{\text{generated}}} \mathbb{1}[e \in G]}{|\text{Entities}_{\text{generated}}|}$$

- Relationship accuracy:

$$\text{RA} = \frac{\sum_{(s,p,o) \in \text{Triples}_{\text{generated}}} \mathbb{1}[(s, p, o) \in G]}{|\text{Triples}_{\text{generated}}|}$$

- Attribute correctness:

$$\text{AC} = \frac{\sum_{(e,a,v) \in \text{Attributes}_{\text{generated}}} \mathbb{1}[G(e, a) = v]}{|\text{Attributes}_{\text{generated}}|}$$

- Overall factual consistency:

$$\text{FC} = \omega_1 \cdot \text{EC} + \omega_2 \cdot \text{RA} + \omega_3 \cdot \text{AC}$$

where $\omega_1 + \omega_2 + \omega_3 = 1$

Figure 9.4 illustrates a simple example of a factual consistency check using KGs. For the given generated claims, the factual consistency score: $3/4 = 0.75$ as 3 out of 4 claims are correct, based on the information from the KG.

9.1.3 Graph-specific Metrics

These metrics evaluate how effectively the Graph-RAG system leverages the underlying graph structure itself, beyond just retrieving individual elements.

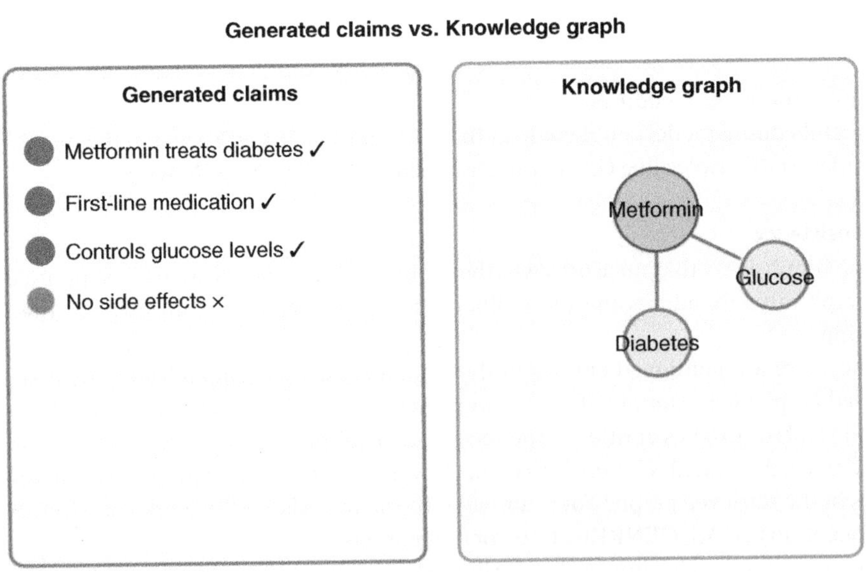

Figure 9.4 Example of factual consistency check in knowledge graphs.

9.1.3.1 Graph Coverage

Graph coverage measures the extent to which the system explores and utilizes relevant portions of the KG. This is important for ensuring the system doesn't get stuck in local optima or miss broader contexts.

Breadth Coverage: The number of distinct graph regions or connected components accessed relative to the total relevant regions. For example: For a query involving multiple sub-domains (e.g., "famous scientists and their discoveries"), does the system access nodes related to different fields (physics, biology) and their associated discoveries?

Depth Coverage: The maximum reasoning depth (number of hops/edges traversed) achieved by the retrieval process, particularly important for multi-hop questions. For example: For a query like "What is the capital of the country where the inventor of the light bulb was born?", this requires several hops: `(Inventor)-[BORN_IN]->(Country)-[HAS_CAPITAL]->(City)`. Depth coverage would measure if all these hops were explored.

Diversity Score: The variety of relationship types (edge labels) traversed and incorporated into the retrieval. A diverse set of relationships often indicates a more nuanced understanding. For example: If a query about a person can involve `HAS_OCCUPATION`, `EDUCATED_AT`, `WORKED_FOR`, `MARRIED_TO`, etc., a higher diversity score means more of these relevant types were used.

Here are the mathematical formula definitions:

- Breadth coverage:

$$\mathrm{BC} = \frac{|\mathrm{UniqueRegions}_{\mathrm{accessed}}|}{|\mathrm{RelevantRegions}_{\mathrm{total}}|}$$

- Depth coverage:

$$\mathrm{DC} = \frac{\max(\mathrm{PathLengths}_{\mathrm{actual}})}{\max(\mathrm{PathLengths}_{\mathrm{required}})}$$

- Diversity score:

$$\mathrm{DS} = -\sum_{r \in \mathrm{RelationTypes}} p(r) \log p(r)$$

where $p(r)$ is the probability of using relation type r.

9.1.3.2 Path Efficiency

Path efficiency evaluates the optimality of graph traversal, particularly for multi-hop reasoning. Systems should ideally find the shortest or most direct relevant path.

Here are the mathematical formula definitions:

- Basic efficiency:

$$\mathrm{Efficiency} = \frac{\mathrm{ShortestPath}_{\mathrm{length}}}{\mathrm{ActualPath}_{\mathrm{length}}}$$

- Weighted path efficiency (WPE):

$$\mathrm{WPE} = \frac{\sum_{e \in \mathrm{ShortestPath}} w(e)}{\sum_{e \in \mathrm{ActualPath}} w(e)}$$

where $w(e)$ is the weight (or cost) of edge e.

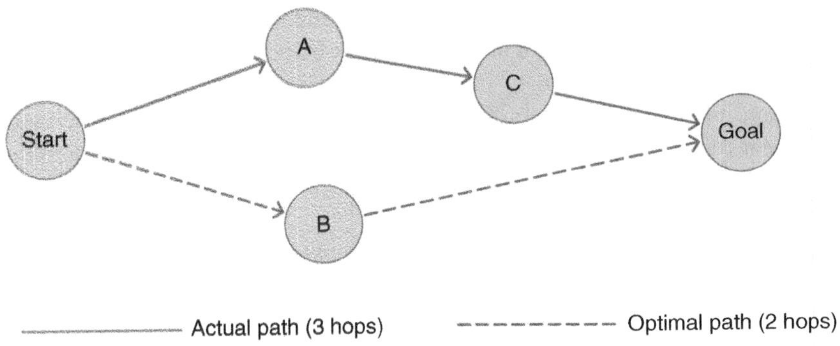

Figure 9.5 Example of path efficiency.

$\text{ShortestPath}_{\text{length}}$ is the length of the shortest path in the ground truth graph that answers the query, and $\text{ActualPath}_{\text{length}}$ is the length of the path retrieved by the system. Values closer to 1 indicate higher efficiency.

Figure 9.5 gives a simple example of computing the path efficiency score, where the length of optimal path is 2, but the length of actual path is 3, so the path efficiency is computed as: Optimal length/actual length = 2/3 = 0.67.

9.1.3.3 Node Importance Score

Based on graph centrality measures, this metric evaluates whether the system prioritizes and retrieves high-value, central, or influential nodes in the KG that are most likely to be relevant.

PageRank-based importance: Nodes with higher PageRank scores are generally more important. The metric would assess if high-PageRank nodes are prioritized in retrieval.

Degree Centrality Weighting: Nodes with a higher degree (more connections) are often more central.

Domain-specific Node Scoring: Some nodes might be intrinsically more important for a specific domain (e.g., "Disease" nodes in a medical KG).

For example: For a query about a scientific field, a Graph-RAG system should ideally retrieve seminal papers or highly cited researchers (high PageRank/degree) rather than obscure works.

Here are the mathematical formula definitions:

- PageRank-based importance:

$$\text{PR}(v) = \frac{1-d}{N} + d \sum_{u \in In(v)} \frac{\text{PR}(u)}{\text{Out}(u)}$$

- Degree centrality:

$$\text{DC}(v) = \frac{\text{degree}(v)}{n-1}$$

- Domain-specific node scoring:

$$\text{NodeScore}(v) = \alpha \cdot \text{PR}(v) + \beta \cdot \text{DC}(v) + \gamma \cdot \text{DomainRelevance}(v)$$

Figure 9.6 gives a simple illustration of the PageRank-based node importance score.

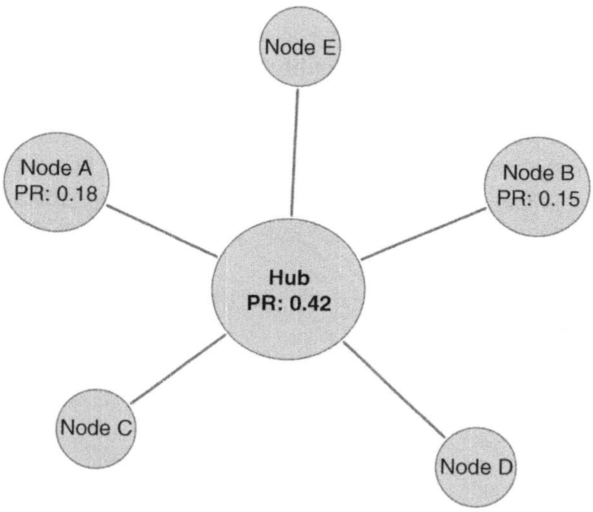

Figure 9.6 Example of PageRank-based node importance.

9.1.3.4 Edge Utilization Rate

Edge utilization rate measures how effectively the system uses different types of relationships (edge labels) within the KG. This indicates the system's ability to leverage the richness of the graph's schema.

Relationship Type Coverage: The percentage of relevant edge types (e.g., `WORKS_FOR`, `HAS_PROPERTY`, `IS_A`) that are utilized in the retrieved subgraph for a given query type. Assessment: Compare the set of edge types in the retrieved subgraph to the set of relevant edge types for the query.

Edge Weight Consideration: For graphs with weighted edges, this assesses if the system correctly prioritizes edges based on their weights (e.g., higher confidence edges).

Temporal Edge Handling: For time-aware graphs, evaluates if the system correctly processes and prioritizes temporal relationships (e.g., `STARTED_ON`, `ENDED_ON`) for time-sensitive queries.

For example: A system answering "What happened to Company X in 2023?" should utilize `ACQUIRED_BY`, `LAUNCHED_PRODUCT`, `REPORTED_EARNINGS` edge types, and potentially temporal edges like `ENDED_ON`.

Here are the mathematical formula definitions:

- Relationship Type Coverage:

$$\mathrm{RTC} = \frac{|\mathrm{UsedRelationTypes}|}{|\mathrm{TotalRelationTypes}|}$$

- Edge Weight Consideration:

$$\mathrm{EWC} = \frac{\sum_{e \in \mathrm{UsedEdges}} w(e) \cdot \mathbb{1}[\mathrm{correct}(e)]}{\sum_{e \in \mathrm{UsedEdges}} w(e)}$$

- Temporal Edge Handling:

$$\mathrm{TEH} = \frac{|\mathrm{TemporallyValidEdges}|}{|\mathrm{TotalTemporalEdges}|}$$

9.1.3.5 KG Completion Metrics

For Graph-RAG systems that also have a component for identifying or proposing missing information (e.g., missing links or entities), these metrics assess that capability.

Link Prediction Accuracy: Measures how accurately the system can predict missing edges (relationships) between existing nodes. This is often evaluated using metrics like Hits@K, Mean Reciprocal Rank (MRR), or AUC for binary classification (whether a link exists). For example: Given `(Entity A)-[?]-(Entity C)`, can the system correctly predict that a `(Entity A)-[WORKS_FOR]->(Entity B)` and `(Entity B)-[LOCATED_IN]->(Entity C)` implies a missing `(Entity A)-[LOCATED_IN]->(Entity C)` link?

Entity Resolution Quality: Measures the system's ability to correctly identify and merge duplicate entities or link mentions to canonical entities in the graph.

Consistency with Existing Graph Structure: Ensures that any proposed additions or modifications to the graph are logically consistent with its existing schema and data.

Here are the mathematical formula definitions:

- Link Prediction Accuracy:

$$\mathrm{LPA} = \frac{|\mathrm{PredictedLinks} \cap \mathrm{TrueLinks}|}{|\mathrm{PredictedLinks}|}$$

- Entity Resolution Quality:

$$\mathrm{ERQ} = \frac{\sum_{(e_1,e_2)\in \mathrm{Pairs}} \mathbb{1}[\mathrm{SameEntity}(e_1, e_2) = \mathrm{Prediction}(e_1, e_2)]}{|\mathrm{Pairs}|}$$

- Triple Classification Accuracy:

$$\mathrm{TCA} = \frac{|\mathrm{CorrectlyClassifiedTriples}|}{|\mathrm{TotalTriples}|}$$

9.1.4 System Performance Metrics

Beyond accuracy, practical deployment of Graph-RAG systems requires attention to computational efficiency, scalability, and resource consumption.

9.1.4.1 End-to-end Latency

End-to-end latency measures the total time from query submission to response generation. This is crucial for user experience and real-time applications. It can be broken down to pinpoint bottlenecks:

Query Processing Time: Time taken to parse the user query, identify entities, and formulate initial graph queries.

Graph Retrieval Latency: Time spent traversing the graph and extracting the relevant subgraph/-paths.

Subgraph Extraction Time: Time to format the retrieved graph information for input to the LLM (e.g., linearization, serialization).

Generation Latency: Time for the LLM to process the prompt and generate the final answer.

Here are the mathematical formula definitions: Latency Breakdown:

$$L_{\mathrm{total}} = L_{\mathrm{query}} + L_{\mathrm{retrieval}} + L_{\mathrm{extraction}} + L_{\mathrm{generation}}$$

Detailed Components:

- L_{query}: Query parsing and embedding time
- $L_{\text{retrieval}}$: Graph search and traversal time
- $L_{\text{extraction}}$: Subgraph extraction and ranking time
- $L_{\text{generation}}$: LLM inference and decoding time

Example Measurement:

```
Query: "What are the side effects of metformin?"
- Query Processing: 12ms
- Graph Retrieval: 45ms
- Subgraph Extraction: 23ms
- Generation: 1,200ms
Total: 1,280ms
```

9.1.4.2 Graph Traversal Efficiency

Graph Traversal Efficiency metrics focus specifically on the efficiency of interacting with the KG.

Nodes Visited Per Query: The average number of nodes explored during graph traversal for a single query. Lower is generally better, indicating more directed search.

Memory Usage During Traversal: The peak memory footprint consumed by the graph traversal algorithm.

Cache Hit Rates for Repeated Patterns: For systems that cache frequently accessed graph patterns or subgraphs, this measures how often a requested item is found in the cache, indicating efficiency gains.

Here are the mathematical formula definitions:

- Nodes Visited Ratio:

$$\text{NVR} = \frac{\text{NodesInFinalSubgraph}}{\text{TotalNodesVisited}}$$

- Memory Efficiency:

$$\text{ME} = \frac{\text{PeakMemoryUsage}}{|V| + |E|}$$

  where $|V|$ and $|E|$ are the numbers of vertices and edges.
- Cache Hit Rate:

$$\text{CHR} = \frac{\text{CacheHits}}{\text{CacheHits} + \text{CacheMisses}}$$

9.1.4.3 LLM Inference Latency

LLM Inference Latency refers to specific latency metrics related to the Large Language Model's performance.

Time for Initial Prompt Processing: Latency incurred by the LLM in processing the input prompt (including the retrieved graph context) before generating the first token.

Token Generation Rate: The speed at which the LLM generates output tokens (words/subwords) per second. Higher is better for faster responses.

Batching Efficiency for Multiple Queries: For systems handling concurrent users, how efficiently the LLM processes multiple queries in batches to maximize throughput.

For example: A powerful LLM might have a low initial prompt processing time (e.g., 500ms) and a high token generation rate (e.g., 30 tokens/second), enabling fast responses.

Here are the mathematical formula definitions:

- Token Generation Rate:

$$\mathrm{TGR} = \frac{\mathrm{TokensGenerated}}{\mathrm{GenerationTime}}$$

- First Token Latency:

$$\mathrm{FTL} = T_{\mathrm{firsttoken}} - T_{\mathrm{queryreceived}}$$

- Batching Efficiency:

$$\mathrm{BE} = \frac{\mathrm{ThroughputBatched}}{\mathrm{ThroughputSingle} \times \mathrm{BatchSize}}$$

9.1.4.4 Resource Utilization

Resource Utilization measurements of hardware and software resources consumed by the Graph-RAG system.

CPU/GPU Usage Patterns: The percentage of CPU or GPU cores utilized during different phases of the query process.

Memory Footprint: The amount of RAM or GPU memory consumed by the system components (graph database, GNN models, LLM).

Network Bandwidth: For distributed graph databases or remote LLM APIs, the amount of data transferred over the network.

Here are the mathematical formula definitions:

- Computational Efficiency:

$$\mathrm{CE} = \frac{\mathrm{QueriesProcessed}}{\mathrm{CPUHours} \times \mathrm{Cost}_{\mathrm{CPU}} + \mathrm{GPUHours} \times \mathrm{Cost}_{\mathrm{GPU}}}$$

- Memory Footprint:

$$\mathrm{MF} = \frac{\mathrm{PeakMemoryUsage}}{\mathrm{GraphSize} + \mathrm{ModelSize}}$$

- Network Efficiency (for distributed systems):

$$\mathrm{NE} = \frac{\mathrm{UsefulDataTransferred}}{\mathrm{TotalDataTransferred}}$$

These comprehensive metrics provide a holistic view of a Graph-RAG system's performance, enabling developers to identify strengths, pinpoint weaknesses, and make informed decisions for continuous improvement and successful deployment.

9.2 Quality Assessment Methodologies

While quantitative metrics provide objective measurements, comprehensive evaluation of Graph-RAG systems requires a multifaceted approach combining human judgment, automated quality checks, and meta-evaluation strategies. This section presents a complete framework for assessing the nuanced aspects of system performance that traditional metrics may overlook.

9.2.1 Human Evaluation Framework

Human evaluation remains the gold standard for assessing aspects of Graph-RAG systems that automated metrics struggle to capture. The complexity of graph-based reasoning and natural language generation demands expert judgment to ensure quality, relevance, and interpretability. Figure 9.7 illustrates the general workflow of human evaluations of the Graph-RAG system outputs.

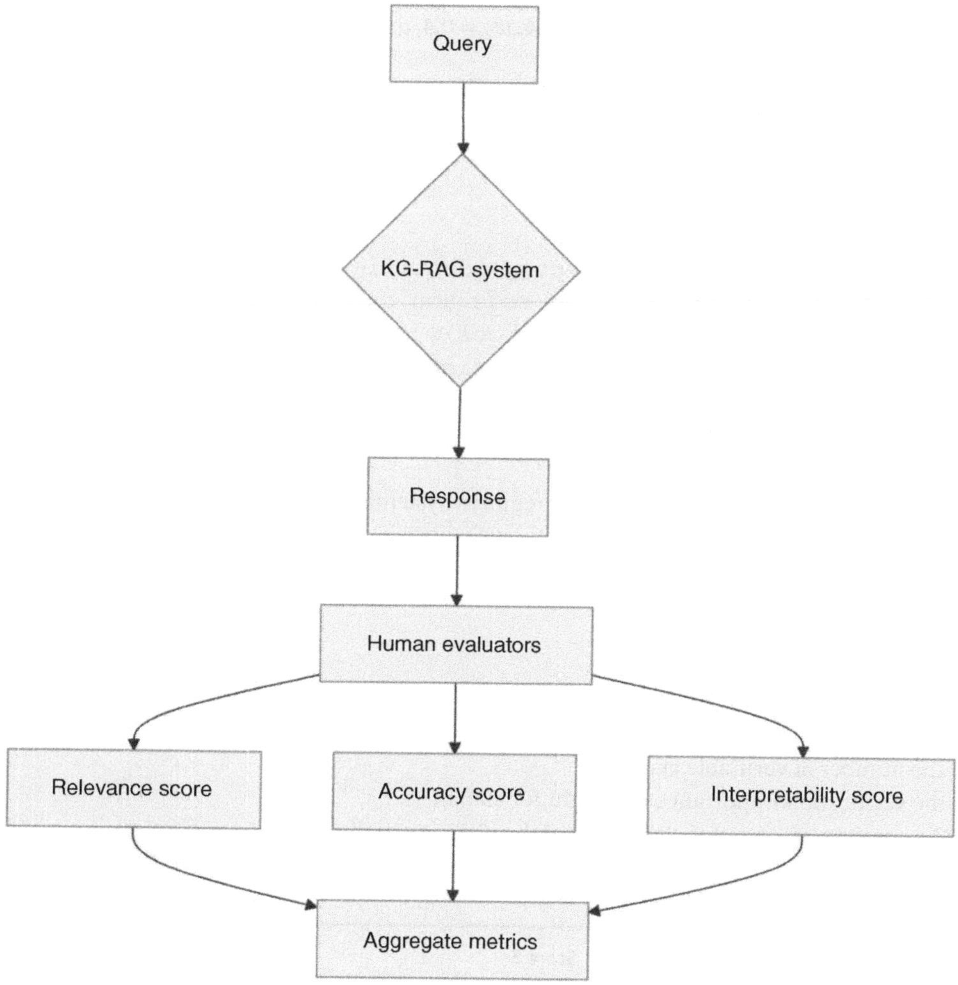

Figure 9.7 Human evaluation workflow.

9.2.1.1 Relevance Assessment

Human evaluators assess whether responses genuinely address the query intent through multiple dimensions:

Query Understanding Score (QUS)

$$\text{QUS} = \frac{1}{n}\sum_{i=1}^{n}(w_1 \cdot \text{Intent}_i + w_2 \cdot \text{Coverage}_i + w_3 \cdot \text{Precision}_i)$$

where

- Intent_i is the score for correct interpretation of query intent (0–1)
- Coverage_i is the degree to which all query aspects are addressed (0–1)
- Precision_i is the absence of irrelevant information (0–1)
- w_1, w_2, w_3 are the weights (typically $w_1 = 0.4$, $w_2 = 0.4$, $w_3 = 0.2$).

Table 9.1 lists an evaluation rubric example:

Example Assessment:

Query: "What are the cardiovascular side effects of diabetes medications that are safe for elderly patients with kidney disease?"

Response Assessment:

```
Intent Understanding: 4/5 (Correctly identifies triple constraint)
Coverage: 3/5 (Addresses diabetes meds and cardiovascular,
        limited on elderly/kidney aspects)
Precision: 5/5 (No irrelevant information)
QUS = 0.4(4) + 0.4(3) + 0.2(5) = 3.8/5
```

9.2.1.2 Factual Accuracy

Expert verification of information correctness requires systematic assessment: Factual Accuracy Score (FAS)

$$\text{FAS} = \frac{\sum_{i=1}^{m} v_i \cdot c_i}{\sum_{i=1}^{m} v_i}$$

where

- m is the number of verifiable claims,
- v_i is the verification importance of weight for claim i
- c_i is the correctness score for claim i (0 = false, 0.5 = partially correct, 1 = correct).

Table 9.1 Example rubric for relevance assessment.

Criterion	Score 5	Score 3	Score 1
Query understanding	Perfect interpretation, captures implicit needs	Correct main intent, misses nuances	Misunderstandings query
Completeness	All facets thoroughly addressed	Main aspects covered, some gaps	Major omissions
Focus	Laser-focused on query	Mostly relevant, minor tangents	Significant irrelevant content

Hallucination detection rate (HDR)

$$\text{HDR} = \frac{|\text{Claims}_{\text{hallucinated}}|}{|\text{Claims}_{\text{total}}|}$$

Source attribution accuracy (SAA)

$$\text{SAA} = \frac{|\text{Claims}_{\text{correctlyattributed}}|}{|\text{Claims}_{\text{requiringattribution}}|}$$

Evaluation Protocol:

- Claim Extraction: Identify all factual statements
- Source Verification: Check against KG
- Hallucination Flagging: Mark unsupported claims
- Attribution Check: Verify source citations

Example Fact-checking: Generated Claim: "Metformin reduces HbA1c by 1–2% on average"
Graph Evidence: Metformin, `reduces_HbA1c_by`, "1.0–2.0%"
Verification: ✓ Correct ($c_i = 1.0$)
Generated Claim: "Safe for all kidney disease stages"
Graph Evidence: Metformin, `contraindicated_in`, "eGFR < 30"
Verification: ✗ Incorrect ($c_i = 0.0$)

9.2.1.3 Coherence and Readability

Linguistic quality assessment using structured evaluation: Coherence Score (CS)

$$\text{CS} = \alpha \cdot \text{LocalCoherence} + \beta \cdot \text{GlobalCoherence} + \gamma \cdot \text{Cohesion}$$

where

- LocalCoherence is the sentence-to-sentence flow (0–1),
- GlobalCoherence is the overall narrative structure (0–1),
- Cohesion is the use of connecting devices (0–1), and
- $\alpha + \beta + \gamma = 1$ (typically $\alpha = 0.3, \beta = 0.5, \gamma = 0.2$)

Readability Metrics:

- Flesch Reading Ease: $206.835 - 1.015 \times \frac{\text{words}}{\text{sentences}} - 84.6 \times \frac{\text{syllables}}{\text{words}}$
- Gunning Fog Index: $0.4 \times \left(\frac{\text{words}}{\text{sentences}} + 100 \times \frac{\text{complexwords}}{\text{words}}\right)$

Technical Appropriateness Score (TAS)

$$\text{TAS} = \frac{1}{n}\sum_{i=1}^{n} \text{match}(\text{TechnicalLevel}_i, \text{AudienceExpectation}_i)$$

Figure 9.8 lists an example illustration of multidimensional assessment of the coherence and readability scores.

9.2.1.4 Graph Path Interpretability

Unique to Graph-RAG systems, this evaluates reasoning transparency:
Path Interpretability Index (PII):

$$\text{PII} = w_1 \cdot \text{Transparency} + w_2 \cdot \text{Justification} + w_3 \cdot \text{Presentation}$$

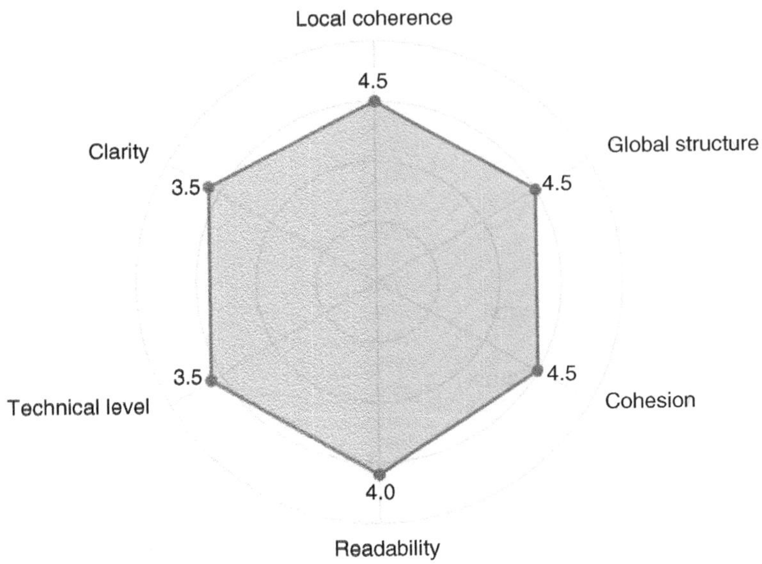

Figure 9.8 Example illustration of multidimensional assessment.

Transparency Score:

$$\text{Transparency} = \frac{|\text{ExplainedSteps}|}{|\text{TotalReasoningSteps}|}$$

Justification Quality:

$$\text{Justification} = \frac{1}{p}\sum_{i=1}^{p}(\text{Logic}_i \times \text{Relevance}_i)$$

where p is the number of path segments.

Example Path Evaluation: Path: `Patient -> has_condition -> Diabetes -> contraindicates -> Drug_X -> alternative -> Drug_Y -> safe_for -> Elderly`

Transparency: 4/5 (Most steps explained)

Justification: 5/5 (Clear medical reasoning)

Presentation: 3/5 (Could be simplified)

$\text{PII} = 0.4(4) + 0.4(5) + 0.2(3) = 4.2/5$

9.2.2 Automated Evaluation Methods

Automated methods complement human evaluation by providing scalable, consistent assessment across large datasets.

9.2.2.1 Model-based Evaluators

Leveraging LLMs as judges has emerged as a powerful evaluation paradigm. Here is an example pseudo code for utilizing llm-as-a-judge:

```
def llm_as_judge(query, response, graph_context, criteria_weights=None):
    """
    Use LLM to evaluate response quality across multiple dimensions.

```

```
    Args:
        query: Original user query
        response: Generated response
        graph_context: Retrieved graph information
        criteria_weights: Dict of criterion -> weight mappings

    Returns:
        Dictionary of scores and explanations
    """
    if criteria_weights is None:
        criteria_weights = {
            'relevance': 0.3,
            'accuracy': 0.4,
            'reasoning': 0.3
        }

    prompt = f"""
    As an expert evaluator, assess the following response:

    Query: {query}

    Graph Context: {graph_context}

    Response: {response}

    Evaluate based on:
    1. Relevance to Query (1-5): Does it address what was asked?
    2. Factual Accuracy (1-5): Is it consistent with the graph context?
    3. Reasoning Validity (1-5): Are the logical steps sound?

    For each criterion:
    - Provide a score (1-5)
    - Give specific evidence for your rating
    - Note any issues or strengths

    Format: JSON with scores, evidence, and overall assessment.
    """

    evaluation = llm.evaluate(prompt)

    # Calculate weighted score
    weighted_score = sum(
        evaluation['scores'][criterion] * weight
        for criterion, weight in criteria_weights.items()
    )

    return {
        'scores': evaluation['scores'],
        'evidence': evaluation['evidence'],
        'weighted_score': weighted_score,
        'recommendations': evaluation.get('improvements', [])
    }
```

Evaluation Consistency Metrics:

$$\text{Consistency}_{\text{LLM}} = 1 - \frac{\sigma_{\text{scores}}}{\mu_{\text{scores}}}$$

where σ_{scores} is the standard deviation of scores for similar examples.

9.2.2.2 Knowledge Consistency Checkers

Automated systems that verify logical consistency: Contradiction Detection Score (CDS)

$$\text{CDS} = 1 - \frac{|\text{Contradictions}|}{|\text{TotalClaims}|}$$

Implementation Example:

```
def detect_contradictions(response, knowledge_graph):
    """
    Detect logical contradictions in generated response.
    """
    claims = extract_claims(response)
    contradictions = []

    for i, claim1 in enumerate(claims):
        for claim2 in claims[i+1:]:
            if are_contradictory(claim1, claim2, knowledge_graph):
                contradictions.append((claim1, claim2))

    # Check against graph constraints
    for claim in claims:
        if violates_graph_constraint(claim, knowledge_graph):
            contradictions.append((claim, "graph_constraint"))

    return {
        'contradiction_score': 1 - len(contradictions) / len(claims),
        'contradictions': contradictions
    }
```

Temporal Consistency Validation:

$$\text{TCV} = \frac{|\text{TemporallyConsistent}|}{|\text{TemporalClaims}|}$$

Example Consistency Check:
Claim 1: "Drug X was approved in 2020"
Claim 2: "Clinical trials for Drug X began in 2021"
Result: Temporal contradiction detected

9.2.2.3 Cross-validation with External Sources

Automated fact-checking against multiple knowledge bases:
Multisource Agreement Score (MSAS)

$$\text{MSAS} = \frac{1}{|C|} \sum_{c \in C} \frac{|\text{Sources}_{\text{agreeing}}(c)|}{|\text{Sources}_{\text{total}}|}$$

Confidence Weighted Agreement (CWA)

$$\text{CWA} = \frac{\sum_{s \in \text{Sources}} \text{confidence}_s \times \text{agreement}_s}{\sum_{s \in \text{Sources}} \text{confidence}_s}$$

Entity Linking Accuracy:

$$\text{ELA} = \frac{|\text{Entities}_{\text{correctlylinked}}|}{|\text{Entities}_{\text{mentioned}}|}$$

9.2.2.4 Graph Structure Validators

Ensuring retrieved subgraphs maintain structural integrity: Structural Validity Score (SVS)

$$\text{SVS} = w_1 \cdot \text{Connectivity} + w_2 \cdot \text{Constraints} + w_3 \cdot \text{Schema}$$

Connectivity Validation:

```
def validate_connectivity(subgraph):
    """
    Ensure subgraph forms connected components as expected.
    """
    components = find_connected_components(subgraph)

    # Single component expected for most queries
    single_component_score = 1.0 if len(components) == 1 else 0.5

    # All query entities should be connected
    query_entities_connected = all(
        are_connected(e1, e2, subgraph)
        for e1, e2 in query_entity_pairs
    )

    return {
        'connectivity_score': 0.7 * single_component_score +
                              0.3 * int(query_entities_connected),
        'num_components': len(components)
    }
```

Constraint Satisfaction:

$$\text{ConstraintScore} = \frac{|\text{Satisfiedconstraints}|}{|\text{Totalconstraints}|}$$

Schema Compliance:

$$\text{SchemaCompliance} = \prod_{e \in \text{Edges}} \text{valid}(\text{type}(e), \text{schema})$$

9.2.3 Meta-evaluation

Evaluating the evaluation process itself ensures reliability and validity of assessments.

9.2.3.1 Inter-annotator Agreement

For human evaluation reliability:

Cohen's Kappa (for binary judgments):

$$\kappa = \frac{p_o - p_e}{1 - p_e}$$

where

- p_o is the observed agreement probability and
- p_e is the expected agreement probability by chance.

Interpretation:

- $\kappa < 0.2$: Poor agreement
- $0.2 \leq \kappa < 0.4$: Fair agreement
- $0.4 \leq \kappa < 0.6$: Moderate agreement
- $0.6 \leq \kappa < 0.8$: Substantial agreement
- $\kappa \geq 0.8$: Almost perfect agreement

Krippendorff's Alpha (for ordinal ratings):

$$\alpha = 1 - \frac{D_o}{D_e}$$

where

- D_o is the observed disagreement and
- D_e is the expected disagreement

Weighted Agreement for Multi-rater Scenarios:

$$W_{\text{agreement}} = \frac{\sum_{i,j} w_{ij} \times \text{agree}_{ij}}{\sum_{i,j} w_{ij}}$$

where w_{ij} represents the expertise weight for rater pair (i, j).

9.2.3.2 Evaluator Calibration

For automated evaluators:

Human–AI Correlation:

$$\rho_{\text{human-AI}} = \frac{\text{cov}(H, AI)}{\sigma_H \sigma_{AI}}$$

Consistency Across Perturbations:

$$\text{Robustness} = 1 - \frac{1}{n}\sum_{i=1}^{n} |\text{score}_{\text{original}} - \text{score}_{\text{perturbed}_i}|$$

Calibration Error:

$$\text{ECE} = \sum_{m=1}^{M} \frac{|B_m|}{n} |\text{acc}(B_m) - \text{conf}(B_m)|$$

Where B_m are confidence bins.

9.2.3.3 Meta-evaluation Framework

Evaluation Quality Index (EQI):

$$\text{EQI} = \sqrt[4]{\text{Reliability} \times \text{Validity} \times \text{Consistency} \times \text{Coverage}}$$

Example Meta-evaluation Report:
Evaluation Method: LLM-as-Judge + Human Verification

Dataset: 1,000 Graph-RAG responses
Reliability Metrics:

- Inter-rater Agreement (Cohen's κ): 0.78
- Test-retest Reliability: 0.85
- Internal Consistency (Cronbach's α): 0.91

Validity Metrics:

- Correlation with Human Experts: 0.82
- Predictive Validity: 0.76
- Content Coverage: 0.95

Overall EQI: 0.84 (High Quality)

9.2.4 Best Practices for Quality Assessment

Ensuring the quality and reliability of Graph-RAG systems requires a robust and multifaceted evaluation framework. While automated metrics can offer fast and scalable insights, they often fall short in capturing the nuanced aspects of factuality, coherence, and domain-specific correctness. Thus, the most effective approach combines both algorithmic and human judgment in a hybrid evaluation strategy.

The hybrid strategy begins with automated methods, which serve as an efficient first pass to screen large volumes of generated responses. These evaluators—often based on classifier models or rule-based heuristics—can flag potential issues in fluency, factuality, or coverage. However, automated evaluations are only as good as their design and training data. To ensure their validity, human evaluators should assess a representative sample of outputs, not only to provide high-quality feedback but also to calibrate and benchmark the automated metrics. This hybrid loop ensures that the automated evaluators remain aligned with real-world expectations and domain standards.

Equally important is the use of domain-specific evaluation rubrics. A one-size-fits-all metric often overlooks what matters most in different use cases. For instance, in medical applications, factual accuracy and safety-related implications are paramount, while in legal domains, correctness of cited precedents and logical consistency of arguments take precedence. In educational systems, clarity of explanation and pedagogical soundness are essential. Tailoring the evaluation framework to the domain ensures that the system is optimized for the criteria that matter most to its end users.

Continuous calibration is necessary to maintain the reliability of evaluations over time. Regular checks on inter-rater agreement among human annotators help surface ambiguities in the rubric or annotation guidelines. Automated evaluators, especially those based on machine learning models, should be retrained periodically to reflect updated knowledge or shifts in usage patterns. An annual review of the overall evaluation framework helps identify blind spots and incorporate new best practices.

Finally, evaluation efficiency should not be overlooked. Given the cost and time associated with human evaluation, intelligent sampling strategies can help optimize resource allocation. For example, an adaptive evaluation strategy can be employed where only those outputs with low confidence scores from the automated evaluator are flagged for human review. Meanwhile, a small portion of high-confidence outputs can still be sampled randomly to serve as a calibration set, ensuring that overconfidence or drift is detected early.

Example pseudo code for adaptive evaluation:

```
def adaptive_evaluation_sampling(responses, confidence_threshold=0.8):
    """
    Adaptively sample responses for human evaluation based on
    automated evaluator confidence.
    """
    auto_evaluations = [auto_evaluate(r) for r in responses]

    human_eval_needed = [
        r for r, e in zip(responses, auto_evaluations)
        if e['confidence'] < confidence_threshold
    ]

    # Also sample high-confidence cases for calibration
    calibration_sample = random.sample(
        [r for r, e in zip(responses, auto_evaluations)
         if e['confidence'] >= confidence_threshold],
        k=int(0.1 * len(responses))
    )

    return human_eval_needed + calibration_sample
```

This holistic assessment strategy ensures that Graph-RAG systems are evaluated not just with numerical scores, but through a thoughtful and evolving framework that reflects the complexities of real-world usage and the diversity of application domains.

9.3 Benchmarking Frameworks

Standardized benchmarks serve as the foundation for fair and reproducible comparison across different Graph-RAG systems. As the field rapidly evolves, comprehensive benchmarking frameworks enable researchers and practitioners to assess progress, identify strengths and weaknesses, and guide future development. This section presents both established and emerging benchmarks, design principles for creating effective evaluation datasets, and frameworks for comparative analysis.

9.3.1 Established Benchmarks

The landscape of Graph-RAG benchmarking has evolved significantly, with several key frameworks emerging as industry standards. Each benchmark targets specific aspects of graph-enhanced retrieval and generation, providing complementary perspectives on system performance.

While the field of Graph-RAG is rapidly evolving, several benchmarks from related areas (knowledge graph question answering, general RAG) have laid the groundwork, and new ones are emerging specifically for Graph-RAG.

9.3.1.1 Semi-structured Knowledge Base Retrieval

Semi-structured knowledge base retrieval (STARK) [1] represents one of the most comprehensive benchmarks for evaluating LLM retrieval on semi-structured knowledge bases. Released in 2024, it addresses the critical gap in evaluating systems that combine structured graph data with unstructured text.

Figure 9.9 illustrates the STARK workflow shown in its original paper "STARK: Benchmarking LLM Retrieval on Textual and Relational Knowledge Bases."

STARK covers various domains to test different aspects of Graph-RAG:

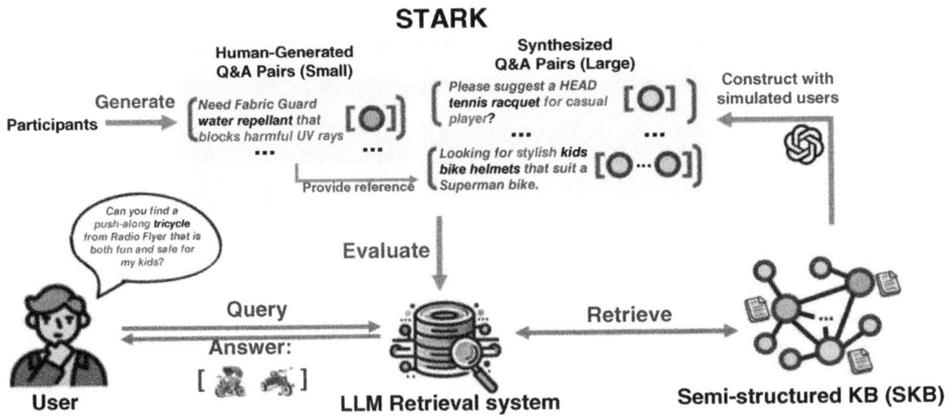

Figure 9.9 Illustration of STARK workflow from its original paper. *Source*: [1]/Neural Information Processing Systems Foundation, Inc.

- Product Search: E-commerce KGs, testing entity matching and attribute retrieval (e.g., "Find phones under $500 with 128GB storage").
- Academic Paper Search: Citation networks and paper metadata, assessing the ability to navigate complex scholarly relationships (e.g., "Which papers by [Author X] cite [Paper Y]?").
- Precision Medicine: Biomedical KGs, evaluating complex relationship reasoning for medical concepts (e.g., "What genes are associated with [Disease A] and what drugs target them?"). Each domain tests different aspects of Graph-RAG capabilities, from simple entity matching to complex multi-hop relationship reasoning and attribute filtering.

9.3.1.2 Graph Reasoning Benchmark

Graph Reasoning Benchmark (GRBENCH) [2] is constructed specifically for augmenting LLMs with graphs, providing a structured dataset for evaluating graph reasoning capabilities in RAG contexts. It has been used in recent research to evaluate LLM reasoning strategies grounded in KGs.

- Scale: Features 1,740 questions across 10 diverse domain graphs, providing a substantial dataset for robust evaluation.
- Reasoning Categorization: Questions are explicitly categorized by reasoning type (e.g., direct fact, multi-hop, aggregation, comparison), allowing for targeted evaluation of different reasoning capabilities.
- Difficulty Levels: Covers a spectrum of difficulty, from single-hop factual queries to complex multi-hop inference questions, assessing a system's ability to handle varying complexities.

9.3.1.3 Comprehensive RAG

Comprehensive RAG (CRAG) [3] provides structured query datasets with mock APIs, enabling controlled experimental conditions for fair comparison across different RAG systems, including those that might leverage KGs.

- Simulated KG Access: Allows researchers to test systems against a predefined, simulated KG environment, abstracting away real-world database complexities.
- Controlled Experimental Conditions: Facilitates A/B testing and systematic analysis of different retrieval and generation strategies by ensuring consistent input data and evaluation setups.
- Fair Comparison: Promotes unbiased evaluation by providing standardized datasets and evaluation scripts, reducing variability in results due to implementation differences.

9.3.1.4 GraphRAG-Bench

GraphRAG-Bench [4] is a large-scale, domain-specific benchmark meticulously designed to rigorously evaluate Graph-RAG models, particularly focusing on challenging domain-specific questions requiring multi-hop reasoning.

- Multi-format QA: Includes multiple-choice, true/false, multi-select, open-ended, and fill-in-the-blank questions to assess various response formats.
- College-level Tasks: Questions are designed to be college-level and domain-specific, ensuring that simple content retrieval is insufficient.
- Rationale Coherence: Evaluates not just the final answer but also the logical coherence of the reasoning process through expert-crafted rationales.
- Full-pipeline Assessment: Provides a comprehensive evaluation framework covering graph construction, knowledge retrieval, and answer generation stages of the Graph-RAG pipeline.

9.3.1.5 Education QA over Graphs

Education QA over graphs (TutorQA) [5, 6] is a benchmark specifically tailored for question answering in educational contexts, leveraging KGs to provide accurate and contextually relevant responses for tutoring systems.

- Multi-hop Reasoning: Questions are designed to require complex reasoning over educational KGs.
- Factual Grounding: Emphasizes the accuracy and verifiability of answers based on the underlying graph data.
- Curriculum-derived Questions: Questions are often derived from real educational curriculums or textbooks, ensuring practical relevance.

9.3.1.6 Factual QA over KGs

Factual QA over KG (FACTKG) [7] is a dataset specifically designed for fact verification using KGs, focusing on assessing factual consistency and hallucination detection in generated text.

- Hallucination Detection: Explicitly designed to test whether claims made in text are supported by facts in a KG.
- Factual Consistency: Evaluates if the generated statements align with the truth encoded in the KG.
- Diverse Reasoning Patterns: Includes various reasoning types such as 1-hop, conjunction, existence, multi-hop, and negation.

9.3.1.7 KQAPro/MetaQA/WebQSP (Classic KGQA)

These are foundational benchmarks in Knowledge Graph Question Answering (KGQA) [8], which, while not exclusively for RAG, are highly relevant for evaluating the core graph reasoning capabilities of Graph-RAG systems. They serve as excellent baselines for measuring improvements.

- KQAPro [9]: A dataset with explicit compositional programs for complex question answering over knowledge bases, emphasizing multi-hop inference, attribute comparison, and set operations.
- MetaQA [10]: A large-scale benchmark for multi-hop question answering over a movie KG, designed to test multi-hop reasoning (1-hop, 2-hop, and 3-hop) and robustness to paraphrasing.
- WebQSP [11]: A benchmark dataset for question answering over Freebase, focusing on structured queries and supporting logical form parsing and knowledge base inference. It is a subset of WebQuestions.

9.3.2 Benchmark Design Principles

Creating effective and meaningful benchmarks for Graph-RAG systems requires careful consideration of various design principles to ensure relevance, fairness, and utility.

9.3.2.1 Test Case Design

Coverage of different reasoning patterns: Benchmarks should include queries that necessitate various types of graph reasoning, such as single-hop lookup, multi-hop path traversal, aggregation (e.g., "total number of employees"), comparison (e.g., "which city is larger?"), temporal reasoning (e.g., "events after 2020"), and inferential reasoning.

Balance between simple and complex queries: A good benchmark includes a mix of straightforward fact retrieval questions and highly complex queries requiring deep graph navigation, allowing systems to be evaluated across their operational range.

Inclusion of adversarial examples: Designing questions that intentionally challenge common retrieval biases or simplistic pattern matching can expose system weaknesses (e.g., queries with negation, implicit knowledge, or distractor information).

Representative domain distribution: Benchmarks should ideally cover multiple knowledge domains (e.g., biomedical, legal, financial, general commonsense) to assess generalizability and domain adaptability.

9.3.2.2 Control Group Setup

To truly understand the value of graph enhancement, benchmarks should facilitate comparative analysis with relevant baselines.

Ablation Studies Removing Graph Components: Evaluating the system's performance when specific graph components (e.g., graph embeddings, relationship types, specific traversal algorithms) are removed or simplified, to understand their individual contributions.

Comparison with Pure LLM Baselines: Direct comparison with an LLM that performs question answering without any external retrieval, highlighting the factual grounding and reasoning benefits of RAG.

Traditional RAG Without Graph Enhancement: Comparing Graph-RAG systems against conventional RAG systems that rely on vector similarity over unstructured text chunks, demonstrating the added value of graph structure.

9.3.2.3 Evaluation Protocols

Consistency in evaluation is paramount for comparability.

Standardized Data Splits: Clearly defined training, validation, and test sets to prevent data leakage and ensure fair evaluation.

Consistent Evaluation Metrics: Utilizing agreed-upon metrics (e.g., those detailed in Section 9.1) and reporting methods across all evaluated systems.

Reproducibility Guidelines: Providing clear instructions, code, and data to allow other researchers to replicate results and verify claims.

Statistical Significance Testing: Employing statistical tests to determine if observed performance differences between systems are truly significant or merely due to chance.

9.3.3 Comparative Analysis Framework

Benchmarking frameworks facilitate structured comparisons across different Graph-RAG paradigms and architectural choices.

9.3.3.1 Traditional RAG vs. Graph-RAG

Key comparison dimensions that highlight the unique advantages of graph-based approaches:

Retrieval Accuracy for Multi-hop Questions: Graph-RAG often excels here due to explicit relational paths.

Factual Consistency in Responses: Graph-grounded responses can reduce hallucinations compared to purely text-based RAG.

Explanation Quality and Transparency: Graph-RAG can potentially offer interpretable reasoning paths, enhancing trust and explainability.

Computational Overhead: Assessing the added latency and resource consumption due to graph construction, storage, and traversal compared to simpler vector retrieval.

9.3.3.2 Graph Retrieval Method Comparison

Evaluating different strategies for extracting relevant graph context:

Embedding-based vs. Traversal-based Retrieval: Comparing GNNs or graph embeddings for similarity search against explicit graph traversal algorithms (e.g., BFS, DFS, shortest path).

Sparse vs. Dense Graph Representations: Assessing the impact of different graph representation techniques on retrieval efficiency and effectiveness.

Static vs. Dynamic Subgraph Extraction: Comparing precomputed subgraphs or graph embeddings versus on-the-fly, query-dependent subgraph extraction.

9.3.3.3 LLM Integration Approaches

Analyzing how the retrieved graph information is presented to and utilized by the LLM:

Prompt-based Graph Injection: Linearizing graph snippets (e.g., triples, paths, summaries) directly into the LLM's prompt. This is the most common and flexible approach.

Fine-tuned Graph-aware Models: LLMs that have been specifically fine-tuned on graph-structured data or tasks, potentially learning to better interpret graph representations.

Hybrid Architectures: Combinations of explicit graph processing modules (e.g., GNNs) with LLMs, where the graph module processes and filters information before passing a refined input to the LLM.

By systematically evaluating Graph-RAG systems across these dimensions, benchmarking frameworks play a critical role in advancing the field, guiding research, and informing the development of robust and efficient Graph-RAG solutions for real-world applications.

9.4 Tools and Platforms for Graph-RAG Evaluations

Evaluating Graph-RAG systems effectively requires leveraging a combination of lightweight prototyping tools, production-ready observability platforms, and specialized benchmark kits. Each tool offers unique strengths depending on the development stage—whether for research, deployment, or benchmarking.

9.4.1 Tools and Platforms Overview

In this section, we'll list the major tools and platforms for Graph-RAG evaluation and compare their usages.

9.4.1.1 List of Tools

Several toolkits and platforms are available to assist in the evaluation of Graph-RAG systems:

9.4.1.1.1 Retrieval Augmented Generation Assessment

Retrieval Augmented Generation Assessment (Ragas) is a flexible Python library ideal for research and rapid prototyping of RAG systems. It provides a suite of metrics to evaluate the quality of generated responses and their retrieved contexts without requiring ground truth answers for every metric. Ragas evaluates response faithfulness, context precision, and answer relevancy by integrating with popular LLM frameworks like LlamaIndex, LangChain, and HuggingFace. Its modular design also supports custom metric integration and benchmarking pipelines, making it highly adaptable for various RAG evaluation needs, including those with graph-based components.

- Repository: https://github.com/explodinggradients/ragas
- Documentation: https://docs.ragas.io/

9.4.1.1.2 Arize AI

Arize AI is designed for production-scale monitoring and observability of machine learning models, including LLMs and RAG systems. It offers detailed LLM tracing, comprehensive metric dashboards, and sophisticated alerting capabilities for issues like model drift, hallucinations, and latency spikes. Arize AI excels in maintaining real-time observability and providing root-cause diagnostics, ensuring the reliability and performance of Graph-RAG systems in live environments. It helps teams understand why a system is performing in a certain way in production.

- Website: https://arize.com/
- Documentation: https://docs.arize.com/

9.4.1.1.3 TruLens

TruLens is an open-source evaluation and explainability framework for LLM applications. It offers a "RAG Triad" of metrics (context relevance, groundedness, and answer relevance) specifically designed to assess the quality of RAG pipelines, helping to identify and mitigate hallucinations. TruLens allows developers to instrument their LLM applications to capture traces, apply feedback functions for programmatic evaluation, and visualize performance over time. While not exclusively for Graph-RAG, its focus on context evaluation and groundedness makes it valuable for understanding how well a Graph-RAG system retrieves and utilizes graph-based information.

- Repository: https://github.com/truera/trulens
- Documentation: https://www.trulens.org/

9.4.1.1.4 LangSmith

LangSmith is a LangChain-native evaluation platform developed by the creators of LangChain. It provides a robust environment for prompt tracing, evaluation tracking, and implementing LLM-as-a-judge pipelines. Designed to facilitate experiment-driven development for RAG systems, LangSmith offers rich visualization tools to understand the flow and performance of complex LLM applications, including those incorporating graph retrieval. Its tight integration with LangChain makes it a natural choice for projects built on that framework.

- Website: https://smith.langchain.com/
- Documentation: https://docs.smith.langchain.com/

9.4.1.1.5 DeepEval

DeepEval is an open-source LLM evaluation framework that emphasizes robust and deterministic evaluation, particularly focused on hallucination detection and factuality assessment. It's built to function similarly to a unit testing framework (like Pytest for LLMs), enabling fine-grained inspection of retrieved and generated content. For Graph-RAG, DeepEval is particularly valuable for verifying the grounding of generated answers in the KG, performing edge/node grounding checks, and providing visual feedback on the factual accuracy of LLM outputs based on structured data.

- Repository: https://github.com/confident-ai/deepeval
- Documentation: https://www.confident-ai.com/deepeval

9.4.1.1.6 LLM Eval Harness

Developed by EleutherAI, the LLM Eval Harness (or lm-evaluation-harness) is a widely used open-source framework for benchmark-driven evaluation of LLM outputs across a diverse range of NLP tasks. While general-purpose for LLMs, its modular design allows it to be extended for graph-based

QA tasks by defining custom tasks and metrics. It supports metric customization, controlled testing environments, and various LLM backends (HuggingFace, OpenAI, local models), making it suitable for rigorous academic and research-oriented benchmarking of Graph-RAG components or end-to-end systems.

- Repository: https://github.com/EleutherAI/lm-evaluation-harness
- Documentation: lm-evaluation-harness/tree/main/docs

9.4.1.1.7 GraphRAG-Bench

GraphRAG-Bench is a recently introduced, large-scale, domain-specific benchmark meticulously designed to rigorously evaluate Graph-RAG models. It aims to overcome the limitations of traditional QA datasets by featuring college-level, domain-specific questions that demand multi-hop reasoning and diverse task coverage (multiple-choice, true/false, multi-select, open-ended, fill-in-the-blank). Critically, it provides a holistic evaluation framework covering the entire Graph-RAG pipeline—including graph construction, knowledge retrieval, and answer generation—and assesses not only final answer correctness but also the logical coherence of the reasoning process through expert-crafted rationales.

- Repository: https://github.com/jeremycp3/GraphRAG-Bench
- Documentation (Paper): GraphRAG-Bench: Challenging domain-specific reasoning for evaluating graph retrieval-augmented generation
- Bench Leaderboard: https://deep-polyu.github.io/RAG/

9.4.1.2 Tools Comparisons and Choices

Table 9.2 lists the comparisons of these evaluation tools.

Choose the right tool based on the stage and goals of your project:

For Research Prototypes: Use Ragas for fast setup and common evaluation metrics to quickly validate system behavior and iterate on design. Try DeepEval when domain-specific hallucination control and granular factuality assessment are crucial, especially for structured or graph-based content where grounding is key. Build Custom Pipelines if developing novel retrieval or evaluation strategies that require deep customization beyond existing frameworks.

For Production Deployments: Use Arize AI for real-time monitoring, comprehensive alerting, and robust observability to ensure consistent system health and performance in live environments, catching issues like data drift or unexpected latency. Adopt TruLens or LangSmith for interpretability, experiment tracking, and auditing, allowing you to debug complex RAG flows, trace prompt executions, and continuously improve your RAG system based on user interactions. Integrate LangChain Benchmarks (via LangSmith) when already leveraging LangChain agents and seeking a tightly integrated development and evaluation workflow that provides insights into agent behavior.

For Benchmarking and Competitions: Use GraphRAG-Bench for standardized, challenging, and domain-specific evaluations that cover the entire Graph-RAG pipeline, providing a rigorous comparison across different Graph-RAG approaches. Leverage LLM Eval Harness for reproducible, academic-grade benchmarking of core LLM capabilities and extend it for graph-based QA tasks, enabling controlled testing environments for model comparison. Combine Ragas with custom metrics for flexible dataset evaluations and leaderboard tracking on unique datasets or when proposing novel Graph-RAG architectures.

Table 9.2 Comparison table of evaluation tools.

Tool	Type and pricing	Best use case	Integration targets
Ragas	Open-source library, free	Research and prototyping, rapid RAG iteration	LlamaIndex, LangChain, HuggingFace, various LLM providers
Arize AI	Production platform, paid (free tier available)	Production monitoring, real-time observability, root-cause analysis	Custom LLM pipelines, observability for production systems
TruLens	Open-source library, free (with commercial offerings from TruEra)	Experiment tracking and explainability, Debugging RAG, Hallucination detection	LangChain, LlamaIndex, OpenAI, HuggingFace, Custom LLMs
LangSmith	Development platform, paid (free tier available)	Experiment tracking and explainability, Iterative RAG development, Debugging LangChain apps	LangChain-native, OpenAI, HuggingFace models, Custom LLMs
DeepEval	Open-source library, free (with commercial platform from Confident AI)	Hallucination detection, Graph-based factuality, Unit testing LLMs, Structured QA evaluation	Custom and biomedical Graph-RAGs, Weaviate, LangChain, LlamaIndex, various LLM providers
LLM Eval Harness	Open-source library, free	Benchmarking for QA systems, academic research, model comparison	HuggingFace models, OpenAI APIs, Custom local APIs, NVIDIA Evals Factory
GraphRAG-Bench	Research benchmark, free	Domain-specific Graph-RAG evaluation, multi-hop reasoning, holistic pipeline assessment	Integrates with various Graph-RAG models/frameworks for evaluation

9.4.2 Integration Best Practices

To maximize evaluation quality and development efficiency in Graph-RAG projects:

Start Simple with Foundational Metrics: Begin by establishing a baseline using lightweight tools like Ragas or LangSmith. These provide quick, high-level feedback on answer relevance, faithfulness, and context precision, allowing for rapid iteration on initial Graph-RAG system designs.

Add Graph-specific Metrics and Grounding Checks: As your Graph-RAG system matures, extend your evaluation to explicitly assess the graph's role. Utilize tools like DeepEval to perform fine-grained hallucination detection and verify that generated answers are truly grounded in the retrieved graph components (nodes, edges, paths). Consider custom metrics to evaluate aspects such as graph coverage (how much of the relevant graph was retrieved), path efficiency (shortest relevant paths), and the quality of graph-to-text verbalization.

Monitor in Production with Observability Platforms: Once deployed, transition to production-grade observability platforms like Arize AI or TruLens. Set up real-time monitoring for key LLM and RAG specific issues, including hallucination rates, context drift, retrieval latency, and overall answer quality. These platforms are crucial for identifying performance regressions and understanding the root cause of issues in a dynamic, real-world environment.

Iterate Based on Comprehensive Feedback: Continuously improve your Graph-RAG system by analyzing insights from your evaluation tools. This feedback should directly inform improvements in prompt design, refinement of graph construction and data modeling, optimization of retrieval algorithms (e.g., using more sophisticated GNN architectures for retrieval), and strategies for better integrating graph knowledge into the LLM's generation process.

9.4.3 Future Directions in Tooling

The Graph-RAG evaluation ecosystem is rapidly evolving, driven by the increasing complexity and adoption of these systems. Anticipated developments in tooling include:

Unified Metric Standards and Frameworks: The emergence of common APIs and standardized metric definitions across different evaluation frameworks will foster greater interoperability and comparability of Graph-RAG systems. This will allow researchers and practitioners to more easily share results, build upon existing work, and accelerate progress.

Enhanced Graph Visualization and Explainability: Future tools will offer more sophisticated, interactive interfaces to visualize the Graph-RAG process. This includes tracing the flow of information through the graph, highlighting which nodes, edges, or paths contributed to specific parts of the generated answer, and visually debugging reasoning chains to pinpoint where errors or inconsistencies occur.

Automated Error Detection and Root Cause Analysis: Advanced AI-powered analysis will go beyond simple metric calculation to automatically detect complex errors like subtle hallucinations, broken reasoning steps, and semantic retrieval gaps. These tools will provide deeper insights into why a Graph-RAG system failed, facilitating faster debugging and more targeted improvements.

Cross-modal Evaluation Capabilities: As Graph-RAG expands into multimodal domains (e.g., integrating text, images, and sensor data with graphs), new evaluation tools will emerge to assess cross-modal alignment and consistency. This will involve evaluating how well information is synthesized from heterogeneous sources and whether the generated responses accurately reflect the combined knowledge from different modalities.

These evolving tools not only help quantify performance but also profoundly support iterative design improvements and reliability audits for building increasingly robust and capable Graph-RAG deployments.

9.5 Practice: Evaluating RAG Pipelines Using Ragas

The objective of this section is to evaluate the RAG pipeline built in previous sections that uses Wikipedia pages to answer questions. We will use Ragas,[1] a powerful open-source library for evaluating applications powered by LLMs, to perform this assessment. The evaluation process involves three key steps:

1. Preparing the evaluation dataset.
2. Defining evaluation metrics.
3. Running the evaluation.

9.5.1 Preparing the Evaluation Dataset

To begin, we initialize a test set generator and provide it with text from the Wikipedia webpage in the form of LlamaIndex `Document` objects. The generator then creates synthetic question-answer pairs, which are converted into a format suitable for evaluation with Ragas.

1 https://docs.ragas.io/en/latest/

```
from ragas.testset import TestsetGenerator

generator = TestsetGenerator.from_llama_index(
    llm=Settings.llm, embedding_model=Settings.embed_model
)
testset = generator.generate_with_llamaindex_docs(
    documents,
    testset_size=3,
)                                                        # [1]
ragas_dataset = testset.to_evaluation_dataset()  # [2]
```

[1] The `Testset` object represents the synthetic data generated for testing, including metadata about its creation.

[2] The `EvaluationDataset` object contains the structured input/output pairs used as the ground truth for evaluating the RAG system using Ragas metrics.

The resulting dataset includes four main columns. Here is an example entry:

{
"reference": "2025 is the 25th year of the 3rd millennium. The context does not provide specific information on how the 3rd millennium has affected global conflict patterns or future projections.",
"reference_contexts": [
"2025 (MMXXV) is the current year, and is a common year starting on Wednesday of the Gregorian calendar, the 2025th year of the Common Era (CE) and Anno Domini (AD) designations, the 25th year of the 3rd millennium and the 21st century, and the 6th year of the 2020s decade."
],
"synthesizer_name": "single_hop_specifc_query_synthesizer",
"user_input": "How the 3rd millennium been affect global conflict patterns and what analyst think about it for future?"
}

In this dataset, `reference_contexts` corresponds to the retrieved context, `user_input` is the query posed by the user, and `reference` is the expected (ground truth) answer. This test case was generated using the `single_hop_specific_query_synthesizer`. Ragas provides several built-in synthesizers: `single_hop_specific_query_synthesizer`, `multi_hop_abstract_query_synthesizer`, and `multi_hop_specific_query_synthesizer`. These synthesizers allow you to simulate different types of queries based on the complexity and depth of information retrieval required.

9.5.2 Defining Evaluation Metrics

Next, we define the evaluation metrics that will be used to assess the performance of our RAG pipeline. These metrics are backed by the LlamaIndex LLM, which serves as the evaluator.

```
from ragas.llms import LlamaIndexLLMWrapper
from ragas.metrics import Faithfulness, AnswerRelevancy, ContextPrecision,
    ContextRecall

evaluator_llm = LlamaIndexLLMWrapper(Settings.llm)
metrics = [
    Faithfulness(llm=evaluator_llm),        # [1]
    AnswerRelevancy(llm=evaluator_llm),     # [2]
    ContextPrecision(llm=evaluator_llm),    # [3]
    ContextRecall(llm=evaluator_llm),       # [4]
]
```

[1] Faithfulness: Measures whether the response is factually consistent with the provided context.
[2] Answer Relevancy: Evaluates how relevant the generated answer is to the user's query.
[3] Context Precision: Assesses the relevance of the retrieved contexts used to generate the answer.
[4] Context Recall: Measures how much of the relevant context was successfully retrieved.

9.5.3 Executing the Evaluation Process

Ragas offers a convenient integration with LlamaIndex through a helper function that enables seamless evaluation of the RAG pipeline. We pass the configured query engine, evaluation metrics, and the prepared dataset to this function. The result includes both aggregated metric scores and detailed per-sample results in a DataFrame.

```
from ragas.integrations.llama_index import evaluate

result = evaluate(
    query_engine=query_engine,
    metrics=metrics,
    dataset=ragas_dataset,
    llm=Settings.llm,
    embeddings=Settings.embed_model,
)
display(result)
display(result.to_pandas())
```

An example output might look like this:

{'faithfulness': 0.6492, 'answer_relevancy': 0.5092, 'context_precision': 0.1667, 'context_recall': 0.8333}

These scores offer a multidimensional view of the pipeline's performance. By analyzing these metrics, developers can identify areas for improvement and iteratively refine their RAG systems for better accuracy, relevance, and robustness.

9.5.4 Key Takeaways

- Ragas is an open-source library used to evaluate RAG pipelines. It simplifies the assessment process by generating synthetic datasets and providing key metrics.
- The evaluation process involves three steps: preparing a dataset (often synthetically generated), defining metrics for assessment, and running the evaluation.
- Four core metrics measure different aspects of a RAG pipeline's performance:
  - Faithfulness: Checks if the generated answer is factually consistent with the retrieved context.
  - Answer Relevancy: Measures if the answer is directly relevant to the user's question.
  - Context Precision: Evaluates the relevance of the retrieved documents.
  - Context Recall: Assesses if all necessary information was retrieved from the source documents.

9.6 Summary

This chapter outlines a comprehensive approach to evaluating RAG pipelines, emphasizing the necessity for both quantitative and qualitative assessments. The evaluation begins with performance metrics, which are categorized into retrieval-specific, generation quality, graph-specific, and system latency metrics. These metrics measure the effectiveness of the RAG system in retrieving relevant information, generating coherent and accurate responses, efficiently traversing KGs, and maintaining optimal performance under various workloads.

The quality assessment process involves a dual approach: human evaluation and automated methods. Human evaluation focuses on aspects such as the relevance and factual accuracy of the generated responses, their coherence, readability, and the interpretability of the graph paths. Automated evaluation employs tools to check consistency and validate the structure of the KGs, ensuring reliability and robustness.

Benchmarking frameworks play a crucial role in standardizing evaluations and facilitating comparisons between different RAG systems. Datasets like STARK, GraphRAG-Bench, GRBENCH, and CRAG provide a foundation for testing across various domains, enabling researchers to design test cases, set up control groups, conduct A/B testing, and compare systems comprehensively. The comparative analysis includes traditional RAG versus Graph-RAG systems, different retrieval methods, and various approaches to integrating LLMs. This section also addresses performance trade-offs, such as balancing accuracy with speed or coverage.

Practical evaluation is demonstrated through the use of tools like Ragas, which enhances the evaluation of applications driven by LLMs. This practical approach involves preparing evaluation datasets, defining appropriate metrics, and executing systematic evaluations, as illustrated with the LlamaIndex RAG pipeline for Wikipedia pages. The summary highlights the significance of aligning theoretical frameworks with real-world applications, ensuring that comprehensive evaluation methods lead to the development of robust, reliable, and efficient RAG systems.

References

1 Shirley Wu, Shiyu Zhao, Michihiro Yasunaga, Kexin Huang, Kaidi Cao, Qian Huang, Vassilis Ioannidis, Karthik Subbian, James Y Zou, and Jure Leskovec. Stark: Benchmarking llm retrieval on textual and relational knowledge bases. *Advances in Neural Information Processing Systems*, 37:127129–127153, 2024.

2 Bowen Jin, Chulin Xie, Jiawei Zhang, Kashob Kumar Roy, Yu Zhang, Zheng Li, Ruirui Li, Xianfeng Tang, Suhang Wang, Yu Meng, et al. Graph chain-of-thought: Augmenting large language models by reasoning on graphs. *arXiv preprint arXiv:2404.07103*, 2024.

3 Xiao Yang, Kai Sun, Hao Xin, Yushi Sun, Nikita Bhalla, Xiangsen Chen, Sajal Choudhary, Rongze Gui, Ziran Jiang, Ziyu Jiang, et al. Crag-comprehensive rag benchmark. *Advances in Neural Information Processing Systems*, 37:10470–10490, 2024.

4 Yilin Xiao, Junnan Dong, Chuang Zhou, Su Dong, Qianwen Zhang, Di Yin, Xing Sun, and Xiao Huang. Graphrag-bench: Challenging domain-specific reasoning for evaluating graph retrieval-augmented generation. *arXiv preprint arXiv:2506.02404*, 2025.

5 Rui Yang, Boming Yang, Aosong Feng, Sixun Ouyang, Moritz Blum, Tianwei She, Yuang Jiang, Freddy Lecue, Jinghui Lu, and Irene Li. Graphusion: A rag framework for knowledge graph construction with a global perspective. *arXiv preprint arXiv:2410.17600*, 2024.

6 Rui Yang, Boming Yang, Sixun Ouyang, Tianwei She, Aosong Feng, Yuang Jiang, Freddy Lecue, Jinghui Lu, and Irene Li. Leveraging large language models for concept graph recovery and question answering in NLP education. *arXiv preprint arXiv:2402.14293*, 2024.

7 Jiho Kim, Sungjin Park, Yeonsu Kwon, Yohan Jo, James Thorne, and Edward Choi. Factkg: Fact verification via reasoning on knowledge graphs. *arXiv preprint arXiv:2305.06590*, 2023.

8 Aleksandr Perevalov, Xi Yan, Liubov Kovriguina, Longquan Jiang, Andreas Both, and Ricardo Usbeck. Knowledge graph question answering leaderboard: A community resource to prevent a replication crisis. *arXiv preprint arXiv:2201.08174*, 2022.

9 Shulin Cao, Jiaxin Shi, Liangming Pan, Lunyiu Nie, Yutong Xiang, Lei Hou, Juanzi Li, Bin He, and Hanwang Zhang. KQA Pro: A dataset with explicit compositional programs for complex question answering over knowledge base. *arXiv preprint arXiv:2007.03875*, 2020.

10 Yuyu Zhang, Hanjun Dai, Zornitsa Kozareva, Alexander Smola, and Le Song. Variational reasoning for question answering with knowledge graph. In *Proceedings of the AAAI Conference on Artificial Intelligence*, volume 32, 2018.

11 Wen-tau Yih, Matthew Richardson, Christopher Meek, Ming-Wei Chang, and Jina Suh. The value of semantic parse labeling for knowledge base question answering. In *Proceedings of the 54th Annual Meeting of the Association for Computational Linguistics* (Volume 2: *Short Papers)*, pages 201–206, 2016.

Part IV

Advanced Implementations and Frontiers

As we enter Part IV, we transition from theoretical foundations to the cutting-edge implementations and future horizons of graph-retrieval-augmented generation (Graph-RAG). This section illustrates how Graph-RAG is being actively deployed in real-world information retrieval systems and how it is evolving to meet the demands of increasingly complex tasks and domains.

Chapter 10 explores a diverse set of Graph-RAG frameworks—such as RAPTOR, G-Retriever, Microsoft-GraphRAG, and HippoRAG—that demonstrate novel strategies for integrating knowledge graphs into search and recommendation engines. These systems showcase innovations ranging from hierarchical tree-based abstraction to biologically inspired memory modeling, and include practical implementation guides for datasets like WANDS across both search and recommender scenarios.

Chapter 11 extends the discussion toward the future of Graph-RAG. It investigates frontier topics such as multimodal knowledge graphs, temporal and dynamic graph structures, and next-generation graph neural networks. The chapter also engages with ethical considerations, including fairness and privacy, and presents advanced learning paradigms such as reinforcement learning and federated learning. Finally, it outlines transformative applications in healthcare, finance, education, and smart cities.

Together, these chapters provide both practical tools for today's challenges and a visionary roadmap for tomorrow's innovations, equipping you to harness the full potential of Graph-RAG in an increasingly complex information landscape.

10

Graph-RAG Frameworks for Enhanced Information Retrieval

In Part III, we systematically introduced the motivation behind graph-based retrieval-augmented generation (Graph-RAG) and provided a deep dive into its architectural components, retrieval methods, generation techniques, and evaluation criteria. We examined why traditional retrieval-augmented generation (RAG) pipelines often fall short in scenarios requiring structured reasoning or multi-hop context, and how Graph-RAG methods address these challenges through graph-based indexing, traversal, and prompt augmentation.

In this chapter, we take the next step: demonstrating how Graph-RAG is applied end-to-end in information retrieval, especially in search and recommendation systems. These two pillars of information access are foundational across industries—from e-commerce and media to scientific research and social networks. We will explore the core architectures and innovations of leading-edge algorithms, demonstrating how they can be applied to customized datasets. By examining their open-source implementations and community support, you will see how these advanced frameworks can be seamlessly adapted to your own projects.

We begin with an overview of state-of-the-art Graph-RAG algorithms powering modern search and recommendation systems. We conclude the chapter with a real-world case study using the Wayfair WANDS dataset, showing how to build a Graph-RAG system that enhances e-commerce search with semantic understanding and recommendation with graph-based personalization.

10.1 Overview of Graph-RAG Frameworks for Search and Recommendations

In this section, we dive deep into cutting-edge algorithms that exemplify the power of graph-enhanced RAG in search and recommendation contexts. While KG-RAG specifically refers to systems using knowledge graphs (KGs) with entities and relationships, the broader family of Graph-RAG includes any approach that leverages graph structures to enhance retrieval and generation.

The algorithms presented here represent different points on the Graph-RAG spectrum:

- RAPTOR: Uses hierarchical tree structures for multilevel text abstraction [1]
- G-Retriever: Operates on entity-relationship KGs [2]
- Microsoft Graph-RAG: Combines KGs with community detection for global sensemaking [3]
- HippoRAG: Implements associative graph structures inspired by human memory [4, 5]

Each approach offers unique advantages and can be combined in hybrid systems. For instance, RAPTOR trees could organize textual content about entities in a KG, while G-Retriever handles structured entity relationships. Understanding these complementary approaches enables practitioners to select and combine the right tools for their specific use cases.

10.1.1 RAPTOR: Recursive Abstractive Processing for Tree-organized Retrieval

RAPTOR is a novel retrieval framework introduced to address a core limitation of traditional RAG: the inability to capture and integrate long-range, hierarchical document context [1]. While typical RAG methods retrieve flat chunks of contiguous text, RAPTOR constructs and retrieves from a multilevel tree structure built via recursive summarization, enabling multiscale semantic access to large documents.

This method is particularly useful for question answering (QA) over long-form narratives, scientific papers, or structured documents, where understanding the global discourse—rather than isolated snippets—is essential. RAPTOR demonstrates strong empirical performance, achieving state-of-the-art results across several QA datasets, especially when paired with advanced language models like GPT-4.

10.1.1.1 The Core Innovation: Hierarchical Abstraction

Traditional retrieval methods operate on a single level of granularity—typically retrieving chunks of 100–500 tokens that match a query. This approach, while effective for finding specific facts, fails catastrophically when answering thematic questions that require synthesizing information across an entire document. Consider the question "How did Cinderella reach her happy ending?" The answer isn't contained in any single paragraph but emerges from the narrative arc spanning the entire story.

RAPTOR's breakthrough lies in constructing a multilayered tree structure that mirrors how humans naturally organize and understand information:

Tree Structure Components:

- **Leaf Nodes**: Original text chunks (typically 100 tokens), preserving all granular details and specific facts
- **Intermediate Nodes**: LLM-generated summaries of semantically related clusters, capturing mid-level themes and connections
- **Root Nodes**: High-level thematic summaries representing the document's core concepts and overarching narrative

As illustrated in Figure 10.1 (cited from the original paper): "RAPTOR recursively clusters chunks of text based on their vector embeddings and generates text summaries of those clusters, constructing a tree from the bottom up."

This recursive abstraction process enables RAPTOR to answer questions at any level of specificity—from "What color were Cinderella's slippers?" (detail) to "What is the moral of the story?" (theme).

10.1.1.2 The RAPTOR Construction Process

Building a RAPTOR tree involves a sophisticated pipeline that balances semantic coherence with computational efficiency:

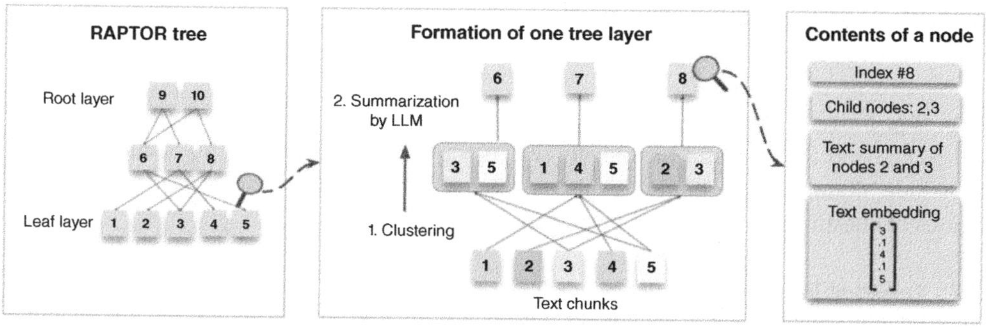

Figure 10.1 Tree construction process of RAPTOR. *Source*: [1]/Open Review.

Step 1: Initial Chunking

For initial chunking, documents are segmented into coherent chunks of ~100 tokens. Sentence boundaries are respected to maintain semantic integrity. Each chunk is embedded using Sentence-bidirectional encoder representations from transformers (SBERT) to create dense vector representations.

Step 2: Intelligent Clustering

RAPTOR employs a unique soft clustering approach using Gaussian mixture models (GMMs):

- Why GMMs? Unlike hard clustering (k-means), GMMs allow chunks to belong to multiple clusters—essential because text often relates to multiple themes.
- UMAP Dimensionality Reduction: High-dimensional embeddings are projected to lower dimensions to improve clustering quality.
- Hierarchical Clustering: The algorithm first identifies global clusters, then performs local clustering within each global cluster.
- Automatic Cluster Selection: Bayesian information criterion (BIC) determines the optimal number of clusters without manual tuning.

Step 3: Recursive Summarization

Nodes within each cluster are sent to a large language model (LLM) (e.g., GPT-3.5-turbo) for summarization. The summarization prompt is carefully designed to preserve key information while abstracting details. Summaries maintain semantic relationships between disparate text segments that traditional retrieval would miss.

Step 4: Tree Building

Summarized nodes are re-embedded and become candidates for the next level of clustering. The process repeats until further meaningful clustering becomes impossible. The result is a balanced tree where height indicates abstraction level and breadth captures thematic diversity.

For computational efficiency, RAPTOR scales linearly with document size in both time and token cost, making it practical for large corpora. A 50,000-token document typically produces a 3–4 level tree with ~100–200 total nodes.

10.1.1.3 Retrieval Strategies: Navigating the Tree

RAPTOR offers two retrieval strategies, each optimized for different query types:

10.1.1.3.1 Tree Traversal Method

This method starts at the root and recursively selects top-k most relevant nodes at each level. Its benefits include: maintaining a hierarchical context and being efficient for focused queries. It is best for cases of questions with clear scope and specificity levels.

10.1.1.3.2 Collapsed Tree Method (Recommended)

This method flattens all tree nodes into a single layer for simultaneous comparison. It selects nodes based on relevance until reaching a token budget (typically 2,000 tokens). It has the benefits of maximum flexibility and better performance on diverse query types. It is best for cases where open-ended questions require information at multiple abstraction levels.

Figure 10.2 (cited from the original paper) compares two retrieval approaches employed within hierarchical retrieval frameworks. In the first approach, tree traversal begins at the root and incrementally selects the top-k nodes at each level based on their cosine similarity to the query embedding—effectively navigating the hierarchy layer by layer. In contrast, the collapsed tree strategy flattens the entire tree structure into a single layer and ranks all candidate nodes globally, selecting those that best match the query until a token budget is met.

Empirical evaluation showed the collapsed tree method outperformed tree traversal across multiple benchmarks, as it dynamically selects the appropriate level of abstraction for each query.

10.1.1.4 Implementation Considerations

RAPTOR excels with document collections containing rich thematic structures—technical documentation, research papers, or narrative content—particularly when handling large documents (>10k tokens) where traditional chunking loses critical context. It's ideal for queries requiring both granular details and high-level insights, but comes with trade-offs: tree construction requires preprocessing time, storage increases by ~1.5–2×, and the recursive summarization can introduce minor information loss (~4% hallucination rate). The system works best with relatively static corpora since frequent updates require expensive tree reconstruction. For example, use small initial chunks (~100 tokens), domain-specific embeddings when available, GPT-3.5-turbo for cost-effective summarization, a 2,000-token retrieval budget, and aim for 3–4 tree levels.

Most exciting is RAPTOR's potential in hybrid architectures with traditional KG-RAG systems. Imagine KG nodes linking to RAPTOR trees containing relevant documents about each entity—a

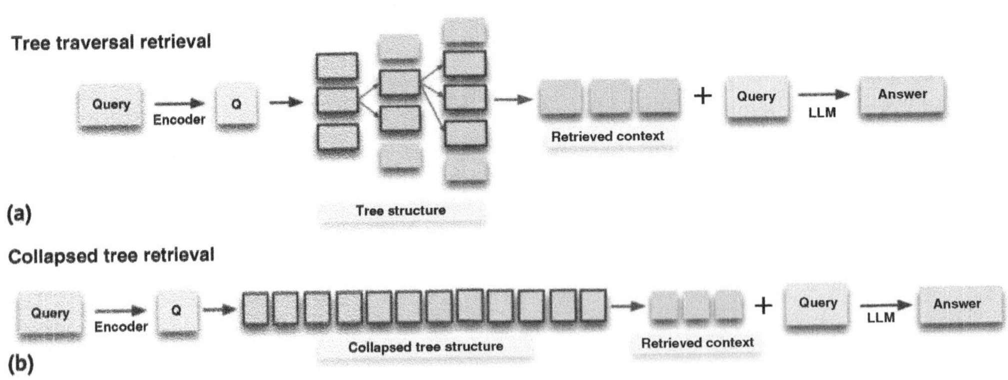

Figure 10.2 Illustration of the tree traversal and collapsed tree retrieval mechanisms in RAPTOR. *Source:* [1]/Open Review.

query about "Tesla's sustainability initiatives" could traverse the KG for structured facts (environmental certifications) while RAPTOR retrieves thematic insights from sustainability reports and news articles. This combination leverages RAPTOR's document understanding alongside KG-RAG's relationship reasoning, creating systems that seamlessly blend structured and unstructured information at every level of abstraction.

In summary, RAPTOR's tree-based architecture aligns closely with the way humans process large documents: summarizing details into themes, navigating structures, and selecting the appropriate abstraction level to answer complex queries. As such, RAPTOR is an important building block in the evolution of Graph-RAG systems and serves as a blueprint for hierarchical knowledge retrieval in LLM applications.

10.1.2 G-Retriever: Joint Text-graph Embeddings

G-Retriever is a novel RAG framework designed to enable large LLMs to reason over real-world graphs with rich textual attributes [2]. While prior work has explored combining LLMs with graph neural networks (GNNs), most methods either target narrow graph tasks (e.g., node classification) or struggle to scale to large, complex graphs—particularly those found in e-commerce environments, such as product graphs, user-item interaction graphs, or knowledge-enriched scene graphs.

10.1.2.1 Architecture and Core Components

G-Retriever combines the strengths of GNNs, LLMs, and RAG to create a powerful search and recommendation system. The framework operates through four main steps: indexing, retrieval, subgraph construction and generation.

Figure 10.3 (cited from He et al. [2]) presents a step-by-step overview of the G-Retriever architecture, highlighting how it integrates retrieval, graph optimization, and language generation to support natural language QA over textual graphs.

Step 1: Indexing

In the first step, G-Retriever prepares the graph for efficient retrieval. The graph consists of nodes (e.g., people, objects) and edges (e.g., relations such as parent, sibling, or profession), where both nodes and edges contain natural language descriptions.

These textual attributes are encoded into vector embeddings using a frozen language model (e.g., SBERT), producing a semantically meaningful representation for every node and edge. The resulting embeddings are stored in a retrieval index that supports fast similarity search.

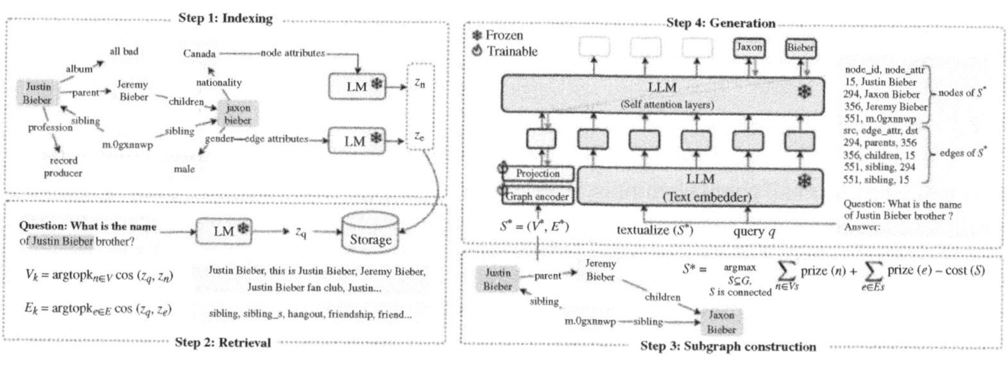

Figure 10.3 Overview of G-Retriever architecture and components.

Step 2: Retrieval

When a user submits a query, the system encodes the query into a vector using the same embedding model as in Step 1.

G-Retriever then computes the cosine similarity between the query embedding and the stored node and edge embeddings computed in Step 1. It retrieves the top-k nodes and edges that are most relevant to the query, forming a candidate set of semantically related components.

For instance, for a query "what is the name of Justin Bieber's brother?," we apply the retrieval process. The retrieved results may include entities like "Jeremy Bieber" and "Jaxon Bieber," and relationships such as "sibling" or "children."

Step 3: Subgraph Construction

To ensure the retrieved nodes and edges form a coherent structure, G-Retriever constructs a connected subgraph from the candidate set. This is achieved through a Prize-collecting Steiner Tree (PCST) optimization. Each retrieved node and edge is assigned a prize based on its similarity to the query.

The PCST algorithm selects a connected subgraph $S^* = (V^*, E^*)$ that maximizes the total prize of the included components while minimizing the cost of connecting them. This balances relevance and connectivity, resulting in a compact subgraph that captures the semantic core of the query without overwhelming the LLM's input capacity.

Step 4: Generation

Finally, the selected subgraph is converted into a textual format and fed into an LLM, alongside the original question. A separate graph encoder processes the structure of the subgraph and produces a graph embedding, which is projected into the same latent space as the LLM's hidden states. This is combined with the textualized graph using soft prompt tuning (PT). The LLM, kept frozen, generates an answer grounded in both the question and the retrieved subgraph.

For instance, given the subgraph containing "Justin Bieber," "Jaxon Bieber," and "sibling" relations, the LLM can accurately answer "Jaxon Bieber."

Overall, G-Retriever offers a robust and explainable pipeline for grounding LLM responses in large, text-rich graphs. By retrieving a focused subgraph tailored to the user query, it improves answer accuracy and reduces hallucination, making it highly applicable for structured domains like e-commerce, knowledge bases, and scene graphs.

10.1.2.2 PCST for Graph Retrieval

G-Retriever's core innovation lies in its use of PCST optimization to retrieve relevant, connected subgraphs from large textual graphs. Below, we break down how PCST works and why it's ideal for Graph-RAG applications.

10.1.2.2.1 What is PCST?

The PCST problem is a generalization of the classic Steiner tree. Unlike the traditional version, where you must connect all given terminals, in PCST each node also carries a prize (or importance). You're allowed to leave some nodes unconnected—paying their prize as a penalty instead. The goal:

- Maximize the total prize collected from nodes and edges included in the subgraph.
- Minus the cost of edges used (each edge has a connection cost).

Given a graph $G = (V, E)$, each node $n \in V$ and edge $e \in E$ has:

- A prize $p(n)$ or $p(e)$ (higher = more relevant to the query).
- A cost $c(e)$ (fixed penalty per edge to limit subgraph size).

The goal is to find a connected subgraph S^* that maximizes:

$$\text{Total Prize}(S^*) - \text{Total Cost}(S^*)$$

10.1.2.2.2 Step-by-step PCST in G-Retriever

1. Prize Assignment: Nodes/edges are scored by their similarity to the query (e.g., cosine similarity of embeddings).
   - Top-k nodes/edges get prizes (e.g., $k = 5$):

$$p(n) = \begin{cases} k - i & \text{if } n \text{ is top-}k \text{ relevant} \\ 0 & \text{otherwise} \end{cases}$$

   - Edge prizes: Analogous to nodes (if edge attributes exist).
2. Virtual Nodes for Edge Prizes If an edge's prize $p(e) >$ cost $c(e)$, it's converted to a virtual node v_e (see Figure 10.3):
   - v_e is connected to the edge's endpoints.
   - Prize of v_e: $p(e) - c(e)$.
   - New edges (dashed) have cost 0.
3. PCST Optimization: A fast approximation algorithm [2] is used to extract the optimal subgraph S^*.

10.1.2.2.3 Why PCST for G-Retriever?

When users ask graph-based questions, not all parts of the graph are relevant. G-Retriever:

- Assigns higher prizes to nodes and edges most semantically related to the user's query (via embedding similarity).
- Penalizes irrelevant or distant parts of the graph through edge costs and low/no node prizes.
- Uses PCST to select a connected subgraph that strikes the optimal balance—capturing high-prize elements while minimizing edge traversal and avoiding less relevant nodes.

This ensures the retrieved subgraph is both informative and compact, ready for prompting the LLM. Computing PCST is NP-hard, but G-Retriever uses efficient approximations (e.g., based on the Goemans-Williamson algorithm or faster heuristics) (Table 10.1).

Table 10.1 Advantages of PCST over alternatives.

Method	Connected?	Scalable?	Relevance-aware?
Top-k nodes	✗ No	✔ Yes	✔ Yes
Random walk	✔ Yes	✗ No (slow)	✗ No
PCST	✔ Yes	✔ Yes	✔ Yes

10.1.3 Microsoft GraphRAG: Global Sensemaking Through Community Summaries

While RAPTOR builds hierarchical summaries, Microsoft Graph-RAG tackles a different challenge: enabling "global sensemaking" over entire document collections [3]. Traditional RAG excels at retrieving specific facts but fails on questions like "What are the main themes in this dataset?" Microsoft Graph-RAG addresses this limitation through a novel combination of KG construction and hierarchical community detection.

10.1.3.1 The Global Sensemaking Problem

Traditional vector RAG systems work well for queries answerable with localized information:

- "When was Tesla founded?" → Retrieve chunks mentioning Tesla's founding
- "What is the capital of France?" → Find passages about Paris

However, they fail catastrophically on global questions requiring corpus-wide understanding:

- "What are the emerging trends across all research papers?"
- "How do different departments' priorities align with company strategy?"
- "What are the key themes connecting seemingly disparate news events?"

Microsoft Graph-RAG solves this by precomputing hierarchical summaries of thematically related content, enabling comprehensive answers that would require reading the entire corpus.

10.1.3.2 The Microsoft GraphRAG Architecture

Figure 10.4 (cited from the original paper) illustrates the offline indexing and online querying pipelines of Microsoft GraphRAG.

There are four major phases of the Microsoft GraphRAG architecture:

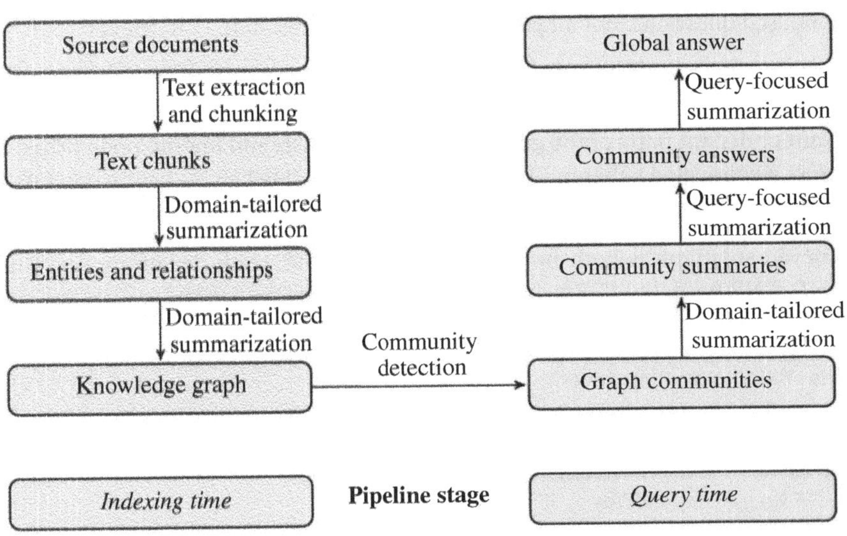

Figure 10.4 Microsoft Graph-RAG offline indexing and online query pipelines [3].

10.1.3.2.1 Phase 1: Rich KG Construction

Unlike simple entity extraction, Microsoft GraphRAG extracts richer semantic information:

- Entity Extraction with Descriptions: Not just "NeoChip" but "NeoChip is a publicly traded company specializing in low-power processors for wearables"
- Relationship Extraction with Context: Not just "acquired_by" but "Quantum Systems acquired NeoChip in 2016 and held ownership until NeoChip went public"
- Factual Claim: Factual statements like "NeoChip"s shares surged during their first week of trading on the NewTech Exchange"

This richer extraction creates a more semantically meaningful graph where even individual elements carry substantial context.

10.1.3.2.2 Phase 2: Hierarchical Community Detection

Microsoft GraphRAG's key innovation lies in applying community detection algorithms (specifically Leiden algorithm) hierarchically:

- Initial Partitioning: Detect communities of closely related entities
- Recursive Sub-partitioning: Within each community, detect sub-communities
- Hierarchical Structure: Create multiple levels (C0→C1→C2→C3) from global to granular

This produces a multiresolution view of the corpus where:

- C0 (Root): Highest-level themes across the entire corpus
- C1–C2: Intermediate themes and sub-themes
- C3 (Leaf): Detailed topic clusters

10.1.3.2.3 Phase 3: Bottom-up Community Summarization

Each community receives an LLM-generated summary following a carefully designed priority system:

- For Leaf Communities:
  - Prioritize edges by combined node degree (prominence)
  - Add node descriptions, edge descriptions, and claims until context limit
  - Generate a comprehensive summary of the community
- For Higher-level Communities:
  - If all sub-elements fit: Summarize directly
  - Otherwise: Substitute subcommunity summaries for their elements
  - Recursively incorporate lower-level insights

This bottom-up approach ensures higher-level summaries capture the essence of their subcommunities while maintaining computational efficiency.

10.1.3.3 Query-time Processing: Map-reduce for Global Answers

Microsoft GraphRAG's query processing employs a sophisticated map-reduce approach:

The map phase includes: (1) shuffle community summaries randomly (distribute relevant information); (2) chunk summaries to fit context windows; (3) generate partial answers in parallel; (4) score each answer's helpfulness (0–100).

The reduce phase includes: (1) filter out unhelpful answers (score = 0); (2) sort by helpfulness score; (3) iteratively combine top answers; (4) generate final global response.

This approach elegantly handles the challenge of synthesizing information that may be scattered across hundreds of community summaries.

10.1.3.4 Performance and Applications

Microsoft GraphRAG evaluation on million-token corpora revealed compelling results. It achieves 72–83% comprehensiveness vs. vector RAG on podcast/news corpora (1M tokens), and root-level (C0) summaries use 97% fewer tokens than full-text summarization.

Microsoft GraphRAG excels in scenarios requiring holistic corpus understanding. These are some ideal applications of Microsoft GraphRAG:

- Strategic Analysis: "What are the main competitive themes across industry reports?"
- Research Synthesis: "How do different research groups' findings relate to each other?"
- Trend Detection: "What emerging patterns connect seemingly unrelated news events?"
- Organizational Intelligence: "What are the key concerns across all customer feedback?"

While the term Graph-RAG broadly refers to any retrieval-augmented generation system leveraging graph-based retrieval or graph-structured indices, Microsoft GraphRAG distinguishes itself by:

- Building a global entity-centric KG with LLM-generated entity and relation summaries;
- Applying hierarchical community detection to construct multilevel graph partitions;
- Performing query-focused summarization across communities in a map-reduce fashion, designed specifically for sensemaking over large corpora;
- Supporting open-ended, multi-hop, and thematic queries, as opposed to factoid retrieval.

Its approach combines the strengths of KG-based retrieval and document summarization in a way that scales with both corpus size and query generality—paving the way for scalable global sensemaking in enterprise and scientific applications.

Microsoft GraphRAG's advantages over other approaches include: It handles global questions impossible for local retrieval; The community detection method finds natural semantic boundaries rather than similarity-based clusters; The precomputed summaries enable faster global queries without graph traversal.

Though there are also some limitations to consider: It has relatively high indexing cost (graph construction + community detection + summarization) and is less suitable for rapidly changing corpora. Also it may miss nuanced details preserved in original text and the performance depends on quality of community detection.

Microsoft GraphRAG represents a paradigm shift in how we approach corpus-wide understanding, transforming an impossible task for traditional RAG into a tractable problem through clever use of graph structure and hierarchical summarization.

10.1.4 HippoRAG: Neurobiologically-inspired Long-term Memory

While RAPTOR organizes information hierarchically and G-Retriever uses graph structures, HippoRAG takes inspiration from how the human brain stores and retrieves memories. Introduced in 2024 [4] and enhanced in 2025 with HippoRAG 2 [5], this framework mimics how our hippocampus and neocortex work together—offering powerful solutions for tasks that require connecting information across multiple documents.

10.1.4.1 The Neuroscience Behind the Algorithm

Think about how you instantly recognize connections—like realizing your friend's new colleague is your old roommate's cousin. This ability comes from our brain's hippocampal memory system, which HippoRAG translates into three computational components.

10.1.4.1.1 The Artificial Neocortex (LLM)

Just as our neocortex processes sensory input into abstract ideas, HippoRAG uses an LLM to process text. The LLM extracts entities (people, places, concepts) and relationships between them using Open Information Extraction (OpenIE). This creates structured knowledge from unstructured text.

10.1.4.1.2 The Parahippocampal Regions (Retrieval Encoders)

In our brain, these regions help connect related concepts. HippoRAG's retrieval encoders do the same by finding similar concepts and linking them. When two concepts are similar enough (above 0.8 cosine similarity), the system connects them with "synonym edges"—recognizing, for example, that "car" and "automobile" mean the same thing.

10.1.4.1.3 The Hippocampus (KG + PageRank)

The hippocampus acts as the brain's index, rapidly connecting different memories. HippoRAG implements this through a KG that stores relationships between concepts. It uses Personalized PageRank (PPR)—an algorithm that simulates how activation spreads through neural networks—to find relevant information even when it's several connections away from the starting point.

10.1.4.2 The HippoRAG Process

HippoRAG works in two phases that mirror human memory: encoding (storing information) and recall (retrieving it).

10.1.4.2.1 Offline Indexing: Building the Memory Network

During indexing, HippoRAG builds its knowledge representation step by step. First, the LLM reads each passage and identifies key entities—the important people, places, and concepts. Then it extracts relationships as (subject, predicate, object) triples, like "(Einstein, worked at, Princeton)." The retrieval encoders add connections between similar concepts, and each node remembers which passages it came from.

10.1.4.2.2 Online Retrieval: Following the Memory Trail

When you ask a question, HippoRAG activates its memory network. For a query like "Which Stanford professor works on Alzheimer's?", it first identifies the key concepts: "Stanford" and "Alzheimer's." These become starting points in the KG.

The system then runs PPR from these starting points, spreading activation through the graph like ripples in water. This activation follows the edges, discovering relevant information that might be several hops away. Finally, it ranks passages based on how much activation they received from the spreading process.

10.1.4.2.3 *Node Specificity: A Brain-inspired Innovation*

Traditional search systems use global statistics to identify important terms. But our brains don't have access to global statistics—they work with local information. HippoRAG mimics this by calculating "node specificity": how unique a concept is based on how many passages contain it. Common concepts get lower weights, while specific ones get higher weights, helping the system focus on what's most relevant.

10.1.4.3 Why HippoRAG Excels at Knowledge Integration

HippoRAG shines when answering questions that require connecting information from multiple sources. To understand why, consider two types of questions:

10.1.4.3.1 *Path-following vs. Pathfinding*

Path-following questions have a clear route: "Where was Einstein born?" just requires following Einstein → birthplace. Traditional systems can sometimes handle these if lucky enough to find a passage mentioning both concepts.

Pathfinding questions are harder: "Which Stanford professor works on Alzheimer's?" requires finding the specific path among thousands of possible connections. Traditional RAG systems fail because they look at passages in isolation. HippoRAG succeeds because it maintains a web of connections that can be explored from multiple angles.

10.1.4.3.2 *Single-step Multi-hop Magic*

While other systems like IRCoT [6] need multiple expensive rounds of retrieval and reasoning, HippoRAG connects information in one graph traversal. This makes it 10–30 times cheaper and 6–13 times faster while often performing better.

10.1.4.3.3 *Real-world Performance*

Testing on challenging benchmarks shows HippoRAG's strength. On MuSiQue [7], it improves retrieval by 3–6% over strong baselines. On 2WikiMultiHopQA [8], the improvements jump to 20% for top-5 retrieval and 38% for finding all relevant passages.

10.1.4.3.4 *Best Practices for Implementation*

Through extensive testing, optimal settings have emerged. GPT-3.5 provides good cost-performance balance, while Llama-3.1-70B offers an open-source option. A PageRank damping factor of 0.5 works well, balancing local and distant exploration. The 0.8 threshold for synonym detection proves effective across domains. Combining HippoRAG with other methods like IRCoT can yield an additional 4–20% improvements.

10.1.4.3.5 *When to Use HippoRAG*

HippoRAG excels in domains where relationships matter. Scientific literature, biographical databases, and any corpus where connections between entities provide value are ideal use cases.

Its ability to add new knowledge by simply inserting edges makes it perfect for growing knowledge bases.

However, be aware that OpenIE quality affects performance. For simple lookups where answers typically appear in single passages, traditional dense retrieval might be more efficient.

10.1.4.4 HippoRAG 2: Achieving Robust Long-term Memory

While the original HippoRAG excelled at associative reasoning, it struggled with other retrieval tasks. HippoRAG 2 (published in 2025) fixes these limitations while keeping the associative strengths, creating a system that handles all types of questions well.

10.1.4.4.1 Key Innovations

HippoRAG 2 introduces three major improvements:

First, it combines sparse and dense representations in one graph. Phrase nodes represent abstract concepts (sparse), while passage nodes store full contexts (dense). This mirrors how our brain uses different neural systems for concepts vs. detailed memories.

Second, it matches entire queries to knowledge triples instead of just extracting entities. Rather than pulling out "Stanford" and "Alzheimer's," it finds triples like "(Professor Smith, researches, Alzheimer's at Stanford)." This deeper understanding improves retrieval by 12.5%.

Third, it adds "recognition memory"—after initial retrieval, an LLM filters out irrelevant results. This two-stage process prevents the system from being misled by keyword matches that don't actually answer the question.

10.1.4.4.2 The Enhanced Process

Figure 10.5 shows how HippoRAG 2 works. During indexing, it creates a richer graph with both phrase and passage nodes connected by context edges. During retrieval, it matches queries to triples, filters with recognition memory, then runs PageRank from both phrase nodes (for associations) and passage nodes (for context).

10.1.4.4.3 Comprehensive Performance

Unlike its predecessor, HippoRAG 2 excels across all retrieval types. It matches the best systems on factual QA (63.3% on NaturalQuestions), improves multi-hop reasoning by 7–20%, and shows remarkable gains on narrative understanding (12.9% on NarrativeQA). This versatility comes from flexibly combining different retrieval strategies based on the question type.

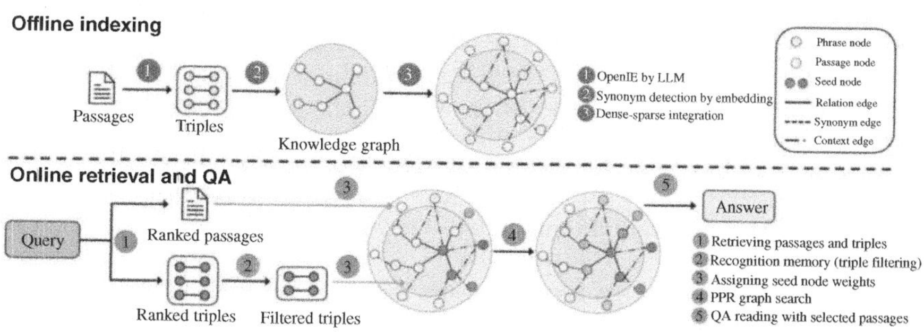

Figure 10.5 HippoRAG 2 methodology showing the dual-node architecture and recognition memory filtering. *Source*: [5]/Proceedings of Machine Learning Research.

10.1.4.4.4 A Practical Example

Consider "What county is Erik Hort's birthplace a part of?"

HippoRAG 1 would extract "Erik Hort" and "county," missing the crucial birthplace connection. It might return passages about various counties associated with Erik Hort.

HippoRAG 2 matches the full query to find triples like "(Erik Hort, born in, Montebello)" and "(Montebello, located in, Rockland County)." Recognition memory filters out irrelevant geographic information, leading to the correct two-hop answer.

Table 10.2 summarizes the key differences between versions.

10.1.4.4.5 Choosing the Right Version

HippoRAG 1 remains useful for entity-focused domains with limited computational resources. HippoRAG 2 is the better choice for production systems needing robust performance across diverse queries, especially those mixing factual lookups, complex reasoning, and document understanding.

10.1.4.4.6 Looking Forward

The progression from HippoRAG 1 to HippoRAG 2 represents a significant step toward AI systems that can truly learn continuously. By mimicking biological memory systems, these frameworks offer a path beyond static models to systems that grow and adapt their knowledge over time.

As we've explored through RAPTOR's hierarchies, G-Retriever's graph enhancement, Microsoft Graph-RAG's global analysis, and HippoRAG's brain-inspired approach, each framework contributes unique insights. The future lies not in choosing one approach but in understanding how to combine their strengths—much like how our brains use multiple memory systems working together to create human intelligence.

Table 10.2 Comparison of HippoRAG 1 vs. HippoRAG 2.

Feature	HippoRAG 1	HippoRAG 2	Improvement in v2
Core innovation	Hippocampal indexing theory for associative memory	Unified factual, associative, and sense-making memory	Holistic memory system
Retrieval method	Entity-centric (NER-based seed nodes)	Triple-centric (query-to-triple + passage nodes)	Deeper contextualization
Knowledge graph (KG)	Phrase nodes + synonym edges	Phrase nodes + passage nodes + context edges	Dense-sparse integration
Multi-hop reasoning	Single-step PPR over phrase nodes	PPR with balanced reset probabilities (phrase + passage nodes)	Better pathfinding
Key components	– LLM (OpenIE) – Dense retriever - PPR	Adds: – Recognition memory (triple filtering) – Passage-context edges	Robustness to noise
Performance	–+20% R@5 over ColBERTv2 (2Wiki) – Weak in sense-making	– +7% F1 over NV-Embed-v2 (multi-hop) – +12.9% F1 (NarrativeQA)	Cross-task robustness
Neurobiological basis	Mimics hippocampus-neocortex interaction	Adds dense-sparse coding (brain-like context integration)	Closer to human memory

10.2 Graph-RAG Tools and Software Overview

As the complexity of Graph-RAG systems increases, so does the demand for robust tools and frameworks that streamline their development and deployment. In Chapter 7, we explored a KG-enhanced RAG implementation using the LlamaIndex framework. Building on that foundation, this chapter introduces key tools and frameworks essential for constructing, integrating, and evaluating Graph-RAG models.

These tools not only simplify the technical challenges associated with combining retrieval mechanisms and KGs but also enable developers to build scalable and high-performance systems. Among the most prominent frameworks are LangChain [9] and Microsoft GraphRAG [3], both of which offer powerful capabilities for integrating KGs into RAG workflows. We will begin by exploring how LangChain provides an intuitive interface for connecting LLMs with KGs, thereby enhancing both the retrieval and generation phases of RAG. Following that, we will examine how Microsoft GraphRAG facilitates the seamless incorporation of graph structures into retrieval processes, offering developers a suite of tools to efficiently build and deploy graph-augmented RAG systems.

Throughout this section, code examples and practical demonstrations will illustrate the application of these frameworks in real-world scenarios. Whether your goal is to implement sophisticated KGs or optimize RAG model training pipelines, this chapter will equip you with the essential tools and techniques to accelerate your development process.

This section presents an overview of the current landscape of Graph-RAG frameworks available in the market. It explores both general-purpose and specialized frameworks that support the integration of KGs into RAG systems, highlighting their key features, functional capabilities, and typical use cases.

(Note: To clearly distinguish the general concept of "graph-based RAG" (or "Graph-RAG") from Microsoft's specific implementation, we will consistently refer to the latter as "Microsoft GraphRAG.")

10.2.1 What Is Available in the Current Market for Graph-RAG?

The ecosystem of Graph-RAG frameworks can be broadly categorized into two groups:

- **General RAG Frameworks with KG Support:** These are versatile platforms primarily designed for building RAG applications, which also offer extensions or modules to incorporate KGs into the retrieval and generation pipeline.
- **KG-focused RAG Frameworks:** These frameworks are specifically designed to leverage KGs as a core component of the retrieval process, providing optimized tools for graph-based information retrieval and reasoning within RAG systems.

10.2.1.1 General RAG Frameworks with KG Support

In the domain of general RAG frameworks, two prominent tools—LangChain and LlamaIndex—stand out due to their robust support for KGs. These frameworks have been thoroughly discussed in Chapter 4, and to avoid redundancy, we will not revisit them in detail here.

Below are notable open-source projects and frameworks that follow a general RAG paradigm while incorporating KG capabilities:

- infiniflow/ragflow: **RAGFlow** is an open-source RAG engine focused on deep document understanding. Drawing inspiration from Microsoft's Graph-RAG, it provides developers with tools to build sophisticated RAG systems enhanced with KG integration.
- HKUDS/LightRAG: **LightRAG** is a lightweight and high-speed retrieval-augmented generation framework designed for efficiency and simplicity. It is particularly well-suited for applications where low computational overhead and fast execution are critical.
- intel/ipex-llm: The **IPEX-LLM** project offers performance optimizations for running local RAG workflows, including support for Microsoft's Graph-RAG on Intel GPUs. This makes it a valuable tool for deploying KG-enhanced RAG systems on Intel hardware platforms.

10.2.1.2 KG-focused RAG Frameworks

The following frameworks are specifically designed to integrate KGs directly into RAG workflows. These tools were identified using a systematic survey approach:

- **Scope:** Open-source projects available on GitHub.
- **Inclusion Criteria:** Projects with keywords such as "RAG," "graph," "graph-rag," or "graphrag."
- **Evaluation Metrics:**
  - By Stars: Indicates broader interest and recognition from diverse stakeholders, including product managers and engineers.
  - By Forks: Reflects active development, customization potential, and community contributions.

After filtering out general-purpose RAG frameworks, excluding projects with fewer than 2,000 stars, merging related repositories, and organizing them hierarchically, the following prominent KG-focused RAG frameworks were identified.

microsoft/graphrag A robust framework designed for enterprise environments, leveraging Microsoft's cloud ecosystem—such as Azure Cognitive Services—for seamless integration with KGs. It supports advanced querying and reasoning over large-scale graphs, making it ideal for organizations seeking a fully integrated solution. We will refer to this tool as "Microsoft GraphRAG" throughout the remainder of this book.

- gusye1234/nano-graphrag: A compact and efficient reimplementation of Microsoft's GraphRAG, prioritizing code readability, portability, and performance. With approximately 1,100 lines of core code, it supports asynchronous execution and integrates seamlessly with tools like FAISS, Neo4j, and Ollama.
- severian42/GraphRAG-Local-UI: A local adaptation of Microsoft's GraphRAG that adds support for interactive user interfaces and visualization tools. Key features include: (1) real-time graph visualization (2D/3D using Plotly), (2) customizable querying with local LLMs, and (3) cost-effective operation by eliminating reliance on cloud-based models.

neo4j-labs/llm-graph-builder This framework transforms unstructured data (e.g., PDFs, DOCs, TXT files, YouTube videos, web pages) into structured KGs stored in Neo4j. It utilizes powerful LLMs (such as OpenAI and Gemini) to extract nodes, relationships, and their properties using the LangChain framework. Users can upload files from local machines, GCS, S3 buckets, or web sources, select an LLM, and generate a KG.

circlemind-ai/fast-graphrag Optimized for high-performance graph search, this framework is particularly well-suited for applications requiring rapid query execution, such as real-time recommendation systems or large-scale graph exploration tasks.

pingcap/autoflow AutoFlow is an open-source KG-enhanced RAG tool built on top of TiDB Vector, LlamaIndex, and DSPy. It aims to provide scalable and efficient knowledge retrieval and generation capabilities within a database-centric architecture.

getzep/graphiti Graphiti focuses on constructing dynamic, temporally aware KGs. It supports querying complex, evolving relationships through a combination of time-based analysis, semantic embeddings, and graph algorithms.

OSU-NLP-Group/HippoRAG Inspired by human long-term memory mechanisms, HippoRAG introduces a novel approach to continuously integrating new knowledge into LLMs. This enables more adaptive, context-aware reasoning and supports continual learning without retraining the model.

10.2.2 When to Choose Which Framework?

Selecting the appropriate Graph-RAG framework depends heavily on the specific goals, architectural needs, and performance requirements of your project. The following guidelines can help streamline the decision:

10.2.2.1 For Applications Beyond KGs

When your application requires flexibility beyond graph-based retrieval or involves integrating multiple data types, consider using **general RAG frameworks with KG support**. These tools offer broader functionality while still allowing for graph-enhanced retrieval when needed.

Example: `LlamaIndex` provides robust backend storage options, customizable retrieval pipelines, and modular generation components, making it an excellent choice for hybrid architectures that combine structured and unstructured data sources.

10.2.2.2 For KG-centric Applications

If your system relies heavily on structured relationships, semantic reasoning, or complex graph queries, opt for **KG-focused RAG frameworks**. These are specifically designed to maximize the utility of graph structures in retrieval and reasoning tasks.

Example: `Microsoft-GraphRAG` is ideal for enterprise environments already integrated with the Microsoft ecosystem (e.g., Azure services), offering scalable, production-ready capabilities for large-scale KG processing and advanced querying.

By aligning the choice of framework with your application's focus—whether general-purpose flexibility or deep graph integration—and considering performance, scalability, and ecosystem compatibility, developers can fully harness the power of Graph-RAG technologies.

10.3 Practice: Run Graph-RAG Frameworks

As we've dived into the overview and theoretical foundations of a few advanced Graph-RAG frameworks in Section 10.1, this section demonstrates the implementation examples of these frameworks on various datasets.

10.3.1 RAPTOR

The implementation of RAPTOR (introduced in Section 10.1.1) is available at: https://github.com/parthsarthi03/raptor.

10.3.1.1 Initializing RAPTOR

The objective of this section is to demonstrate how to customize RAPTOR and run RAPTOR on an example dataset. To get started with RAPTOR, first clone the GitHub repository and install the required dependencies:

```
conda create -n raptor
conda activate raptor

git clone https://github.com/parthsarthi03/raptor.git
cd raptor
pip install -r requirements.txt
```

All subsequent code should be executed from the root directory of this repository to ensure proper import of the `raptor` module.

RAPTOR reuses the chat and embedding model servers described in Appendix A, along with their LlamaIndex clients. The following custom classes define the summarization, QA, and embedding models used in RAPTOR. These are designed to interface with the server endpoints defined in the appendix.

10.3.1.1.1 Custom LLM Model Classes

RAPTOR utilizes two types of conversational models: one for summarization and another for QA.

The `CustomSummarizationModel` class extends `BaseSummarizationModel` and implements logic for generating document summaries using a preconfigured chat model:

```
from raptor import BaseSummarizationModel
from openai import OpenAI

class CustomSummarizationModel(BaseSummarizationModel):
    def summarize(self, context, max_tokens=1024, model="o4-mini"):
        response = OpenAI(
            base_url=api_base,
            api_key="sk-fake-api-key-abcdefghijklmnopqrstuvwxyz-0123456789",
            timeout=timeout,
        ).chat.completions.create(
            model=model,
            messages=[
                {
                    "role": "user",
                    "content": f"Write a summary of the following, including as many
    key details as possible:\n\n\n{context}",
                },
            ],
            max_tokens=max_tokens,
        )
        summary = response.choices[0].message.content
        return summary
```

The `CustomQAModel` class extends `BaseQAModel` to answer user questions based on provided context:

```
from raptor import BaseQAModel

class CustomQAModel(BaseQAModel):
    def answer_question(self, context, question, model="o4-mini"):
        response = OpenAI(
            base_url=api_base,
            api_key="sk-fake-api-key-abcdefghijklmnopqrstuvwxyz-0123456789",
            timeout=timeout,
        ).chat.completions.create(
            model=model,
            messages=[
                {
                    "role": "user",
                    "content": f"Given Context:\n{context}\n\n\nGive the best full
    answer amongst the option to question:\n{question}",
                },
            ],
        )
        answer = response.choices[0].message.content
        print("answer_question", type(answer), answer)
        return answer
```

10.3.1.1.2 Custom Embedding Model Class

The `CustomEmbeddingModel` class wraps an existing embedding model to generate vector representations of text:

```
from raptor import BaseEmbeddingModel

class CustomEmbeddingModel(BaseEmbeddingModel):
    def create_embedding(self, text, model="text-embedding-ada-002"):
        embedding = (
            OpenAI(
                base_url=api_base,
                api_key="sk-fake-api-key-abcdefghijklmnopqrstuvwxyz-0123456789",
                timeout=timeout,
            )
            .embeddings.create(input=[text], model=model)
            .data[0]
            .embedding
        )
        return embedding
```

10.3.1.1.3 Registering Custom Models with RAPTOR

Once the custom models are defined, they can be registered with the RAPTOR framework by instantiating a `RetrievalAugmentationConfig` object:

```
from raptor import RetrievalAugmentation, RetrievalAugmentationConfig

custom_summarizer = CustomSummarizationModel()
custom_qa = CustomQAModel()
custom_embedding = CustomEmbeddingModel()

custom_config = RetrievalAugmentationConfig(
    summarization_model=custom_summarizer,
    qa_model=custom_qa,
    embedding_model=custom_embedding,
    tb_summarization_length=32 * 1024,
    tb_max_tokens=32 * 1024,
)

RA = RetrievalAugmentation(config=custom_config)
```

10.3.1.1.4 Running RAPTOR on an Example Dataset

Start by loading a sample text file included in the repository:

```
with open(RAPTOR_GIT_ROOT_DIR + "/demo/sample.txt", "r") as file:
    text = file.read()

print(text[:100])
```

Next, construct the hierarchical tree structure using RAPTOR's retrieval augmentation pipeline:

```
RA.add_documents(text)
```

Finally, ask a question and retrieve a generated answer:

```
question = "How did Cinderella reach her happy ending ?"
answer = RA.answer_question(question=question)
print("Answer: ", answer)
```

10.3.1.2 Running RAPTOR on the WANDS Search Dataset

The objective of this section is to demonstrate how to run RAPTOR using the WANDS dataset for search applications. The full WANDS dataset is too large for demonstration purposes, so we will sample a small subset to illustrate how RAPTOR can be applied in a search-based retrieval setting. We follow the same pattern as before: prepare a sample dataset, index it using RAPTOR, and then perform QA over the indexed documents using a RAG approach.

10.3.1.2.1 Data Preparation

The goal is to generate a list of text strings, referred to as `docs`, using the three preprocessed dataframes—`df_product`, `df_query`, and `df_label`—introduced in Section 1.3.3.

To construct a manageable test set:

1. Select a few query IDs from `queries.csv` to form a list called `query_id_list` (e.g., `query_id_list = [0, 1]`).
2. For each selected query:
   - Sample `n` positive (relevant) products labeled "Exact."
   - Sample `n` negative (irrelevant) products labeled "Irrelevant."
3. Combine both sets into a document collection that will be fed into RAPTOR.

Below is the code to sample positive (gold) examples:

```
df_label_gold = df_label.head(0)
for query_id in query_id_list:
    df_label_exact = df_label.loc[
        (df_label["query_id"] == query_id)
        & (df_label["label"] == "Exact")
        & (~df_label["product_id"].isin(df_label_gold["product_id"]))
    ].sample(n=n, random_state=random_state)
    df_label_gold = pd.concat([df_label_gold, df_label_exact], ignore_index=True)

df_gold = df_label_gold.merge(
    df_query,
    on="query_id",
    how="left",
).merge(
    df_product,
    on="product_id",
    how="left",
)
```

Next, we sample negative (irrelevant) examples:

```
df_label_negative = df_label.head(0)
for query_id in query_id_list:
    df_label_irrelevant = df_label.loc[
        (df_label["query_id"] == query_id)
        & (df_label["label"] == "Irrelevant")
        & (~df_label["product_id"].isin(df_label_negative["product_id"]))
        & (~df_label["product_id"].isin(df_gold["product_id"]))
    ].sample(n=n, random_state=random_state)
    df_label_negative = pd.concat(
        [df_label_negative, df_label_irrelevant], ignore_index=True
    )

df_negative = df_label_negative.merge(df_product, on="product_id", how="left")
```

Finally, combine gold and negative samples into a single list of documents:

```
df_product_columns = df_product.columns.to_list()
docs = (
    pd.concat(
        [
            df_gold[df_product_columns],
            df_negative[df_product_columns],
        ],
        ignore_index=True,
    )
    .to_json(orient="records", lines=True)
    .splitlines()
)
```

10.3.1.2.2 Indexing with RAPTOR

By default, RAPTOR uses generic text chunking logic which may not work well with structured data like CSVs. To better align with our input format, we patch the chunking function to treat each line (i.e., each JSON object) as a separate chunk:

```
from unittest.mock import patch

def custom_split_text(
    text: str,
    tokenizer,
    max_tokens: int,
    overlap: int = 0,
):
    return text.splitlines()

with patch("raptor.tree_builder.split_text", new=custom_split_text):
    RA.add_documents(docs="\n".join(docs))
```

10.3.1.2.3 Querying with RAPTOR

When performing QA, we prepend a prompt prefix such as "search product" to guide the model toward the expected task:

```
for i, q in enumerate(queries):
    print("+" * 20, i)
    answer = RA.answer_question(question=f"search product {q}")
    print("Answer: ", answer)
```

10.3.1.3 Run RAPTOR on WANDS Recommendation Dataset

Following the same workflow as in Section 10.3.1.2, our objective is now to apply RAPTOR to the WANDS recommendation dataset. We first sample and prepare data based on the dataset setup described in Appendix B. Then, we index the documents using RAPTOR and perform retrieval-augmented QA. In this example, we use the `synthetic_sequences.csv` file prepared in Appendix B. As outlined in Section 10.3.1.2, we generate gold (relevant) documents and additional non-gold (irrelevant) documents, then combine them into a single document set. Unlike the previous search-based example, this task includes user purchase history embedded directly into the query prompt—simulating a personalized product recommendation scenario.

The following helper function converts a list of `product_ids` into a list of JSON-formatted strings representing product details:

```
def product_jsonl(
    product_id_list,
    df_product=df_product,
    columns=df_product.columns.to_list(),
):
    return (
        df_product.loc[
            df_product["product_id"].isin(product_id_list),
            columns,
        ]
        .to_json(orient="records", lines=True)
        .splitlines()
    )
```

Select a small number of sequence IDs from `synthetic_sequences.csv` to form a list called `sequence_id_list`, e.g., `sequence_id_list = ['seq_0', 'seq_1']`. Using these sequences, we construct queries that include user purchase history and ask the model to predict the next likely product to be purchased.

```
queries, gold_docs, gold_answers = [], [], []
visited = set()                                          # [1]
for sequence_id in sequence_id_list:
    product_sequence = df_sequences.loc[
        df_sequences["sequence_id"] == sequence_id, "product_sequence"
    ].item()                                             # [2]
    queries.append(
        "Based on the list of candidate products provided above and the user's purchase history below, "
        "predict the product that the user is most likely to buy next. "
        "Only output the `product_name` of the predicted product.\n"
        "Do not recommend any product that has already been purchased.\n"
        "The purchase history is given as a sequence of previously purchased products:\n\n"
        + "\n\n".join(product_jsonl(product_sequence))
    )                                                    # [3]

    next_product = df_sequences.loc[
        df_sequences["sequence_id"] == sequence_id, "next_product"
    ].item()                                             # [4]
    gold_docs.append(product_jsonl([next_product]))      # [5]
    gold_answers.append(
        df_product.loc[df_product["product_id"] == next_product, "product_name"]
    )                                                    # [6]

    visited.update(product_sequence + [next_product])    # [1]
```

[1] A set to track all seen product IDs, used later to filter irrelevant products.
[2] Extract the list of purchased product IDs for the current sequence.
[3] Embed the user's purchase history into the query prompt.
[4] Get the ground truth label—the expected next product.
[5] Append the corresponding gold document.
[6] Record the gold answer for evaluation purposes (see Section 10.3.4.3).

Now, we sample some irrelevant (non-gold) products to include in the document set, ensuring diversity in the retrieval context:

```
other_docs = product_jsonl(
    df_product[~df_product["product_id"].isin(visited)].sample(
        n=len(sequence_id_list), random_state=random_state
    )["product_id"]
)

docs = [y for x in gold_docs for y in x]
docs.extend(other_docs)
```

Once the document set is ready, the indexing and querying steps follow the same process as described in Section 10.3.1.2.

10.3.1.4 Key Takeaways

- RAPTOR is a RAG framework that builds a hierarchical tree structure from documents to support efficient, context-aware information retrieval and QA.
- You can customize RAPTOR by defining custom classes for summarization, QA, and embedding models, allowing it to interface with various LLM and embedding service endpoints.
- The framework can be applied to diverse tasks, including general QA on text documents and specialized applications like product search and recommendation on structured datasets like WANDS.
- For structured data, the default chunking behavior can be patched to treat each data entry (e.g., a JSON object) as a single chunk, optimizing the indexing process.

10.3.2 G-Retriever

The objective of this section is to demonstrate how to run G-Retriever in various modes using the example dataset. The G-Retriever codebase is publicly available at: https://github.com/XiaoxinHe/G-Retriever. G-Retriever offers three distinct model architectures:

- `inference_llm`: A standard LLM
- `pt_llm`: LLM combined with PT
- `graph_llm`: LLM integrated with GNN components

Additionally, it supports three fine-tuning configurations (shown in Table 10.3):

- Inference-only: Uses a frozen LLM for direct question answering without any training.
- Frozen LLM with PT: The LLM parameters remain frozen while only the prompt embeddings are adapted.
- Tuned LLM: The LLM is fine-tuned using Low-rank Adaptation (LoRA), allowing parameter updates.

Table 10.3 Mapping between G-Retriever models and fine-tuning configurations.

Scenario	Inference_llm	pt_llm	graph_llm
Inference-Only LLM	Yes	No	No
Frozen LLM + PT	No	Yes	Yes
Tuned LLM	Yes	Yes	Yes

10.3.2.1 Setup and Configuration

To set up the environment:

```
conda create --name g_retriever python=3.12 -y
conda activate g_retriever

# Check your CUDA version using `nvidia-smi`, then install PyTorch with CUDA 12.8
    support
pip3 install torch==2.7.1 torchvision==0.22.1 torchaudio==2.7.1 --index-url https://
    download.pytorch.org/whl/cu128

# Verify installation
python -c "import torch; print(torch.__version__)"
# Expected output: 2.7.1+cu128
python -c "import torch; print(torch.version.cuda)"
# Expected output: 12.8
```

Install additional dependencies:

```
# Install PyG dependencies
pip install pyg_lib torch_scatter torch_sparse torch_cluster torch_spline_conv -f
    https://data.pyg.org/whl/torch-2.7.1+cu128.html

# Install other required packages
pip install peft pandas ogb transformers wandb sentencepiece torch_geometric datasets
    pcst_fast gensim scipy==1.12 protobuf
```

Clone the repository:

```
git clone git@github.com:GraphRAGBook/G-Retriever.git
cd G-Retriever/
```

Due to changes in newer versions of PyTorch, all calls to `torch.load()` must include the argument `weights_only=False`. For example:[1]

```
torch.load(..., weights_only=False)
```

If you want to use a custom LLM—such as a smaller model to better fit your GPU VRAM constraints—edit the model path in `src/model/__init__.py`:

```
llama_model_path = {
    ...
    "TinyLlama": "TinyLlama/TinyLlama_v1.1",
}
```

10.3.2.2 Dataset Preparation

From the root directory of the repository, preprocess the dataset as follows:

```
python -m src.dataset.preprocess.expla_graphs
python -m src.dataset.expla_graphs
```

This will convert the input CSV file into three folders (`nodes`, `edges`, and `graphs`) that can be recognized by G-Retriever. The default dataset, located at `dataset/expla_graphs`, contains 2,766 rows/graphs, with an average of 5.14 nodes and 4.25 edges per graph. The task involves comparing two sentences and predicting whether they are `'support'` or `'counter'`:

```
Question: Do argument 1 and argument 2 support or counter each other? Answer in one
    word in the form of 'support' or 'counter'.
```

1 See an example patch here: https://github.com/GraphRAGBook/G-Retriever/pull/1

10.3.2.3 Inference Only Mode

In this mode, no training is required before performing inference. Use the script `inference.py` to run the LLM on the dataset and obtain accuracy results. There are two inference types:

- Question-only

```
python inference.py --model_name llm --llm_model_name TinyLlama --max_txt_len 0
```

  Resulting test accuracy: < 0.1

- Textual Graph + Question

```
python inference.py --model_name llm --llm_model_name TinyLlama
```

  Resulting test accuracy: < 0.01

10.3.2.4 Frozen LLM with PT

After training, models are saved in `output/expla_graphs`. Training scripts automatically evaluate performance after completion.

- Prompt Tuning `pt_llm`

```
python train.py --model_name pt_llm --llm_model_name TinyLlama
```

  Test Accuracy: 0.4927

- G-Retriever `graph_llm`

```
python train.py --model_name graph_llm --llm_model_name TinyLlama
```

  Test Accuracy: 0.5758

10.3.2.5 Fine-tuned LLM (Full Parameter Tuning)

This setup is similar to the previous one, but the LLM is not frozen during training. Instead, LoRA is used for efficient fine-tuning.

- Prompt Tuning with Fine-Tuning

```
python train.py --model_name pt_llm --llm_model_name TinyLlama \
    --llm_frozen False
```

  Test Accuracy: 0.5632

- G-Retriever with Fine-Tuning

```
python train.py --model_name graph_llm --llm_model_name TinyLlama \
    --llm_frozen False
```

  Test Accuracy: 0.5703

10.3.2.6 Key Takeaways

- G-Retriever is a framework that combines LLMs with GNNs for question-answering on textual graphs.
- It offers three model architectures: a standard LLM, an LLM with prompt tuning, and an LLM integrated with GNNs.
- The framework supports three fine-tuning configurations: inference-only, prompt tuning with a frozen LLM, and full fine-tuning using LoRA.
- Performance results demonstrate that the `graph_llm` model with prompt tuning or fine-tuning significantly outperforms the `inference_llm` model, highlighting the benefit of incorporating graph structure.

10.3.3 Microsoft GraphRAG

The objective of this section is to demonstrate how to run Microsoft GraphRAG for a question-answering task on data from Wikipedia web pages. Microsoft GraphRAG, described in Section 10.1.3, is documented at https://microsoft.github.io/graphrag/get_started/, and its source code is available on GitHub at https://github.com/microsoft/graphrag. The following provides a simple end-to-end example demonstrating how to use the Microsoft GraphRAG system. This walkthrough illustrates the process of indexing a collection of text documents and then leveraging the resulting KG to answer domain-specific questions. To begin, install the package using `pip`:

```
pip install graphrag
```

10.3.3.1 Indexing Phase

This phase includes the following steps: setting up the working directory and configuration, preparing input files, running the index construction command, examining the graph visualization, and analyzing the LLM conversation.

10.3.3.1.1 Set Up Working Directory and Configuration

The Microsoft GraphRAG library includes a command-line interface (CLI) that offers a no-code approach to setting up and running the system. To initialize your workspace, run the following command to create a working directory:

```
graphrag init --root ./microsoft-graphrag
```

The working directory will be created at `./microsoft-graphrag`. All subsequent commands will be executed within this directory in your command line. This command creates two essential configuration files in the `./microsoft-graphrag` directory: `.env` and `settings.yaml`.

- `.env`: This file contains environment variables required by the Graph-RAG pipeline, such as the OpenAI base URL and API key. You can modify it to point to a local LLM server by using a placeholder API key:

```
OPENAI_BASE_URL=http://localhost:4000/v1
```

- `settings.yaml`: This file holds the configuration settings for the indexing pipeline, including model parameters, output formats, and more. For later graph visualization, enable GraphML snapshot generation by updating the configuration:

```
sed -i 's/graphml: false/graphml: true/g' ./microsoft-graphrag/settings.yaml
```

10.3.3.1.2 Input Text Files and Indexing Process

Place your input text files (`*.txt`) into the `./microsoft-graphrag/input/` directory. For this demonstration, we use sample text extracted from the Wikipedia page summary on "Tariffs in the Second Trump Administration," available at: https://en.wikipedia.org/wiki/ Tariffs_in_the_second_Trump_administration. Once the files are in place, run the indexing command using the following instruction:

```
graphrag index --verbose --root ./microsoft-graphrag
```

The duration of this step depends on several factors, including the size of your dataset, the LLM being used, and the chunking and processing strategies configured in `settings.yaml`. Upon completion, a new folder named `./microsoft-graphrag/output/` will be created. This directory contains multiple Parquet files [10], along with the generated KG file.

10.3.3.1.3 Graph Visualization

During the indexing process, Microsoft GraphRAG generates a GraphML file named `graph.graphml` in the `./microsoft-graphrag/output` directory. This file contains the structured representation of the KG extracted from your input documents. To visualize the graph, follow the official (Graph-RAG visualization guide) (https://microsoft.github.io/ graphrag/visualization_guide/) using tools such as Gephi [11]. An example of a rendered KG visualization is shown in Figure 10.6.

10.3.3.1.4 LLM Conversation Analysis

The LLM conversation analysis involves two key steps:

1. Entity and Relationship Extraction: Identify entities (with properties: `entity_name`, `entity_type`, `entity_description`) and relationships (with properties: `source_entity`, `target_entity`, `relationship_description`, `relationship_strength`).
2. Community Summarization and Insight Generation: Analyze grouped communities to produce structured reports containing summaries, impact ratings, and detailed findings grounded in the input data.

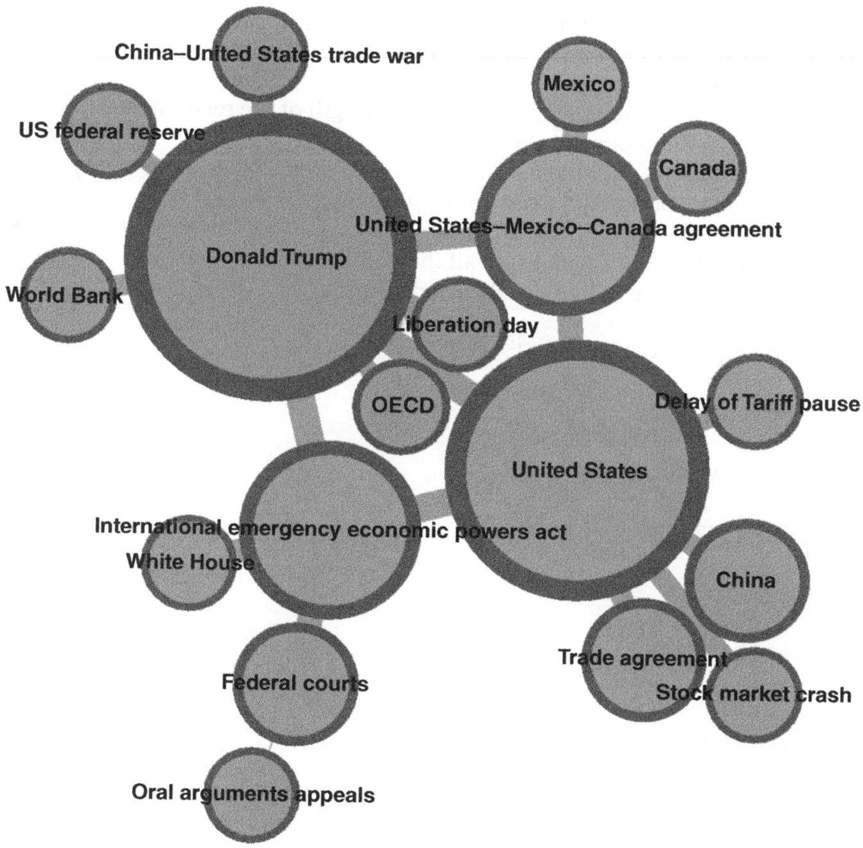

Figure 10.6 Visualize `graph.graphml`.

The first stage of the LLM conversation focuses on extracting structured knowledge from raw text. Below is an example prompt and response used for this purpose:

user

```
–Goal–
Given a text document that is potentially relevant to this activity and a list of entity types, identify all entities of those types from the text and all relationships among the identified entities.

–Steps–
1. Identify all entities. For each identified entity, extract the following information:
– entity_name: Name of the entity, capitalized
– entity_type: One of the following types: [organization,person,geo,event]
– entity_description: Comprehensive description of the entity's attributes and activities
Format each entity as ("entity"<|><entity_name><|><entity_type><|>
<entity_description>)

2. From the entities identified in step 1, identify all pairs of (source_entity, target_entity) that are *clearly related* to each other.
For each pair of related entities, extract the following information:
– source_entity: name of the source entity, as identified in step 1
– target_entity: name of the target entity, as identified in step 1
– relationship_description: explanation as to why you think the source entity and the target entity are related to each other
– relationship_strength: a numeric score indicating strength of the relationship between the source entity and target entity
Format each relationship as ("relationship"<|><source_entity><|><target_entity
><|><relationship_description><|><relationship_strength>)

3. Return output in English as a single list of all the entities and relationships identified in steps 1 and 2. Use **##** as the list delimiter.

4. When finished, output <|COMPLETE|>

######################
–Examples–
######################
Example 1:
Entity_types: ORGANIZATION,PERSON
Text:
...

######################
–Real Data–
######################
Entity_types: organization,person,geo,event
Text: ...
######################
Output:
```

assistant

("entity"<|>DONALD TRUMP<|>PERSON<|>Donald Trump is the former United States president who enacted protective tariffs and invoked the IEEPA to impose broad tariffs<|>9)
##
...
##
("relationship"<|>DONALD TRUMP<|>UNITED STATES<|>Donald Trump is the president of the United States who enacted tariffs<|>9)
##
...
<|COMPLETE|>

user MANY entities and relationships were missed in the last extraction. Remember to ONLY emit entities that match any of the previously extracted types. Add them below using the same format:

assistant

("entity"<|>FEDERAL COURTS<|>ORGANIZATION<|>Federal courts ruled the use of IEEPA for tariffs unconstitutional<|>7)
##
...
##
("relationship"<|>FEDERAL COURTS<|>INTERNATIONAL EMERGENCY ECONOMIC POWERS ACT<|>Federal courts ruled the use of IEEPA for tariffs unconstitutional<|>8)
##
...
<|COMPLETE|>

This iterative process ensures completeness and accuracy in identifying all relevant entities and their interconnections within the domain-specific context.

Once entities and relationships are extracted and organized into communities, the second stage generates a structured report for each community. Below is the prompt and sample response used for summarizing one such community:

user You are an AI assistant that helps a human analyst to perform general information discovery. Information discovery is the process of identifying and assessing relevant information associated with certain entities (e.g., organizations and individuals) within a network.

Goal
Write a comprehensive report of a community, given a list of entities that belong to the community as well as their relationships and optional associated claims. The report will be used to inform decision–makers about information associated with the community and their potential impact. The content of this report includes an overview of the community's key entities, their legal compliance, technical capabilities, reputation, and noteworthy claims.

Report Structure

The report should include the following sections:

– TITLE: community's name that represents its key entities – title should be short but specific. When possible, include representative named entities in the title.
– SUMMARY: An executive summary of the community's overall structure, how its entities are related to each other, and significant information associated with its entities.
– IMPACT SEVERITY RATING: a float score between 0–10 that represents the severity of IMPACT posed by entities within the community. IMPACT is the scored importance of a community.
– RATING EXPLANATION: Give a single sentence explanation of the IMPACT severity rating.
– DETAILED FINDINGS: A list of 5–10 key insights about the community. Each insight should have a short summary followed by multiple paragraphs of explanatory text grounded according to the grounding rules below. Be comprehensive.

Return output as a well–formed JSON–formatted string with the following format:

```
{
   "title": <report_title>,
   "summary": <executive_summary>,
   "rating": <impact_severity_rating>,
   "rating_explanation": <rating_explanation>,
   "findings": [
     {
        "summary":<insight_1_summary>,
        "explanation": <insight_1_explanation>
     },
     {
        "summary":<insight_2_summary>,
        "explanation": <insight_2_explanation>
     }
   ]
}
```

Grounding Rules

Points supported by data should list their data references as follows:

"This is an example sentence supported by multiple data references [Data: <dataset name> (record ids); <dataset name> (record ids)]."

Do not list more than 5 record ids in a single reference. Instead, list the top 5 most relevant record ids and add "+more" to indicate that there are more.

For example:
"Person X is the owner of Company Y and subject to many allegations of wrongdoing [Data: Reports (1), Entities (5, 7); Relationships (23); Claims (7, 2, 34, 64, 46, +more)]."

where 1, 5, 7, 23, 2, 34, 46, and 64 represent the id (not the index) of the relevant data record.

Do not include information where the supporting evidence for it is not provided.

Limit the total report length to 2000 words.

Example Input

...

Real Data

Use the following text for your answer. Do not make anything up in your answer.

Text:
------Entities------
human_readable_id,title,description,degree
2,INTERNATIONAL EMERGENCY ECONOMIC POWERS ACT,"IEEPA is a U.S. law invoked by Trump to justify broad, universal tariffs on imports",4
17,FEDERAL COURTS,,2
12,WHITE HOUSE,The White House sent letters to countries regarding new tariff rates,1
13,ORAL ARGUMENTS ON APPEALS,"Scheduled for July 31, 2025, regarding the IEEPA tariffs",1

------Relationships------
human_readable_id,source,target,description,combined_degree
14,INTERNATIONAL EMERGENCY ECONOMIC POWERS ACT,FEDERAL COURTS,Federal courts ruled the use of IEEPA for tariffs unconstitutional,6
15,INTERNATIONAL EMERGENCY ECONOMIC POWERS ACT,WHITE HOUSE, The White House used IEEPA to justify tariffs,5
19,ORAL ARGUMENTS ON APPEALS,FEDERAL COURTS,Oral arguments are scheduled for the IEEPA tariffs case,3

...

Output:

assistant

{
"title": "IEEPA Tariff Dispute and Legal Challenges",
"summary": "The community centers around the International Emergency Economic Powers Act (IEEPA), which the White House invoked to justify tariffs. Federal courts ruled this application unconstitutional, prompting appeals. The upcoming oral arguments on July 31, 2025, represent a critical legal milestone in challenging the executive's authority under IEEPA.",
"rating": 7.5,

"rating_explanation": "The high impact rating reflects the legal conflict over executive authority and potential policy shifts in international trade, with unresolved judicial outcomes.",
"findings": [
{
"summary": "IEEPA's central role in tariff justification",
"explanation": "The International Emergency Economic Powers Act (IEEPA) is the foundational entity in this community, invoked by the White House to justify broad tariffs. These tariffs, framed as a response to national emergencies, have sparked significant legal scrutiny. The IEEPA's application here demonstrates its dual role as both a legislative tool and a potential overreach of executive power [Data: Entities (2); Relationships (15)]. The law's broad language enables its misuse, as seen in this case where it was used to impose universal tariffs without clear emergency justification [Data: Entities (2)]."
},
...
]
}

This round of prompts and responses is repeated for each detected community in the KG, enabling comprehensive insight generation across multiple domains or clusters of interest.

10.3.3.2 Query Phase

The following examples demonstrate how to use the Graph-RAG query engine to answer both global and local questions using the KG generated during the indexing phase.

10.3.3.2.1 Global Search

Global search enables high-level reasoning over the entire KG by aggregating insights from community summaries. Below is an example of querying for key policy arguments made by the Trump administration in favor of its tariffs:

```
graphrag query \
    --root ./microsoft-graphrag \
    --method global \
    --query "What are the main arguments made by the Trump administration in favor of
    its tariffs?"
```

This query triggers a two-step LLM conversation process within the Graph-RAG framework. In the first step, community summaries are sent to the LLM to extract relevant key points that address the user's question. An example prompt-response interaction is shown below:

system

---Role---

You are a helpful assistant responding to questions about data in the tables provided.

---Goal---

Generate a response consisting of a list of key points that responds to the user's question, summarizing all relevant information in the input data tables.

You should use the data provided in the data tables below as the primary context for generating the response.
If you don't know the answer or if the input data tables do not contain sufficient information to provide an answer, just say so. Do not make anything up.
Each key point in the response should have the following element:
– Description: A comprehensive description of the point.
– Importance Score: An integer score between 0–100 that indicates how important the point is in answering the user's question. An 'I don't know' type of response should have a score of 0.
The response should be JSON formatted as follows:
{
  "points": [
    {"description": "Description of point 1 [Data: Reports (report ids)]", "score": score_value},
    {"description": "Description of point 2 [Data: Reports (report ids)]", "score": score_value}
  ]
}
The response shall preserve the original meaning and use of modal verbs such as "shall", "may" or "will".
Points supported by data should list the relevant reports as references as follows:
"This is an example sentence supported by data references [Data: Reports (report ids)]"
Do not list more than 5 record ids in a single reference. Instead, list the top 5 most relevant record ids and add "+more" to indicate that there are more.
For example:
"Person X is the owner of Company Y and subject to many allegations of wrongdoing [Data: Reports (2, 7, 64, 46, 34, +more)]. He is also CEO of company X [Data: Reports (1, 3)]"
where 1, 2, 3, 7, 34, 46, and 64 represent the id (not the index) of the relevant data report in the provided tables.
Do not include information where the supporting evidence for it is not provided.
Limit your response length to 1000 words.

---Data tables---

id|title|occurrence weight|content|rank
3|IEEPA Tariff Dispute and Legal Challenges|1.0|"..."|7.5
2|Donald Trump's Tariff Policies and Economic Impact Community|1.0|"..."|7.5
0|USMCA Trade Agreement and Mexico–Canada Exemptions|1.0|"..."|7.5
1|U.S.–China Trade Conflict and Tariff Agreements|1.0|"..."|6.5

...

user What are the main arguments made by the Trump administration in favor of its tariffs?

assistant
{
  "points": [
    {

```
            "description": "The Trump administration argued that protective tariffs were
necessary to safeguard domestic industries from foreign competition, particularly from
China, which was subject to tariffs peaking at 145% on Chinese goods [Data: Reports
(2, 1)]",
            "score": 90
        },
        ...
    ]
}
```

The second step synthesizes the highest-ranked key points into a coherent, multi-paragraph summary. Analyst reports—ranked by importance score—are combined into a final structured response.

```
system
    ---Role---
    You are a helpful assistant responding to questions about a dataset by synthesizing
    perspectives from multiple analysts.

    ---Goal---
    Generate a response of the target length and format that responds to the user's
    question, summarize all the reports from multiple analysts who focused on different
    parts of the dataset.
    Note that the analysts' reports provided below are ranked in the **descending order of
    importance**.
    If you don't know the answer or if the provided reports do not contain sufficient
    information to provide an answer, just say so. Do not make anything up.
    The final response should remove all irrelevant information from the analysts' reports
    and merge the cleaned information into a comprehensive answer that provides
    explanations of all the key points and implications appropriate for the response length
    and format.
    Add sections and commentary to the response as appropriate for the length and
    format. Style the response in markdown.
    The response shall preserve the original meaning and use of modal verbs such as
    "shall", "may" or "will".
    The response should also preserve all the data references previously included in the
    analysts' reports, but do not mention the roles of multiple analysts in the analysis
    process.
    **Do not list more than 5 record ids in a single reference**. Instead, list the top 5 most
    relevant record ids and add "+more" to indicate that there are more.
    For example:
    "Person X is the owner of Company Y and subject to many allegations of wrongdoing
    [Data: Reports (2, 7, 34, 46, 64, +more)]. He is also CEO of company X [Data: Reports
    (1, 3)]"
    where 1, 2, 3, 7, 34, 46, and 64 represent the id (not the index) of the relevant data
    record.
    Do not include information where the supporting evidence for it is not provided.
    Limit your response length to 2000 words.
```

---Target response length and format---
Multiple Paragraphs

---Analyst Reports---

----Analyst 1----
Importance Score: 90
...

----Analyst 1----
Importance Score: 85
...

----Analyst 1----
...

----Analyst 1----
...

...
Add sections and commentary to the response as appropriate for the length and format. Style the response in markdown.

user What are the main arguments made by the Trump administration in favor of its tariffs?

assistant

Main Arguments Made by the Trump Administration in Favor of Tariffs

The Trump administration's advocacy for tariffs was rooted in a multifaceted strategy to reshape U.S. trade policy, emphasizing economic protectionism, national security, and renegotiation of trade agreements. Below is a synthesis of the key arguments, supported by data from the provided reports.

---

1. Protection of Domestic Industries from Foreign Competition
The administration argued that tariffs were essential to shield U.S. industries from what it characterized as unfair foreign competition, particularly from China. Tariffs on Chinese goods reached as high as **145%**, framed as a necessary measure to counteract practices like intellectual property theft and currency manipulation [Data: Reports (2, 1)]. This approach aimed to reduce reliance on imported goods and incentivize domestic production, thereby preserving American jobs and industrial capacity.

Implication: By imposing steep tariffs, the administration sought to level the playing field for domestic manufacturers, which it claimed were disadvantaged by foreign subsidies and dumping practices.

---

2. National Security and Emergency Powers Justification

...

Conclusion
The Trump administration's tariff policies were underpinned by a combination of economic, legal, and symbolic arguments. By framing tariffs as essential for protecting domestic industries, national security, and fair trade practices, the administration sought to justify a protectionist agenda that reshaped U.S. trade relationships. These measures reflected a strategic shift toward prioritizing domestic economic interests over traditional multilateral trade frameworks, with lasting implications for global trade dynamics.

10.3.3.2.2 Local Search

The following demonstrates how to use local search in Microsoft GraphRAG to answer a specific question involving detailed entities and events. In this case, we query for actions taken by Donald Trump on a specific date and their economic consequences.

```
graphrag query \
    --root ./microsoft-graphrag \
    --method local \
    --query "What specific actions did Donald Trump take on April 2, 2025, concerning
    tariffs, and what was the immediate consequence of these actions on the stock
    market?"
```

This query triggers a single-round LLM interaction that draws from entity, relationship, and source tables derived from the KG. Below is the structured prompt and final response.

user ---Role---

You are a helpful assistant responding to questions about data in the tables provided.

---Goal---

Generate a response of the target length and format that responds to the user's question, summarizing all information in the input data tables appropriate for the response length and format, and incorporating any relevant general knowledge.
If you don't know the answer, just say so. Do not make anything up.
Points supported by data should list their data references as follows:
"This is an example sentence supported by multiple data references [Data: <dataset name> (record ids); <dataset name> (record ids)]."
Do not list more than 5 record ids in a single reference. Instead, list the top 5 most relevant record ids and add "+more" to indicate that there are more.
For example:
"Person X is the owner of Company Y and subject to many allegations of wrongdoing [Data: Sources (15, 16), Reports (1), Entities (5, 7); Relationships (23); Claims (2, 7, 34, 46, 64, +more)]."
where 15, 16, 1, 5, 7, 23, 2, 7, 34, 46, and 64 represent the id (not the index) of the relevant data record.
Do not include information where the supporting evidence for it is not provided.

---Target response length and format---

Multiple Paragraphs

```
---Data tables---

-----Reports-----
id|title|content
2|Donald Trump's Tariff Policies and Economic Impact Community|"..."
1|U.S.–China Trade Conflict and Tariff Agreements|"..."

-----Entities-----
id|entity|description|number of relationships
1|UNITED STATES|The United States implemented steep tariffs and engaged in trade conflicts under President Trump|7
0|DONALD TRUMP|Donald Trump is the former United States president who enacted protective tariffs and invoked the IEEPA to impose broad tariffs|8
...

-----Relationships-----
id|source|target|description|weight|links
0|DONALD TRUMP|UNITED STATES|Donald Trump is the president of the United States who enacted tariffs|9.0|1
1|DONALD TRUMP|INTERNATIONAL EMERGENCY ECONOMIC POWERS ACT|Trump invoked the IEEPA to justify tariffs|9.0|3
...

-----Sources-----
id|text
0|...

...
Add sections and commentary to the response as appropriate for the length and format. Style the response in markdown.
```

user What specific actions did Donald Trump take on April 2, 2025, concerning tariffs, and what was the immediate consequence of these actions on the stock market?

user
```
# Donald Trump's April 2, 2025, Tariff Actions and Market Impact

## **April 2, 2025: Declaration of "Liberation Day" and Tariff Announcements**
On April 2, 2025, ... [Data: **Entities** (3); **Relationships** (2); **Sources** (0)].
Trump also ... [Data: **Entities** (4); **Relationships** (3, 10, 18)].

---
## **Immediate Consequence: Stock Market Crash and Tariff Pause**
...

---
## **Contextual Analysis**
Trump's actions ... [Data: **Entities** (6, 7, 8); **Relationships** (5, 6, 7); **Sources** (0)].
The use of IEEPA to ... [Data: **Relationships** (14, 15); **Entities** (13)].

---
**References**
```

– [Data: Entities (3, 4, 5, 6, 7, 8, 14, 15); Relationships (2, 3, 4, 5, 6, 7, 10, 11, 12, 13, 14, 15, 16, 17, 18); Sources (0)]

10.3.3.3 Key Takeaways

- **Microsoft GraphRAG** is a RAG framework that uses a KG for document-based QA.
- The indexing phase involves converting raw text into a structured KG by extracting entities and relationships. This process is iterative to ensure completeness.
- The query phase supports both global search, which synthesizes high-level insights from the entire graph, and local search, which provides specific answers by referencing detailed entities and events.
- The system leverages LLM conversations throughout the process to perform tasks like entity extraction, community summarization, and answer generation.

10.3.4 HippoRAG

The objective of this section is to demonstrate how to run HippoRAG on three datasets: HippoRAG example dataset, WANDS search dataset, and WANDS recommendation dataset. HippoRAG 1 and HippoRAG 2, as detailed in Section 10.1.4, are publicly accessible at https://github.com/OSU-NLP-Group/HippoRAG.

10.3.4.1 Running HippoRAG on a Simple String List

To install HippoRAG, follow the steps in the official installation guide:

```
conda create -n hipporag python=3.10
conda activate hipporag
pip install -q hipporag==2.0.0a3
```

Next, instantiate the `HippoRAG` object. We reuse the LLM endpoints configured in Appendix A:

```
from hipporag import HippoRAG

hipporag = HippoRAG(
    save_dir="hipporag-outputs",
    llm_model_name="o1-mini",
    llm_base_url="http://localhost:4000/v1",
    embedding_model_name="text-embedding-ada-002",
    embedding_base_url="http://localhost:4000/v1",
)
```

We use a small example dataset to demonstrate how HippoRAG structures LLM conversations during the indexing and querying phases:

10.3.4.1.1 Indexing

Prepare a demo set of documents as a list of strings.[2] Then, index them using HippoRAG:

```
docs = [
    "Oliver Badman is a politician.",
    "George Rankin is a politician.",
    "Thomas Marwick is a politician.",
    "Cinderella attended the royal ball.",
    "The prince used the lost glass slipper to search the kingdom.",
    "When the slipper fit perfectly, Cinderella was reunited with the prince.",
    "Erik Hort's birthplace is Montebello.",
    "Marina is bom in Minsk.",
```

2 Document source: https://github.com/OSU-NLP-Group/HippoRAG/blob/main/README.md#openai-models

```
    "Montebello is a part of Rockland County.",
]

hipporag.index(docs=docs)
```

HippoRAG builds an index with two main steps using LLM assistance: named entity recognition (NER) extraction and triple generation.

Step 1: NER Extraction First, HippoRAG extracts named entities from each document. Below is an example of the LLM prompt and response for NER extraction:

system
Your task is to extract named entities from the given paragraph.
Respond with a JSON list of entities.

user Radio City
Radio City is India's first private FM radio station and was started on 3 July 2001.
It plays Hindi, English and regional songs.
Radio City recently forayed into New Media in May 2008 with the launch of a music portal – PlanetRadiocity.com that offers music related news, videos, songs, and other music–related features.

assistant
{"named_entities":
["Radio City", "India", "3 July 2001", "Hindi", "English", "May 2008", "PlanetRadiocity.com"]
}

user George Rankin is a politician.

assistant
{"named_entities": ["George Rankin"]}

Step 2: Triple Extraction Next, HippoRAG constructs an RDF graph by extracting triples based on the text and previously extracted named entities.

system
Your task is to construct an RDF (Resource Description Framework) graph from the given passages and named entity lists.
Respond with a JSON list of triples, with each triple representing a relationship in the RDF graph.

Pay attention to the following requirements:
– Each triple should contain at least one, but preferably two, of the named entities in the list for each passage.
– Clearly resolve pronouns to their specific names to maintain clarity.

user Convert the paragraph into a JSON dict, it has a named entity list and a triple list.
Paragraph:
```

Radio City
```

Radio City is India's first private FM radio station and was started on 3 July 2001.
It plays Hindi, English and regional songs.
Radio City recently forayed into New Media in May 2008 with the launch of a music portal – PlanetRadiocity.com that offers music related news, videos, songs, and other music–related features.
```

{"named_entities":
["Radio City", "India", "3 July 2001", "Hindi", "English", "May 2008", "PlanetRadiocity.com"]
}

**assistant**
{"triples": [
["Radio City", "located in", "India"],
["Radio City", "is", "private FM radio station"],
["Radio City", "started on", "3 July 2001"],
["Radio City", "plays songs in", "Hindi"],
["Radio City", "plays songs in", "English"],
["Radio City", "forayed into", "New Media"],
["Radio City", "launched", "PlanetRadiocity.com"],
["PlanetRadiocity.com", "launched in", "May 2008"],
["PlanetRadiocity.com", "is", "music portal"],
["PlanetRadiocity.com", "offers", "news"],
["PlanetRadiocity.com", "offers", "videos"],
["PlanetRadiocity.com", "offers", "songs"]
]
}

**user** Convert the paragraph into a JSON dict, it has a named entity list and a triple list.
Paragraph:
```
George Rankin is a politician.
```

{"named_entities": ["George Rankin"]}

**assistant**
{"named_entities": ["George Rankin"], "triples": [["George Rankin", "is", "politician"]]}

#### 10.3.4.1.2 Query

Using the previously indexed demo documents, we evaluate HippoRAG on the following example queries:[3]

```
queries = [
    "What is George Rankin's occupation?",
    "How did Cinderella reach her happy ending?",
    "What county is Erik Hort's birthplace a part of?",
]
```

3 Query source: https://github.com/OSU-NLP-Group/HippoRAG/blob/main/README.md#openai-models
```

Query Steps Queries in HippoRAG are processed in two stages: retrieval and generation. Here's how you can execute them:

```
retrieval_results = hipporag.retrieve(queries=queries, num_to_retrieve=2)
print(retrieval_results)

qa_results = hipporag.rag_qa(retrieval_results)
qa_results
```

During retrieval, an LLM filters relevant facts from a list of candidate triples based on the query. An example of this interaction looks like:

system

Your input fields are:
1. `question` (str): Query for retrieval
2. `fact_before_filter` (str): Candidate facts to be filtered

Your output fields are:
1. `fact_after_filter` (Fact): Filtered facts in JSON format

All interactions will be structured in the following way, with the appropriate values filled in.

[[## question ##]]
{question}

[[## fact_before_filter ##]]
{fact_before_filter}

[[## fact_after_filter ##]]
{fact_after_filter} # note: the value you produce must be pareseable according to the following JSON schema: {"type": "object", "properties": {"fact": {"type": "array", " description": "A list of facts, each fact is a list of 3 strings: [subject, predicate, object]", " items": {"type": "array", "items": {"type": "string"}}, "title": "Fact"}}, "required": ["fact"], " title": "Fact"}

[[## completed ##]]

In adhering to this structure, your objective is:
You are a critical component of a high–stakes question–answering system used by top researchers and decision–makers worldwide. Your task is to filter facts based on their relevance to a given query, ensuring that the most crucial information is presented to these stakeholders. The query requires careful analysis and possibly multi–hop reasoning to connect different pieces of information. You must select up to 4 relevant facts from the provided candidate list that have a strong connection to the query, aiding in reasoning and providing an accurate answer. The output should be in JSON format, e.g., {"fact": [["s1", "p1", "o1"], ["s2", "p2", "o2"]]}, and if no facts are relevant, return an empty list, {"fact": []}. The accuracy of your response is paramount, as it will directly impact the decisions made by these high–level stakeholders. You must only use facts from the candidate list and not generate new facts. The future of critical decision–making relies on your ability to accurately filter and present relevant information.

user [[## question ##]]
Are Imperial River (Florida) and Amaradia (Dolj) both located in the same country?

[[## fact_before_filter ##]]
{"fact": [["imperial river", "is located in", "florida"], ["imperial river", "is a river in", " united states"], ["imperial river", "may refer to", "south america"], ["amaradia", "flows through", "ro ia de amaradia"], ["imperial river", "may refer to", "united states"]]}

Respond with the corresponding output fields, starting with the field `[[ ## fact_after_filter ## ]]` (must be formatted as a valid Python Fact), and then ending with the marker for `[[ ## completed ## ]]`.

assistant
[[## fact_after_filter ##]]
{"fact":[["imperial river","is located in","florida"],["imperial river","is a river in","united states"],["amaradia","flows through","ro ia de amaradia"]]}

[[## completed ##]]

user [[## question ##]]
What is George Rankin's occupation?

[[## fact_before_filter ##]]
{"fact": [["george rankin", "is", "politician"], ["thomas marwick", "is", "politician"], ["oliver badman", "is", "politician"], ["erik hort", "birthplace", "montebello"], ["prince", " searched", "kingdom"]]}

Respond with the corresponding output fields, starting with the field `[[ ## fact_after_filter ## ]]` (must be formatted as a valid Python Fact), and then ending with the marker for `[[ ## completed ## ]]`.

assistant
[[## fact_after_filter ##]]
{"fact":[["george rankin","is","politician"]]}

[[## completed ##]]

An example of the returned `retrieval_results` object is:

```
{
  "question": "What is George Rankin's occupation?",
  "answer": null,
  "gold_answers": null,
  "docs": [
    "George Rankin is a politician.",
    "Thomas Marwick is a politician."
  ],
  "doc_scores": [
    0.1014,
    0.0336
  ],
  "gold_docs": null
}
```

In the generation step, the LLM uses the retrieved documents to produce a final answer. A sample dialog during this phase is shown below:

system

As an advanced reading comprehension assistant, your task is to analyze text passages and corresponding questions meticulously. Your response start after "Thought: ", where you will methodically break down the reasoning process, illustrating how you arrive at conclusions. Conclude with "Answer: " to present a concise, definitive response, devoid of additional elaborations.

user Wikipedia Title: The Last Horse

The Last Horse (Spanish:El ultimo caballo) is a 1950 Spanish comedy film directed by Edgar Neville starring Fernando Fernan Gomez.

Wikipedia Title: Southampton

The University of Southampton, which was founded in 1862 and received its Royal Charter as a university in 1952, has over 22,000 students. The university is ranked in the top 100 research universities in the world in the Academic Ranking of World Universities 2010. In 2010, the THES – QS World University Rankings positioned the University of Southampton in the top 80 universities in the world. The university considers itself one of the top 5 research universities in the UK. The university has a global reputation for research into engineering sciences, oceanography, chemistry, cancer sciences, sound and vibration research, computer science and electronics, optoelectronics and textile conservation at the Textile Conservation Centre (which is due to close in October 2009.) It is also home to the National Oceanography Centre, Southampton (NOCS), the focus of Natural Environment Research Council–funded marine research.

Wikipedia Title: Stanton Township, Champaign County, Illinois

Stanton Township is a township in Champaign County, Illinois, USA. As of the 2010 census, its population was 505 and it contained 202 housing units.

Wikipedia Title: Neville A. Stanton

Neville A. Stanton is a British Professor of Human Factors and Ergonomics at the University of Southampton. Prof Stanton is a Chartered Engineer (C.Eng), Chartered Psychologist (C.Psychol) and Chartered Ergonomist (C.ErgHF). He has written and edited over a forty books and over three hundred peer–reviewed journal papers on applications of the subject. Stanton is a Fellow of the British Psychological Society, a Fellow of The Institute of Ergonomics and Human Factors and a member of the Institution of Engineering and Technology. He has been published in academic journals including "Nature." He has also helped organizations design new human–machine interfaces, such as the Adaptive Cruise Control system for Jaguar Cars.

Wikipedia Title: Finding Nemo

Finding Nemo Theatrical release poster Directed by Andrew Stanton Produced by Graham Walters Screenplay

by Andrew Stanton Bob Peterson David Reynolds Story by Andrew Stanton Starring Albert Brooks Ellen DeGeneres Alexander Gould Willem Dafoe Music by Thomas Newman Cinematography Sharon Calahan Jeremy Lasky Edited by David Ian Salter Production company Walt Disney Pictures Pixar Animation Studios Distributed by Buena Vista Pictures Distribution Release date May 30, 2003 (2003 – 05 – 30) Running time 100 minutes Country United States Language English Budget $94 million Box office $940.3 million

Question: When was Neville A. Stanton's employer founded?

Thought:

assistant
The employer of Neville A. Stanton is University of Southampton. The University of Southampton was founded in 1862.
Answer: 1862.

user Wikipedia Title: George Rankin is a politician.

Wikipedia Title: Thomas Marwick is a politician.

Question: What is George Rankin's occupation?
Thought:

assistant
The Wikipedia title explicitly states that George Rankin is a politician.
Answer: Politician.

After running `rag_qa()`, the `qa_results` includes a populated `answer` field derived from the LLM's final response:

```
{
  "question": "What is George Rankin's occupation?",
  "answer": "Politician.",
  "gold_answers": null,
  "docs": [
    "George Rankin is a politician.",
    "Thomas Marwick is a politician."
  ],
  "doc_scores": [
    0.1014,
    0.0336
  ],
  "gold_docs": null
}
```

All-in-one Query Alternatively, you can use the all-in-one query method to combine retrieval and generation into a single step:

```
rag_results = hipporag.rag_qa(queries=queries)
rag_results
```

This simplifies the benchmarking process by eliminating the need to explicitly separate retrieval and generation steps.

Benchmarking For benchmarking, HippoRAG requires two additional inputs:

- `gold_docs`: Reference documents used to evaluate the retrieval step.
- `gold_answers`: Ground-truth answers used to assess the quality of generated responses.

Below is an example setup for benchmarking:[4]

```
gold_docs = [
    ["George Rankin is a politician."],
    [
```

4 Query and answer source: https://github.com/OSU-NLP-Group/HippoRAG/blob/main/README.md#openai-models

```
        "Cinderella attended the royal ball.",
        "The prince used the lost glass slipper to search the kingdom.",
        "When the slipper fit perfectly, Cinderella was reunited with the prince.",
    ],
    [
        "Erik Hort's birthplace is Montebello.",
        "Montebello is a part of Rockland County.",
    ],
]
answers = [
    ["Politician"],
    ["By going to the ball."],
    ["Rockland County"],
]

rag_results_gold = hipporag.rag_qa(
    queries=queries, gold_docs=gold_docs, gold_answers=answers
)
(
    queries_solutions, all_response_message, all_metadata,
    overall_retrieval_result, overall_qa_results,
) = rag_results_gold
```

After running the benchmark, you will receive performance metrics for both retrieval and QA (generation) steps. An example of retrieval evaluation results:

```
{'Recall@1': 0.6111,
 'Recall@2': 0.8889,
 'Recall@5': 1.0,
 'Recall@10': 1.0,
 'Recall@20': 1.0,
 'Recall@30': 1.0,
 'Recall@50': 1.0,
 'Recall@100': 1.0,
 'Recall@150': 1.0,
 'Recall@200': 1.0}
```

And the QA evaluation results:

```
{'ExactMatch': 0.6667, 'F1': 0.6667}
```

These metrics provide insight into how well the system retrieves relevant documents and generates accurate answers.

10.3.4.2 Running HippoRAG on the WANDS Search Dataset

In the previous section, we saw that HippoRAG accepts three main inputs: `queries`, `gold_docs`, and `gold_answers`. To use the WANDS search dataset with HippoRAG, the key task is to convert the three WANDS dataframes (`df_product`, `df_query`, and `df_label` loaded from CSV files in Section 10.3.1.2) into these three corresponding Python lists. We reuse the `df_gold` dataframe prepared in Section 10.3.1.2, and construct the required lists: `queries`, `gold_docs`, and `gold_answers`, as follows:

```
queries, gold_docs, gold_answers = [], [], []
for (query_id, query, query_class), df in df_gold.groupby(
    ["query_id", "query", "query_class"]
):
    queries.append(
        "get product_name: " + json.dumps({"query": query, "query_class": query_class
    })
    )
    gold_docs.append(
        df[df_product_columns].to_json(orient="records", lines=True).splitlines()
    )
    gold_answers.append(df["product_name"].tolist())
```

Once this preprocessing is complete, the indexing and querying steps follow exactly the same procedure as described in the previous section.

```
hipporag.index(docs=docs)
rag_results = hipporag.rag_qa(
    queries=queries,
    gold_docs=gold_docs,
    gold_answers=gold_answers,
)
(
    queries_solutions,
    all_response_message,
    all_metadata,
    overall_retrieval_result,
    overall_qa_results,
) = rag_results
```

The retrieval evaluation results `overall_retrieval_result` are shown below:

```
{'Recall@1': 0.3333,
 'Recall@2': 0.6667,
 'Recall@5': 1.0,
 'Recall@10': 1.0,
 'Recall@20': 1.0,
 'Recall@30': 1.0,
 'Recall@50': 1.0,
 'Recall@100': 1.0,
 'Recall@150': 1.0,
 'Recall@200': 1.0}
```

And the overall QA (generation) evaluation results `overall_qa_results` are:

```
{'ExactMatch': 0.5, 'F1': 0.6818}
```

10.3.4.3 Running HippoRAG on the WANDS Recommendation Dataset

We reuse the data prepared in Section 10.3.1.3, including the lists `docs`, `queries`, `gold_docs`, and `gold_answers`. The indexing and querying steps are executed using the same code as in the previous section. The retrieval evaluation results are as follows:

```
{'Recall@1': 0.5,
 'Recall@2': 1.0,
 'Recall@5': 1.0,
 'Recall@10': 1.0,
 'Recall@20': 1.0,
 'Recall@30': 1.0,
 'Recall@50': 1.0,
 'Recall@100': 1.0,
 'Recall@150': 1.0,
 'Recall@200': 1.0}
```

And the overall QA evaluation results are:

```
{'ExactMatch': 1.0, 'F1': 1.0}
```

10.3.4.4 Key Takeaways

- HippoRAG is a two-stage RAG framework. It operates by first building a KG from text and then using that graph to answer queries.
- Indexing involves two key steps:
  - Named Entity Recognition (NER): The system extracts key entities from documents.
  - Triple Generation: It then creates subject-predicate-object triples to form an RDF graph.

- Querying also consists of two stages:
  - Retrieval: An LLM filters relevant facts (triples) from the KG based on the query.
  - Generation: Another LLM uses the retrieved facts to formulate the final answer.
- Performance is evaluated using specific metrics. Retrieval performance is measured by Recall@k, while the quality of generated answers is assessed using ExactMatch and F1 scores.
- The process is flexible. It can be run on various datasets, including simple string lists and complex datasets like WANDS, with a single all-in-one query command simplifying the process.

10.4 Summary

This chapter provides a comprehensive overview of Graph-RAG frameworks for enhanced information retrieval. We began by exploring RAPTOR, a recursive abstractive processing technique for tree-organized retrieval, highlighting its construction process through initial chunking, intelligent clustering, recursive summarization, and tree building. We then discussed various retrieval strategies, including the tree traversal method and the recommended collapsed tree method, along with important implementation considerations.

Next, we dived into G-Retriever, a joint text-graph embeddings approach. We examined its architecture and core components, detailing the indexing, retrieval, subgraph construction, and generation steps. A key component, the PCST, was explained in depth, along with its role in G-Retriever.

The discussion then shifted to Microsoft GraphRAG, emphasizing its global sensemaking capability through community summaries. We outlined the sensemaking problem and the core construction phases: rich KG construction, hierarchical community detection, and bottom-up community summarization. The chapter also touched upon the query-time processing, specifically mapping-reduce for global answers, and its broader implications and applications.

Following this, we introduced HippoRAG, a neurobiologically inspired long-term memory system, exploring the neuroscience behind its algorithm, including the artificial neocortex, parahippocampal regions, and the hippocampus. We detailed the HippoRAG construction process, covering offline indexing, building the memory network, and online retrieval. We discussed the hypothesis that HippoRAG excels at knowledge integration, highlighting path-following vs. pathfinding, single-step multi-hop magic, and real-world performance, along with best practices for implementation. We concluded the HippoRAG section by discussing its role in achieving robust long-term memory, key innovations, the enhanced process, comprehensive performance, and practical examples.

Finally, we explored the current landscape of Graph-RAG tools and software, providing an overview of available general RAG frameworks with KG support, specifically focusing on KG-focused RAG frameworks. We then provided practical instructions for each framework: initializing and running RAPTOR, including custom LLM models, custom embedding model classes, and running on datasets. For G-Retriever, we covered setup and configuration, dataset preparation, inference-only mode, and fine-tuning. For Microsoft GraphRAG, we detailed the indexing phase (working directory, input files, indexing process, graph visualization, LLM conversation analysis) and the query phase (global search, local search). Lastly, for HippoRAG, we provided steps for running on a simple string list (NER extraction, triple extraction, query), and also on the WANDS search and recommendation datasets. This chapter aimed to provide a comprehensive understanding of these advanced Graph-RAG frameworks, equipping readers with the knowledge to leverage them for enhanced information retrieval.

References

1 Parth Sarthi, Salman Abdullah, Aditi Tuli, Shubh Khanna, Anna Goldie, and Christopher D Manning. Raptor: Recursive abstractive processing for tree-organized retrieval. In *The Twelfth International Conference on Learning Representations*, 2024.

2 Xiaoxin He, Yijun Tian, Yifei Sun, Nitesh Chawla, Thomas Laurent, Yann LeCun, Xavier Bresson, and Bryan Hooi. G-retriever: Retrieval-augmented generation for textual graph understanding and question answering. *Advances in Neural Information Processing Systems*, 37:132876–132907, 2024.

3 Darren Edge, Ha Trinh, Newman Cheng, Joshua Bradley, Alex Chao, Apurva Mody, Steven Truitt, Dasha Metropolitansky, Robert Osazuwa Ness, and Jonathan Larson. From local to global: A graph rag approach to query-focused summarization. *arXiv preprint arXiv:2404.16130*, 2024.

4 Bernal Jiménez Gutiérrez, Yiheng Shu, Yu Gu, Michihiro Yasunaga, and Yu Su. Hipporag: Neurobiologically inspired long-term memory for large language models. In *The Thirty-eighth Annual Conference on Neural Information Processing Systems*, 2024.

5 Bernal Jiménez Gutiérrez, Yiheng Shu, Weijian Qi, Sizhe Zhou, and Yu Su. From rag to memory: Non-parametric continual learning for large language models. *arXiv preprint arXiv:2502.14802*, 2025.

6 Harsh Trivedi, Niranjan Balasubramanian, Tushar Khot, and Ashish Sabharwal. Interleaving retrieval with chain-of-thought reasoning for knowledge-intensive multi-step questions. *ArXiv preprint arXiv:2212.10509*, 2022.

7 Harsh Trivedi, Niranjan Balasubramanian, Tushar Khot, and Ashish Sabharwal. MuSiQue: Multihop questions via single-hop question composition. *Transactions of the Association for Computational Linguistics*, 10:539–554, 2022.

8 Xanh Ho, Anh-Khoa Duong Nguyen, Saku Sugawara, and Akiko Aizawa. Constructing a multi-hop QA dataset for comprehensive evaluation of reasoning steps. In *Proceedings of the 28th International Conference on Computational Linguistics*, pages 6609–6625. Barcelona, Spain: International Committee on Computational Linguistics, December 2020. https://www.aclweb.org/anthology/2020.coling-main.580.

9 LangChain. Langchain documentation: Introduction, 2025. https://python.langchain.com/docs/introduction/. Accessed: 2025-6-22.

10 Apache Parquet. Apache Parquet documentation, 2025. https://parquet.apache.org/docs/. Accessed: 2025-6-29.

11 Mathieu Bastian, Sebastien Heymann, and Mathieu Jacomy. Gephi - The open Graph Viz platform, 2025. https://gephi.org/. Accessed: 2025-6-29.

11

Frontiers of Graph-RAG

In the fast-evolving landscape of artificial intelligence (AI) and machine learning, the frontiers of research hold the key to unprecedented possibilities. This chapter embarks on a forward-looking exploration of the future trajectories of graph-based retrieval-augmented generation (Graph-RAG) and its applications. We examine emerging trends, potential challenges, and the implications of integrating graph-based methodologies across various domains. By analyzing cutting-edge research and theoretical frameworks, we aim to spark curiosity and inspire innovation in the ongoing development of intelligent systems. As we peer into the future, we invite readers to envision a world where Graph-RAG not only enhances existing technologies but also serves as a catalyst for breakthroughs yet to be imagined.

11.1 Emerging Trends in Graph-RAG

The evolution of Graph-RAG is accelerating as researchers push the boundaries of what's possible when combining structured knowledge with generative AI. These emerging trends represent not just incremental improvements, but fundamental shifts in how we conceptualize and implement intelligent information systems.

11.1.1 Multimodal Knowledge Graphs

The future of Graph-RAG lies not in text alone, but in the seamless integration of all forms of human knowledge. Multimodal knowledge graphs (MMKGs) represent a paradigm shift from single-modality systems to comprehensive frameworks that mirror the richness of human perception and understanding. Just as humans naturally process information through multiple senses—sight, sound, touch—to form cohesive understanding, MMKGs represent a paradigm shift toward AI systems that mirror this richness of perception [1].

Traditional knowledge graphs, while powerful for textual relationships, capture only a fraction of real-world information. The future lies in comprehensive frameworks that seamlessly integrate diverse data modalities—text, images, audio, video, and structured data—enabling AI to develop more nuanced, context-aware understanding of complex scenarios. For example, consider a medical diagnosis scenario (refer to Figure 6.6 multimodal knowledge graph in Chapter 6): a doctor doesn't rely solely on written reports but integrates patient history (text), medical imaging (visual), lab results (tabular), and even patient voice patterns (audio) to form a complete picture.

The MMKGs have become increasingly popular in research and industry. Recent studies [2] have provided a comprehensive survey of the large-scale MMKGs such as MMpedia[3], TIVA-KG[4], and VTKGs[5], which have become central, with examples like ImageNet and VisualGenome supporting applications across domains. Figure 11.1 illustrates the knowledge graph and multimodal learning synergy as cited from the paper.

11.1.1.1 Handling Diverse Node Types

MMKGs fundamentally reimagine what constitutes a "node." Beyond simple entity representation, nodes now encompass rich, heterogeneous data types, each requiring specialized processing while maintaining semantic coherence within the unified graph structure.

As illustrated in Figure 11.2, these are the major node types:

11.1.1.1.1 Text Nodes: The Semantic Foundation

Text remains the lingua franca of multimodal systems, serving as the connective tissue between modalities:

- Entity Descriptions: Rich semantic representations using state-of-the-art language models, capturing nuanced meanings through contextual embeddings (bidirectional encoder representations from transformers [BERT], generative pretrained transformer [GPT] architectures).
- Contextual Narratives: Complete documents and passages providing comprehensive background, maintaining discourse-level understanding.

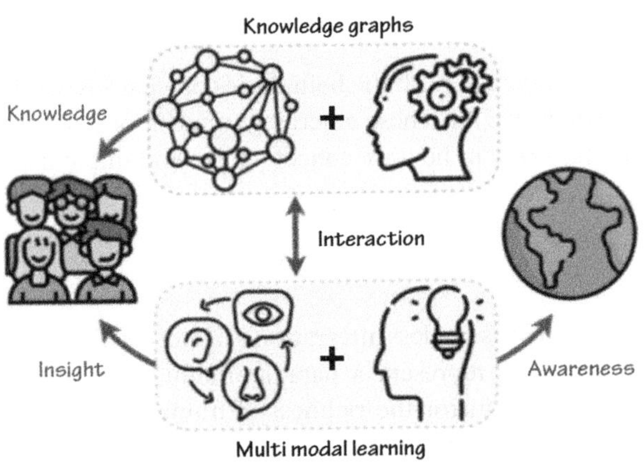

Figure 11.1 Knowledge graphs meet multimodal learning [2].

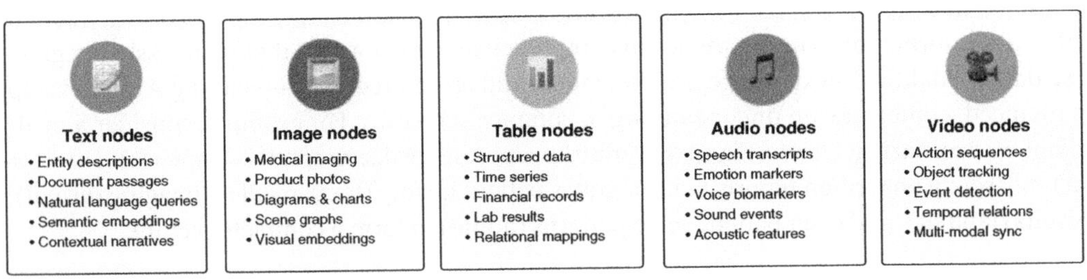

Figure 11.2 Multimodal node type architecture.

- Cross-modal Bridges: Natural language descriptions that link and explain relationships between other modalities.
- Query Interfaces: Enabling users to access any modality through natural language queries.

11.1.1.1.2 Image Nodes: Visual Intelligence Integration

Visual data introduces spatial reasoning capabilities that complement textual understanding:

- Medical Imaging Integration: X-rays, MRIs, and CT scans linked to diagnostic entities, with spatial annotations preserving critical visual features through CNN-extracted representations.
- Scene Graph Decomposition: Complex images automatically parsed into entity-relationship structures, enabling structured queries over visual content.
- Multiresolution Hierarchies: Images stored at multiple resolutions, with attention mechanisms focusing on relevant regions based on query context.
- Cross-modal Alignment: Visual features mapped to shared embedding spaces using techniques like contrastive language-image training (CLIP) [6], enabling seamless image-text reasoning.

11.1.1.1.3 Audio Nodes: Capturing the Acoustic Dimension

Audio data enriches graphs with temporal and acoustic intelligence:

- Speech Analytics: Beyond transcription, capturing prosodic features, emotional undertones, and speaker characteristics through spectrogram analysis and transformer architectures.
- Environmental Context: Acoustic scene understanding linking sounds to locations, events, and activities.
- Multimodal Synchronization: Temporal alignment with other modalities, particularly crucial for video understanding.
- Clinical Voice Biomarkers: Subtle vocal changes linked to health conditions, enabling early disease detection.

11.1.1.1.4 Table Nodes: Preserving Structural Integrity

Structured data maintains its relational properties while enabling graph-based reasoning:

- Financial Time Series: Market data with temporal dependencies, calculation relationships, and statistical properties preserved.
- Scientific Datasets: Experimental results maintaining units, precision, statistical significance, and methodological context.
- Relational Mappings: Database foreign keys naturally translated into graph edges, enabling SQL-like queries over graph structures.
- Aggregation Hierarchies: Precomputed summaries at various granularities for efficient analytical queries.

11.1.1.1.5 Video Nodes: Spatio-temporal Reasoning

Video represents the frontier of multimodal integration, combining visual, audio, and temporal dimensions:

- Action Recognition: Decomposition into semantic event sequences with temporal boundaries.
- Object Tracking: Persistent entity identification across frames, maintaining identity through occlusions and scene changes.

- Multimodal Alignment: Synchronizing visual, audio, and textual streams for comprehensive scene understanding.
- Hierarchical Representation: From frame-level features to scene-level summaries, enabling queries at multiple temporal scales.

11.1.1.2 Cross-modal Reasoning

Cross-modal reasoning represents the pinnacle achievement of multimodal Graph-RAG—the ability to seamlessly traverse and integrate information across modalities to answer complex queries that no single modality could address alone.

11.1.1.2.1 Query Translation and Expansion

Multimodal systems excel at accepting queries in one modality and retrieving relevant information across all others. This capability transforms user interaction patterns:

Here is an example of Python pseudocode of visual query with multimodal results:

```
user_query = upload_image("skin_condition.jpg")
results = multi_modal_graph.query(
    input_modality="image",
    retrieve_modalities=["text", "image", "table", "audio"],
    query=user_query
)
# Returns:
# - Medical literature descriptions (text)
# - Similar case images from clinical databases (image)
# - Treatment efficacy statistics (table)
# - Patient testimonials (audio)
```

11.1.1.2.2 Semantic Bridge Construction

The system automatically identifies and leverages conceptual bridges between modalities:

- Visual-textual Grounding: Linking visual elements to their textual descriptions, enabling bidirectional reasoning.
- Audio-visual Synchronization: Aligning speech with visual events for comprehensive scene understanding.
- Cross-modal Attention: Attention mechanisms that dynamically weight different modalities based on query relevance.

11.1.1.2.3 Complementary Information Fusion

Different modalities provide unique, often complementary perspectives that, when combined, yield superior understanding:

- Medical Diagnosis: Integrating symptoms (text), imaging (visual), lab results (tables), and clinical observations (audio/video) for holistic assessment.
- Product Intelligence: Combining reviews (text), images (visual), specifications (structured), and demonstration videos (multimodal) for comprehensive product understanding.
- Security Analysis: Correlating incident reports (text), surveillance footage (video), sensor readings (structured), and audio recordings for threat detection.

11.1.1.2.4 Dynamic Fusion Architectures

- Early Fusion: Raw modality concatenation for deep integration
- Late Fusion: Modality-specific processing with decision-level integration
- Hybrid approaches like cross-modal memory networks

11.1.1.2.5 Disambiguation Through Multimodal Context

When single-modality information proves ambiguous, other modalities provide crucial disambiguating context:

- Text: "Python" + Code snippet → Programming language ✔ (not snake)
- Text: "Bank" + River image → Riverbank ✔ (not financial institution)
- Audio: "Cell" + Microscope image → Biological cell ✔ (not prison cell)

This contextual disambiguation happens automatically through learned cross-modal associations, dramatically improving query understanding accuracy.

11.1.1.3 Implementation Challenges and Solutions

The implementation of multimodal Graph-RAG systems presents significant technical challenges requiring innovative solutions across multiple dimensions. The major challenges in multimodal learning are illustrated in Figure 11.3 [7].

Here we list a few challenges and their potential solutions.

11.1.1.3.1 Data Alignment: Bridging Modality Gaps

Data alignment represents one of the fundamental challenges in multimodal Graph-RAG systems. Different modalities possess inherently different structures, dimensionalities, and semantic properties, making their meaningful integration nontrivial.

Consider image captioning: aligning pixel-level visual information with word-level textual descriptions requires bridging a significant semantic gap. The relationship between an image region and its corresponding textual description is often implicit, contextual, and many-to-many.

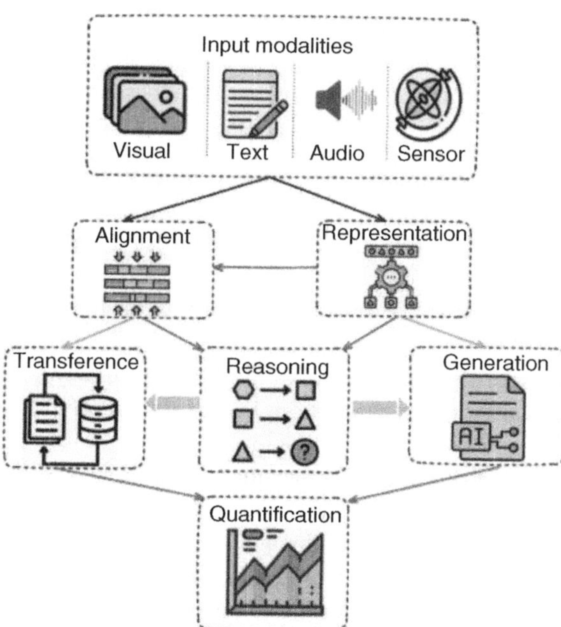

Figure 11.3 The core technical challenges in multimodal learning [7].

A single word might relate to multiple image regions, while a single visual element might be described by various textual expressions.

Creating shared representational spaces where diverse modalities can be meaningfully compared has emerged as the cornerstone solution:

1. Contrastive Learning Frameworks
   - CLIP [6]: Aligns visual and textual representations through large-scale contrastive learning, creating a joint embedding space where semantically similar concepts cluster together regardless of modality.
   - ALIGN [8]: Extends the contrastive approach to noisy, web-scale data, demonstrating robustness in real-world scenarios.
   - ImageBind: Pushes beyond vision-language to create unified spaces for six modalities including audio, depth, thermal, and IMU data.
2. Universal Architecture Approaches
   - Perceiver IO [9]: Processes arbitrary modalities through a single, unified transformer architecture, eliminating the need for modality-specific encoders.
   - Unified Transformer Networks: Share parameters across modalities while maintaining modality-specific tokenization strategies.
3. Learned Projection Networks: Modality-specific encoders project into shared spaces through learned transformations that preserve semantic relationships.

Here is the integration pipeline for MMKGs, where all modalities are mapped to common semantic representations:

Raw Data Input → Modality-specific Encoding → Unified Embedding Space → Graph Construction

11.1.1.3.2 *Reasoning Complexity: Navigating Heterogeneous Graphs*

Cross-modal reasoning in Graph-RAG systems requires sophisticated algorithms capable of traversing paths that span different modalities while maintaining semantic coherence. These are some advanced navigation strategies:

1. Modality-aware Attention Mechanisms: These mechanisms dynamically weight the importance of different modalities based on the query context.
2. Typed Graph Neural Networks (GNNs): GNNs that explicitly model different edge types for intra-modal vs. cross-modal connections:
   - Heterogeneous Graph Transformers (HGT): Apply type-specific transformations for different node and edge types.
   - Relational GCNs (R-GCNs): Use relation-specific weight matrices for different edge types.
   - Modality-aware Message Passing: Customize aggregation functions based on source and target modalities.
3. Reinforcement Learning (RL) for Optimal Traversal: RL agents learn query-specific navigation strategies:
   - State Representation: Current node features + query embedding + traversal history
   - Action Space: Available edges from current node, filtered by modality constraints
   - Reward Design: Balances relevance, diversity, and computational efficiency
   - Policy Learning: Deep q-networks or policy gradient methods for path selection

4. Hybrid Symbolic-neural Reasoning: Combining the interpretability of symbolic logic with the flexibility of neural approaches:
   - Neural Theorem Provers: Learn to apply logical rules while navigating multimodal graphs
   - Differentiable Logic: Soft logic operations that maintain gradient flow
   - Explainable Pathways: Generate human-readable explanations for reasoning chains

11.1.1.3.3 Scalability: Managing Complex Multimodal Graphs

Multimodal graphs face exponential growth in both data volume and computational complexity. Each additional modality multiplies the potential connections and storage requirements. To address this, researchers are adopting the following strategies:

1. Hierarchical Indexing: Multilevel indexing schemes that first partition data by modality and then organize it semantically allow for targeted retrieval and efficient storage. Figure 11.4 illustrates an example for hierarchical indexing for the multimodal graph:
2. Intelligent Loading and Caching: Optimizes resource usage through strategic data management. Lazy loading defers heavy media retrieval until needed, while predictive prefetching analyzes query patterns to anticipate and preload likely-needed data. The caching layer employs modality-aware eviction policies that consider data size, access frequency, and modality importance beyond simple LRU strategies. Progressive resolution loading balances responsiveness with efficiency by serving low-resolution previews first, then loading full quality on demand.
3. Distributed Processing Architecture: Exploits modality characteristics for optimal performance. Text and metadata reside on high-IOPS SSDs for rapid access, while images and videos are distributed across object storage with CDN integration. Memory-optimized instances cache hot embeddings for microsecond retrieval. This enables federated query processing—

```
MultiModalGraph/
├── Modality_Index/
│   ├── Text/
│   │   ├── Entities/
│   │   ├── Documents/
│   │   └── Embeddings/
│   ├── Images/
│   │   ├── Objects/
│   │   ├── Scenes/
│   │   └── Features/
│   └── Audio/
│       ├── Speech/
│       ├── Music/
│       └── Environmental/
└── Cross_Modal_Index/
    ├── Text_Image_Alignments/
    ├── Audio_Visual_Sync/
    └── Multimodal_Clusters/
```

Figure 11.4 Hierarchical indexing of multimodal graph.

decomposing complex queries for parallel execution across distributed partitions—while edge computing processes data near its source for real-time applications.

4. Progressive Refinement Strategies: Balance search completeness with computational efficiency. Coarse-to-fine search begins broadly to identify promising graph regions, then narrows for precise results. Sampling-based exploration estimates properties of million-node regions without exhaustive traversal. Approximate nearest neighbor algorithms trade marginal accuracy for massive speedups, while dynamic pruning eliminates unpromising paths early, focusing resources on high-potential directions.

Performance optimization through these integrated techniques enables scaling to billions of nodes with interactive response times. This scalable foundation maintains consistent performance as graphs grow, supporting both real-time applications and deep analytical queries across the same infrastructure.

11.1.2 Temporal and Dynamic Graphs

Real-world knowledge is not static—it evolves, updates, and sometimes contradicts itself over time. Temporal Graph-RAG represents a fundamental shift from snapshot-based knowledge to living, breathing information systems that understand the flow of time [10]. This temporal awareness transforms RAG systems from static repositories into dynamic knowledge ecosystems that capture the evolution of facts, relationships, and understanding.

11.1.2.1 The Temporal Dimension

Traditional RAG systems operate on a fixed snapshot of knowledge, unable to reason about temporal dynamics. This limitation becomes critical when dealing with queries that require temporal understanding, such as:

- Historical Analysis: "How did Amazon's business model evolve from an online bookstore in 1995 to a cloud computing giant in 2025?"
- Trend Detection: "What are the emerging topics in AI research over the past five years?"
- Policy Evolution: "How have data privacy regulations changed across different jurisdictions from GDPR (2018) to present?"
- Scientific Progress: "Track the development of COVID-19 treatment protocols from March 2020 to December 2024."

Temporal Graph-RAG addresses these challenges through sophisticated temporal modeling that preserves the historical context while enabling dynamic updates.

Figure 11.5 illustrates an example of the Amazon's knowledge graph evolved from 1995 to 2005, demonstrating node appearing, changing types, and relationships evolving over time.

11.1.2.1.1 Versioned Knowledge Elements

Every piece of information in a temporal graph carries rich temporal metadata:

- Node Versions: Each entity maintains temporal variants (e.g., "Amazon_1995" vs. "Amazon_2023").
- Temporal Edges: Relationships with validity windows.
- Confidence Decay: Fact reliability scores that decrease over time (e.g., $\Delta t = -0.02$/week in Bloomberg's financial KG).

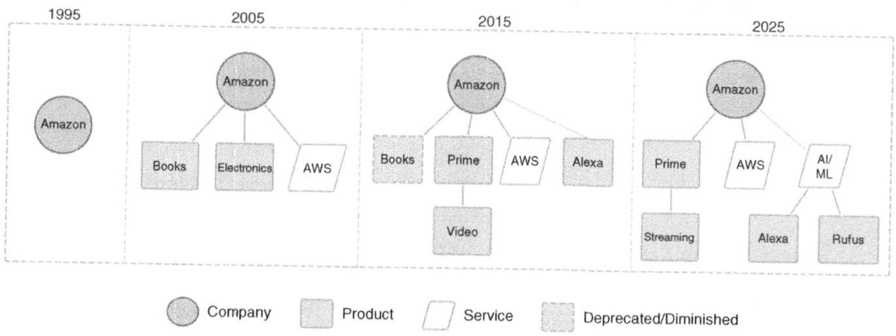

Figure 11.5 Temporal graph evolution: How knowledge graphs evolve over time with changing entities and relationships.

Specifically for temporal edges, it incorporates temporal relationship modeling, where relationships between entities are inherently temporal, capturing how connections form, evolve, and dissolve. Some examples include:

- Employment History:
  "Sundar Pichai [CEO_OF] Google (2015–present), previously [SENIOR_VP_OF] Google (2014–2015)"
- Corporate Evolution:
  "Instagram [SUBSIDIARY_OF] Facebook (2012–2021), then [SUBSIDIARY_OF] Meta (2021–present)"
- Scientific Consensus:
  "Pluto [CLASSIFIED_AS] Planet (1930–2006), then [CLASSIFIED_AS] Dwarf Planet (2006–present)"

11.1.2.1.2 Temporal Reasoning Operators

Temporal Graph-RAG implements sophisticated operators based on Allen's interval algebra [11]:

Here is the Python pseudo code for the temporal reasoning operators:

```
class TemporalOperators:
    @staticmethod
    def before(interval1, interval2):
        """Check if interval1 occurs entirely before interval2"""
        return interval1.end < interval2.start

    @staticmethod
    def overlaps(interval1, interval2):
        """Check if intervals have temporal overlap"""
        return interval1.start < interval2.end and interval2.start < interval1.end

    @staticmethod
    def during(point_or_interval, interval):
        """Check if point/interval occurs during another interval"""
        if hasattr(point_or_interval, 'start'):
            return (point_or_interval.start >= interval.start and
                    point_or_interval.end <= interval.end)
        else:
            return interval.start <= point_or_interval <= interval.end

    @staticmethod
    def meets(interval1, interval2):
        """Check if interval1 immediately precedes interval2"""
        return interval1.end == interval2.start
```

Table 11.1 Temporal reasoning primitives.

Operator	Description	Example use case queries
BEFORE/AFTER	Causal relationship analysis	"What events preceded the 2008 financial crisis?"
DURING	Concurrent event detection	"Which companies were developing COVID vaccines simultaneously?"
FIRST/LAST	State transition / change points	"When did public sentiment on climate change reach a tipping point?"
INTERVAL	Duration calculations	"How long has Taiwan Semiconductor maintained market leadership?"

These operators enable complex temporal queries. Table 11.1 lists examples of the use cases for each operator.

11.1.2.1.3 *Versioning and Contradiction Resolution*

In temporal knowledge graphs, facts evolve, overlap, or even contradict each other over time. Resolving these inconsistencies while maintaining historical accuracy is critical for reliable reasoning. Temporal Graph-RAG employs several advanced techniques to manage versioning and contradiction resolution.

1. Versioned Knowledge Representation
   Every entity, relationship, or attribute in a temporal graph is associated with metadata that captures its temporal validity. This includes:
   - Temporal validity intervals (e.g., valid_from, valid_to)
   - Source provenance (e.g., origin dataset, extraction timestamp)
   - Confidence scores (decaying over time or updated with new evidence)
     For example:
     A company's CEO transition would be stored as:

```
{
 "entity": "Google",
 "relationship": "has_CEO",
 "target": "Sundar Pichai",
 "valid_from": "2015–10–01",
 "valid_to": null, // Indicates ongoing
 "source": "SEC Filing 2015",
 "confidence": 0.98,
 "decay_rate": "0.001/month" // Confidence decreases slightly over time
}
```

2. Contradiction Detection and Resolution
   Contradictions in temporal knowledge graphs arise when two or more conflicting statements overlap in their time frames or logical assertions. For instance: "Pluto is a planet" (1930–2006); "Pluto is a dwarf planet" (2006–present). While these assertions conflict semantically, they are temporally consistent, and can coexist when versioned properly. Problems arise when conflicting facts overlap temporally, or when a source claims both in the same time interval.
   To handle this, several mechanisms are implemented:
   Temporal Precedence Heuristics: When two contradictory facts exist for the same time interval, the system favors:
   - More recent sources with higher trust scores

- Explicit timestamps over inferred dates
- Data with higher confidence and lower uncertainty

Source Credibility Modeling: Assigns weighted trust scores to sources based on historical reliability, domain expertise, and consistency over time. Contradictions are resolved by computing credibility-weighted voting.

Conflict Annotation and Soft Reasoning: Instead of discarding one fact, both are retained and marked as conflicting. A reasoning engine considers both sides and may:

- Alert users to the ambiguity
- Ask for user input
- Apply probabilistic or logic-based inference to assess likelihood

Versioned Fact Provenance: Each fact stores a traceable path of its origin—documents, timestamps, source identity, and context. This enables auditable explanations and transparent fact tracking.

Here is an implementation example with pseudo code:

```
def resolve_conflict(fact1, fact2):
    if fact1.source_authority > fact2.source_authority:
        return fact1
    elif fact1.timestamp > fact2.timestamp and fact1.confidence > threshold:
        return fact1
    else:
        return fact2  # Fallback to consensus or manual review
```

11.1.2.2 Dynamic Graph Evolution

Temporal knowledge graphs become truly dynamic when they are equipped with mechanisms for autonomous adaptation—continually evolving with the influx of new data, feedback, and contextual shifts. Unlike static graphs that require manual refreshes, dynamic Graph-RAG systems self-update, reindex, and relearn in response to changes in the external world.

11.1.2.2.1 Self-updating Mechanisms

Modern dynamic graphs are equipped with continuous ingestion pipelines capable of detecting new information and integrating it incrementally. The update process includes several key components:

1. Change Detection and Monitoring
   Real-time monitors track a wide range of structured and unstructured data sources—news streams, research publications, APIs, social media, enterprise logs, etc.—to detect changes that may affect existing knowledge. Techniques such as named entity drift detection, relation frequency shifts, and event extraction are used.
2. Incremental Knowledge Integration
   Rather than rebuilding the entire graph with each update, the system performs delta updates, modifying only the affected nodes and edges. Embeddings are updated incrementally, using online learning or streaming GNNs [12]. For example:

```
new_fact = ("OpenAI", "released", "GPT-5", "2025-03-01")
temporal_graph.update(new_fact)
```

3. Confidence Decay Models
   To model the degradation of factual reliability over time, facts are assigned a decay function. These functions vary by domain:
   - Financial Events: Fast decay (e.g., half-life of weeks)
   - Scientific Knowledge: Slow decay, step-function drops (e.g., retractions)
   - Historical Records: Minimal decay

A simple decay model:

$$\text{Confidence}_t = \text{Confidence}_0 \cdot e^{-\lambda \cdot (t - t_0)}$$

where λ is domain-dependent, and $t - t_0$ is the time since fact insertion.

4. Automated Deprecation
Information no longer relevant or superseded by newer data is systematically archived or removed. For example, if a company changes its name, the old name might be deprecated, with the new name becoming active, while the old fact is retained for historical reference.

This prevents stale or invalid information from influencing future queries. Archived facts are retained in cold storage layers and can be reinstated if corroborated by new sources. This aligns with privacy and compliance needs, where the system implements "forget-me" pipelines and entity-level TTL policies for GDPR and HIPAA compliance.

Automated Deprecation and Archival

- Time-based Archival: Facts beyond a validity window (e.g., outdated product prices) are moved to cold storage.
- Usage-based Retention: Rarely queried data is compressed or purged (e.g., obsolete smartphone models).
- Provenance Tracking: Retains lineage for auditability (e.g., "This fact was sourced from WHO 2020-03 report").

Example Policy:

```
{
  "retention_policy": {
    "financial_data": "7 years (SEC compliance)",
    "scientific_claims": "10 years unless superseded",
    "news_headlines": "6 months"
  }
}
```

Implementation Pipeline:

[Change Detection] → [Temporal Alignment] → [Conflict Resolution] → [Graph Rewiring] → [Index Update]

11.1.2.2.2 *Adaptive Learning Systems*

Beyond data updates, the graph itself adapts its structure and representation based on usage patterns and feedback.

Query-driven Reinforcement: Frequently accessed subgraphs are prioritized in caching, and edge weights may be updated to reflect user engagement. This results in a dynamic reranking of relevance over time.

Feedback-informed Refinement: User feedback loops (thumbs-up/down, click-throughs, time-on-node) train the system to adjust traversal strategies, retrain entity-linking models, and reweight conflicting sources.

Anomaly Detection: Graph evolution can also highlight unexpected or suspicious changes:

- Sudden formation or deletion of dense subgraphs
- Emerging echo chambers in media graphs
- Divergence in consensus across previously aligned sources
- Facts violating previously held temporal constraints

Tools such as temporal GNN-based outlier detection and Bayesian changepoint models are employed to flag and investigate anomalies.

Example Anomaly Detection Rule:

```
def detect_anomaly(node):
    if node.type == "Stock" and node.volatility > 3 * historical_avg:
        trigger_alert("Possible market manipulation", node)
```

These sophisticated mechanisms ensure that temporal Graph-RAG systems not only capture the evolution of knowledge but actively participate in that evolution, learning and adapting to provide increasingly accurate and relevant information over time.

11.1.2.3 Applications and Implications

Temporal and dynamic Graph-RAG systems unlock powerful capabilities across a wide range of domains by capturing the evolution of knowledge over time and enabling time-aware reasoning.

Table 11.2 illustrates how temporal and dynamic Graph-RAG transforms various industries by capturing time-sensitive patterns and evolutionary dynamics. Each domain leverages temporal awareness differently—from microsecond precision in financial markets to multiyear paradigm shifts in scientific research.

The table demonstrates that while the core temporal graph technology remains consistent, its application varies dramatically by domain—from tracking millisecond market movements to monitoring decade-long scientific paradigm shifts. This flexibility makes temporal Graph-RAG a powerful tool for any field where understanding change over time is crucial.

The integration of temporal and dynamic capabilities into Graph-RAG systems represents a crucial evolution in AI's ability to understand and reason about our ever-changing world. By capturing not just what is true but when it was true and how it changed, these systems provide a foundation for more nuanced, context-aware, and historically informed AI applications. As we continue to generate information at unprecedented rates, temporal Graph-RAG becomes not just useful but

Table 11.2 Domain-specific applications of temporal Graph-RAG systems.

Domain	Key applications	Temporal characteristics	Example use cases
Financial markets and trading	Risk monitoring, regulatory compliance, market sentiment analysis, audit trail maintenance	Microsecond precision, real-time updates, historical pattern matching, continuous compliance tracking	Track portfolio exposure evolution; Correlate news events with price movements; maintain regulatory audit information
Scientific research and discovery	Literature evolution, citation dynamics, reproducibility tracking, paradigm shift detection	Multiyear timescales, knowledge emergence patterns, consensus evolution, influence propagation	COVID-19 treatment efficacy evolution (2020–2021); track research topic lifecycles; identify scientific consensus shifts
Healthcare and epidemiology	Disease progression modeling, treatment protocol evolution, epidemic dynamics, drug safety monitoring	Patient timeline tracking, evidence-based updates, real-time case integration, long-term safety patterns	Model patient symptom journeys; Update treatment guidelines with new evidence; track adverse events post-market
News and media analysis	Fact-checking, narrative analysis, echo chamber detection, misinformation tracking	Story mutation tracking, source reliability decay, information cascade timing, narrative evolution patterns	Trace claim origins across sources; Identify information silos; track misinformation spread

essential for maintaining accurate, up-to-date knowledge systems that can adapt as quickly as the world they seek to understand.

11.1.3 Advanced GNNs for RAG

RAG systems traditionally fetch facts or passages from knowledge sources to assist a language model. By integrating GNNs into RAG, we move beyond simple rule-based traversal or keyword search toward learned, intelligent navigation of knowledge structures. In Part III, we introduced GNN-based RAG and noted that this integration is a quantum leap: a GNN can learn which nodes, edges, or subgraphs are relevant to a query, enabling multi-hop reasoning that static retrieval often misses. For example, a GNN-based retriever can take a complex question and propagate influence through a knowledge graph, automatically highlighting a chain of relationships that leads to the answer. This section dives deeper into advanced GNN architectures and learning paradigms that empower RAG systems, along with emerging techniques to make these systems more explainable and trustworthy.

11.1.3.1 Next-generation GNN Architectures

Recent research has produced GNN architectures tailored for richer, more complex graph data and tasks. These "next-generation" GNNs enhance RAG by allowing the model to reason over heterogeneous data, handle temporal knowledge, and incorporate memory for long-term dependencies. We discuss three key advancements—HGT, continuous-time graph networks, and memory-augmented GNNs—and how each can bolster RAG.

11.1.3.1.1 Heterogeneous Graph Transformers

Real-world knowledge graphs are often heterogeneous, containing multiple types of nodes and relations. For instance, a scholarly graph might include papers, authors, institutions, etc., with edges like writes, cites, or affiliated_with. Unlike standard GNNs that assume one node type, HGT [13] introduces attention mechanisms that account for different node and edge types. These models have type-specific parameters or projections for each relation, so the attention weight between an author node and a paper node is computed differently than between two paper nodes. This allows the model to learn the significance of each edge type in context—for example, a cites link might be more relevant for a question about influence, whereas an authored_by link might matter for questions about an author's work. An example heterogeneous graph with two node types (User, Game) and two edge types (follows, plays). HGT uses specialized attention for each relation type, so they can weigh "follows" edges differently from "plays" edges when reasoning. Beyond handling multiple node types, some transformers are designed for multimodal or cross-domain graphs. Cross-modal attention allows the GNN-RAG to integrate information from different data modalities—for example, linking text nodes to image nodes in a knowledge graph. Imagine a question-answering system that reasons over a graph of encyclopedia articles (text nodes) and their associated images (image nodes). A heterogeneous transformer can attend to both modalities: if the query is "What does the flag of California look like?", the model can focus on the image node of the flag while also considering text descriptions, thanks to cross-modal attention mechanisms that align visual and textual information. This flexibility is crucial in RAG scenarios where knowledge isn't purely textual. Another frontier is hierarchical attention for multilevel reasoning. In complex graphs, information may be structured in layers or communities. Hierarchical transformers apply attention at multiple levels—for example, first aggregating information within local neighborhoods or type-specific groups, then at a higher level across groups. A recent Hierarchical Heterogeneous Graph

Transformer (HHGT) exemplifies this: it uses a "type-level" transformer to aggregate nodes of the same type within a neighborhood, and a "ring-level" transformer to combine information from neighbors at different hop distances. Such two-stage attention lets the model reason about both fine-grained details (within a group of related nodes) and coarse-grained structure (how far something is in the graph), which can improve multi-hop question answering. If a query requires reasoning through several hops, hierarchical attention ensures the GNN gives appropriate focus at each step of the path, rather than treating all neighbors uniformly. Importantly, attention weights can be made dynamic based on the query context. Traditional graph neural nets compute a single set of weights for all tasks, but in a RAG system the query can influence what's important. Advanced architectures allow injecting the query embedding into the attention mechanism, effectively making the graph traversal query-aware. For instance, if the user asks "Which biologist discovered a chemical element?", the model can bias attention toward scientist—discovery—element relations in a science knowledge graph, rather than, say, author-paper relations. This dynamic reweighting means the GNN doesn't just blindly follow graph connections—it steers through the graph in a way that is contextually relevant to the question. This capability moves us closer to intelligent retrieval: the system learns to navigate the graph as a human expert might, focusing on paths likely to yield an answer. In summary, the key components of HGT are:

- Heterogeneous Mutual Attention: Uses multi-head attention (h=8 heads) with node-type dependent projections (K-Linear$^{\tau(s)}: R^d \rightarrow R^{d_h}$, Q-Linear$^{\tau(t)}: R^d \rightarrow R^{d_h}$) and edge-type matrix $W^{\text{ATT}}_{\phi(e)} \in R^{d_h \times d_h}$, scaled by prior tensor $\mu \in R^{a \times r \times a}$.
- Heterogeneous Message Passing: Passes information using M-Linear$^{\tau(s)}: R^d \rightarrow R^{d_h}$ and $W^{\text{msg}}_{\phi(e)} \in R^{d_h \times d_h}$.
- Target-specific Aggregation: Aggregates using attention weights, followed by A-Linear$^{\tau(t)}$ and residual connection, with $L = 3$ layers.
- Relative Temporal Encoding (RTE): Models dynamics using sinusoid functions, adding temporal information to node representations.
- HGSampling: Samples sub-graphs with n nodes per type, L depth, using importance sampling for scalability.

11.1.3.1.2 Continuous-time Graph Networks (CTGNs)

Many knowledge graphs and retrieval scenarios involve a temporal dimension. New facts emerge over time, and some relationships are transient. Continuous-time Graph Networks (CTGNs) are designed to handle graphs that evolve continuously, rather than in discrete snapshots [14]. Traditional dynamic GNNs might divide time into steps (e.g., a separate graph for each year) and run a normal GNN on each snapshot. In contrast, continuous-time models use techniques like neural ordinary differential equations (Neural ODEs) to model smooth evolution of node embeddings over real time. The key idea is that the change in a node's state is governed by a differential equation learned by a neural network, enabling the model to integrate these changes over any time interval. One benefit of this approach is the ability to make fine-grained updates triggered by events rather than fixed intervals. In a RAG context, imagine a live knowledge base (such as a news graph or a social network) where edges (interactions or new info) stream in at irregular times. A continuous-time GNN can update the relevant node representations at the moment an event occurs (e.g., when a new article is published or when a user posts a comment), instead of waiting for a batch update. This event-based updating preserves temporal fidelity—it can capture the sequence and gaps between events more naturally than snapshot-based methods. Such networks also offer predictive capabilities for future states. Because the model learns the time-derivative of node features, we can extrapolate the graph state forward. For example, a continuous-time GNN might predict how a citation network will grow: given the trend of publications, it could forecast which fields will be connected next, or which author might collaborate with whom in the near future. For RAG, this means the system isn't limited to answering questions about the current knowledge graph—it could potentially answer "what-if" questions or predict future relations (with appropriate uncertainty). In fact, uncertainty quantification is another strength of some continuous-time models. By formulating graph evolution in a probabilistic or ODE framework, it's possible to estimate confidence in predictions over time. For instance, if we ask, "Will researcher A and researcher B collaborate within the next year?," a continuous-time GNN could give an answer along with a confidence score that accounts for the sparsity of past interactions or volatility of the field. A well-calibrated continuous-time GNN provides probabilities that reflect true likelihoods, helping the RAG system decide when to trust its predictions. This is crucial in time-sensitive queries—if the model is unsure about a future connection, it can signal uncertainty rather than making a flimsy guess. Overall, continuous-time graph networks empower RAG systems to deal with dynamic knowledge. They can integrate facts as they arrive, reason about temporal relationships ("Which events lead up to this outcome?"), and even look ahead. This smooth handling of time contrasts with earlier static approaches, bringing RAG closer to real-world applications like news assistants, financial analytics, or any domain where "when" is as important as "what." In summary, the key components of CTGNs are:

- Uses link starting timestamps and durations as evolving information, integrating neural layers over continuous time.
- Workflow: TGN encoder learns latent states, Neural ODE module models node representation with link duration as integration variable, LSTM decoder computes interaction probabilities, and memory updates as encoder input.
- Supports event-based and contact sequence graphs, effective in transductive and inductive tasks.

11.1.3.1.3 Memory-augmented GNNs

Standard GNNs compute node representations by aggregating information from neighbors, which is inherently a short-term memory—the network's state is typically reset or overwritten each time you run it on a new query or new graph. However, complex reasoning (especially in a multi-hop Q&A) may require long-term memory or the ability to remember intermediate results. Memory-augmented GNNs incorporate external or internal memory modules to store and retrieve information, giving RAG systems a form of "brain" that persists beyond a single pass of message-passing [15]. One approach uses external memory banks for long-term dependencies. These are akin to a scratch pad where the GNN can write useful facts or partial conclusions and read them back when needed. For example, consider a question: "In what year did the author of GraphX (a paper) join Stanford University?" A GNN might need to find the author of GraphX, then find that person's employment info. A memory-augmented GNN could explicitly store the result of the first hop ("Author of GraphX is Dr. Y") in an external memory, then use it to guide the second hop ("Dr. Y is a professor at Stanford since 2015"). Without memory, the intermediate fact might be lost or only implicitly stored in a node embedding, whereas an external memory allows the model to cache important results and reuse them directly. A related concept is the Differentiable Neural Computer (DNC), a neural network with a trainable memory matrix and read/write controller[16]. DNCs and similar neural Turing machines have been shown to excel at tasks like graph traversal and planning because they can store complex data structures and sequences of steps. In the context of GNN-RAG, a DNC-augmented GNN might, for instance, store the path it has traversed in the memory and use that to decide the next step, essentially learning algorithms on the graph. Researchers have demonstrated that such memory-augmented networks can navigate transportation networks and other graphs more effectively than standard recurrent networks. By plugging this capability into a RAG system, we enable the system not only to retrieve an answer but to internally simulate reasoning pathways (like following a chain of reasoning in a stored memory) and backtrack or explore alternatives if needed. Memory in GNNs can also be tailored to specialized needs:

- Episodic memory modules are inspired by how humans remember specific events or experiences. In a GNN, an episodic memory could store embeddings or subgraphs related to past queries or tasks. This enables few-shot learning, where the model quickly adapts to a new type of query by recalling analogous past episodes. Suppose our system has answered only a couple of questions about medical research in the past. With episodic memory, when a new medical question comes, the GNN might retrieve those past case memories (e.g., a subgraph of a biomedical knowledge base it traversed before) to inform the new reasoning process. Such meta-learning approaches allow transfer of knowledge across graph domains, so the system doesn't start from scratch on each new subject area.
- Working memory for multi-hop reasoning is another angle: this refers to a temporary storage that holds intermediate computations within a single complex query. When a GNN has to do, say, a 5-hop reasoning (A → B → C → D → E → answer), working memory helps it not "forget" the early hops by the time it reaches the later hops. Concretely, a GNN could be designed to accumulate information over hops in a dedicated vector (or set of vectors) rather than compressing everything into the last node's embedding. This might involve recurrent mechanisms or stacking multiple reasoning layers that pass along a state. The result is that the model can carry forward partial results (like "A is related to B and C") and refine them as it goes further ("A is related to B who knows C, who in turn works at D"). By the end, the working memory contains a trace of the whole reasoning chain, which can improve accuracy, especially for very complex queries requiring many hops.

The key components are:

- General Interest Module: Uses matrix factorization ($p_u^T \cdot q_j$).
- Short-term Interest Module: GNN aggregates neighborhood information, e.g., $h_i = \tanh(W^{(1)} \cdot [\sum_{k \in N_i} e_k A_{i,k}; e_i])$, with sliding window strategy ($|L| = 5$, $|T| = 3$).
- Long-term Interest Module: Shared key-value memory network, e.g., $H_{u,l} := H_{u,l} + \mathrm{PE}(H_{u,l})$, attention $S_{u,l} = \mathrm{softmax}(W_a^{(3)} \tanh(W_a^{(1)} H_{u,l} + (W_a^{(2)} p_u) \otimes 1_\phi))$, memory output $o_{u,l} = \sum_i s_i v_i$.
- Interest Fusion: Gating mechanism $g_{u,l} = \sigma(W_g^{(1)} \cdot \frac{1}{|L|} \sum_{i \in L_{u,l}} h_i + W_g^{(2)} \cdot p_{u,l}^H + W_g^{(3)} \cdot p_u)$.
- Item Co-occurrence: Bilinear function $e_i^T W_r q_j$.

In summary, memory-augmented GNNs endow RAG systems with recollection and persistence. They mitigate the limitation of pure message-passing, which is usually bounded by local neighborhoods and ephemeral computations. With memory, the system can draw on past knowledge (even from previous queries), handle long-range dependencies in graphs that would otherwise be impractical to traverse, and methodically tackle multistep problems. This significantly improves the robustness of RAG on tasks like long-form question answering, dialog (where context from earlier turns must be remembered), and any scenario where reasoning is not a single-shot operation but a chain of thought.

11.1.3.2 Learning Paradigms

Advances in GNN architectures go hand-in-hand with new learning paradigms that train these models effectively. For GNN-RAG, we need models that can learn from unlabeled graph data, adapt quickly to new domains or questions, and even learn strategies to retrieve information optimally. Three important paradigms are self-supervised learning on graphs, few-shot/zero-shot learning, and RL integration. Each contributes to making GNN-based retrieval more powerful and general.

11.1.3.2.1 Self-supervised Graph Learning

Much like language models benefit from unsupervised pretraining on large text corpora, GNNs can be pretrained on graphs via self-supervised objectives [17]. The idea is to create learning tasks from the graph structure itself, so the GNN can learn general patterns without needing manual labels. Several approaches have emerged:

- Graph Reconstruction Tasks: The GNN is asked to reconstruct parts of the graph that were hidden or corrupted, essentially learning the structure. For example, a Graph Auto-Encoder (GAE) might take a graph, remove some edges, and train the GNN to predict which edges were missing by reconstructing the adjacency matrix. Another variant is to mask out some node features or even entire nodes/edges and have the GNN predict them—analogous to masked language modeling in NLP. By doing so, the GNN learns to embed nodes in a way that can regenerate the local neighborhood, capturing the underlying semantics. As a concrete case, suppose we hide the relationship between two proteins in a biomedical network and ask the GNN to guess if they interact; to succeed, it must learn the connectivity patterns of such interaction networks. This pretraining can greatly improve downstream performance when the GNN is later fine-tuned on a supervised task (like link prediction or node classification).
- Contrastive Learning on Graphs: Here the model learns by comparing different views or substructures of the graph. One popular technique is to generate two random perturbations of the same graph (e.g., remove some edges or features in each) and train the GNN such that the representations of the two perturbed graphs are close, while representations of very different graphs are far apart. This teaches the GNN to capture global graph properties and invariances. For instance, in a knowledge graph, one could take subgraphs around a topic and use contrastive loss to ensure the GNN produces similar embeddings for two subgraphs that essentially contain the same facts, even if some details differ. This way, the GNN learns a robust representation of a knowledge context, which helps RAG because the retriever can recognize when two questions are asking about a similar underlying concept (even if worded differently or some facts missing). Contrastive methods like GraphCL and DGI have shown that such pretraining yields strong transferable embeddings for graphs, nodes, or edges.
- Masked Node and Edge Prediction: A more fine-grained self-supervision is to mask individual elements. For example, GPT-GNN is a framework where the model masks one node and all its connections, then tries to sequentially predict what that node is and which edges it had [18]. This is analogous to language models predicting the next word, but on graphs—predicting the existence of a node or relationship given its context. Another approach is GraphMAE, which masks some attributes of nodes (like setting some features to zero) and trains the GNN to impute them correctly [19]. These tasks force the GNN to understand the structural context deeply: if a node is removed, the model must infer it from neighbors (e.g., "there's likely a missing author node connecting these two papers"). Successfully predicting masked nodes/edges indicates the GNN has captured complex dependencies in the graph.

- Graph-level Property Prediction: Sometimes we can set up tasks where the entire graph has a pseudo-label derived from its structure. A fun example is to pick two nodes at random and ask the GNN to predict whether they are connected within *k* hops or not. This requires the GNN to develop a notion of connectivity and distance. Another example: in molecular graphs, one might generate a label like "has motif X" (where motif X could be a particular functional group in a molecule) and train the GNN to detect that—even if those labels are automatically extracted by an algorithm rather than human annotations. Such property prediction tasks at the graph level encourage the model to encode high-level information. In RAG, this could translate to the ability to quickly judge if a candidate subgraph contains the answer. For instance, given a subgraph of a knowledge base, a pretrained GNN could have an internal neuron for "contains a person's birthdate" because it has learned to predict that property; then when asked "When was person Y born?", it can immediately spot which subgraphs (documents, nodes) likely contain a birthdate property.

Through self-supervised learning, GNNs acquire a form of background knowledge about graphs before seeing any specific Q&A pairs or labels. This pretraining can be done on large unlabeled graphs (e.g., all of Wikipedia as a huge graph of entities). When the GNN is later used in a RAG pipeline, it starts with a rich understanding of how nodes and edges relate, making it far more effective at retrieval.

11.1.3.2.2 Few-shot and Zero-shot Capabilities

Few-shot and zero-shot learning allow GNNs to generalize with limited or no labeled examples. In many realistic scenarios, we don't have a large training set of Q&A pairs or labeled data for every new domain of knowledge. GNN-augmented RAG systems therefore benefit from few-shot learning (learning from a very small number of examples) and zero-shot generalization (performing tasks with no direct examples, by leveraging prior experience). Achieving this in GNNs involves techniques like meta-learning, transfer learning, and compositional reasoning. Meta-learning for graphs often means training the GNN on a variety of tasks so that it can adapt to a new task with minimal fine-tuning—essentially "learning to learn" on graphs. One meta-learning approach is to simulate many small training tasks on subgraphs (like classifying subgraphs or predicting some property) and train a model that can quickly adjust its weights to optimize any of those tasks. When a new task comes (say, a new type of relationship question), the model's internal representation is already suited to adapt with just a few gradient steps (this is analogous to techniques like MAML, but applied to GNNs). For example, a GNN might be meta-trained on a set of knowledge graphs from different domains—biology, history, geography—each with its own query-answer pairs. At meta-test time, the GNN is given a new domain (say, literature) with only a handful of Q&A examples. Because it has meta-learned how to quickly form a reasoning strategy, it can adapt to this new domain much faster than a regular model. Researchers have found that such rapid adaptation is possible by learning good initialization parameters that are sensitive to new training signals. Transfer learning across graph domains is closely related: it means using knowledge learned on one graph for another. If our RAG has been thoroughly trained on a large general knowledge graph (like Freebase or ConceptNet), we'd like to transfer that to a more specific graph (like a medical ontology) without retraining from scratch. GNNs can transfer in various ways. One is by freezing or reusing parts of the network (say, the first few message-passing layers) that capture generic graph reasoning skills (like following a path or counting connections). Another is by attaching new embeddings for new node types but using the same update rules learned previously. For instance, a model trained on a social network might know concepts like "friend-of-a-friend" or "in the same community." If we introduce it to a protein interaction network, those concepts might translate to "proteins in the same pathway" or "close in the interaction graph." With minimal fine-tuning, the GNN can repurpose its relational reasoning to the new context—this is essentially what happened with the DNC memory model, which was trained on one transit system and could apply to another with little extra training, indicating that the high-level reasoning carried over. Compositional generalization is a fancy way of saying the model can handle new combinations of known pieces. In language, for example, if you know "red" and "ball," you can understand "red ball" even if you've never heard it. For GNNs, compositionality might mean understanding a new multi-hop query by composing known one-hop or two-hop patterns. Suppose a GNN has seen how to answer "X is married to Y" and separately learned "Y works at Z." A compositional generalization would be answering a query "Is X's spouse an employee of Z?" by combining those reasoning steps. Few-shot learning techniques encourage this by exposing the model to many small tasks where it must recombine knowledge. Also, designing the GNN to have modular components (like separate sub-networks for different relation types or for different reasoning steps) can help—the model can swap in/out modules and still function. In RAG, this is invaluable for answering novel queries that are longer or more complex than those seen in training. The system won't be explicitly trained on every possible question; instead, it must compose reasoning skills on the fly (e.g., combining "located in" + "bordered by" relations to answer a geography query that needs both). Finally, cross-lingual and cross-domain reasoning is an exciting aspect of zero-shot GNN capabilities. If the knowledge graph (or a set of graphs) spans multiple languages or domains, a well-trained GNN might answer questions in a new language without direct training, purely because the structure of the graph is the same. For example, if a GNN-based RAG knows how to find a scientist's notable work in an English knowledge graph, and we then give it a French knowledge graph with a similar structure (scientists, works, dates, etc.), it could transfer its method by aligning the embeddings or using a shared representation space. Cross-lingual transfer can be aided by using language-agnostic node features (like Wikidata IDs or multilingual embeddings for names). One can also train with translational invariance—e.g., during training ask questions in different languages about the same graph, forcing the GNN to develop a language-independent way of reasoning. As a result, the system achieves zero-shot cross-lingual QA: answer in language B having only trained in language A. In essence, few-shot and zero-shot techniques strive to make GNN-RAG universally adaptable. We don't want to train a new model from scratch for every possible knowledge base or every specialized domain. Instead, we train one (or a few) powerful GNNs that can generalize their graph reasoning ability broadly. This dramatically expands the usability of RAG systems—one model could handle everything from Wikipedia to a company's internal knowledge graph to a medical database, given just a bit of quick tuning or even no examples at all, leveraging its prior experience.

11.1.3.2.3 RL Integration

RL with GNNs learns optimal policies [20]. So far, we've discussed GNNs mostly as passive networks that learn from static data. But retrieval can be viewed as a decision-making process: the system must decide which nodes or paths to explore in the graph to find the answer. This is where RL comes into play. Integrating RL with GNN-RAG means the model learns a retrieval policy—a strategy of traversal—by trial and error, getting rewards for correct answers or efficient searches. Consider the task of multi-hop question answering on a knowledge graph. The GNN at any step has a "state" (perhaps the current node or subgraph it's focusing on, plus the question context) and can take an "action" (follow one of the outgoing edges, or stop if it thinks it found the answer). We can train a policy network (which could itself utilize a GNN to encode state) using reinforcement signals: for example, a reward of +1 for arriving at the correct answer node and perhaps a small negative reward for each step taken (to encourage short, efficient paths). Over time, the agent learns an optimal retrieval policy that balances finding the correct information with not wandering too much—essentially learning how to explore the graph efficiently for answering questions. One notable early example is DeepPath, where an RL agent traverses a knowledge graph to find multi-hop relationship paths [21]. The agent's policy was guided by rewards that explicitly combined accuracy (finding the right answer) with efficiency (shorter paths) and even diversity (finding novel paths). This kind of reward shaping is exactly what we need in RAG: we might reward the agent for retrieving supporting facts that lead to the correct answer and penalize it for extraneous hops that don't add value. By tuning the reward, we can get the GNN+RL system to be very precise (if we heavily reward accuracy) or very fast (if we penalize long retrievals), or a balanced trade-off. In practice, one would set these based on the application—a live question-answering system might favor efficiency to keep response times low, whereas a research assistant might favor thoroughness, willing to traverse a bit more to gather complete evidence. Another application of RL is dynamic subgraph selection for cases where retrieving the whole graph is infeasible. Instead of the GNN taking the entire knowledge graph as input, an RL agent can learn to construct a relevant subgraph on the fly. Think of it as the agent "walking" the graph and including nodes that seem promising into a subgraph that will be passed to the language model for generation. The agent gets positive reward when the subgraph it built contains the correct answer or enough evidence for the language model to answer confidently. This is akin to a scavenger hunt: the agent must pick out only the useful pieces from a huge graph. Over time, it learns patterns like "if the query is about a person's work, first go to that person's node, then follow profession or known_for edges, ignore unrelated branches, etc." Instead of manually programming these heuristics, the RL training discovers them. In fact, a 2025 study (DrKGC) used a learned retrieval method to extract query-specific subgraphs which improved a language model's performance [22]. While DrKGC used rules and a GCN rather than deep RL, one can envision an RL-driven variant that continuously improves its retrieval strategy with experience. Moreover, RL allows query-specific traversal strategies. Not all questions should be answered the same way. A "definition" question might only require one hop to a definition node, whereas a "comparison" question might require gathering info from two different parts of the graph then joining them. A non-RL GNN might treat these similarly unless explicitly designed; an RL agent, on the other hand, can have different policies for different query types, because the reward structure can be made conditional on the task. We could train the agent with different scenarios (single-hop tasks, multi-hop tasks, boolean questions, etc.) and it will learn to navigate accordingly. For example, it might learn that for yes/no questions, it should quickly locate a fact that directly affirms or contradicts the statement (often just one or two hops), whereas for list questions, it should do a broader traversal to gather multiple items. Finally, combining RL with GNN opens up the possibility of continual improvement of retrieval. In a deployed RAG system, one could have the model attempt to answer questions and get feedback (from user clicks or an evaluator) on whether the supporting info was relevant. That feedback is essentially a reward signal. Over time, the retrieval policy would get better—perhaps avoiding certain paths that often led to wrong answers, or learning to favor certain trustworthy information sources in the graph. This means the system learns from its own usage, a powerful feedback loop that plain supervised learning can't easily replicate. In summary, RL integration turns GNN-RAG into an active searcher. Instead of passively encoding a subgraph given to it, the model decides where to look next in the graph, much like a human researcher formulating a step-by-step strategy to find information. The end result is a more targeted and efficient retrieval, which is particularly vital when dealing with gigantic graphs where exhaustive search is impossible. By learning optimal policies and pathfinding strategies, RL-augmented GNNs can retrieve answers with minimal steps and maximal confidence.

11.1.3.3 Explainable GNN-RAG

Explainability ensures transparency, crucial for trust in RAG systems. The methods can be categorized into instance-level (gradients/features-based, perturbation-based, decomposition, surrogate) and model-level (graph generation) [23]. As GNNs for RAG become more sophisticated—navigating huge graphs with complex attention and memory—it's increasingly important that they remain transparent and explainable. Users (and developers) will justifiably ask: Why did the system choose this piece of information? How did it derive that answer from the graph? Explainability is key for trust, debugging, and aligning the system's reasoning with human expectations. In this section, we outline approaches to make GNN-based RAG systems interpretable, including visualizing attention, highlighting reasoning paths, performing counterfactual analysis, and calibrating confidence.

11.1.3.3.1 Attention Visualization

Many advanced GNNs (especially transformers and attentional GNNs) produce attention weights that indicate how strongly one node or edge influenced another. These weights can be leveraged to explain the model's focus. For instance, in a HGT answering a question, we can inspect the attention coefficients to see which neighbors of a node were most influential. If the question was "What university did Alan Turing attend?", and the model retrieved the answer from a knowledge graph, an attention visualization might show that the edges from Alan Turing to University of Cambridge and to Princeton University had high weights, while other edges (like Turing's birth place, or profession) had near-zero weights. This tells us the model homed in on the education-related connections. We can present this to the user as: "The system considered Alan Turing's university affiliations most relevant to your query." Technically, one can overlay edge highlightings on a graph diagram or list the top-weighted relations. In graphs with multi-hop attention, we might even stack these visualizations layer by layer, showing how the focus narrows. Attention weights aren't a perfect explanation of model internals, but they provide an intuitive heuristic: big weights = important link for the model's decision. Many frameworks now include tools to color-code graph edges by attention strength, making it easier to interpret GNN decisions.

11.1.3.3.2 Path Highlighting

GNN-RAG often involves following paths through the graph to connect a question to an answer. An explicit reasoning path can be extracted and highlighted to users as an explanation. For example, if the system answered "Alan Turing studied at University of Cambridge," it might have effectively followed a path: Alan Turing → educated_at → University of Cambridge. If the question was more complex like "Did Alan Turing ever collaborate with someone at Cambridge?", the path might be Alan Turing → worked_with → Person X → affiliated_with → Cambridge. By tracing through the GNN's computation or using techniques like GNNExplainer to find critical subgraphs, we can identify which sequence of edges led to the answer [24, 25]. Presenting this as a chain of facts is hugely beneficial: it not only justifies the answer but also provides additional context. In user studies, showing a path through the knowledge graph (especially one that aligns with human reasoning) significantly boosts trust in the system's answer. Path highlighting can be done automatically by taking the nodes/edges that, if removed, most reduce the model's confidence in the answer—these are presumably the supporting facts. Essentially, we are opening the black box and showing a white box rationale: "The model believes this because A is connected to B, and B is connected to C, which answers your question."

11.1.3.3.3 Counterfactual Analysis

A powerful way to test and explain GNN reasoning is to ask "What if...?" questions about the graph. For instance, "What if this edge didn't exist? Would the answer change?" In practice, one can remove or alter certain edges/nodes and see how the model's output or confidence shifts. This yields insights into which connections are truly pivotal. If removing the Alan Turing—Cambridge edge causes the model to no longer find an answer, that edge was obviously critical for that query—a point to highlight in explanation. On the other hand, if the model still answers correctly, perhaps it found an alternate route (maybe Alan Turing had another connection to Cambridge via a colleague). This method can uncover redundant vs. crucial evidence. In a RAG system, such counterfactual tests might be used behind the scenes to double-check an answer's robustness. If multiple disjoint paths yield the same answer, the system can be more confident (and explain "Even without fact X, the answer still holds via fact Y"). Conversely, if the answer hinges on a single edge, the system might flag that as a point of uncertainty or interest: "The answer relies heavily on Alan Turing's Cambridge affiliation; if that is incorrect, the conclusion might be different." This is related to research on counterfactual explanations for GNNs, where one aims to find the minimal change in the graph that flips the model's prediction. For an explainable RAG, we can use this to identify a minimal subgraph that supports the answer—a succinct explanation set.

11.1.3.3.4 Confidence Calibration

Lastly, an explainable system should convey how sure it is about the answer. GNNs, especially when integrated into larger pipelines with language models, can produce a probability or confidence score for each answer. However, raw model probabilities are not always trustworthy—models can be overconfident or under-confident. Calibration is the process of adjusting these so that, say, an "80% confidence" truly means an 80% chance of being correct [25]. In practice, one might use techniques like temperature scaling or isotonic regression on the GNN's output scores to better align them with actual accuracy observed on a validation set. A well-calibrated GNN-RAG will, for example, only display "I'm 99% sure" if historically 99% of such cases were right, whereas a less calibrated one might say "99%" but be wrong 20% of the time (very misleading!). Providing calibrated confidence alongside answers helps users make informed judgments. For critical applications (medical, legal, etc.), the system might even refuse to give a high-confidence answer if it's not at least, say, 95% sure—instead, it could respond with uncertainty or ask for clarification. From an explainability perspective, showing confidence can be as simple as a probability bar or as elaborate as: "I'm fairly confident in this answer because the supporting evidence is strong and consistent." Under the hood, this may involve the GNN quantifying uncertainty (like using Monte Carlo dropout or Bayesian neural network techniques to get a distribution over answers). Recent surveys on GNN uncertainty stress that improving calibration makes GNNs more trustworthy and robust [25]. In conclusion, explainable GNN-RAG systems combine these approaches to not only find the correct answer but also show their work. By visualizing attention, we see what the model deemed important. By highlighting paths, we see how pieces of knowledge connect to form the answer. Through counterfactuals, we understand the dependency of the answer on each fact. And with confidence estimates, we know how much to trust the result. Together, these make the advanced GNN-driven retrieval not a magic oracle, but a transparent assistant—one that can articulate the reasoning behind its answers in a human-friendly manner. This fosters user trust and paves the way for collaboration between humans and AI in exploring knowledge graphs.

11.2 Ethical Considerations and Bias Mitigation

As Graph-RAG systems become increasingly powerful and pervasive across industries, their ethical implications demand careful consideration. The structured nature of knowledge graphs presents a double-edged sword: while it can amplify existing societal biases through explicit encoding of relationships, it also offers opportunities for detecting, measuring, and mitigating these biases through transparent and auditable mechanisms.

11.2.1 The Dual Nature of Bias in Graph-RAG

Knowledge graphs, by their very nature of encoding explicit relationships between entities, can inadvertently perpetuate and even amplify societal biases. Research has identified several critical forms of bias that emerge in knowledge graph-based systems, each with distinct characteristics and implications for Graph-RAG applications.

Representation bias emerges when certain entities or demographic groups are systematically over- or under-represented in the knowledge graph. This imbalance in representation directly impacts the quality and fairness of retrieval results. For instance, analyses of major public knowledge bases reveal striking disparities: DBpedia, extracted from Wikipedia, predominantly represents Western and industrialized nations, while Wikidata exhibits significant gender imbalance with approximately 70% of person entries being male and only 20% female. These representation gaps have cascading effects on Graph-RAG systems built upon such foundations.

Consider a medical knowledge graph constructed primarily from Western medical literature. Such a system might inadequately represent diseases prevalent in tropical regions or traditional medicine practices that have proven effective in non-Western contexts. When a Graph-RAG system built on this biased foundation is queried about treatment options, it may systematically underserve populations whose medical conditions or cultural contexts are underrepresented in the source data.

Relationship bias manifests in the types and strengths of connections encoded between entities. Even when entities from diverse groups are present in the graph, the relationships between them may reflect historical prejudices or systematic omissions. Professional knowledge graphs might overemphasize connections within certain demographic groups while missing equally valid relationships in underrepresented communities. Historical data encoded in graphs often shows disproportionate links between male individuals and leadership positions, reflecting past gender imbalances rather than current realities or future possibilities.

This relationship bias creates a self-reinforcing cycle: if a knowledge graph associates certain professions predominantly with specific genders or ethnicities based on historical patterns, the Graph-RAG system may perpetuate these stereotypes in its responses. For example, when asked about notable figures in technology, the system might predominantly retrieve paths connecting innovation with Western male figures, not because they are the only innovators, but because the graph's relationship structure reflects historical documentation biases.

Temporal bias represents perhaps the most insidious form of bias in knowledge graphs. Information encoded at a particular point in time can become outdated, yet continue to influence system outputs. This is especially problematic in rapidly evolving fields such as technology, medicine, or social sciences. A knowledge graph that hasn't been updated might still contain debunked medical beliefs, outdated social classifications, or obsolete technological information. When such temporal biases persist, they can lead Graph-RAG systems to provide answers that are not merely incomplete but actively harmful.

11.2.2 Comprehensive Mitigation Strategies

Addressing bias in Graph-RAG systems requires a multifaceted approach that intervenes at every stage of the system life cycle. Drawing from recent research in fair AI and knowledge graph construction, we identify four key pillars of bias mitigation.

As an overview, Table 11.3 summarizes the mitigation strategies and their applications.

11.2.2.1 Diverse Data Sources and Inclusive Construction

The foundation of a fair Graph-RAG system lies in the diversity and representativeness of its underlying data. This extends beyond simply adding more data sources; it requires intentional efforts to seek out and integrate knowledge from marginalized perspectives. Successful implementation involves several complementary approaches.

Active outreach to underrepresented communities ensures their knowledge and experiences are captured in the graph. For medical applications, this might mean partnering with healthcare providers in rural areas, integrating research from global south institutions, and including traditional medicine practices that have empirical support. The Yusuf Hamied Department of Chemistry at Cambridge has demonstrated this approach by creating knowledge graphs that intentionally include research from diverse geographic regions, resulting in more comprehensive coverage of global scientific knowledge.

Community-driven knowledge contribution platforms, similar to Wikipedia but with stronger diversity initiatives, enable broader participation in graph construction. These platforms must be designed with accessibility in mind, providing interfaces in multiple languages and accommodating varying levels of technical expertise. Regular diversity audits help identify gaps in representation. These audits go beyond simple demographic counts to examine the depth and quality of information about different groups. For instance, even if a knowledge graph includes entries for scientists from all continents, an audit might reveal that entries for Western scientists contain richer biographical information and more numerous achievement links.

11.2.2.2 Algorithmic Fairness in Graph Operations

Beyond diversifying data, the algorithms that operate on knowledge graphs must be designed with fairness as a core objective. This involves both modifying existing graph algorithms and developing new fairness-aware approaches.

Debiasing techniques for knowledge graph embeddings adapt methods from natural language processing to remove unwanted correlations in vector representations. These techniques work by

Table 11.3 Mitigation strategies and their applications.

Strategy	Description	Example application
Diverse data sources	Include multiple perspectives and marginalized voices	Add data from diverse medical studies for fair treatment recommendations
Algorithmic fairness	Use debiasing and fairness-aware models	Apply fairness constraints in loan approval systems
Transparency and accountability	Track provenance, explain decisions, and allow feedback	Document sources in medical graphs, explain treatment choices
Privacy-preserving techniques	Use differential privacy, federated learning, and encryption	Protect patient data in healthcare graphs with federated learning

identifying subspaces in the embedding that correlate with protected attributes and constraining the model to avoid using these subspaces for unrelated predictions.

Fairness-aware GNNs incorporate explicit constraints during training to ensure equitable treatment across different groups. For instance, when predicting missing links in a professional network graph, these models can be regularized to ensure that the probability of predicting high-value connections (such as mentorship or collaboration links) does not systematically differ based on the gender or ethnicity attributes of nodes.

During the retrieval phase, balanced sampling ensures diverse results. Rather than simply returning the top-k most relevant results, which might all come from overrepresented groups, the system can use diversity-aware ranking algorithms. These algorithms optimize for both relevance and representational fairness, ensuring that users are exposed to a broader range of perspectives.

11.2.2.3 Transparency and Accountability Mechanisms

The structured nature of knowledge graphs provides unique opportunities for transparency that are difficult to achieve in black-box neural models. Every fact in a Graph-RAG system can carry provenance information, creating an audit trail from question to answer.

Comprehensive provenance tracking involves recording not just the source of each fact but also when it was added, by whom (human or automated system), and what validation processes it underwent. Advanced systems might even track the confidence level of each fact and any disputes or corrections over time. For example, if a medical Graph-RAG states that "Treatment X is effective for Condition Y," it can also provide that this information comes from a specific clinical trial published in a particular journal, with links to the full study and any subsequent research that confirms or challenges the finding.

Explainable retrieval paths make the system's reasoning process transparent. Unlike traditional search engines that return results without explanation, Graph-RAG systems can show the exact path through the knowledge graph that led to each answer. This allows users to verify the logic and identify potential biases in the reasoning chain. Visual tools can make these paths accessible to nontechnical users, showing how the system connected different pieces of information to arrive at its conclusion.

User feedback mechanisms for bias reporting create a continuous improvement loop. When users identify biased or incorrect outputs, they can flag them with specific categories (representation bias, outdated information, missing context, etc.). These reports feed into both immediate fixes and longer term system improvements.

11.2.2.4 Privacy-preserving Techniques

As Graph-RAG systems often aggregate vast amounts of potentially sensitive data, privacy preservation becomes paramount. This is particularly critical in domains like healthcare, finance, and education where individual privacy rights must be balanced with the collective benefit of shared knowledge.

Differential privacy techniques add carefully calibrated noise to statistical queries on the graph, ensuring that no individual's data can be inferred from the outputs. For instance, when a Graph-RAG system answers "What percentage of patients with Condition X respond to Treatment Y?" differential privacy ensures that the answer doesn't reveal whether any specific patient is in the database.

Federated learning approaches enable multiple organizations to collaboratively build and query knowledge graphs without sharing raw data. In healthcare, this allows hospitals to benefit from collective knowledge while maintaining patient privacy. The MELLODDY project in pharmaceutical research demonstrates this at scale, with 10 pharmaceutical companies collaborating on a federated knowledge graph for drug discovery without sharing proprietary compound structures [26].

11.2.3 Ethical AI Governance Framework

Technical measures alone cannot ensure ethical Graph-RAG systems. Organizations must embed ethics into their governance structures through comprehensive frameworks that address the full lifecycle of these systems.

11.2.3.1 Establishing Ethics Review Boards

Ethics review boards for Graph-RAG systems should include diverse stakeholders: domain experts who understand the application context, ethicists who can identify moral implications, legal advisors familiar with relevant regulations, technologists who understand system capabilities and limitations, and crucially, representatives from communities that will be affected by the system.

These boards must have genuine authority to influence system design and deployment decisions. Their responsibilities include reviewing knowledge graph schemas for potential biases, evaluating data sources for representativeness and consent, assessing the fairness of retrieval algorithms, and establishing policies for handling sensitive information. For example, when designing a Graph-RAG system for criminal justice applications, the ethics board might mandate that arrest records without convictions be excluded from the graph to avoid perpetuating presumption of guilt.

11.2.3.2 Implementing Regular Bias Audits

Bias audits for Graph-RAG systems must be both comprehensive and actionable. Quantitative audits analyze the distribution of entities and relationships across demographic groups, measuring metrics such as coverage parity (are all groups equally represented?), relationship density (do all groups have equally rich connections?), and retrieval fairness (do queries return balanced results across groups?).

Qualitative audits involve domain experts reviewing system outputs for subtle biases that metrics might miss. For instance, a Graph-RAG system for career counseling might pass quantitative fairness tests but still exhibit bias if it consistently describes leadership qualities using masculine-coded language. These audits should use standardized test sets that probe for known biases, such as asking about professionals in various fields and analyzing whether the gender distribution in results matches reality or perpetuates stereotypes.

When biases are identified, remediation must be systematic. This might involve reweighting relationships in the graph to counteract historical biases, adding missing connections based on evidence rather than historical patterns, or annotating biased relationships with contextual warnings. The New York City Department of Education's knowledge graph for student resources underwent such remediation after audits revealed that advanced learning opportunities were disproportionately connected to schools in affluent neighborhoods.

11.2.3.3 Continuous Monitoring and Adaptation

Ethical governance of Graph-RAG systems requires continuous vigilance. Automated monitoring systems track key metrics in real time, including fairness indicators across different queries and user groups, accuracy metrics with breakdowns by demographic categories, and usage patterns that might indicate biased behavior. Anomaly detection can flag sudden changes that might indicate new biases introduced through automated updates.

Human oversight remains crucial. Regular reviews by diverse teams can catch issues that automated systems miss. These reviews should examine not just system outputs but also user feedback, emerging societal concerns, and changes in the regulatory landscape. The governance framework must be adaptive, ready to respond to new ethical challenges as they emerge.

11.3 Future Research Directions

The frontier of Graph-RAG research presents numerous opportunities for breakthrough innovations that could fundamentally transform how AI systems understand and reason with knowledge. These research directions address current limitations while opening new possibilities for applications across every sector of society.

11.3.1 Explainable Graph-RAG: Making AI Reasoning Transparent

Current RAG systems often operate as black boxes, providing answers without clear explanations of their reasoning process. Explainable Graph-RAG aims to leverage the inherent structure of knowledge graphs to make every step of the reasoning process transparent and interpretable. This transparency is especially crucial in high-stakes domains such as healthcare, finance, and legal applications where users must understand and verify the rationale behind AI recommendations.

11.3.1.1 Natural Language Explanations from Graph Traversals

The primary challenge in explainable Graph-RAG is translating graph-based reasoning into natural language explanations that are meaningful to end users. While it's technically straightforward to highlight the nodes and edges used in answering a query, this falls far short of what users need. Instead, systems must generate coherent narratives that explain not just what facts were used, but why they are relevant and how they connect to form a conclusion.

This translation requires sophisticated natural language generation that understands the semantic meaning of relationships in the graph. Template-based approaches provide a starting point, but neural generation models conditioned on graph structures show more promise for producing fluent, context-appropriate explanations. The key challenge is maintaining fidelity—ensuring that the natural language explanation accurately represents the underlying graph-based reasoning without oversimplification or distortion.

11.3.1.2 Causal Reasoning and Counterfactual Explanations

Moving beyond correlation to causation represents a fundamental advance in explainable Graph-RAG. While knowledge graphs often encode associations, users frequently need to understand causal relationships for decision-making. This requires augmenting graphs with causal information and developing reasoning mechanisms that can distinguish correlation from causation.

The integration of causal inference with knowledge graphs is exemplified by the CausalKG framework [27], which enriches knowledge graphs with causal relationships extracted from randomized controlled trials and observational studies. When a user asks why a particular treatment is recommended, the system can explain not just that the treatment is associated with positive outcomes, but that clinical trials have established a causal relationship.

Counterfactual reasoning adds another dimension to explainability. By simulating alternative scenarios, Graph-RAG systems can help users understand the boundaries and dependencies of recommendations. For instance, a financial Graph-RAG advising on investment strategies might explain: "This recommendation assumes stable interest rates. If rates were to increase by 2%, the optimal strategy would shift to shorter-term bonds because..." This type of counterfactual explanation helps users understand not just what to do, but under what conditions the advice remains valid.

11.3.1.3 Interactive Exploration and Dialog-based Explanations

The future of explainable Graph-RAG lies in interactive systems that engage users in dialog about their reasoning. Rather than providing static explanations, these systems allow users to probe deeper, ask follow-up questions, and explore alternative scenarios. This transforms explanation from a one-way information dump into a collaborative exploration of knowledge.

Consider a physician using a Graph-RAG system for diagnosis support. After the system suggests a diagnosis, the physician might ask: "What if the patient also had symptom X?" The system would retraverse the graph with this additional information and explain how it changes the diagnostic reasoning. Or the physician might challenge: "Why didn't you consider diagnosis Y?" The system could then show the path to that alternative diagnosis and explain why it was considered less likely given the available evidence.

Implementing such interactive explanation requires sophisticated dialog management that maintains context across multiple turns of conversation. The system must track what aspects of its reasoning have already been explained, what the user has understood or challenged, and how to provide additional detail without overwhelming the user. Research in conversational AI provides foundations, but the integration with graph-based reasoning presents unique challenges and opportunities.

11.3.2 Automated Knowledge Graph Construction and Maintenance

The manual construction of knowledge graphs remains one of the most significant bottlenecks in deploying Graph-RAG systems at scale. While current approaches require extensive human curation, the future lies in automated systems that can continuously construct, update, and refine knowledge graphs from diverse sources while maintaining quality and trustworthiness.

11.3.2.1 Continuous Knowledge Extraction from Streaming Data

The vision of "living" knowledge graphs that update themselves in real time is becoming reality through advances in automated extraction. These systems monitor continuous streams of information—research papers, news articles, social media, sensor data, transaction logs—and automatically extract structured knowledge. The technical challenges are formidable: identifying relevant information in noisy streams, extracting entities and relationships with high precision, resolving entities across different sources, and integrating new information while maintaining consistency.

The COVID-19 pandemic demonstrated both the potential and challenges of rapid knowledge graph construction. The COVID-19 Open Research Dataset (CORD-19) knowledge graph grew from zero to millions of facts within months, integrating information from thousands of research papers. Automated extraction systems identified new virus variants, tracked emerging symptoms, and connected treatment reports from around the world. However, the rapid pace also led to quality issues—contradictory information, premature conclusions, and difficulty distinguishing peer-reviewed research from preprints.

Future systems must balance speed with accuracy. Advanced natural language processing models, particularly large language models fine-tuned for knowledge extraction, show promise. These models can understand context and nuance in ways that rule-based extractors cannot. For instance, when extracting information about drug efficacy, they can distinguish between preliminary observations ("may be effective") and established facts ("proven effective in Phase 3 trials").

11.3.2.2 Quality Assurance Through Automated Validation

As knowledge graphs grow autonomously, ensuring information quality becomes paramount. Automated quality assurance must catch errors, resolve contradictions, and maintain consistency without constant human oversight. This requires multiple complementary approaches working in concert.

Cross-source validation leverages the principle that facts appearing in multiple independent sources are more likely to be accurate. When the system extracts a new fact, it can query other sources to find corroborating or contradicting evidence. For instance, if one source claims "Company X acquired Company Y for $10 billion" while another reports "$10 million," the system flags this discrepancy for resolution. Advanced systems might use source reliability scores, giving more weight to established, peer-reviewed sources over informal reports.

Logical consistency checking ensures that new facts don't contradict existing knowledge in the graph. This goes beyond simple contradiction detection to include more subtle inconsistencies. For example, if the graph knows that "Person A was born in 1990" and a new source claims "Person A graduated from university in 1995," the system should flag this as unlikely (though not impossible). Constraint-based reasoning and description logics provide formal frameworks for such consistency checking.

Confidence scoring mechanisms assign reliability estimates to facts based on multiple factors: source credibility, extraction confidence, corroboration from other sources, and consistency with existing knowledge. These scores propagate through the graph during reasoning, allowing the Graph-RAG system to convey uncertainty in its answers. Rather than stating facts as absolute truths, the system might say: "According to three reliable sources, the merger value was $10 billion, though one source reports a different figure."

11.3.2.3 Dynamic Schema Evolution

Knowledge domains are not static—new concepts emerge, relationships evolve, and entire fields of knowledge can spring into existence. Traditional knowledge graphs with fixed schemas struggle to accommodate such changes. Future Graph-RAG systems must support dynamic schema evolution, automatically detecting when new concepts or relationships are needed and updating the graph structure accordingly.

The emergence of cryptocurrency provides an illustrative example. A financial knowledge graph designed in 2010 would have no concept of Bitcoin, blockchain, or DeFi. By 2025, these concepts are central to financial technology. An adaptive system would notice patterns in the data—frequent mentions of unfamiliar terms, new types of transactions, novel relationships between entities—and infer the need for schema updates.

Automated ontology learning techniques can identify emerging concepts by clustering entities with similar properties or relationships. When many entities share characteristics not captured by existing categories, the system can propose new classes. For instance, noticing many drugs with similar mechanisms targeting the same pathway might lead to creating a new drug class in a medical knowledge graph.

The challenge lies in making such updates without breaking existing functionality. Version-controlled schemas, similar to database migration systems, allow gradual evolution while maintaining backward compatibility. The system might maintain mappings between old and new concepts, ensuring that queries using outdated terminology still return relevant results.

11.3.2.4 Human–AI Collaborative Construction

Despite advances in automation, human expertise remains invaluable for knowledge graph construction. The future lies not in replacing humans but in creating powerful collaborative frameworks where AI handles scale while humans provide nuance, validation, and domain expertise.

Collaborative platforms might present human experts with AI-extracted facts for validation, using active learning to focus human attention where it's most needed. The AI learns from human corrections, improving its extraction accuracy over time. For instance, in building a legal knowledge graph, AI might extract regulatory relationships from thousands of documents, while legal experts validate critical interpretations and resolve ambiguities that require domain expertise.

Crowdsourcing mechanisms, similar to Wikipedia but AI-assisted, can democratize knowledge graph construction. Contributors without technical knowledge of graph schemas can provide information in natural language, which AI converts to structured format. The system might prompt contributors with specific questions to fill knowledge gaps: "We noticed Company X has no listed subsidiaries. Can you provide information about its corporate structure?"

Gamification and incentive mechanisms encourage quality contributions. Contributors might earn reputation points for accurate additions, with expert validators rating contributions. Blockchain-based systems have been proposed to create tamper-evident logs of contributions, ensuring attribution and building trust in collaboratively constructed graphs.

11.3.3 RL for Intelligent Graph Traversal

Current Graph-RAG systems typically use simple, predetermined strategies for navigating knowledge graphs—breadth-first search, fixed-depth exploration, or basic heuristics. RL offers a paradigm shift: agents that learn optimal traversal strategies through experience, adapting to query types, graph structures, and user preferences.

11.3.3.1 Adaptive Exploration Strategies

By framing graph traversal as a sequential decision-making problem, RL agents can learn sophisticated exploration policies. Each state represents a position in the graph, actions correspond to following specific edges, and rewards reflect the value of information gathered. Through trial and error across many queries, agents learn which paths are most promising for different types of questions.

The DeepPath system pioneered this approach for knowledge-base reasoning. When answering complex queries requiring multi-hop reasoning, DeepPath's RL agent learned to navigate efficiently through massive graphs. For the query "Which proteins interact with those targeted by Drug X?", the agent learned to first find Drug X's targets, then explore protein–protein interaction edges, rather than wandering through irrelevant relationships.

Advanced RL approaches can balance exploration and exploitation dynamically. For well-understood query types, the agent exploits known successful paths. For novel queries, it explores more broadly, potentially discovering new reasoning patterns. This adaptability is crucial for handling the diverse queries that real-world Graph-RAG systems encounter.

11.3.3.2 Multiobjective Optimization in Graph Traversal

Real-world graph traversal involves multiple, often competing objectives: finding accurate information, minimizing computational cost, ensuring diverse perspectives, and meeting response time constraints. Multi-objective RL provides frameworks for navigating these trade-offs intelligently.

Consider a medical Graph-RAG with strict response time requirements for emergency queries. The RL agent must balance thoroughness with speed. For a query about emergency drug interactions, it might learn to check only the most critical interaction paths first, returning a partial but timely answer. For nonurgent research queries, the same system might explore more extensively, trading speed for comprehensiveness.

Reward shaping becomes crucial in multiobjective settings. Rewards might include positive components for finding relevant information, negative components for computational cost, bonuses for diversity of sources, and penalties for exceeding time limits. The agent learns to optimize this complex reward landscape, developing nuanced strategies for different contexts.

11.3.3.3 Meta-learning for Cross-domain Adaptation

A key challenge in applying RL to Graph-RAG is generalization across domains. An agent trained on biomedical graphs may struggle with financial graphs due to different structures and relationship types. Meta-learning approaches address this by training agents to quickly adapt to new domains.

Meta-RL for graph traversal involves training across multiple graphs or query types to extract transferable strategies. The agent learns not just specific paths but general principles: follow high-connectivity nodes for broad context, prioritize recent relationships for time-sensitive queries, or check inverse relationships when direct paths fail. These meta-strategies transfer across domains, allowing rapid adaptation to new knowledge graphs.

Recent work demonstrates promising results. Agents trained on diverse knowledge graphs show superior performance on entirely new graphs compared to those trained on single domains. The key insight is that while specific facts differ across domains, structural patterns and reasoning strategies often generalize.

11.3.3.4 Integration with Human Feedback

RL from Human Feedback (RLHF) has revolutionized large language models by aligning them with human preferences. Similar principles apply to graph traversal in Graph-RAG systems. Rather than defining rigid reward functions, systems can learn from human judgments of result quality.

In practice, users might rate the helpfulness of answers, and these ratings train the traversal agent. If users consistently prefer answers that cite multiple sources, the agent learns to explore more diverse paths. If users in a specific domain value recency, the agent learns to prioritize newer relationships. This creates a virtuous cycle where the system becomes increasingly aligned with user needs.

The integration of RLHF with graph traversal presents unique opportunities. Users might provide feedback not just on final answers but on the reasoning process itself. By exposing traversal paths and allowing users to indicate which steps were helpful or irrelevant, the system can learn more nuanced strategies. This fine-grained feedback is particularly valuable in expert domains where users can evaluate the quality of reasoning paths.

11.3.4 Federated Learning for Privacy-preserving Graph-RAG

As Graph-RAG systems handle increasingly sensitive information across healthcare, finance, and personal data, the traditional approach of centralizing all knowledge in a single graph becomes untenable. Federated learning offers a revolutionary alternative: enabling powerful collective intelligence while keeping sensitive data distributed and private.

11.3.4.1 Distributed Knowledge Graph Architecture

The federated approach reimagines knowledge graphs not as monolithic structures but as interconnected networks of local graphs. Each participant—whether a hospital, bank, or research institution—maintains its own knowledge graph with sensitive data. Federated learning techniques enable these distributed graphs to function as a coherent whole without raw data ever leaving its source.

The MELLODDY consortium demonstrates this at scale in pharmaceutical research [26]. Ten competing pharmaceutical companies collaborated on drug discovery without sharing proprietary compound structures. Each company maintained its own knowledge graph of molecular structures and bioactivity data. Federated learning algorithms trained predictive models across all graphs while preserving trade secrets. The resulting models benefited from 10× more data than any single company possessed, accelerating drug discovery while maintaining competitive advantages.

Technically, this requires sophisticated coordination protocols. When processing a query that spans multiple private graphs, the system must decompose it into sub-queries that reveal minimal information. Secure multiparty computation ensures that intermediate results can be combined without any party seeing others' raw data. For instance, finding treatment patterns across hospitals might involve each hospital computing local statistics, encrypting them, and combining them in a way that only the aggregate result is revealed.

11.3.4.2 Privacy-preserving Query Processing

Processing queries across federated knowledge graphs while maintaining privacy requires novel cryptographic and algorithmic approaches. Differential privacy provides mathematical guarantees that query results don't reveal information about specific individuals or entities in any participant's graph.

Consider a federated medical Graph-RAG where researchers want to study treatment effectiveness across multiple hospitals. Each hospital's graph contains sensitive patient data that cannot be shared. The query "What percentage of patients with Condition X responded to Treatment Y?" must be answered without revealing whether any specific patient is in any hospital's database.

Differential privacy achieves this by adding carefully calibrated noise to query results. If Hospital A has 100 eligible patients with 70% response rate and Hospital B has 50 patients with 60% response rate, the system might report "approximately 66% response rate across 150 patients," with noise added to prevent reverse engineering of individual hospital statistics. The key innovation is maintaining useful accuracy while providing strong privacy guarantees.

Recent advances in private federated learning for graphs show promising results, such as the FedGraph framework [28]. This makes federated Graph-RAG practical for sensitive applications where privacy regulations like GDPR or HIPAA apply.

11.3.4.3 Incentive Mechanisms for Participation

Creating successful federated Graph-RAG systems requires solving not just technical but also economic challenges. Why should organizations contribute their valuable knowledge graphs to a federated system? How can the system ensure fair participation and prevent free-riding?

Blockchain-based incentive mechanisms provide one solution. Smart contracts can automatically track contributions and query usage, ensuring fair compensation. When Organization A's data helps answer Organization B's query, A receives credits that can be used for its own queries or converted to monetary compensation. The Ocean Protocol demonstrates this concept for data marketplaces, which could extend to knowledge graphs [29].

Reputation systems provide nonmonetary incentives. Organizations that contribute high-quality data or computational resources earn reputation scores, gaining priority access to collective insights. This creates positive feedback loops where active contributors receive the most value from the system.

Privacy-preserving contribution tracking ensures that the incentive system itself doesn't leak sensitive information. Zero-knowledge proofs can verify that an organization contributed valuable data without revealing what that data contains. This maintains the privacy guarantees while enabling fair attribution and rewards.

11.3.5 Transformative Industry Applications

The convergence of advanced Graph-RAG capabilities—explainability, automated construction, intelligent traversal, and privacy preservation—enables transformative applications across every major industry. These applications go beyond incremental improvements to fundamentally reshape how knowledge powers decision-making and innovation.

11.3.5.1 Healthcare: From Reactive to Predictive Medicine

The integration of Graph-RAG into healthcare promises a shift from reactive treatment to predictive, personalized medicine. By combining individual patient data with vast medical knowledge, these systems enable unprecedented precision in diagnosis and treatment.

The PrimeKG medical knowledge graph exemplifies current capabilities [30]. With over 17,000 diseases and 4 million relationships spanning genetic, molecular, and clinical data, it enables complex reasoning about disease mechanisms and treatment options. When integrated into a Graph-RAG system, clinicians can ask sophisticated questions like: "What off-label drugs might benefit this patient given their specific genetic mutations and comorbidities?"

Future healthcare Graph-RAG systems will integrate real-time data streams from wearable devices, environmental sensors, and population health monitoring. Imagine a system that continuously updates its knowledge graph with the latest research while monitoring individual patient trajectories. It might alert physicians: "Three patients with similar genetic profiles to your patient developed adverse reactions to the prescribed medication. Recent research suggests an alternative treatment pathway with better outcomes for this genotype."

Epidemic modeling and public health response represent another transformative application. During the COVID-19 pandemic, knowledge graphs proved valuable for tracking virus mutations, treatment effectiveness, and transmission patterns. Future systems will provide real-time epidemic intelligence, integrating genomic surveillance, population movement data, clinical outcomes, and social determinants of health.

Such systems could answer complex queries like: "Given the current mutation patterns of Influenza strain X and population immunity levels, which regions are at highest risk for severe outbreaks in the next month?" The Graph-RAG would traverse connections between viral genetics, population demographics, historical outbreak patterns, and current vaccination rates to provide actionable intelligence for public health interventions.

11.3.5.2 Financial Services: Systemic Risk Intelligence

The interconnected nature of modern finance makes it an ideal domain for Graph-RAG applications. By representing financial entities and their relationships as knowledge graphs, these systems can provide unprecedented visibility into systemic risks and opportunities.

Current financial knowledge graphs capture basic relationships: banks, loans, derivatives, and counterparty exposures. Advanced Graph-RAG systems will integrate real-time market data, news sentiment, regulatory changes, and macroeconomic indicators. This enables dynamic risk assessment that adapts to changing conditions.

Consider a query to a future financial Graph-RAG: "What would be the cascade effects if Commercial Real Estate prices drop 20% in major cities?" The system would traverse paths through property valuations, mortgage-backed securities, bank exposures, derivative positions, and interconnected financial institutions. It might respond: "A 20% CRE price drop would initially impact Regional Banks A, B, and C with $X billion in direct exposure. Secondary effects through CMBS markets would affect Investment Funds D and E. Tertiary impacts through derivative positions could trigger margin calls at Hedge Fund F, potentially forcing asset sales that further depress prices."

Regulatory compliance represents another transformative application. Financial institutions struggle with thousands of regulations across multiple jurisdictions. A Graph-RAG system maintaining a regulatory knowledge graph can provide real-time compliance intelligence. When new regulations are published, natural language processing extracts requirements and updates the graph. The system can then traverse connections between regulations, business processes, and control mechanisms to identify compliance gaps.

11.3.5.3 Education: Personalized Learning Ecosystems

Graph-RAG systems promise to revolutionize education by creating truly personalized learning experiences. Each student's learning journey becomes a path through a vast knowledge graph, with the system continuously optimizing for individual needs, interests, and goals.

Current educational knowledge graphs capture curriculum standards, learning objectives, and prerequisite relationships. Future Graph-RAG systems will integrate individual student profiles, learning styles, real-time performance data, and connections to real-world applications. This enables unprecedented personalization.

A student struggling with calculus might query: "Why do I need to learn integration?" The Graph-RAG system, knowing the student's interest in video game design, might respond: "Integration is crucial for physics engines in games. When calculating realistic motion paths for characters or projectiles, integration determines positions from velocities. Here's how the game engine for your favorite racing game uses integration..." The system would then provide personalized examples and exercises connecting calculus to game development.

Beyond individual learning, Graph-RAG enables intelligent collaboration. The system might notice that Student A excels at visual thinking while Student B is strong in mathematical formulation. When both struggle with a physics problem requiring both skills, it could suggest collaboration and provide complementary resources to each student.

Career development and lifelong learning benefit from Graph-RAG systems that map evolving skill requirements to learning resources and opportunities. A professional might ask: "What skills should I develop to transition from traditional manufacturing to Industry 4.0 roles?" The system would analyze the knowledge graph connecting current skills, emerging job requirements, available training programs, and success stories of similar transitions. It might recommend: "Based on your mechanical engineering background, focusing on IoT sensor integration and data analytics would position you well. Here's a personalized learning path with estimated timeline and expected salary progression..."

11.3.5.4 Smart Cities: Urban Intelligence Networks

Cities generate vast amounts of interconnected data: traffic flows, energy consumption, public safety incidents, citizen requests, and environmental conditions. Graph-RAG systems can transform this data deluge into actionable urban intelligence.

Current smart city initiatives often treat different systems in isolation. Future Graph-RAG platforms will integrate all urban systems into a unified knowledge graph, enabling holistic optimization. City planners could query: "What would be the multiyear impact of converting Main Street to pedestrian-only?"

The system would traverse connections between traffic patterns, local business revenues, property values, air quality, public health outcomes, and citizen satisfaction. Its response might include: "Converting Main Street would initially increase traffic on parallel routes by 15%, but models predict 30% increase in foot traffic benefiting local businesses. Air quality improvements could prevent an estimated five hospitalizations annually. Similar conversions in comparable cities led to 12% property value increases over three years. Recommended phasing plan and mitigation strategies..."

Emergency response showcases the real-time capabilities of urban Graph-RAG. During a major incident, the system integrates sensor data, resource locations, traffic conditions, and historical patterns. First responders could query: "Optimal deployment strategy for a chemical spill at Location X?" The system would instantly compute evacuation routes considering wind patterns, identify available hazmat teams, predict hospital capacity needs, and coordinate traffic management—all by traversing the urban knowledge graph in real time.

Citizen engagement transforms from one-way communication to intelligent dialog. Citizens could ask complex questions about their city: "Why is my neighborhood park maintenance delayed?" The Graph-RAG might explain: "The park maintenance crew was redirected to emergency storm damage repairs. Based on the current schedule and resource allocation, maintenance should resume next Tuesday. Alternative nearby parks with similar amenities include..." This transparency builds trust and enables informed civic participation.

11.4 Summary

This chapter explored the cutting-edge frontiers of Graph-RAG systems, emphasizing their evolution from static, text-centric architectures to dynamic, multimodal reasoning engines. We began by examining MMKGs, which integrate diverse data types—text, images, audio, tables, and video—into unified representations. This integration enables richer, context-aware reasoning through cross-modal alignment and attention, supporting sophisticated applications such as medical diagnostics, product intelligence, and security analysis. We then turned to temporal and dynamic Graph-RAG, which infuses knowledge systems with an awareness of time. By incorporating versioned entities, time-stamped relationships, and temporal reasoning operators, these systems can answer complex historical and trend-based queries while adapting in real time to new information streams.

The chapter also introduced advanced GNN architectures tailored for Graph-RAG, including heterogeneous transformers, continuous-time networks, and memory-augmented models. These innovations allow for intelligent, efficient navigation of complex knowledge spaces, while emerging RL strategies offer adaptive, query-specific graph traversal.

These advances bring profound ethical responsibilities. Graph structures can both expose and amplify societal biases, demanding proactive mitigation through diverse data sources, algorithmic fairness, and transparency. We discussed the dual nature of bias in graph systems and proposed mitigation strategies—such as fairness-aware algorithms, provenance tracking, and privacy-preserving learning—to ensure accountability and inclusivity. Finally, we outlined future research directions and transformative industry applications across healthcare, finance, education, and smart governance.

Key insights for the future:

- Integration over Isolation: The most powerful systems will seamlessly combine multiple modalities, temporal reasoning, and advanced learning paradigms.
- Augmentation, Not Replacement: Graph-RAG evolves toward true human–AI collaboration, creating capabilities neither could achieve alone.
- Ethics as Foundation: Current decisions about fairness and accountability will determine whether this technology amplifies the best or worst of human knowledge.
- Continuous Evolution: Like knowledge itself, these systems must continuously learn and adapt, never reaching a final state.

The research frontiers—from explainable Graph-RAG to automated knowledge construction—represent opportunities to fundamentally reimagine how AI acquires and reasons with knowledge. Applications spanning healthcare, finance, education, and governance demonstrate that these technologies can address society's most pressing challenges.

As you continue your Graph-RAG journey, remember that you're not just building technology but shaping how future AI systems understand human knowledge. The frontier is vast, the challenges significant, but the potential to create AI that truly augments human intelligence has never been greater. The future of Graph-RAG will be written by those who dare to push boundaries and imagine new possibilities.

11.5 Conclusion and Final Thoughts

As we stand at the frontier of Graph-RAG technology, we see not just an evolution of information retrieval but a fundamental transformation in how AI systems understand and reason about the world. The convergence of graph structures with large language models has opened possibilities that seemed like science fiction just years ago.

The journey from simple RAG to sophisticated graph-based reasoning systems represents more than technical progress—it reflects our growing understanding of how knowledge itself is structured and how intelligence emerges from the connections between ideas. As these systems become more capable, they promise to augment human intelligence in ways that preserve our agency while expanding our capabilities.

Yet with great power comes great responsibility. The ethical considerations discussed in this chapter are not merely academic exercises but urgent imperatives that will shape the societal impact of these technologies. Privacy, fairness, transparency, and accountability must be built into the foundation of Graph-RAG systems, not added as afterthoughts.

Looking ahead, the research directions outlined here offer tantalizing glimpses of what's possible: AI systems that not only retrieve information but truly understand it, that can explain their reasoning in human terms, and that can collaborate across organizational boundaries while preserving privacy. The industry applications we've explored are just the beginning—the true potential of Graph-RAG will likely manifest in ways we haven't yet imagined.

As practitioners, researchers, and stakeholders in this technological revolution, we have the opportunity and responsibility to shape its direction. The choices we make today about how to build, deploy, and govern Graph-RAG systems will reverberate through the decades to come. May we choose wisely, with both ambition for what's possible and humility about what we don't yet understand.

The frontier beckons. The future of intelligent systems lies not in isolated models but in connected knowledge, not in static databases but in living graphs, not in narrow AI but in systems that mirror the rich, interconnected nature of human understanding. Graph-RAG is not the destination but a powerful vehicle for the journey ahead. Where we go from here is limited only by our imagination and our commitment to beneficial AI for all.

As you close this chapter and this book, we invite you to become part of this frontier. Whether through research, development, deployment, or thoughtful critique, your contribution to the Graph-RAG ecosystem will help shape the future of AI. The graph of knowledge continues to grow, and every node—every idea, every connection, every innovation—matters. The story of Graph-RAG has just begun. What chapter will you write?

References

1 Tadas Baltrušaitis, Chaitanya Ahuja, and Louis-Philippe Morency. Multimodal machine learning: A survey and taxonomy. *IEEE Transactions on Pattern Analysis and Machine Intelligence*, 41(2):423–443, 2018.

2 Zhuo Chen, Yichi Zhang, Yin Fang, Yuxia Geng, Lingbing Guo, Xiang Chen, Qian Li, Wen Zhang, Jiaoyan Chen, Yushan Zhu, et al. Knowledge graphs meet multi-modal learning: A comprehensive survey. *arXiv preprint arXiv:2402.05391*, 2024.

3 Yinan Wu, Xiaowei Wu, Junwen Li, Yue Zhang, Haofen Wang, Wen Du, Zhidong He, Jingping Liu, and Tong Ruan. Mmpedia: A large-scale multi-modal knowledge graph. In *International Semantic Web Conference*, pages 18–37. Springer, 2023.

4 Xin Wang, Benyuan Meng, Hong Chen, Yuan Meng, Ke Lv, and Wenwu Zhu. TIVA-KG: A multimodal knowledge graph with text, image, video and audio. In *Proceedings of the 31st ACM International Conference on Multimedia*, pages 2391–2399, 2023.

5 Jaejun Lee, Chanyoung Chung, Hochang Lee, Sungho Jo, and Joyce Whang. Vista: Visual-textual knowledge graph representation learning. In *Findings of the Association for Computational Linguistics: EMNLP 2023*, pages 7314–7328, 2023.

6 Alec Radford, Jong Wook Kim, Chris Hallacy, Aditya Ramesh, Gabriel Goh, Sandhini Agarwal, Girish Sastry, Amanda Askell, Pamela Mishkin, Jack Clark, et al. Learning transferable visual models from natural language supervision. In *International Conference on Machine Learning*, pages 8748–8763. PMLR, 2021.

7 Qihang Jin, Enze Ge, Yuhang Xie, Hongying Luo, Junhao Song, Ziqian Bi, Chia Xin Liang, Jibin Guan, Joe Yeong, and Junfeng Hao. Multimodal representation learning and fusion. *arXiv preprint arXiv:2506.20494*, 2025.

8 Chao Jia, Yinfei Yang, Ye Xia, Yi-Ting Chen, Zarana Parekh, Hieu Pham, Quoc Le, Yun-Hsuan Sung, Zhen Li, and Tom Duerig. Scaling up visual and vision-language representation learning with noisy text supervision. In *International Conference on Machine Learning*, pages 4904–4916. PMLR, 2021.

9 Andrew Jaegle, Felix Gimeno, Andy Brock, Oriol Vinyals, Andrew Zisserman, and Joao Carreira. Perceiver: General perception with iterative attention. In *International Conference on Machine Learning*, pages 4651–4664. PMLR, 2021.

10 Seyed Mehran Kazemi, Rishab Goel, Kshitij Jain, Ivan Kobyzev, Akshay Sethi, Peter Forsyth, and Pascal Poupart. Representation learning for dynamic graphs: A survey. *Journal of Machine Learning Research*, 21(70):1–73, 2020.

11 James F Allen. Maintaining knowledge about temporal intervals. *Communications of the ACM*, 26(11):832–843, 1983.

12 Aldo Pareja, Giacomo Domeniconi, Jie Chen, Tengfei Ma, Toyotaro Suzumura, Hiroki Kanezashi, Tim Kaler, Tao Schardl, and Charles Leiserson. Evolvegcn: Evolving graph convolutional networks for dynamic graphs. In *Proceedings of the AAAI Conference on Artificial Intelligence*, volume 34, pages 5363–5370, 2020.

13 Ziniu Hu, Yuxiao Dong, Kuansan Wang, and Yizhou Sun. Heterogeneous graph transformer. In *Proceedings of the Web Conference 2020*, pages 2704–2710, 2020.

14 Jin Guo, Zhen Han, Zhou Su, Jiliang Li, Volker Tresp, and Yuyi Wang. Continuous temporal graph networks for event-based graph data. *arXiv preprint arXiv:2205.15924*, 2022.

15 Guixiang Ma, Vy A Vo, Theodore L Willke, and Nesreen K Ahmed. Memory-augmented graph neural networks: A brain-inspired review. *IEEE Transactions on Artificial Intelligence*, 5(5):2011–2025, 2023.

16 Wikipedia contributors. Differentiable neural computer — Wikipedia, the free encyclopedia, 2025. https://en.wikipedia.org/wiki/Differentiable_neural_computer. Accessed: 2025-7-13.

17 Yaochen Xie, Zhao Xu, Jingtun Zhang, Zhengyang Wang, and Shuiwang Ji. Self-supervised learning of graph neural networks: A unified review. *IEEE Transactions on Pattern Analysis and Machine Intelligence*, 45(2):2412–2429, 2022.

18 Ziniu Hu, Yuxiao Dong, Kuansan Wang, Kai-Wei Chang, and Yizhou Sun. Gpt-gnn: Generative pre-training of graph neural networks. In *Proceedings of the 26th ACM SIGKDD International Conference on Knowledge Discovery & Data Mining*, pages 1857–1867, 2020.

19 Zhenyu Hou, Xiao Liu, Yukuo Cen, Yuxiao Dong, Hongxia Yang, Chunjie Wang, and Jie Tang. Graphmae: Self-supervised masked graph autoencoders. In *Proceedings of the 28th ACM SIGKDD Conference on Knowledge Discovery and Data Mining*, pages 594–604, 2022.

20 Paul Almasan, José Suárez-Varela, Krzysztof Rusek, Pere Barlet-Ros, and Albert Cabellos-Aparicio. Deep reinforcement learning meets graph neural networks: Exploring a routing optimization use case. *Computer Communications*, 196:184–194, 2022.

21 Wenhan Xiong, Thien Hoang, and William Yang Wang. Deeppath: A reinforcement learning method for knowledge graph reasoning. *arXiv preprint arXiv:1707.06690*, 2017.

22 Yongkang Xiao, Sinian Zhang, Yi Dai, Huixue Zhou, Jue Hou, Jie Ding, and Rui Zhang. DrKGC: Dynamic subgraph retrieval-augmented llms for knowledge graph completion across general and biomedical domains. *arXiv preprint arXiv:2506.00708*, 2025.

23 Hao Yuan, Haiyang Yu, Shurui Gui, and Shuiwang Ji. Explainability in graph neural networks: A taxonomic survey. *IEEE Transactions on Pattern Analysis and Machine Intelligence*, 45(5):5782–5799, 2022.

24 Zhitao Ying, Dylan Bourgeois, Jiaxuan You, Marinka Zitnik, and Jure Leskovec. Gnnexplainer: Generating explanations for graph neural networks. *Advances in Neural Information Processing Systems*, 32:9244–9255, 2019.

25 Fangxin Wang, Yuqing Liu, Kay Liu, Yibo Wang, Sourav Medya, and Philip S Yu. Uncertainty in graph neural networks: A survey. *arXiv preprint arXiv:2403.07185*, 2024.

26 Wouter Heyndrickx, Lewis Mervin, Tobias Morawietz, Noé Sturm, Lukas Friedrich, Adam Zalewski, Anastasia Pentina, Lina Humbeck, Martijn Oldenhof, Ritsuya Niwayama, et al. Melloddy: Cross-pharma federated learning at unprecedented scale unlocks benefits in qsar without compromising proprietary information. *Journal of Chemical Information and Modeling*, 64(7):2331–2344, 2023.

27 Utkarshani Jaimini and Amit Sheth. Causalkg: Causal knowledge graph explainability using interventional and counterfactual reasoning. *IEEE Internet Computing*, 26(1):43–50, 2022.

28 Fahao Chen, Peng Li, Toshiaki Miyazaki, and Celimuge Wu. Fedgraph: Federated graph learning with intelligent sampling. *IEEE Transactions on Parallel and Distributed Systems*, 33(8):1775–1786, 2021.

29 Trent McConaghy. Ocean protocol: Tools for the web3 data economy. In *Handbook on Blockchain*, pages 505–539. Springer, 2022.

30 Payal Chandak, Kexin Huang, and Marinka Zitnik. Building a knowledge graph to enable precision medicine. *Scientific Data*, 10(1):67, 2023.

A

Set Up Experiment Servers

This appendix details the configuration of a multi-server environment set up locally using Docker. The architecture supports both programmatic interaction via Jupyter notebooks and user-friendly access through web browsers. All containers and services operate within a shared Docker network, enabling seamless inter-service communication and port mapping to the host machine. Figure A.1 illustrates the data flow between containers, the Docker network, and client tools.

Large Language Model (LLM) Services: The goal of this component is to emulate the OpenAI API endpoint and provide a web-based interface using a suite of open-source tools: vLLM, LiteLLM, and Open-WebUI.

vLLM Servers: There are two vLLM server instances:

- A generation/chat model server that exposes an LLM endpoint optimized for text generation and natural language conversations.
- An embedding model server that provides an embedding endpoint, generating vector representations of text for semantic search, similarity comparisons, or clustering tasks.

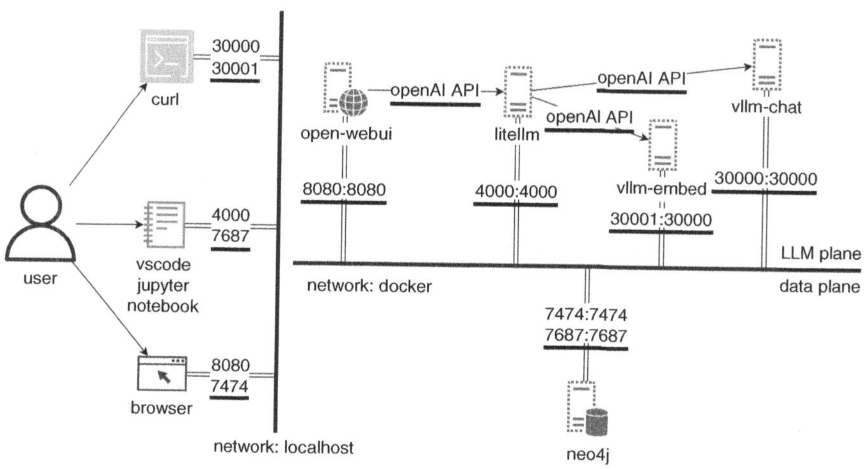

Figure A.1 Server architecture diagram.

These servers are typically accessed indirectly through LiteLLM. Direct access via command-line tools like `curl` is used only for health checks or debugging. Ideally, both models should run on GPUs for optimal performance. If GPU memory is limited, the embedding model can be moved to the CPU as a fallback. However, running chat/generation models on the CPU is not recommended due to performance limitations.

LiteLLM Proxy Server: Acts as a unified interface to both vLLM instances, exposing a single OpenAI-compatible API endpoint. This simplifies integration with Jupyter notebooks or other applications expecting the OpenAI API schema.

Web Interface (Open-WebUI): A lightweight web server that proxies user interactions to the LiteLLM endpoint, providing browser-based access to LLM functionality such as chat interfaces and model queries.

Database Services: Two types of databases are hosted within the system: a vector database and a graph database.

Vector Database (Chroma): Stores dense numerical embeddings generated by the embedding model. It enables fast similarity searches, which are essential for semantic analysis and retrieval tasks.

Graph Database (Neo4j): A Neo4j instance manages relational data and offers two access methods:

Programmatic Access: Provides a Bolt protocol endpoint (bolt://localhost:7687) for direct interaction using libraries such as `neo4j` in Jupyter notebooks.

Web-based Interface: Offers an HTTP endpoint (http://localhost:7474) with a graphical user interface for querying and visualizing relationships within the database.

Docker Network Configuration: All services operate within a shared Docker network, which ensures secure internal communication and port isolation. This setup includes port mapping via NAT, which exposes service ports to the localhost (such as 7474, 7687, and custom ports for vLLM, Litellm, and Open-WebUI), and network isolation, which prevents external access while allowing seamless interaction between services and client tools like browsers, Jupyter, and the command-line interface (CLI).

The services are orchestrated using `docker compose`, with configuration defined in `.yaml` files that specify service dependencies, resource constraints, and runtime parameters. The setup was tested on Ubuntu 24.04 LTS desktop with CUDA 12.8, using two NVIDIA GeForce RTX 4060 Ti GPUs (16 GB VRAM each). Minor adjustments may be necessary when deploying on other operating systems or hardware configurations.

To start or stop the servers running the chat model, use the following Docker Compose commands:

```
docker compose \
    -f docker-compose.vllm.chat.yaml \
    up -d --wait

docker compose \
    -f docker-compose.vllm.chat.yaml \
    down
```

These commands can be adapted to manage other server configurations by replacing `docker-compose.vllm.chat.yaml` with the appropriate Docker Compose file for the target configuration.

A.1 Two Models Hosted by vLLM

vLLM is a fast and user-friendly library for large language model (LLM) inference and serving. It is specifically designed to meet the unique computational and memory demands of LLMs. This guide demonstrates how to deploy two specialized models using vLLM:

1. **Chat Model:** Designed for interactive conversational AI, this model uses quantized variants of Qwen, such as `Qwen/Qwen3-32B-AWQ`.
2. **Embedding Model:** Used for generating vector embeddings, this model leverages `Qwen/Qwen3-Embedding-0.6B`.

Due to current limitations in vLLM—specifically its support for only one model per instance—these models are deployed as separate services running on distinct servers.

A.1.1 Chat Model Service

The Qwen series of models is highly effective for open-source natural language processing tasks. Below is the Docker Compose service definition for deploying the chat model using vLLM:

```
services:
  chat:
    command:
      - '--host=0.0.0.0'
      - '--port=30000'
      - '--model=${MODEL_CHAT}'
      - '--served-model-name=my-o4-mini'
      - '--tensor-parallel-size=2'                              # [1]
      - '--enforce-eager'
      - '--enable-auto-tool-choice'                             # [2]
      - '--tool-call-parser=hermes'                             # [2]
      - '--enable-reasoning'                                    # [3]
      - '--reasoning-parser=deepseek_r1'                        # [3]
      - '--kv-cache-dtype=fp8_e4m3'
    deploy:
      resources:
        reservations:
          devices:
            - capabilities:
                - 'gpu'                                         # [4]
    environment:
      - 'CUDA_LAUNCH_BLOCKING=1'
      - 'VLLM_ATTENTION_BACKEND=FLASHINFER'
    healthcheck:
      interval: '10s'
      retries: 30
      test: 'curl -f http://localhost:30000/health'
      timeout: '5s'
    image: 'vllm/vllm-openai:${RELEASE_VLLM}'
    ipc: 'host'                                                 # [5]
    ports:
      - '30000:30000'
    runtime: 'nvidia'                                           # [4]
    volumes:
      - '${HOME}/.cache/huggingface:/root/.cache/huggingface'   # [6]
```

[1] **Tensor Parallelism:** Distributes the model across multiple GPUs for efficient inference.

[2] **Automatic Tool Selection:** Enables dynamic tool calling; uses the `hermes` parser for interpreting function calls.

[3] **Structured Reasoning:** Supports reasoning output formatting; uses the `deepseek_r1` parser.

[4] **GPU Acceleration:** NVIDIA GPU support is enabled via the `nvidia` runtime, granting access to host GPUs.

[5] **Shared Memory (IPC):** Required for inter-process communication during tensor-parallel execution.

[6] **Model Caching:** Mounts the local Hugging Face cache to avoid re-downloading model weights.

To test the chat service, use the following `curl` command:

```
curl -s http://localhost:30000/v1/chat/completions \
    -H 'Content-Type: application/json' \
    -d '{
        "messages": [
            {"role": "system", "content": "You are a helpful assistant."},
            {"role": "user", "content": "Which weighs more, a pound of water, two
    pounds of bricks, a pound of feathers, or three pounds of air?"}
        ]
    }'
```

A.1.2 Embedding Model Service

The following configuration defines the service for hosting the embedding model using vLLM:

```
services:
  embed:
    command:
      - '--port=30000'
      - '--host=0.0.0.0'
      - '--task=embedding'
      - '--model=${MODEL_EMBED}'
      - '--served-model-name=my-text-embedding-3-small'
      - '--tensor-parallel-size=2'
      - '--enforce-eager'
      - '--kv-cache-dtype=fp8_e4m3'
    deploy:
      resources:
        reservations:
          devices:
            - driver: 'nvidia'
              device_ids:
                - '0'
                - '1'
              capabilities:
                - 'gpu'
    environment:
      - 'VLLM_ATTENTION_BACKEND=FLASHINFER'
    healthcheck:
      interval: '10s'
      retries: 30
      test: 'curl -f http://localhost:30000/health'
      timeout: '5s'
    image: 'vllm/vllm-openai:${RELEASE_VLLM}'
    ipc: 'host'
    ports:
      - '30001:30000'
    restart: 'always'
    runtime: 'nvidia'
    volumes:
      - '${HOME}/.cache/huggingface:/root/.cache/huggingface'
```

This configuration is nearly identical to that of the chat model service, with one key distinction: the `-task=embedding` flag configures the model to run in embedding mode, enabling it to generate vector representations of text inputs. To test the embedding service, use the following `curl` command:

```
curl -s http://localhost:30001/v1/embeddings \
    -H 'Content-Type: application/json' \
    -d '{
        "input": "hello world"
    }'
```

A.2 LLM Gateway

We use LiteLLM as a reverse proxy to unify access to the various model services. It routes incoming requests to the appropriate backend based on the task type (e.g., chat or embedding). Below is the Docker Compose service definition for deploying LiteLLM:

```
configs:
  litellm-config:                                          # [1]
    file: './config-litellm.yaml'                          # [1]
services:
  litellm:
    command:
      - '--config=/app/config.yaml'                        # [1]
      - '--detailed_debug'
    configs:
      - source: 'litellm-config'                           # [1]
        target: '/app/config.yaml'                         # [1]
    environment:
      - 'LANGFUSE_PUBLIC_KEY=lf_pk_1234567890'             # [2]
      - 'LANGFUSE_SECRET_KEY=lf_sk_1234567890'             # [2]
      - 'LANGFUSE_HOST=http://langfuse-web:3000'           # [2]
    healthcheck:
      interval: '10s'
      retries: 30
      test: "python3 -c '\nimport requests;\nr = requests.get(\"http://localhost:4000/
    health/readiness\");\nr.raise_for_status()'\n"
      timeout: '5s'
    image: 'ghcr.io/berriai/litellm:${RELEASE_LITELLM}'
    ports:
      - '4000:4000'
    volumes:
      - './litellm_custom_guardrail.py:/app/custom_guardrail.py'  # [3]
```

[1] **LiteLLM Configuration File:** Specifies routing rules, model mappings, and parameters. Includes references to Langfuse integration and pre-call hooks.

[2] **Langfuse Integration:** Langfuse is an open-source observability platform for LLM applications. It helps teams monitor, debug, and optimize their AI workflows by capturing prompts, traces, and execution flows. Here, it is configured via environment variables pointing to a locally hosted instance. For setup instructions, see the official guide.[1]

[3] **Custom Guardrail Hook:** A custom Python script that runs before each LLM call to inject or modify request parameters.

To verify that the gateway correctly routes requests to both backends, you can run the following commands:

```
curl -s http://localhost:4000/v1/chat/completions \
    -H 'Content-Type: application/json' \
    -d '{
        "messages": [
            {"role": "system", "content": "You are a helpful assistant."},
            {"role": "user", "content": "Please tell a story."}
        ]
    }'

```

1 https://langfuse.com/self-hosting/docker-compose

```
curl -s http://localhost:4000/v1/embeddings \
    -H 'Content-Type: application/json' \
    -d '{
        "input": ["hello world", "world hello"]
    }'
```

A.3 UI for LLM Models

Open-WebUI provides an intuitive web interface for interacting with hosted language models. It enables users to test prompts, view responses, and manage model configurations through a browser. Below is the Docker Compose service definition for deploying Open-WebUI, configured to work with the LiteLLM gateway:

```
services:
  open-webui:
    container_name: 'open-web-ui'
    depends_on:
      litellm:
        condition: 'service_healthy'
    environment:
      - 'WEBUI_AUTH=False'                                  # [1]
      - 'ADMIN_EMAIL=admin@demo.com'
      - 'ENABLE_OLLAMA_API=False'
      - 'OPENAI_API_BASE_URL=http://litellm:4000/v1'        # [2]
      - 'DEFAULT_MODELS=o4-mini'
      - 'GLOBAL_LOG_LEVEL=DEBUG'
      - 'ENABLE_RAG_WEB_SEARCH=True'                        # [3]
      - 'RAG_WEB_SEARCH_ENGINE=duckduckgo'
      - 'RAG_WEB_SEARCH_RESULT_COUNT=20'
    healthcheck:
      interval: '10s'
      retries: 30
      test: 'curl -f http://localhost:8080/health'
      timeout: '5s'
    image: 'ghcr.io/open-webui/open-webui:${RELEASE_OPENWEBUI}'
    ports:
      - '8080:8080'
    restart: 'always'
    volumes:
      - 'open-webui-data:/app/backend/data'
```

[1] **Disable Authentication:** Allows local access without requiring login.

[2] **LiteLLM Integration:** Routes all model requests through the LiteLLM gateway at `http://litellm:4000/v1`.

[3] **Enable RAG Web Search:** Integrates DuckDuckGo search to enrich responses with real-time data during RAG.

A.4 Vector Store

Chroma is a widely used open-source vector database that provides a simple and intuitive API for storing and querying embeddings. It is ideal for applications such as semantic search, recommendation systems, and RAG. Below is the Docker Compose service definition for deploying Chroma:

```
configs:
  chroma-config:
    file: './config-chroma.yaml'
services:
  chroma:
```

```
    configs:
      - source: 'chroma-config'
        target: '/config.yaml'
    container_name: 'chroma-db'
    healthcheck:
      interval: '30s'
      retries: 3
      test:
        - 'CMD'
        - '/bin/bash'
        - '-c'
        - 'cat < /dev/null > /dev/tcp/localhost/8000'
      timeout: '10s'
    image: 'chromadb/chroma:${RELEASE_CHROMA}'
    ports:
      - '8000:8000'
    volumes:
      - 'chroma-data:/data'
```

The corresponding configuration file `config-chroma.yaml` defines important operational settings, such as the maximum payload size for handling larger embedding data, the persistent storage path to keep data safe between container restarts, and the allow-reset permission, which makes it easier to reset the database during development or testing. This setup offers a lightweight and reliable vector storage solution that works well for local development and small-scale deployments. To test the server, run the following Python snippet:

```
python3 -c "
import chromadb
chroma_client = chromadb.HttpClient()
print(chroma_client.heartbeat())
"
```

This script initializes a client connection to the Chroma server and prints the heartbeat response, confirming that the service is running and accessible.

A.5 Graph Database

Neo4j is an open-source graph database designed to efficiently manage and query highly connected data using graph structures. Below is the Docker Compose service definition for deploying Neo4j with essential plugins and configurations for local development:

```
services:
  neo4j:
    container_name: 'neo4j-db'
    environment:
      - 'NEO4J_AUTH=none'                                               # [1]
      - 'NEO4J_PLUGINS=["apoc", "graph-data-science"]'                  # [2]
      - 'NEO4J_dbms_security_procedures_unrestricted=apoc.*,gds.*'      # [3]
      - 'NEO4J_dbms_security_procedures_allowlist=apoc.*,gds.*'         # [4]
    healthcheck:
      interval: '10s'
      retries: 30
      test:
        - 'CMD-SHELL'
        - 'wget --no-verbose --tries=1 --spider localhost:7474 || exit 1'
      timeout: '5s'
    image: 'neo4j:${RELEASE_NEO4J}'
    ports:
      - '7474:7474'                                                     # [5]
      - '7687:7687'                                                     # [6]
    volumes:
      - 'neo4j-data:/data'
      - 'neo4j-logs:/logs'
```

[1] **Authentication Disabled:** Allows access without login credentials, ideal for local development environments.

[2] **Plugins Enabled:** Installs Awesome Procedures on Cypher (APOC) and Graph Data Science (GDS) plugins for advanced analytics and processing.

[3] **Unrestricted Procedure Access:** Grants full access to APOC and GDS procedures during execution.

[4] **Procedure Allowlist:** Explicitly permits specific APOC and GDS functions for safer operation within defined boundaries.

[5] **HTTP Interface:** Exposes the Neo4j web UI and REST API on port 7474.

[6] **Bolt Protocol:** Enables Bolt, a binary protocol used by Neo4j drivers for efficient programmatic interaction.

Command-line Access: To run Cypher queries directly from the terminal:

```
docker exec -t neo4j-db bin/cypher-shell "MATCH (n) RETURN labels(n) AS labels,
    count(*) AS count ORDER BY count DESC;"
```

Web Interface Access: Open the Neo4j browser interface at: http://localhost:7474/ You can also check plugin versions using Cypher:

```
RETURN apoc.version();
RETURN gds.version();
```

To delete all nodes and relationships from the database:

```
MATCH (n) DETACH DELETE n
```

B

Prepare Synthetic Recommendation Data from WANDS

This appendix provides a detailed guide for generating synthetic recommendation datasets from the Wayfair Annotation Dataset (WANDS). WANDS is a large-scale e-commerce product search relevance dataset published by Wayfair, designed to advance research in product search and recommendation systems. The dataset is publicly available on GitHub and contains human-labeled relevance judgments for query-product pairs, making it an ideal foundation for creating realistic recommendation scenarios.

The synthetic datasets generated in this appendix serve multiple purposes:

- Enable reproducible research without privacy concerns.
- Provide controlled experimental conditions for algorithm development.
- Simulate realistic user interaction patterns in e-commerce settings.
- Create benchmark datasets for evaluating recommendation approaches.

B.1 Step 1: Download the WANDS Dataset

The first step involves obtaining the WANDS dataset from the official repository. The dataset is distributed as three separate CSV files that together form a complete picture of search queries, products, and their relevance relationships. We provide two methods for downloading these files:

```
# Option 1: Clone the entire repository
git clone https://github.com/wayfair/WANDS.git

# Option 2: Download individual files directly
wget https://github.com/wayfair/WANDS/raw/main/dataset/product.csv
wget https://github.com/wayfair/WANDS/raw/main/dataset/query.csv
wget https://github.com/wayfair/WANDS/raw/main/dataset/label.csv
```

Option 1 is recommended if you want access to the complete repository, including documentation and any future updates. Option 2 is more suitable for minimal setups where you only need the core dataset files. Both methods will provide you with the three essential CSV files needed for the subsequent steps.

B.2 Step 2: Load the Dataset

Once downloaded, we load the dataset files into memory for processing. The WANDS dataset consists of three CSV files with tab-separated values, each serving a specific purpose:

- **product.csv**: Contains detailed information about products including names, descriptions, categories, and hierarchical classifications
- **query.csv**: Stores search queries along with their classifications and metadata
- **label.csv**: Provides human-annotated relevance judgments linking queries to products

```
import pandas as pd

# Load the dataset files with tab separator
products = pd.read_csv(wands_folder + "product.csv", sep="\t", engine="python")
queries = pd.read_csv(wands_folder + "query.csv", sep="\t", engine="python")
labels = pd.read_csv(wands_folder + "label.csv", sep="\t", engine="python")

# Dataset statistics
print(f"Products: {products.shape[0]} items")
print(f"Queries: {queries.shape[0]} search terms")
print(f"Labels: {labels.shape[0]} relevance judgments")
```

The loaded dataset contains 42,994 products, 480 search queries, and 233,448 labeled query-product pairs. This scale provides sufficient data diversity for generating meaningful synthetic recommendation scenarios while remaining manageable for research purposes.

B.3 Step 3: Prepare the Data for Recommendation Tasks

Before generating synthetic datasets, we need to prepare and integrate the raw data. This step involves several important transformations:

1. **Column Standardization**: We rename columns to follow consistent naming conventions, making the data easier to work with programmatically
2. **Data Integration**: We merge the three separate datasets to create a unified view of all relationships
3. **Quality Filtering**: We filter the data to focus on relevant query-product pairs, excluding irrelevant matches

```
# Rename columns for consistency
products = products.rename(columns={"category hierarchy": "category_hierarchy"})
labels = labels.rename(columns={"id": "label_id"})

# Merge data for easier processing
data = labels.merge(queries, on="query_id").merge(products, on="product_id")

# Filter to only include relevant products (Exact or Partial matches)
relevant_data = data[data["label"].isin(["Exact", "Partial"])]
```

The merged dataset provides a comprehensive view of query-product relationships, containing all metadata needed for our synthetic recommendation tasks. By focusing on "Exact" and "Partial" matches, we ensure that our synthetic data reflects genuine relevance relationships rather than random associations.

B.4 Step 4: Generate Synthetic Recommendation Data

This section presents the core algorithms for generating three distinct types of synthetic recommendation datasets. Each dataset type is designed to support different recommendation scenarios and evaluation methodologies commonly used in recommendation systems research.

Using the WANDS dataset, we created three types of synthetic recommendation datasets:

```
# 1. Session-Based Recommendation Data Generation
def generate_query_sessions(
    num_sessions: int = 1000, min_len: int = 3, max_len: int = 5
) -> List[Dict]:
    """
    Generate synthetic query sessions using correct column names
    """
    # Group products by query (using 'query' column from queries)
    query_to_products = (
        relevant_data.groupby("query")["product_id"].apply(list).to_dict()
    )

    sessions = []
    all_queries = list(query_to_products.keys())

    for _ in range(num_sessions):
        session_len = random.randint(min_len, max_len)
        seed_query = random.choice(all_queries)

        # Use 'query_class' for related queries
        seed_class = data[data["query"] == seed_query]["query_class"].iloc[0]
        candidate_queries = list(
            data[data["query_class"] == seed_class]["query"].unique()
        )

        if len(candidate_queries) >= session_len:
            session_queries = random.sample(candidate_queries, session_len)
        else:
            session_queries = candidate_queries + random.choices(
                all_queries, k=session_len - len(candidate_queries)
            )

        target_query = session_queries[-1]
        context_queries = session_queries[:-1]

        if target_query in query_to_products:
            target_products = query_to_products[target_query]
            sessions.append(
                {
                    "session_id": f"sess_{len(sessions)}",
                    "context_queries": context_queries,
                    "target_query": target_query,
                    "target_products": target_products,
                    "query_class": seed_class,  # Added for better filtering
                }
            )

    return sessions


# 2. User-Query Synthetic Profiles (updated column names)
def generate_user_profiles(
    num_users: int = 500, min_queries: int = 2, max_queries: int = 5
) -> List[Dict]:
    """
    Generate user profiles using correct product class column ('product_class')
    """
    query_to_products = (
        relevant_data.groupby("query")["product_id"].apply(list).to_dict()
    )
    product_to_class = (
        data.drop_duplicates("product_id")
        .set_index("product_id")["product_class"]
        .to_dict()
    )

    profiles = []
    all_queries = list(query_to_products.keys())
```

```

    for user_num in range(num_users):
        num_queries = random.randint(min_queries, max_queries)
        user_queries = random.sample(all_queries, num_queries)

        engaged_products = []
        for q in user_queries:
            engaged_products.extend(query_to_products.get(q, []))

        if not engaged_products:
            continue

        # Use product_class for recommendations
        engaged_classes = {
            product_to_class[p] for p in engaged_products if p in product_to_class
        }
        candidate_products = [
            p
            for p in product_to_class
            if product_to_class[p] in engaged_classes and p not in engaged_products
        ]

        recommended = (
            random.sample(candidate_products, min(3, len(candidate_products)))
            if candidate_products
            else []
        )

        profiles.append(
            {
                "user_id": f"user_{user_num}",
                "query_history": user_queries,
                "engaged_products": engaged_products,
                "recommended_products": recommended,
                "preferred_classes": list(engaged_classes),  # Added for analysis
            }
        )

    return profiles


# 3. Next-Item Prediction via Product Sequences (updated merge keys)
def generate_product_sequences(
    num_sequences: int = 2000, min_len: int = 2, max_len: int = 4
) -> List[Dict]:
    """
    Generate sequences using correct label-to-product relationships
    """
    query_to_products = (
        relevant_data.groupby("query_id")["product_id"].apply(list).to_dict()
    )

    sequences = []

    for _ in range(num_sequences):
        while True:
            query_id = random.choice(list(query_to_products.keys()))
            products = query_to_products[query_id]
            if len(products) >= min_len + 1:
                break

        seq_len = random.randint(min_len, min(max_len, len(products) - 1))
        for i in range(len(products) - seq_len):
            sequence = products[i : i + seq_len]
            next_product = products[i + seq_len]

            sequences.append(
                {
```

```
                    "sequence_id": f"seq_{len(sequences)}",
                    "product_sequence": sequence,
                    "next_product": next_product,
                    "query_id": query_id,  # Added for traceability
                }
            )

            if len(sequences) >= num_sequences:
                return sequences

    return sequences
```

B.4.1 Session-based Recommendation Data

Session-based recommendation is crucial for understanding user intent within a single browsing session. In e-commerce, users often refine their searches through multiple queries, gradually honing in on their desired products. Our synthetic session data captures this behavior by simulating realistic query progressions.

We simulated user sessions as sequences of queries where the goal is to predict products relevant to the final query based on previous queries in the session:

```
# Generate 1,000 synthetic query sessions
synthetic_sessions = generate_query_sessions(1000)
session_cols = [
    "session_id",
    "context_queries",
    "target_query",
    "target_products",
    "query_class",
]
pd.DataFrame(synthetic_sessions)[session_cols].to_csv(
    save_folder + "synthetic_sessions.csv", index=False
)
```

Each session contains:

- Three to five consecutive queries grouped by query class
- Context queries (earlier in session)
- Target query (final query in session)
- Target products (relevant products for the target query)

The session generation algorithm ensures coherence by preferentially selecting queries from the same query class, mimicking how users naturally refine their searches. For example, a user searching for "comfortable accent chair" might follow up with "sancroft armchair" before settling on "turquoise chair"—all within the accent chairs category.

Given the inputs `session_id`, `context_queries`, and `target_query` (search keywords), the objective is to predict the corresponding `target_products`. Formally, the task can be expressed as:

$$\texttt{target\_products} = f(\texttt{context\_queries}, \texttt{target\_query})$$

where f represents the recommendation model to be learned from the session data.

B.4.2 User-query Synthetic Profiles

User profiling is essential for personalized recommendations. Unlike session-based approaches that focus on short-term intent, user profiles capture long-term preferences and interests. Our synthetic profiles simulate diverse user behaviors and preference patterns found in real e-commerce settings.

We created simulated user profiles with query histories and product interactions:

```
# Generate 500 synthetic user profiles
synthetic_profiles = generate_user_profiles(500)
profile_cols = [
    "user_id",
    "query_history",
    "engaged_products",
    "recommended_products",
    "preferred_classes",
]
pd.DataFrame(synthetic_profiles)[profile_cols].to_csv(
    save_folder + "synthetic_profiles.csv", index=False
)
```

Each user profile contains:

- Two to five diverse queries
- Engaged products (products relevant to their queries)
- Recommended products from the same product classes
- Preferred product classes inferred from engagement

The profile generation process models realistic user behavior by:

1. Assigning each user a varied number of historical queries
2. Collecting all products shown for those queries as potential engagements
3. Inferring user preferences from engaged product classes
4. Generating recommendations from similar but unexplored products

This approach creates profiles that exhibit both focused interests (users who primarily search within specific categories) and diverse exploration patterns (users who browse across multiple product types).

Given the observed user data `user_id`, `query_history`, and `engaged_products`, the objective is to predict the appropriate `recommended_products`. Formally, this task is defined as:

$$\texttt{recommended\_products} = f(\texttt{query\_history}, \texttt{engaged\_products})$$

where f denotes the recommendation model to be trained, aiming to generalize from past behavior to suggest relevant, novel items aligned with the user's inferred preferences.

B.4.3 Next-item Prediction Sequences

Sequential recommendation represents one of the most challenging tasks in recommendation systems. The goal is to predict the next item a user will interact with based on their sequential interaction history. This dataset type is particularly valuable for evaluating algorithms that can capture temporal dependencies and sequential patterns in user behavior.

We generated product sequences for next-item prediction tasks:

```
# Generate 2,000 product sequences
synthetic_sequences = generate_product_sequences(2000)
sequence_cols = ["sequence_id", "product_sequence", "next_product", "query_id"]
pd.DataFrame(synthetic_sequences)[sequence_cols].to_csv(
    save_folder + "synthetic_sequences.csv", index=False
)
```

Each sequence contains:

- Two to four consecutive products from the same query
- Next product to predict
- Original query that produced these products

The sequence generation leverages the natural ordering of products within search results. By extracting subsequences from products returned for the same query, we create training data that reflects how users might browse through search results sequentially. This approach ensures that the sequences maintain semantic coherence—products in a sequence are related through their relevance to a common query.

Given the input `sequence_id` and `product_sequence` (an ordered list of previously interacted items), the goal is to predict the `next_product`. Formally, this can be expressed as:

$$\texttt{next\_product} = f(\texttt{product\_sequence})$$

where f represents the sequential recommendation model trained to capture transition patterns and contextual dependencies in user behavior.

B.5 Step 5: Dataset Statistics and Sample Entries

Understanding the characteristics of the generated datasets is crucial for their effective use. This section provides detailed statistics and concrete examples from each dataset type.

The synthetic datasets contain rich recommendation patterns:

- Session-based Dataset: 1,000 sessions with an average of 4 queries per session
- User Profiles Dataset: 500 profiles engaging with an average of 12 products each
- Product Sequences Dataset: 2,000 sequential patterns derived from relevant query results

These statistics reflect realistic usage patterns:

- Sessions typically contain three to five queries, matching observed e-commerce search refinement behavior.
- User profiles show varied engagement levels, from casual browsers to active shoppers.
- Product sequences capture different browsing depths, from quick scans to detailed explorations.

Sample entries from each dataset:

Sample Session

```
session_id                                                        sess_0
context_queries   ['king poster bed', 'maryford queen tufted bed...
target_query                                  aiden black velvet bed
target_products   [5017, 26789, 1103, 26795, 40521, 41249, 26790...
query_class                                                       Beds
```

This sample illustrates a coherent search session where a user progressively refines their search for beds, moving from broad size-based queries toward specific style and color preferences.

Sample User Profile

```
user_id                                                           user_0
query_history           ['barstool patio sets', 'champagne velvet desk...
engaged_products        [10283, 12494, 11549, 11550, 23179, 12222, 301...
recommended_products                                [26348, 2169, 23479]
preferred_classes       ['Dining Chairs', 'Kids Chairs|Licensed Produc...
```

This profile represents a user with diverse interests spanning furniture and kitchen appliances, demonstrating how the synthetic data captures multi-category shopping behavior common in large e-commerce platforms.

Sample Product Sequence

```
sequence_id                 seq_0
product_sequence    [4771, 37624]
next_product                37623
query_id                      263
```

This sequence shows a browsing pattern within products returned for query 263, where the user views products 4771 and 37624 before the model predicts the next product, 37623.

As summarized in Table B.1, these datasets enable the development and evaluation of various recommendation algorithms, including session-based recommenders, personalized product recommendation systems, and sequential recommendation models. Each dataset type addresses different aspects of the recommendation problem:

1. **Session-based data** helps develop algorithms that can understand short-term user intent and provide contextually relevant recommendations within a single browsing session.
2. **User profile data** supports the development of personalization algorithms that can model long-term user preferences and provide tailored recommendations based on historical behavior.
3. **Product sequence data** enables the training of sequential models that can predict future user actions based on ordered interaction patterns.

Together, these synthetic datasets provide a comprehensive testbed for recommendation system research while maintaining the privacy and ethical advantages of synthetic data generation.

Table B.1 Summary of synthetic recommendation datasets.

Dataset type	Size	Main fields	File output
Session-based	1,000	session_id, context_queries, target_query, target_products, query_class	synthetic_sessions.csv
User query profiles	500	user_id, query_history, engaged_products, recommended_products, preferred_classes	synthetic_profiles.csv
Product sequences	2,000	sequence_id, product_sequence, next_product, query_id	synthetic_sequences.csv

Index

m

S

u

Printed and bound by CPI Group (UK) Ltd, Croydon, CR0 4YY
17/09/2026
14964237-0008